Advances in
ORGANOMETALLIC CHEMISTRY

VOLUME 56

Advances in Organometallic Chemistry

The Organotransition Metal Chemistry of Poly(pyrazolyl)borates. Part 1

EDITED BY

ROBERT WEST
ORGANOSILICON
RESEARCH CENTER
DEPARTMENT OF CHEMISTRY,
UNIVERSITY OF WISCONSIN,
MADISON, WI, USA

ANTHONY F. HILL
RESEARCH SCHOOL OF CHEMISTRY,
INSTITUTE OF ADVANCED STUDIES,
AUSTRALIAN NATIONAL UNIVERSITY
CANBERRA, ACT,
AUSTRALIA

MARK J. FINK
DEPARTMENT OF CHEMISTRY
TULANE UNIVERSITY,
NEW ORLEANS, LOUISIANA, USA

FOUNDING EDITOR
F. GORDON A. STONE

VOLUME 56

Amsterdam • Boston • Heidelberg • London • New York • Oxford • Paris
San Diego • San Francisco • Singapore • Sydney • Tokyo
Academic Press is an imprint of Elsevier

Academic Press is an imprint of Elsevier
84 Theobald's Road, London WC1X 8RR, UK
Radarweg 29, PO Box 211, 1000 AE Amsterdam, The Netherlands
Linacre House, Jordan Hill, Oxford OX2 8DP, UK
30 Corporate Drive, Suite 400, Burlington, MA 01803, USA
525 B Street, Suite 1900, San Diego, CA 92101-4495, USA

First edition 2008

ISBN: 978-0-12-374273-5
ISSN: 0065-3055

For information on all Academic Press publications
visit our website at books.elsevier.com

Printed and bound in USA

08 09 10 11 12 10 9 8 7 6 5 4 3 2 1

Contents

The Organometallic Chemistry of Group 9 Poly(pyrazolyl)borate Complexes

IAN R. CROSSLEY

Contributors

Numbers in parentheses indicate the pages on which the authors' contributions begin.

EVA BECKER (155), Institute of Applied Synthetic Chemistry, Vienna University of Technology, Getreidemarkt 9, A-1060 Vienna, Austria

LORRAINE M. CALDWELL (1), Research School of Chemistry, Institute of Advanced Studies, Australian National University, Canberra, ACT 0200, Australia

IAN R. CROSSLEY (199), Research School of Chemistry, Institute of Advanced Studies, Australian National University, Canberra, ACT 0200, Australia

T. BRENT GUNNOE (95), Department of Chemistry, North Carolina State University, Raleigh, NC 27695-8204, USA

KARL KIRCHNER (155), Institute of Applied Synthetic Chemistry, Vienna University of Technology, Getreidemarkt 9, A-1060 Vienna, Austria

MARTY LAIL (95), Department of Chemistry, North Carolina State University, Raleigh, NC 27695-8204, USA

SONJA PAVLIK (155), Institute of Applied Synthetic Chemistry, Vienna University of Technology, Getreidemarkt 9, A-1060 Vienna, Austria

KARL A. PITTARD (95), Department of Chemistry, North Carolina State University, Raleigh, NC 27695-8204, USA

Preface

In 1966, a truly seminal paper (*J. Am. Chem. Soc.* **1966**, *88*, 1842) entitled 'Boron-Pyrazole Chemistry' began with the words 'Boron-pyrazole chemistry, which deals with compounds containing boron bonded to nitrogen of a pyrazole nucleus, is a new and fertile field of remarkable scope.' It is likely that even the single author of this paper, Swiatoslaw Trofimenko, could not have foreseen the true immensity of the field that was to spring from his pioneering discovery of the poly(pyrazolyl)borates ('scorpionates'). A quick database search reveals some 3000 papers that have appeared in the intervening years concerned with the coordination chemistry of this versatile class of ligands. A closer inspection shows that a very substantial component of this literature relates to their use as scaffolds for organometallic chemistry. Indeed much early work explored a perceived analogy between the tris(pyrazolyl)borate and cyclopentadienyl ligands. More recently, it is the features that distinguish these ligands from cyclopentadienyls that are appreciated and widely exploited.

With this ever-growing wealth of scorpionate-supported organometallic chemistry, it seemed a single volume that brought all this wonderful chemistry together would provide a valuable 'one-stop' resource for organometallic chemists. This idea proved to be naïve. The enormity of the field has defied comprehensive coverage in a single volume and accordingly will fill this and a subsequent volume of *Advances in Organometallic Chemistry*. The material is arranged by metal according to group, from which 'flavours' become apparent, either because of the fundamental differences between metals or because the foci of particular research groups have seen different aspects receive particular attention. In addition, a chapter is devoted specifically to alkylidyne complexes – one area that has especially benefited from the special properties offered by these ligands. This volume, Part 1, includes this chapter in addition to chapters dealing with the chemistry of groups 7, 8 and 9. The following volume will cover groups 4–6 and 10, the lanthanides and actinides.

In surveying this literature, one notes a recurrent feature – numerous papers that carry an acknowledgement to Jerry Trofimenko for generously providing samples of his ligands to help others initiate work. This generosity has no doubt played a role in the wider embrace of these ligands and stands as an example. Sadly, Jerry Trofimenko passed away in February 2007. These volumes are dedicated to his memory and to his example.

Anthony F. Hill

Alkylidyne Complexes Ligated by Poly(pyrazolyl)borates

LORRAINE M. CALDWELL*

Research School of Chemistry, Institute of Advanced Studies, Australian National University, Canberra, ACT 0200, Australia

I

INTRODUCTION

The first alkylidyne–metal complexes were prepared by Fischer *et al.* more than 30 years ago.[1] Since this pioneering event, the chemistry of the transition metal–carbon triple bond present in such complexes has developed into a major field of research and though the poly(pyrazolyl)borate ligands were discovered 7 years prior to the synthesis of the first alkylidyne complexes,[2] their importance and significance in this field has only more recently been truly appreciated.

There are a number of reviews that detail the various facets of alkylidyne chemistry,[3–12] but the majority of these pre-date the enormous growth in the field that has been possible with the appreciation of the kinetic and thermodynamic stability conferred upon alkylidyne complexes by the inclusion of the bulky pyrazolylborate ligands. Furthermore, though a general overview of the chemistry of pyrazolylborate derivatives has appeared,[13] there currently exists no comprehensive source of information specific to the chemistry of alkylidyne complexes co-ligated by poly(pyrazolyl)borates, a situation which this review attempts to

*Corresponding author. Tel.: +61 2 6125 8577.
E-mail: caldwell@rsc.anu.edu.au

ADVANCES IN ORGANOMETALLIC CHEMISTRY
VOLUME 56 ISSN 0065-3055/DOI 10.1016/S0065-3055(07)56001-6

remedy. Very few examples of such complexes have been isolated outside of Group 6 but the (rare) examples of Group 7 and 8 alkylidyne complexes have been included. Alkylidyne complexes co-ligated by the closely related pyrazolylmethane ligands, where relevant, have also been discussed. An effort has been made to comprehensively cover the literature till the end of 2005.

The terms carbyne and alkylidyne are used interchangeably in the literature and in this review. For the scorpionate ligand derivatives, the abbreviated nomenclature adopted by Curtis *et al.*[14,15] will be utilised (Chart 1). As such, for the hydrotris(pyrazolyl)borate ligand, $[HB(pz)_3]^-$ (pz = pyrazolyl), the abbreviation Tp will be employed. For the commonly employed 3,5-dimethylpyrazolyl-substituted derivative, $[HB(pz^*)_3]^-$, Tp* will be used. The abbreviation pzTp represents the homoscorpionate tetrakis(pyrazolyl)borate ligand, $[B(pz)_4]^-$, and Bp refers to the heteroscorpionate dihydrobis(pyrazolyl)borate anion, $[H_2B(pz)_2]^-$. Expansion of this system of nomenclature according to the rules of creating these abbreviations suggested by Trofimenko[13] allows encompassment of all of the variously substituted pyrazolyl derivatives, but very few examples of alkylidyne complexes exist that are not based on these preceding ligands. For a general poly(pyrazolyl)borate ligand the abbreviation Tp^x will be used.

The incorporation of facially capping tripodal ligands such as the cyclopentadienyl anion (η-$C_5H_5^-$, Cp) and its substituted derivatives (Cp^x, x = H, alkyl or aryl, Chart 1) into the molecular architecture of alkylidyne complexes is well established. In their coordination chemistry, the tridentate Tp^x ligand systems display similarity to the two cyclopentadienyl ligands, Cp and Cp* (η-C_5Me_5); both ligand systems, when considered uninegative, provide six electrons (charged formalism), being considered to occupy three coordination sites. Accordingly, a particular effort has been made to compare and contrast the chemistry of the poly(pyrazolyl)borate complexes with their cyclopentadienyl analogues. The topology of both the Tp^x and Cp^x ligands pre-disposes their complexes towards pseudo-octahedral geometry, however, the Tp^x ligands present a significantly greater steric profile than the cyclopentadienyl systems (cone angles of 239° and 183° for Tp* and Tp, respectively, cf. 100° and 146° for Cp and Cp*).[16] Furthermore, in contrast to the cyclopentadienyl ligand where the bonds to Cp^x substituents (x = H, Me, other

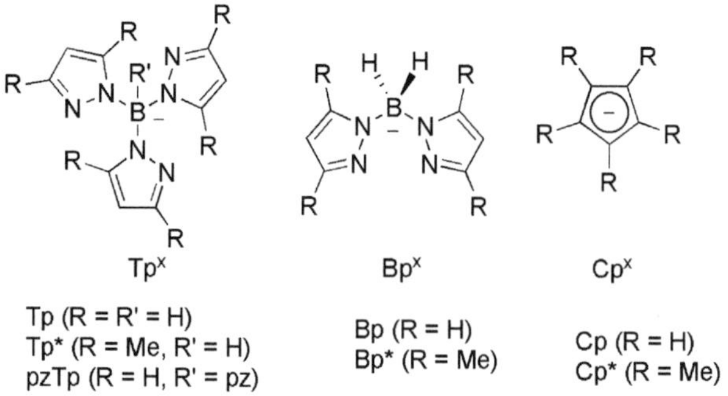

Tp^x Bp^x Cp^x

Tp (R = R' = H)
Tp* (R = Me, R' = H)
pzTp (R = H, R' = pz)

Bp (R = H)
Bp* (R = Me)

Cp (R = H)
Cp* (R = Me)

CHART 1. Commonly encountered poly(pyrazole)borates and cyclopentadienyls.

alkyl or aryl groups) point away from the coordinated metal, the bonds in a 3,5-substituted Tp^x ligand emanate at an angle which makes the pyrazolyl substituents protrude in space past the metal, forming a protective pocket.[13] Thus, in many cases the tris(pyrazolyl)borate metal cages exhibit thermal and chemical properties notably different from their Cp analogues.

Chemistry involving the 'first-generation' pyrazolylborate ligands, Tp and Tp*, forms the foundation of the initial 20 years of scorpionate research[13] and unsurprisingly these ligands have been particularly popular in the chemistry of alkylidyne complexes, whereas the related bis and tetrakis analogues have received much less attention. The merits and special features of the poly(pyrazolyl)borate co-ligand have traditionally been exploited to provide robust molecular platforms that support ligand transformations under very demanding conditions.[17–22] Moreover, the enhanced kinetic stability of the alkylidyne complexes that is imparted by the limited access to the metal centre and metal–carbon multiple bond makes the poly(pyrazolyl)borate anions ideal capping ligands in metal–alkylidyne cluster chemistry, steric factors notwithstanding.

II
SYNTHESIS

A. *General Comments*

The first reported examples of hydrotris(pyrazolyl)borate alkylidyne complexes were the mercaptoalkylidyne complexes prepared by Greaves and Angelici in 1981.[23] Alkylation of the nucleophilic sulfur atom in the anionic thiocarbonyl complex $[TpW(CO)_2(CS)]^-$ resulted in formal reduction of the C–S multiple bond to a C–S single bond, furnishing the thiomethylidyne derivatives $TpW(\equiv CSR)(CO)_2$ ($R = C_6H_4(NO_2)_2$-2,4, Me, Et) (Scheme 1). The mercarptocarbyne complex $TpW(\equiv CSMe)(CO)_2$ could similarly be generated by treatment of $TpWI(CO)_2(CS)$ with the nucleophile LiMe, though the intimate mechanism is different and probably involves single electron transfer processes.[23] Analogous reactions for the Cp-ligated complexes were also described.

The most commonly employed synthetic strategy for inclusion of the bulky poly(pyrazolyl)borate ligand into the coordination sphere of alkylidyne complexes

$[TpW(CO)_2(C{\equiv}S)]^- \xrightarrow{RX} TpW({\equiv}C{-}SR)(CO)_2$

$TpWI(CO)_2(C{\equiv}S) \xrightarrow{LiMe} TpW({\equiv}C{-}SMe)(CO)_2$

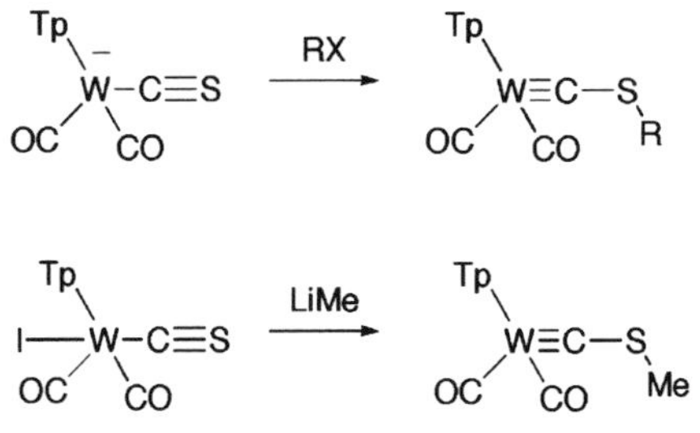

SCHEME 1. Preparation of $TpW(\equiv CSR)(CO)_2$ ($R = C_6H_4(NO_2)_2$-2,4, Me, Et; X = Cl, I).

involves late incorporation of the Tp^x anion into a more highly functionalised organometallic substrate in much the same way that Cp^x derivatives are utilised to generate cyclopentadienyl-containing alkylidyne complexes.[24] Frequently, this involves modification of the metal–ligand framework of a pre-existing alkylidyne complex, which can be achieved by one of three ways: altering the alkylidyne ligand (e.g., replacement of R in M≡CR with R′), ligand substitution, or oxidation/reduction of the metal centre. Thus, the common preparative routes to both Fischer-[1,25–30] and Schrock-type alkylidynes[31–35] are generally employed *en route* to Tp^x alkylidyne complexes. It should be noted, however, that the strategies for constructing and functionalising the metal–carbon triple bond in Lalor's halo-alkylidyne complexes $Tp^*M(\equiv CCl)(CO)_2$ (M = Mo, W)[36,37] (*vide infra*) are unique to the pyrazolylborate systems.

B. *Alkyl, Aryl, Alkenyl and Alkynyl Substituents*

1. Synthesis by Ligand Substitution

A typical synthesis of poly(pyrazolyl)borate alkylidyne complexes involves displacement of labile ligands from pre-existing alkylidynes prepared according to the classic oxide–abstraction Fischer preparation (Scheme 2), which has been developed by Mayr and co-workers[30,38] into an efficient approach to Group 6 alkylidynes.

Treatment of the thermolabile bromo-alkylidyne complexes *trans*-$M(\equiv CR)Br(CO)_4$ with the tris(pyrazolyl)borate salts $K[Tp^x]$ leads to displacement of not only the anionic ligand but also the additional elimination of two carbonyl ligands, affording the *cis*-dicarbonyl complexes $Tp^xM(\equiv CR)(CO)_2$ (M = Cr, Mo, W; Tp^x = Tp, pzTp, Tp*, Tp^{Ph}; R = Me or aryl) (Scheme 3).[39–44] Depending on the Lewis acid utilised in the transformation of the acyl or carbene ligands into the alkylidyne, different anions may be present, i.e., X = halide with oxalyl halides or phosphine dihalides, or the better leaving group $CF_3CO_2^-$ with trifluoroacetic anhydride. This protocol has been similarly applied by Stone and

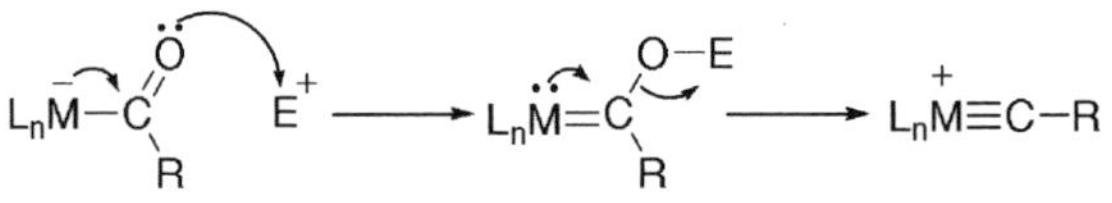

$$L_nM^{-}-C(=\ddot{O})R + E^+ \longrightarrow L_nM=C(O-E)R \longrightarrow L_n\overset{+}{M}\equiv C-R$$

SCHEME 2. Oxide–abstraction as a route to alkylidyne synthesis.

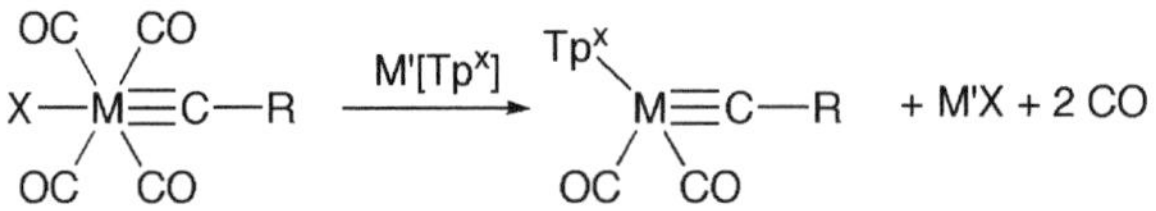

$$X-M(CO)_4\equiv C-R \xrightarrow{M'[Tp^x]} Tp^xM(CO)_2\equiv C-R + M'X + 2\,CO$$

SCHEME 3. Ligand substitution from a tetracarbonyl alkylidyne precursor (M = Cr, Mo, W; M′ = Na, K; Tp^x = Tp, Tp*, pzTp; R = alkyl, aryl).

co-workers in the preparation of a series of cyclopentadienyl complexes $CpM(\equiv CAr)(CO)_2$ (Ar = substituted phenyl; M = Cr, Mo, W).[45]

The bidentate Bp anion is sterically less demanding than the tridentate congeners, and alkylidyne complexes ligated by Bp thus require an additional ancillary ligand in order to obtain a complete coordination sphere. Consequently, treatment of the tetracarbonyl precursors with salts of the bidentate Bp ligand result in displacement of a single carbonyl ligand along with the anionic leaving group to afford the tricarbonyl alkylidyne complexes $BpM(\equiv CR)(CO)_3$ (M = W, R = Me, C_6H_4Me-4)[46] (Scheme 4).

Due to the *trans*-disposition of the four π-acidic carbonyls in the precursor complex $M(\equiv CR)X(CO)_4$, thermal decomposition may compete with the ligand substitution process and, accordingly, the synthesis of Tp-alkylidyne complexes is often achieved indirectly *via* the intermediacy of more stable carbonyl-substitution products. Bis-substituted metal–alkylidyne complexes of the form *trans,cis,cis*-$M(\equiv CR)X(CO)_2(L')_2$ (X = anionic ligand) are obtained upon the treatment of the tetracarbonyl compounds with donor ligands, L′ [L′ = py, ½ 2,2′-bipyridyl (bipy), γ-picoline (pic), ½ tmeda (*N,N,N′,N′*-tetramethylethylenediamine), isocyanide (CNR)]. The coordinative lability of the *N*-donors renders these complexes highly susceptible to ligand displacement reactions; however, the greater electron density at the metal centre confers an enhanced thermal stability when compared with the $M(\equiv CR)X(CO)_4$ derivatives. Displacement of L′ in *trans,cis,cis*-$M(\equiv CR)X(CO)_2(L')_2$ with the stronger *N*-donor poly(pyrazolyl)borate ligands readily affords the desired *cis*-dicarbonyl complexes $Tp^xM(\equiv CR)(CO)_2$ (L′ = py, $Tp^x = Tp^*$, M = Cr, R = C_6H_4Me-4;[47] L′ = pic, Tp^x = Tp, M = Mo, R = C_6H_4OMe-4) (Scheme 5).[48]

As mentioned above, in the case of the bidentate Bp ligand only the anionic ligand and one additional ligand is displaced in subsequent substitution reactions. Thus, for the Bp analogues, complexes of the type $BpM(\equiv CR)(CO)_2(L')$ (M = W, R = $C_6H_3Me_2$-2,6, $C_6H_2Me_3$-2,4,6; L′ = pic, CNR′) are generated, which retain one isonitrile or *N*-donor ligand.[49] Moreover, the bidentate nature of the Bp ligands allows for the formation of two possible geometric isomers of these alkylidyne

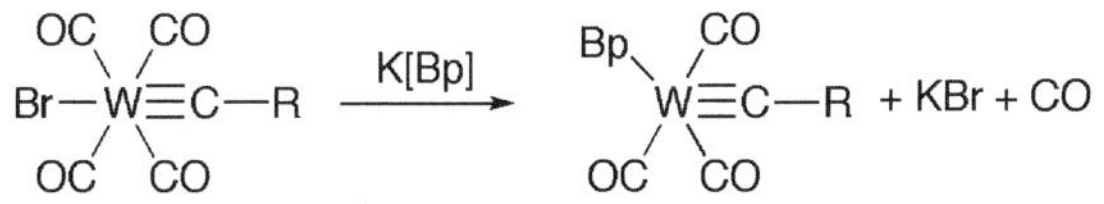

SCHEME 4. Ligand substitution from a tetracarbonyl alkylidyne precursor (R = Me, C_6H_4Me-4).

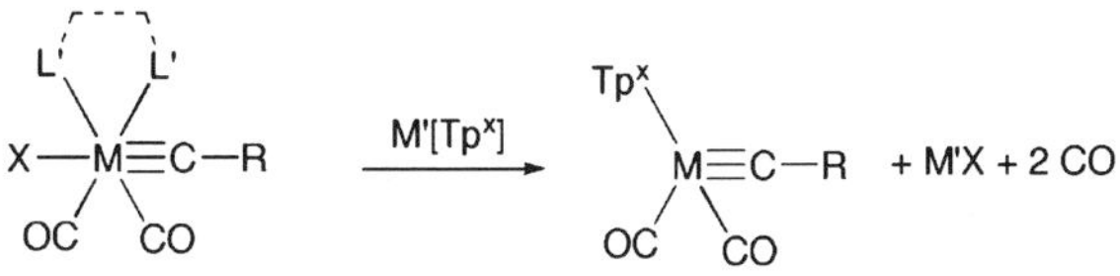

SCHEME 5. Ligand substitution from carbonyl-substitution products (M = Cr, Mo, W; X = anionic ligand; M′ = Na, K; Tp^x = Tp, Tp*, pzTp; R = alkyl, aryl; L′ = py, pic, CNR, ½ bipy, ½ tmeda).

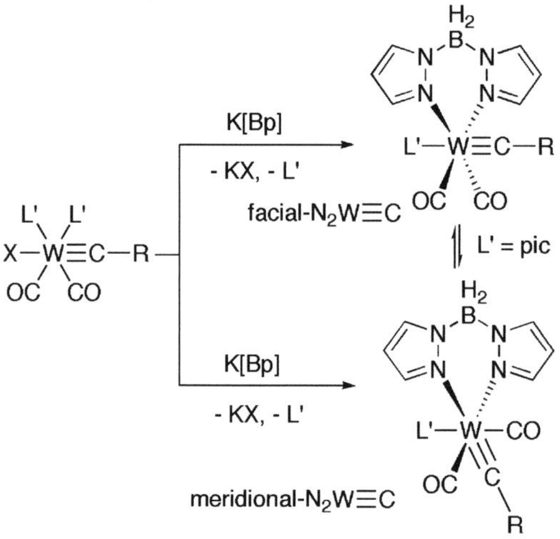

SCHEME 6. Meridional vs. facial coordination of Bp/alkylidyne fragments (M = Mo, W; R = alkyl, aryl; L′ = py, pic, CNR, PMe_2Ph).

complexes, i.e., the alkylidyne fragment may be oriented meridionally or facially with respect to the chelated pyrazolyl arms and the favoured geometry depends upon the relative π-basicity of the remaining ligands. The *trans*-disposition of the weakest π-acid ligand appears to strengthen M$\equiv$C bonds and this geometry predominates (Scheme 6).[49] For L′ = pic, both geometric isomers were formed in unequal proportions, with interconversion between the two evident in solution. Nevertheless, $^{13}C\{^1H\}$ NMR spectroscopic data suggested that the dominant isomeric form featured the more symmetric geometry in which the picoline ligand was *trans* to the alkylidyne substituent (i.e., facial Bp/alkylidyne coordination).[49] This orientation was also found in the solid-state structure. Spectroscopic data for the complexes BpW($\equiv$CR)$(CO)_2$(CNR′), containing a π-acidic isonitrile ligand, are consistent with the alternate geometry in which the alkylidyne ligand is *trans* to a pyrazolyl arm.[49]

Enhanced thermal stability for the precursor complexes M($\equiv$CR)X$(CO)_2$$(L')_2$ (X = anionic ligand) employed in ligand substitution reactions is likewise achieved when L′ is a phosphorus donor. Direct reaction of the dicarbonyl complex *trans,cis,cis*-W($\equiv$CR)Br$(CO)_2(PPh_3)_2$ (R = C_6H_4Me-4) with K[Tp] provides TpW($\equiv$CR)$(CO)_2$.[50] However, initial photolysis of this same substrate leads to a *trans*-disposition of the PR_3 ligands and subsequent reaction with K[Tp] initially provides the thermally unstable alkylidyne–carbonyl coupling ketenyl product[50] TpW(η^2-*C,C*-OCCR)(CO)(PPh_3), which in solution undergoes slow CO extrusion to afford TpW($\equiv$CR)(CO)(PPh_3).[50]

The tris(phosphine) alkylidyne derivatives similarly serve as precursors for ligand displacement reactions. The alkylidyne complex *mer*-Mo($\equiv$CC_6H_4OMe-2) Cl(CO)$\{P(OMe)_3\}_3$ reacts with K[Bp] leading to *trans*-BpMo($\equiv$CC_6H_4OMe-2) (CO)$\{P(OMe)_3\}_2$ (Scheme 7). The dimethylphenyl phosphine analogue *trans*-BpW($\equiv$CC$_6$H$_4$Me-4)(CO)(PMe$_2$Ph)$_2$[49] can be prepared similarly *via* the intermediacy of a bis(picoline) complex W($\equiv$CC$_6$H$_4$Me-4)Br(CO)$_2$(pic)$_2$ that reacts with three equivalents of phosphine to provide, *en route*, W($\equiv$CC$_6$H$_4$Me-4) Br(CO)(PMe$_2$Ph)$_3$.[51]

A variety of alkyl-, aryl- and alkynyl-substituted alkylidynes are accessible *via* the ligand substitution pathway. An illustrative example involves the complex M($\equiv$CC$\equiv$C^{t}Bu)(O$_2$CCF$_3$)(CO)$_2$(L′)$_2$ (M = Mo, L′ = ½ tmeda; M = W, L′ = py), which reacts with Tp^x salts (but not with Cp sources) to produce Tp^{x}M($\equiv$CC$\equiv$C^{t}Bu)(CO)$_2$ (M = Mo, Tp^x = Tp, Tp*) *via* displacement of the chelating tmeda ligand and the excellent leaving group O$_2$CCF$_3^-$. Due to stronger binding of tmeda to tungsten than molybdenum, the Tp tungsten complex may instead be prepared from the more labile bis(pyridine) precursor.[52]

More exotic examples may be obtained in which the alkylidyne substituent is itself a coordinated arene. Treatment of the alkylidyne–metal complex CrMo (μ-η^6:σ-CC$_6$H$_4$OMe-2)(O$_2$CCF$_3$)(CO)$_5$(tmeda) with K[Tp] in dichloromethane affords TpCrMo(μ-η^6:σ-CC$_6$H$_4$OMe-2)(CO)$_5$, in which the alkylidyne ligand is η^6-coordinated to a Cr(CO)$_3$ group.[53] The cymantrenylmethylidyne complex TpMoMn(μ-σ:η^5-CC$_5$H$_4$)(CO)$_5$[54] and the ferrocene-substituted bis(alkylidyne-molybdenum) complex Tp$_2$Mo$_2$Fe(μ-σ,σ':η^5-CC$_5$H$_4$)$_2$(CO)$_4$[55] can be similarly prepared in good yields (Chart 2). Not surprisingly, the coordination of metal fragments to the arene ring results in a significant shielding of the alkylidyne carbon resonance.[41,42,44]

The preparation of high-valent, Schrock-type alkylidyne complexes containing poly(pyrazolyl)borate ligands may also be achieved by ligand substitution reactions. For example, the (dry) air-stable tungsten complexes Tp*W($\equiv$CR)X$_2$

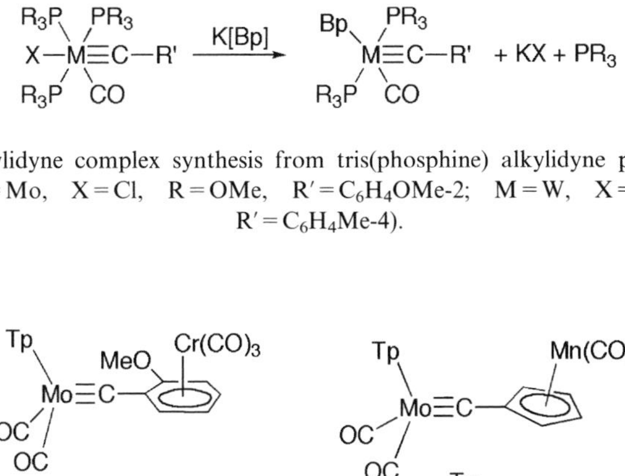

SCHEME 7. Bp-alkylidyne complex synthesis from tris(phosphine) alkylidyne precursors M($\equiv$CR′) X(CO)(PR$_3$)$_3$ (M = Mo, X = Cl, R = OMe, R′ = C$_6$H$_4$OMe-2; M = W, X = Br, PR$_3$ = PMe$_2$Ph, R′ = C$_6$H$_4$Me-4).

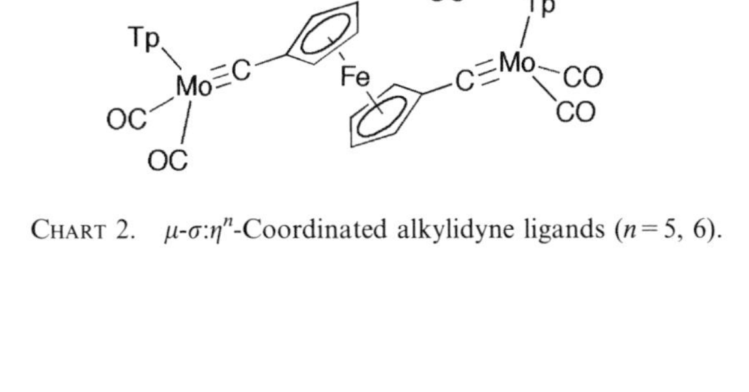

CHART 2. μ-σ:η^n-Coordinated alkylidyne ligands (n = 5, 6).

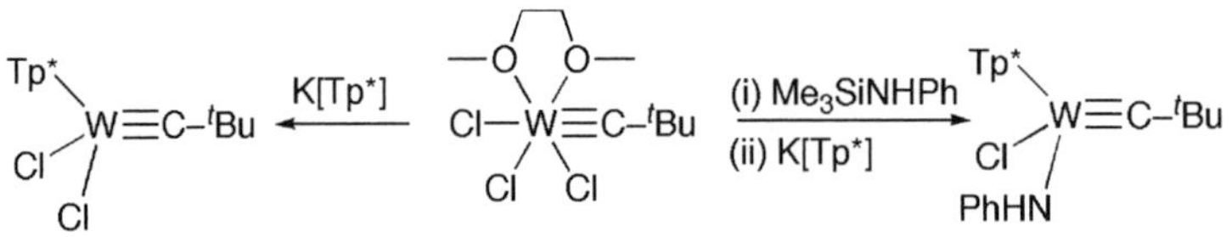

SCHEME 8. High-valent Tp^x-alkylidyne synthesis from ligand substitution.

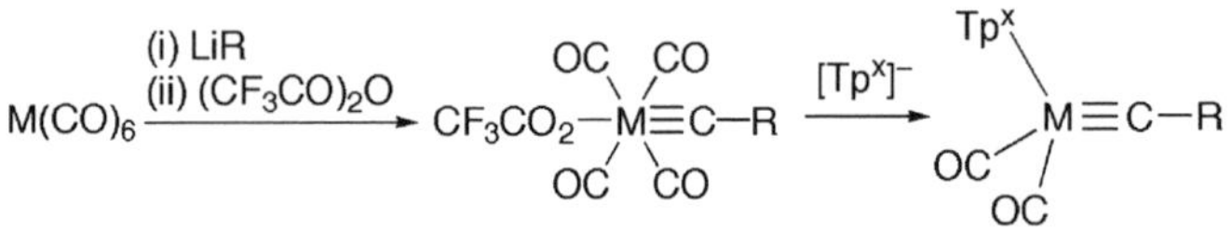

SCHEME 9. One-pot carbyne synthesis (M = Mo, W; Tp^x = Tp, Tp*; R = $SiMe_3$, Ph).

(R = ^{t}Bu, X = Cl;[56,57] R = Ph, X = Br[57,58]) were synthesised from W(≡C^{t}Bu)Cl_3(DME) (DME = dimethoxyethane) or W(≡CPh)Br_3(DME) with K[Tp*] in yields of 50 and 10%, respectively. Similarly, the complex Tp*W(≡C^{t}Bu)Cl(NHPh) was prepared by reaction of Me_3SiN(H)Ph with W(≡C^{t}Bu)Cl_3(DME), to displace one halide, followed by the addition of K[Tp*] to displace the bidentate DME and a second chloride ligand (Scheme 8).[59]

For the complexes Tp^xW(≡C^{t}Bu)Cl(NHR), the sterically demanding Tp^x ligand and strong N–W bond disfavour rotational isomerism.[59] In contrast, the related Mo complex, TpMo(≡$CCMe_2$Ph)(OMe)($NHC_6H_3{}^iPr_2$-2,6), exists as a mixture of two rotamers due to π-donation from the methoxy group which competes with the amido N–Mo π-interaction, thereby sufficiently lowering the barrier to rotation about the N–Mo bond to allow the interconversion of *syn* and *anti* isomers.[60]

More recently, a modification of the Mayr procedure in which the weakly coordinated trifluoroacetato ligand is directly replaced by the poly(pyrazolyl)borate anion (with concomitant CO ligand displacement) has allowed development of a 'one-pot' synthetic procedure. An early example of this involved stepwise preparation of the (alkynylmethylidyne)tungsten complexes Tp^xW(≡CC≡CR)$(CO)_2$ (Tp^x = Tp*, Tp; R = $SiMe_3$, Ph) from $W(CO)_6$, Li[C≡CR], $(CF_3CO)_2O$ and M[Tp^x] (M = Na, K) directly, as shown in Scheme 9.[61] The complex TpMo(≡CC_4H_3S-2)$(CO)_2$ was similarly obtained directly from $Mo(CO)_6$, 2-thienyllithium, $(CF_3CO)_2O$ and K[Tp].[48]

2. Synthesis *via* Alkylidyne Elaboration

Synthesis from pre-existing alkylidyne complexes L_nM(≡CR) can also occur *via* the modification of the alkylidyne ligand, involving the elaboration of R or the replacement of R by R′. These modifications typically involve alkylidyne complexes which already incorporate the poly(pyrazolyl)borate ligand, and consequently there is overlap between this section and Section IV, and such syntheses will be expanded upon therein.

3. Synthesis from Acyl Complexes

Under strongly basic conditions in ethanol, the molybdenum η^2-acyl complexes $Tp^*Mo(\eta^2\text{-OCR})(CO)_2$ (R = Me, Et) were converted to molybdenum alkylidyne complexes $Tp^*Mo(\equiv CR)(CO)_2$ in 20% yield (Scheme 10).[62] In the infrared spectrum, the average CO stretching frequency of the carbyne thus produced is ca. 30 cm^{-1} higher than that of the corresponding η^2-acyl complex, consistent with the strong π-acidity of the alkylidyne ligand.[62]

4. Synthesis from Alkylidenes

Base-catalysed proton transfer results upon the addition of excess potassium methoxide to the complex $TpMo(=CHCMe_2Ph)(OTf)(NHC_6H_3{}^iPr_2\text{-}2,6)$, affording the corresponding alkylidyne compound $TpMo(\equiv CCMe_2Ph)(OMe)(NHC_6H_3{}^iPr_2\text{-}2,6)$ (Scheme 11).[63] The reverse reaction is negligible due to the high stability of the metal–carbon triple bond.[60] Restricted rotation around the molybdenum–amide bond results in the presence of two rotamers, as evident in the ^1H and $^{13}C\{^1H\}$ NMR spectra.[60]

Similarly, photolysis of the complex $TpW(=CH^tBu)Cl(NPh)$ induces rotational isomerism of the alkylidene ligand and proton transfer to generate $TpW(\equiv C^tBu)Cl(NHPh)$. The *syn* orientation of the amido substituent and the alkylidyne ligand positions the amide proton distal to the alkylidyne ligand, disfavouring the reverse tautomerisation reaction without prior amide rotation or catalysis.[59]

5. Synthesis *via* Alkyne Insertion

In a reaction that parallels the one described by Geoffroy and co-workers for the formation of $[CpW(\eta^3{=}CPhCPh{=}CHR)(CO)_2][BF_4]$,[64] protonation of (aryloxy) carbynes generates an agostic carbene complex which undergoes insertion of

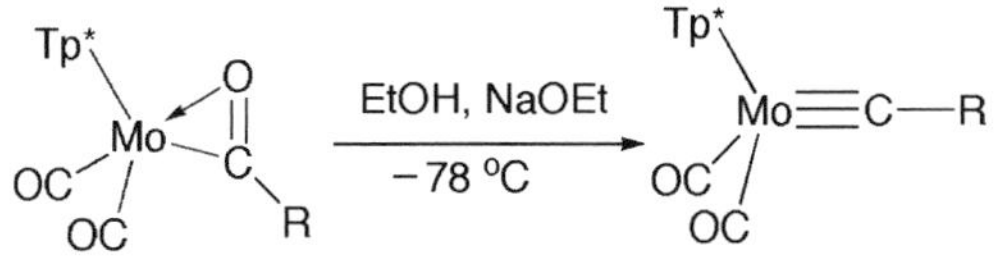

SCHEME 10. Alkylidyne preparation from $[Tp^*Mo(\eta^2\text{-OCR})(CO)_2]$ (R = Et, Me).

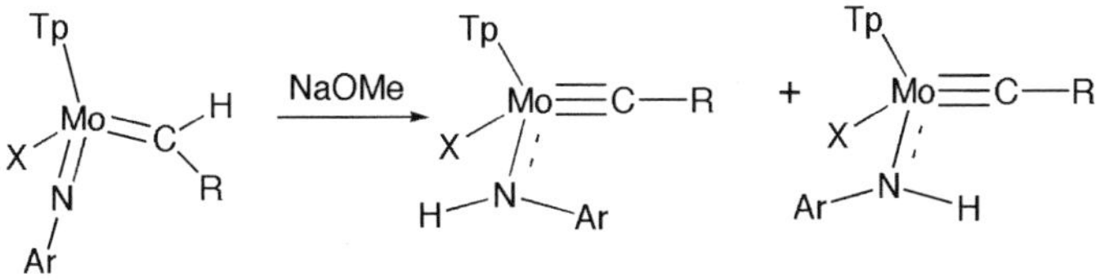

SCHEME 11. Alkylidene/imido–amido/alkylidyne conversion involving $TpMo(=CHR)(X)(NHAr)$ ($R = CMe_2Ph$; $Ar = C_6H_3{}^iPr_2\text{-}2,6$; X = OTf, OMe).

PhC$\equiv$CH to afford the cationic vinyl carbene complexes Tp*W(η^3=CPhCH=CHOAr)$(CO)_2^+$ (Ar = Ph, C_6H_4Me-4). Subsequent treatment with base affords the corresponding vinyl carbynes Tp*W($\equiv$CCPh=CHOAr)$(CO)_2$ in low yield.[65] The major product of this synthesis is, however, the metallafuran complex Tp*W(κ^2-*C*,*O*-=CPhCH=CHO)$(CO)_2$ (Scheme 12).[65]

cis-Insertion of alkynes into photolytically generated 'Tp*WH$(CO)_2$' species may produce alkylidyne complexes, η^2-vinyl, η^2-acyl, or metallafuran complexes, depending on the alkyne employed.[66] Initial *cis*-2,1-insertion is favoured by terminal alkynes, leading to the η^2-vinyl and η^3-allyl products. In contrast, alkynes with large substituents such as *t*-butyl acetylene instead favour *cis*-1,2-insertion. However, in this case, the η^2-vinyl complex is destabilised by steric interactions between the C_β-substituent and the Tp* methyls and further reactions occur, with a 1,2-H-migration from C_α affording the alkylidyne complexes or the (re)addition of CO providing η^2-acyls. In the absence of *trans* substituents, 1,2-migration to afford the carbyne is faster than CO trapping (Scheme 13).[66]

A departure from these generalisations is provided by the reaction with ethynyltrimethylsilane, which yields an η^2-vinyl complex Tp*W{η^2-C($SiMe_3$)=CH_2}$(CO)_2$ in addition to the alkylidyne complex Tp*W($\equiv$$CCH_2SiMe_3$)$(CO)_2$ in a 45:37 ratio, consistent with competing 1,2- and 2,1-insertion pathways. The vinyl complex, however, slowly undergoes 1,2-silyl migration ($>$5 days) to generate the alkylidyne complex.[66]

With *tert*-butyl acetylene, the η^2-acyl Tp*W{*trans*-η^2-OCCH=CH^{t}Bu}$(CO)_2$ forms as the major product, along with small amounts of the η^2-vinyl and alkylidyne complexes, Tp*W(η^2-C^{t}Bu=CH_2)$(CO)_2$ and Tp*W($\equiv$$CH_2$ ^{t}Bu)$(CO)_2$, respectively. With $Me_3SiC\equiv CCH_3$, the allyl complex Tp*W(η^3-*syn*-$CH_2CHCHSiMe_3$)$(CO)_2$ is produced in addition to a significant amount of the alkylidyne complex Tp*W($\equiv$CEt)$(CO)_2$, which is perhaps produced by hydrolysis of an η^2-vinyl intermediate. Thermal isomerisation of the η^3-allyl complex to the

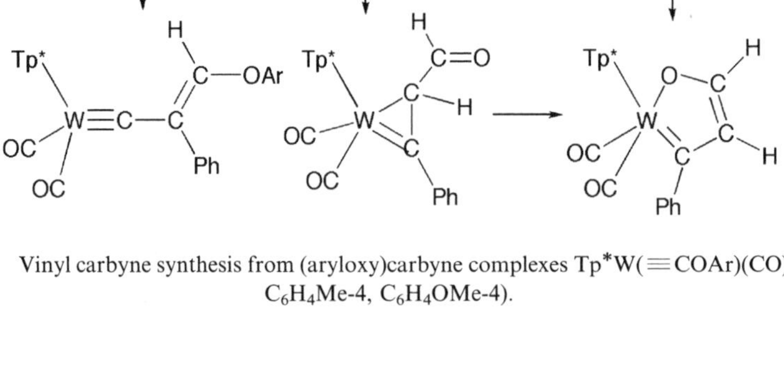

SCHEME 12. Vinyl carbyne synthesis from (aryloxy)carbyne complexes Tp*W($\equiv$COAr)$(CO)_2$ (Ar = Ph, C_6H_4Me-4, C_6H_4OMe-4).

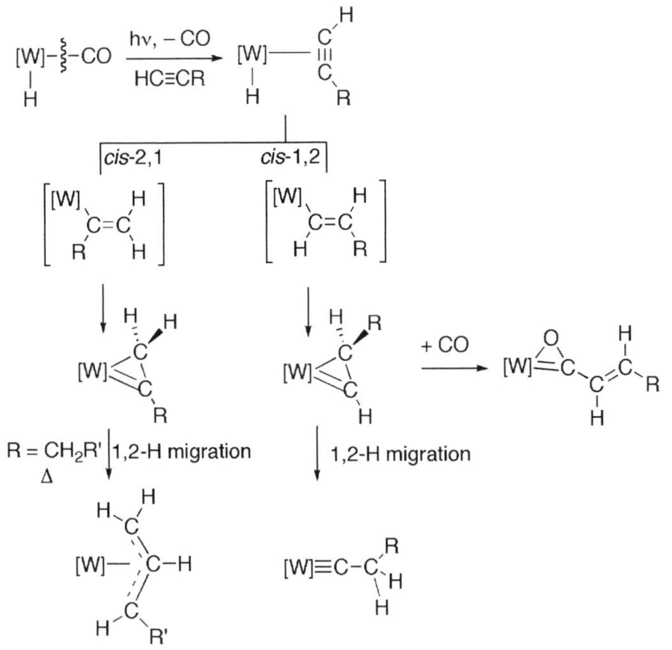

SCHEME 13. Alkyne insertion into the photogenerated 'Tp*WH(CO)$_2$' fragment ([W] = Tp*W(CO)$_2$).

η^2-vinyl species followed by a 1,2-H-shift produces the alkylidyne complex Tp*W(≡CCHMeSiMe$_3$)(CO)$_2$.

The equivalent alkyne-insertion reaction with PhC≡CR (R = H or Ph) gives only η^2-vinyl products, whereas RCH$_2$C≡CH (R = H or ^{n}Pr) gives η^2-vinyl and metallafuran products, though the same substrate with HC≡CH yields exclusively the ethylidyne complex Tp*W(≡CMe)(CO)$_2$.[66] For the latter, an identical reaction course resulted from stoichiometric addition of Li[HBEt$_3$] to a tetrahydrofuran solution of [Tp*W(HC≡CH)(CO)$_2$][OTf][66] and this complex may also be prepared by an alternative route involving elaboration of an alkylidyne ligand (*vide infra*).[67]

C. *Heteroatom Substituents*

1. Haloalkylidynes

The halomethylidyne complexes Tp*M(≡CX)(CO)$_2$ (M = Mo, W; X = Cl, Br and I) are prepared from the displacement of three mutually *cis* (*fac*) terminal carbonyls from metal hexacarbonyls and subsequent oxidation of the poly (pyrazolyl)borate complexes [Tp*M(CO)$_3$]$^-$ (M = Mo, W) as their [Et$_4$N]$^+$ salts with either [ArN$_2$]$^+$ (Ar = aryl) or [Ph$_2$I]$^+$ in the presence of haloalkanes (dichloromethane, bromoform or iodoform) (Scheme 14). The iodomethylidyne complex is unstable and formed in very low yield,[36,37] though an alternative route to this species was later demonstrated by Templeton.[68] In acetonitrile, the η^2-aroyl complexes $Tp^*M(\eta^2\text{-}OCAr)(CO)_2$ ($Ar = C_6H_4X\text{-}4$, $X = NO_2$, CN, COMe, CF_3, H, Me, OMe, NMe_2; $C_6H_4X\text{-}3$, $X = NO_2$, OMe) are instead produced.[36]

The relief of steric crowding and propensity for pseudo-octahedral geometries for the Tp^*M fragment are believed to provide the driving force behind formation of the chloromethylidyne complexes and the proposed synthetic mechanism is depicted in Scheme 15. A seven-coordinate dichloromethyl compound is suggested as an intermediate, with rearrangement to the η^2-dichloroacetyl complex being disfavoured in view of the less crowded coordination sphere which can be obtained *via* simultaneous loss of a proton from the dichloromethyl group and expulsion of CO with formation of anionic $[Tp^*Mo(=CCl_2)(CO)_2]^-$. Loss of chloride anion would then lead directly to the more cylindrically compact σ-chloromethylidyne ligand.[36]

Through comparative studies for a range of $[Tp^xM(CO)_3]^-$ analogues with different substituents in the pyrazolyl ring, Lalor demonstrated that the oxidation by arenediazonium cations occurred in response to the steric rather than the electronic effect of the 3-methyl substituents. However, further steric crowding in either the hydrotris(pyrazolyl)borate ligand or the diazonium cation promoted a reversion to the carbonyl-substitution pathway,[36] producing aryldiazenido complexes $Tp^xM(NNAr)(CO)_2$, which are also the products observed for the

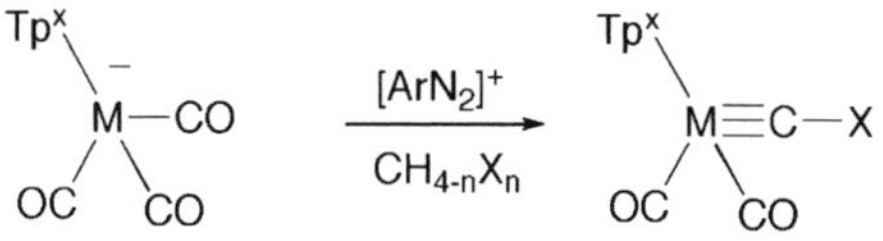

SCHEME 14. Synthesis of haloalkylidyne complexes $Tp^*M(\equiv CX)(CO)_2$ ($M = Mo$, W; $n = 2$, 3; $X = Cl$, Br, I).

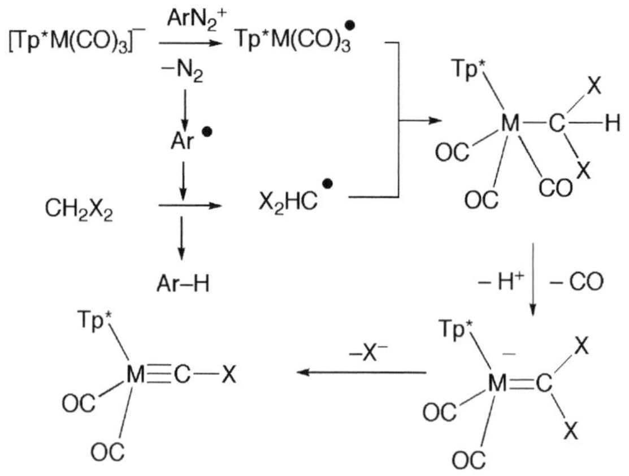

SCHEME 15. Proposed mechanism for haloalkylidyne synthesis.

parallel reaction with the more sterically modest Cp and Tp analogues (Scheme 16).[37]

A general and powerful approach to heteroatom-substituted molybdenum and tungsten alkylidyne complexes is based upon chloride displacement in these haloalkylidyne complexes by a variety of anionic as well as neutral nucleophiles.[69–71] Although some nucleophilic attack at the metal centre in $L_nM(\equiv CCl)$ might be expected to afford chloroketenyl derivatives by analogy with other alkylidynes (*vide infra*), the bulky Tp* ligand sphere seems to inhibit approach of the incoming nucleophile to the coordinatively saturated metal, resulting in clean reactions at C_α.[72]

2. Organothio-, Seleno- and Telluroalkylidynes

Nucleophilic displacement at the carbyne carbon of the terminal halomethylidynes $Tp^*M(\equiv CCl)(CO)_2$ (M = Mo, W) with alkyl and aryl thiolates[73] and phenylselenolate[73] produces thiolatomethylidyne and selenolatomethylidyne complexes (Scheme 17).

Methylthiocarbyne, methylselenocarbyne and methyltellurocarbyne complexes are also available by reacting the chlorocarbyne with Li_2S, Li_2Se or Na_2Te and subsequent methylation of the anionic carbon monochalcogenide complexes.[73] As mentioned earlier, the related tungsten complexes have also been prepared in much lower overall yields from the reaction of the anionic thiocarbonyl complex $[TpW(CO)_2(CS)]^-$ with reactive organic halides and the treatment of *trans*-$TpWI(CO)_2(CS)$ with organolithium compounds (Scheme 1).[23] The requisite

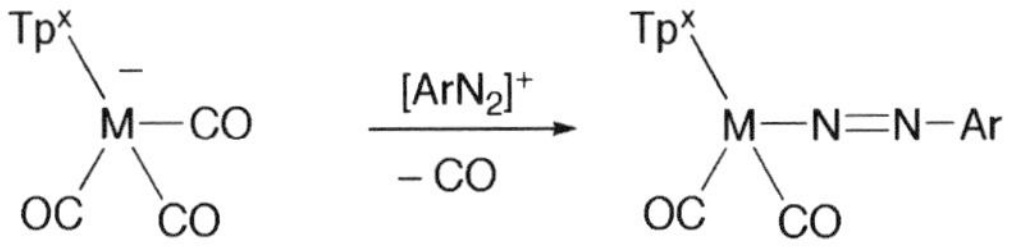

SCHEME 16. Carbonyl-substitution pathway favoured by increased steric crowding of N_2Ar or Tp^x (Ar = aryl).

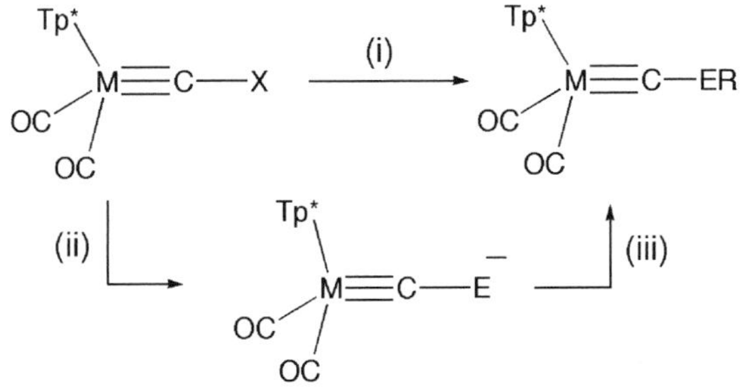

SCHEME 17. Reagents: (i) ER^- (E = S, Se; R = alkyl, aryl), (ii) Li_2E (E = S, Se, Te), and (iii) MeI (M = Mo, W).

$W(CO)_5(CE)$ (E = Se or Te) have yet to be reported, making this route unavailable for the heavier chalcogenolatoalkylidynes.

Angelici has reported a variety of novel synthetic routes to the thiocarbyne $TpW(\equiv CSMe)(CO)_2$. The reagents NaH, $NaBH_4$, NaOMe, NaOPh, PPN[SH], $NaSCH_2Ph$, NaSePh, LiMe, NEt_3, K_2CO_3, NH_2NH_2, $NHMeNH_2$ and NHMeNHMe react with the thiocarbene complex $[TpW(\eta^2\text{-}MeSCH)(CO)_2]^+$ not as nucleophiles but as Brønsted bases to give the thiocarbyne (in 10–20% yields) and $TpW(\eta^2\text{-}MeSCHSMe)(CO)_2$ (ca. 5–40%) as a side product.[74,75] The ylide complex $[TpW\{\eta^2\text{-}MeSCH(PEt_3)\}(CO)_2]^+$ can also be deprotonated at the methylidyne hydrogen with NaH, accompanied by phosphine dissociation to give the thiocarbyne compound in good yield (90%).[76] Although these transformations do not represent useful routes to $TpW(\equiv CSMe)(CO)_2$ being as it is the precursor to $[TpW(\eta^2\text{-}MeSCH)(CO)_2]^+$, they do illustrate fundamental transformations that might be extendable to other systems.

Similarly, the dithiocarbene cation $[TpW(\eta^2\text{-}MeSCSMe)(CO)_2]^+$, which is produced by addition of SMe^+ ($[MeSSMe_2]BF_4$) to $TpW(\equiv CSMe)(CO)_2$, reacts with the reducing agent sodium naphthalenide or the bases LiPh, $LiPPh_2$ and NaSePh to produce a 1:1 mixture of the MeS^- alkylidene adduct $TpW\{\eta^2\text{-}MeSC(SMe)_2\}(CO)_2$ and the regenerated thiocarbyne complex.[77] In reactions with thiolate nucleophiles (RS^-), release of the thiocarbyne occurs in addition to adduct formation, suggesting perhaps a competing electron transfer pathway, the operation of which increases with the increasing size of RS^-.[77] The dithiocarbene can also be used to regenerate Angelici's thiomethylidyne complex (along with $CpMo(SMe)(CO)_3$ or $Mn_2(\mu\text{-}SMe)_2(CO)_8$) *via* sulfenium ion transfer reactions with $[CpMo(CO)_3]^-$ or $[Mn(CO)_5]^-$, respectively (Scheme 18).[77]

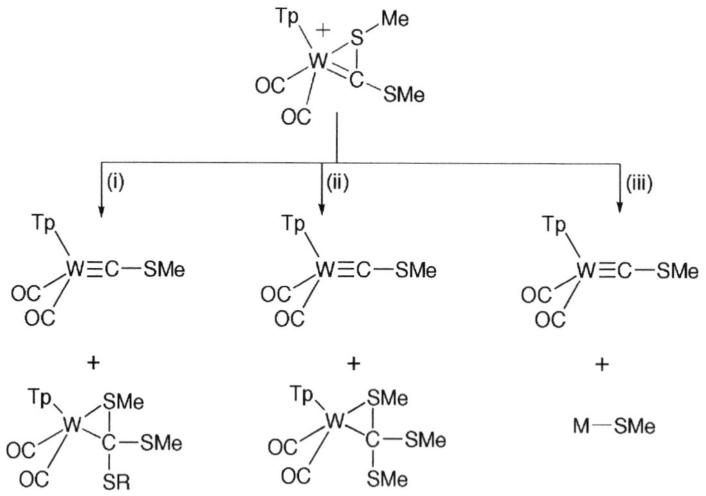

SCHEME 18. Reagents: (i) RS^- (R = Me, Et), (ii) $Na[C_{10}H_8]$, $LiPPh_2$, LiPh or NaSePh, and (iii) $[M]^-$ ($[M]^- = [Mo(CO)_3Cp]^-$ or $[Mn(CO)_5]^-$).

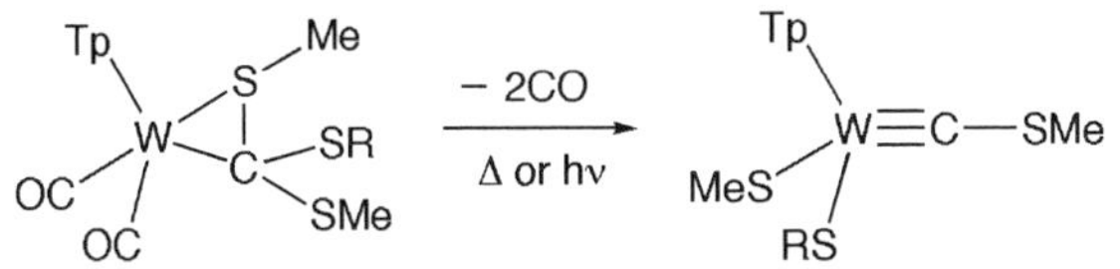

SCHEME 19. Double α-thiolate elimination to generate TpW(≡CSMe)(SMe)(SR) (R = alkyl, aryl).

Angelici's synthesis of TpW(≡CSMe)$(SMe)_2$ *via* the thermally induced double α-tungsten–thiolate elimination reaction of the complex TpW{η^2-MeSC $(SMe)_2$}$(CO)_2$ was the first example of the general class of alkylidyne complex $Tp^xW(\equiv CR)X_2$ (X = anionic ligand; Tp^x = Tp, Tp*).[19] α-Thiolate migration and concomitant CO loss from dithiocarbene complexes of the type TpW{η^2-MeSC (SR)(SMe)}$(CO)_2$ (R = Me, Et, Ph, C_6H_4Me-4) may be thermally (R = alkyl) or photolytically (R = aryl) induced (Scheme 19), though photolysis is usually employed at the expense of the yield. The resulting alkylidyne complexes TpW(≡CSMe)(SMe)(SR) are spectroscopically and crystallographically similar to the related dicarbonyl complex TpW(≡CSMe)$(CO)_2$, despite the difference in oxidation state and formal electron count of the metal centre.[19] Crystallographic evidence suggests a π-donor stabilising role for the thiolato ligands in the formally 16-electron complexes, with W–S bonds that are shorter than expected for simple thiolato ligands bound to an 18-electron metal centre. Under thermal conditions, the less nucleophilic and less basic arylthiolato derivatives eliminate disulfide and produce the known thiocarbyne complex TpW(≡CSMe)$(CO)_2$, proposed to proceed *via* concerted α-elimination of the disulfide.[19]

Photolysis of TpW{η^2-MeSCMe(SMe)}$(CO)_2$ similarly yields TpW(≡CMe) $(SMe)_2$; however, this reaction does not proceed under thermal conditions.[19] The proposed mechanism for the photolytic reactions involves UV-promoted CO loss followed by C–SR bond cleavage and complete transfer of the SR group to tungsten centre.[19]

3. Phosphonioalkylidynes

The SMe substituent is likewise a good nucleofugic leaving group and addition of excess PR_3 (R = Et, Me) to Tp*W(≡CSMe)$(CO)_2$ in dichloromethane yields the cationic phosphonium alkylidyne and bis(phosphonio)carbene complexes, $[Tp^*W(\equiv CPEt_3)(CO)_2]^+$ and $[Tp^*W\{=C(PMe_3)_2\}(CO)_2]^+$, respectively, which are isolated as their PF_6 salts by metathesis with NH_4PF_6.[72] The bis(trimethylphosphonio)carbene complex is in equilibrium with the monophosphonioalkylidyne complex in solution and the addition of MeI allows the isolation of the monophosphinio analogue $[Tp^*W(\equiv CPMe_3)(CO)_2]^+$ by trapping liberated Me_3P as $[Me_4P]I$.[72] This differs from the work of Angelici and Kim involving the Tp-substituted carbynes, which participated in nucleophile-induced carbonyl–carbyne coupling and ketenyl formation upon reaction with PEt_3.[76] The increased steric congestion of the Tp* system means that the reactions more closely resemble the nucleophilic substitution reactions described by Lalor, involving Tp*Mo(≡CCl)$(CO)_2$ wherein Cl^- is the nucleofuge.

Tertiary phosphines (PMe_2Ph, PPh_3, PCy_3) displace chloride from Tp*M(≡CCl)$(CO)_2$ (M = Mo, W) to give the cationic phosphoniocarbynes [Tp*W(≡CPR_3)$(CO)_2$][PF_6] in the presence of K[PF_6] and in a polar solvent (acetonitrile) (Scheme 20).[70] The structure of [Tp*W(≡$CPMe_2Ph$)$(CO)_2$][PF_6] has been crystallographically determined. High-energy CO stretching frequencies (ca. 2022 and 1934 cm^{-1}) reflect the cationic nature of the dicarbonyl products, from which it may be inferred that the phosphonioalkylidyne ligand is a particularly potent π-acid.[70]

4. Aryloxyalkylidynes

Aryloxides similarly displace chloride from Tp*M(≡CCl)$(CO)_2$ (M = Mo, W) with the formation of (aryloxy)carbyne complexes Tp*M(≡COAr)$(CO)_2$ (Ar = Ph, C_6H_4Me-4, C_6H_4OMe-4),[70] which are of interest in that they provide models for the elusive hydroxycarbyne ligand that is implicated in Fischer–Tropsch processes. The neutral PR_3 substituent in the cationic phosphoniocarbyne salts [Tp*W(≡CPR_2Ph)$(CO)_2$][PF_6] (R = Me, Ph) has proven to be an excellent leaving group, with electron-rich aryloxides (ArO^-) replacing the phosphine to afford (aryloxy)alkylidyne products Tp*W(≡COAr)$(CO)_2$. With electron-poor aryloxide nucleophiles, η^2-ketenyl complexes are also formed (Scheme 21).[78]

The product ratio of substitution at the carbyne carbon vs. carbonyl–carbyne coupling can be tuned by variation of the aryloxide *para*-substituent, which

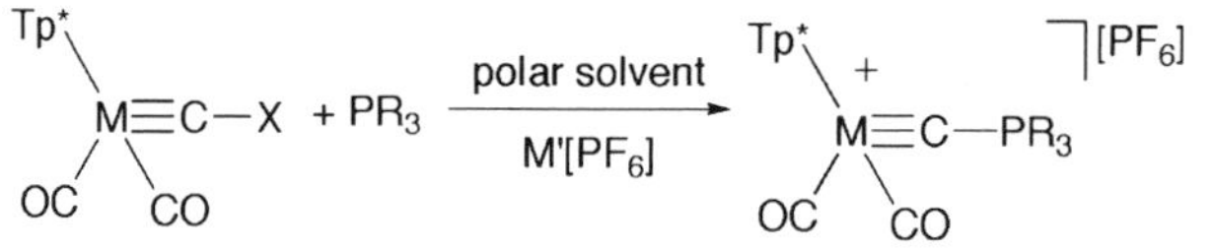

SCHEME 20. General synthesis of phosphonioalkylidynes $[Tp^*M(\equiv CPR_3)(CO)_2]^+$ (M = Mo, W; PR_3 = PMe_2Ph, PPh_3, PCy_3; X = leaving group: SMe, Cl; M′ = NH_4, Na).

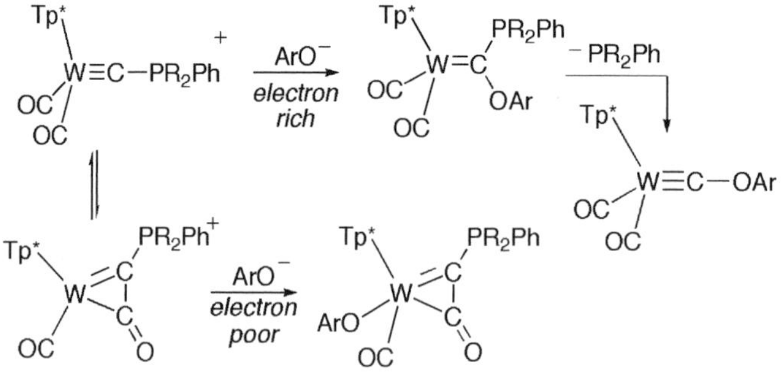

SCHEME 21. Available reaction pathways for the treatment of $[Tp^*W(\equiv CPR_2Ph)(CO)_2Tp^*]^+$ (R = Me, Ph) with aryloxide nucleophiles (ArO^-).

augments the nucleophilic strength.[78] A larger proportion of the alkylidyne product is formed with the dimethylphenylphosphonio cationic reagents than with the triphenylphosphonio reagents, perhaps due to the greater steric bulk of the '$L_nW(\equiv CPPh_3)^+$' fragment, which inhibits nucleophilic attack at the alkylidyne carbon atom.[78]

Addition of methoxide to the cationic phosphonioalkylidyne complexes $[Tp^*W(\equiv CPPh_3)(CO)_2]^+$ likewise allows the isolation of the thermal- and light-sensitive methoxycarbyne complex $Tp^*W(\equiv COMe)(CO)_2$ in 48% yield.[79] It is necessary to quench the reaction with excess MeI to prevent subsequent dealkylative conversion to the tricarbonyl anion $[Tp^*W(CO)_3]^-$. The electron-rich methoxy group results in ν_{CO} frequencies at 1958 and 1862 cm^{-1}, which are towards the low end of the range observed for neutral Tp* dicarbonyl tungsten complexes.[79]

5. Aminoalkylidynes

Group 6 aminocarbyne complexes $Tp^xM(\equiv CN^iPr_2)(CO)_2$ can be synthesised from *trans*-$M(\equiv CN^iPr_2)(O_2CCF_3)(CO)_3(PPh_3)$ (M = Cr, Mo, W)[80] or *trans*-$Cr(\equiv CN^iPr_2)X(CO)_4$ (X = Cl, Br)[81,82] in a manner analogous to the formation of aryl and alkyl alkylidynes. The thermally labile tetracarbonyl complexes are obtained from $M(CO)_6$ using the nucleophile LiN^iPr_2 (LDA) in a Mayr carbyne synthesis. Filippou and co-workers reported that thermal decarbonylation of the tetracarbonyl complex with γ-picoline resulted in the quantitative formation of $Cr(\equiv CN^iPr_2)X(CO)_2(pic)_2$ (X = Cl, Br) in 60–75% overall isolated yield. Treating the bromo complex with $Na[Cp^x]$ or $K[Tp^*]$ affords $LCr(\equiv CN^iPr_2)(CO)_2$ (L = Cp, Cp*, Tp*). The displacement reaction with Tp* occurs at a higher temperature but is lower yielding than that of the analogous cyclopentadienyl aminocarbyne syntheses.[81,82] The complexes $M(\equiv CN^iPr_2)(O_2CCF_3)(CO)_3(PPh_3)$ (M = Cr, Mo, W) reported by Hill equally provide convenient precursors for the synthesis of $TpM(\equiv CN^iPr_2)(CO)_2$ through a ligand exchange reaction with K[Tp].[80]

The isocyanide-substituted complexes can be correspondingly obtained following thermal decarbonylation of the bromotetracarbonyl species with *tert*-butyl isonitrile to produce the cationic aminocarbyne complex $[Cr(\equiv CN^iPr_2)(CO)(CN^tBu)_4]^+$. Subsequent treatment with Na[Cp] or $K[Tp^*]$ in tetrahydrofuran at 50 °C results in formation of $LCr(\equiv CN^iPr_2)(CO)(CN^tBu)$ (L = Cp or Tp* in 61 and 58% yield, respectively). In both reactions the minor product $[Cr(\equiv CN^iPr_2)(CN^tBu)_5]^+$ is formed by a competitive carbonyl-substitution reaction of the tetra(isonitrile) complex with liberated CN^tBu.[81,82]

Electrophilic addition to anionic molybdenum isocyanide derivatives $[Na][Tp^*Mo(CNR)(CO)_2]$ is dominated by direct attack at nitrogen to form alkylidyne products; however, alkylation at the metal followed by isonitrile insertion can lead to η^2-iminoacyl complexes in some cases. The aminocarbyne products $Tp^*Mo(\equiv CNRMe)(CO)_2$ are formed in high yield for methyl and phenyl isocyanide reagents. However, the formation of the corresponding *tert*-butylamino-alkylidyne complex is accompanied by formation of η^2-acyl and η^2-iminoacyl complexes (Scheme 22).[83]

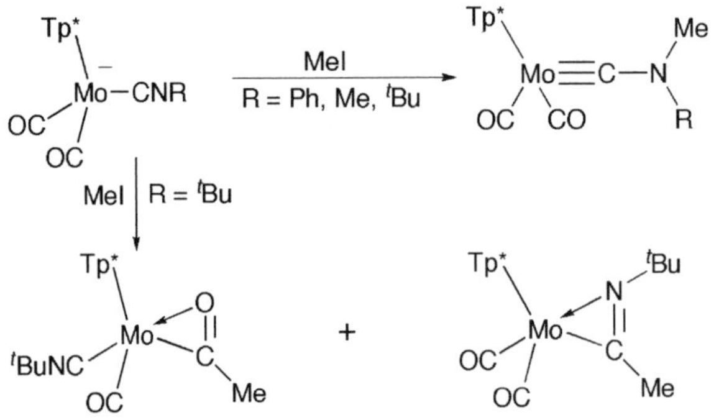

SCHEME 22. Synthesis of aminoalkylidynes $Tp^*Mo(\equiv CNRMe)(CO)_2$ from electrophilic addition to coordinated isonitriles (R = ^{t}Bu, Ph, Me).

Increased nucleophilicity at nitrogen and steric hindrance at the metal centre promote formation of the aminoalkylidyne complexes. It is suggested that the bulky *tert*-butyl substituent serves to shield the nitrogen atom.[83] In a similar fashion, addition of HCl to the anionic $[Tp^*Mo(CN^tBu)(CO)_2]^-$ generates the neutral alkylidyne complex $Tp^*Mo(\equiv CNH^tBu)(CO)_2$, rather than the hydrido tautomer $Tp^*MoH(CO)_2(CN^tBu)$.[83] A large-scale, high-yield, stepwise synthesis of the aminoalkylidyne complexes $Tp^*W(\equiv CNREt)(CO)_2$ (R = Me, Et) from $Tp^*WI(CO)_3$ has been reported. Thermal decarbonylation of $Tp^*WI(CO)_3$ in the presence of CNEt gives *cis*-$Tp^*WI(CNEt)(CO)_2$. Subsequent reduction with Na/Hg provides the metalate $Na[Tp^*W(CNEt)(CO)_2]$, which can be alkylated with RI (R = Me, Et) exclusively at the isocyanide nitrogen to give the aminoalkylidyne complexes.[84] In contrast, the metalates $Na[Cp^xW(CNEt)(CO)_2]$ (Cp^x = Cp, Cp*) undergo alkylation with RI at the metal centre to afford the W(II) alkyl complexes *cis*/*trans*-$Cp^xWR(CNEt)(CO)_2$. The difference in reactivity is ascribed to the steric demands of the Tp* ligand, which shields the metal centre from the incoming electrophile.[84]

By following a related stepwise procedure, aminocarbyne complexes can be used to prepare rare examples of mononuclear bis(aminocarbyne) complexes.[85] Oxidative decarbonylation of $Tp^*W(\equiv CNREt)(CO)_2$ (R = Me, Et) with Br_2 or I_2 affords the six-coordinate dihalido-aminocarbyne complexes $Tp^*W(\equiv CNREt)X_2$ (X = Br, I) in high yield. Subsequent reductive dehalogenation by Na/Hg in the presence of CNEt gives the electron-rich mono-aminocarbyne complexes $Tp^*W(\equiv CNREt)(CNEt)_2$. An unusually low-field chemical shift for the metal-bonded isonitrile carbons as well as low-frequency ν_{CN} absorptions confirms the presence of a very electron-rich metal centre. Alkylation with $[Et_3O][BF_4]$ at one isocyanide nitrogen yields the bis(aminocarbyne) salts $[Tp^*W(\equiv CNREt)(\equiv CNEt_2)(CNEt)][BF_4]$ (Scheme 23).[85]

The introduction of amino substituents on both carbyne carbons stabilises the bis(carbyne) form relative to the carbyne–carbyne coupled alkyne isomer due to strong interaction of the π-type lone pair at the amino nitrogen with the two vacant π-orbitals at the carbyne carbon atoms.[85] Calculations reproduce the preference for the bis(alkylidyne) formulation.

The thiocarbene $[TpW(\eta^2\text{-CHSMe})(CO)_2]^+$ reacts with the primary amines NH_2R to produce the related aminoalkylidyne compounds $TpW(\equiv CNHR)(CO)_2$ in 25–35% yield, which are in equilibrium with their hydride–isocyanide tautomers $TpWH(CNR)(CO)_2$ (R = Me, Et, CH_2CH_2OH, ^{i}Pr, ^{t}Bu) (Scheme 24).[75] The related Cp complex $CpMoH(CNMe)(CO)_2$ shows no evidence for the alkylidyne tautomer,[86] which suggests that it may be the preference of the Tp complexes for six- rather than seven-coordination[13] or the greater electron donor ability of Tp that favours the aminoalkylidyne structure in contrast to the Cp system. If there exists a similar equilibrium for tautomerisation in the Tp* aminoalkylidyne complexes, it greatly favours the alkylidyne version, presumably due to further steric congestion at the metal centre.[83]

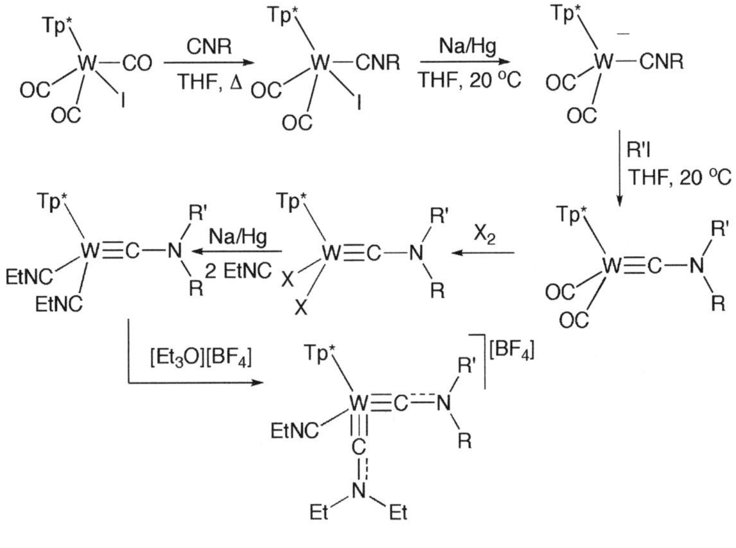

SCHEME 23. Stepwise preparation of mono- and bis(aminoalkylidyne) complexes of tungsten (R, R′ = Me or Et; X = I or Br).

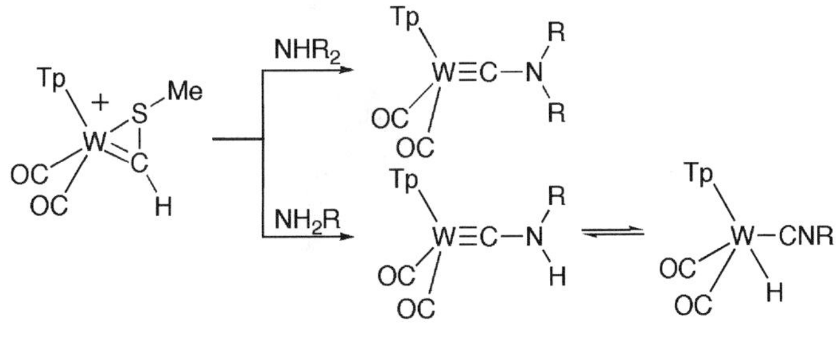

SCHEME 24. Preparation of $TpW(\equiv CNHR)(CO)_2$ (R = Et, Me, CH_2CH_2OH, ^{i}Pr, ^{t}Bu).

The reaction of the thiocarbene $[TpW(\eta^2\text{-}CHSMe)(CO)_2]^+$ with *N*,*N*-dimethyl hydrazine (NH_2NMe_2) produces the dimethylaminocarbyne complex $TpW(\equiv CNMe_2)(CO)_2$ (28% yield) along with $TpW(\eta^2\text{-}SMeCHSMe)(CO)_2$ (28% yield) and a third product not sufficiently stable to be isolated.[75] With secondary amines ($HNMe_2$, $HNEt_2$), the same complex $[TpW(\eta^2\text{-}CHSMe)(CO)_2]^+$ reacts to form air-stable aminocarbyne complexes $TpW(\equiv CNR_2)(CO)_2$ in approximately 30% yield.[75]

6. Phospha- and Arsaalkenyl Alkylidynes

Condensation of Lalor's chlorocarbyne with *P*-silylated phosphaalkenes $Me_3SiP{=}C(NR_2)_2$ (R = Me, Et) produces phosphaalkenyl-functionalised carbyne complexes *via* nucleophilic displacement of the chloride atom (loss of $ClSiMe_3$).[87] In an analogous reaction requiring the exclusion of light, the complexes $Tp^*M\{\equiv CAs{=}C(NMe_2)_2\}(CO)_2$ (M = Mo, W) are prepared by condensation with the corresponding arsaalkene $Me_3SiAs{=}C(NMe_2)_2$. Three mesomeric structures are considered to contribute to a full description of the bonding in these alkylidyne complexes (Scheme 25).[88]

The π-basicity of the organophosphorus and organoarsenic ligands is reflected in their highly deshielded C_α signals (δ_C 318–334 for the phosphaalkenyls; δ_C 330–350 for the arsaalkenyls) compared to those of the starting chloromethylidyne complexes (δ_C 206–209). This property is also reflected in the infrared ν_{CO} stretches at ca. 1950 and 1850 cm^{-1}, which are at much lower frequencies than for the precursor alkylidyne complexes at ca. 2005 and 1921 cm^{-1}.[87,88]

7. Silyl Alkylidynes

By employing the dimethylphenylsilyl anion as nucleophile to initiate Mayr's multistep Fischer-carbyne synthesis,[30,38] and later using the Tp* anion as a capping tridentate ligand, new alkylidyne complexes in which a silyl group is directly attached to the methylidyne carbon can be obtained, i.e., $Tp^*M(\equiv CSiMe_2Ph)(CO)_2$ (M = Mo, W).[89] Alternatively, deprotonation of the parent methylidyne complex $Tp^*W(\equiv CH)(CO)_2$ and subsequent silylations ($ClSiMe_3$) also generates

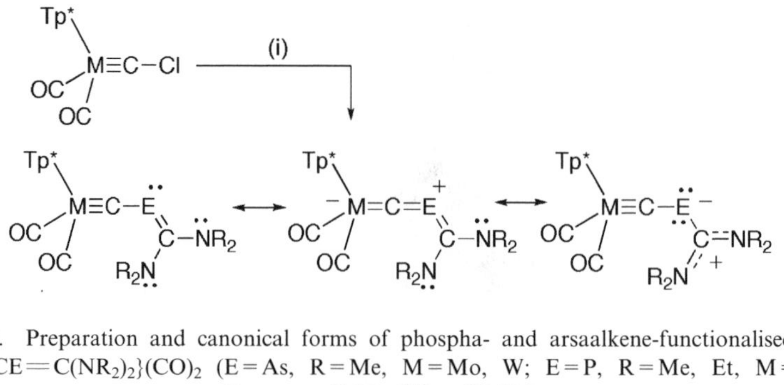

Tp*(OC)₂M≡C–Cl —(i)→ Tp*(OC)₂M≡C–Ë=C(NR₂)(NR₂) ⟷ Tp*(OC)₂M⁻=C=E⁺–C(NR₂)(NR₂) ⟷ Tp*(OC)₂M≡C–Ë⁻–C(=NR₂)(NR₂)⁺

SCHEME 25. Preparation and canonical forms of phospha- and arsaalkene-functionalised carbynes $Tp^*M\{\equiv CE{=}C(NR_2)_2\}(CO)_2$ (E = As, R = Me, M = Mo, W; E = P, R = Me, Et, M = Mo, W). Reagents: (i) $Me_3SiE{=}C(NR_2)_2$.

the silyl alkylidyne complex of the type Tp*W(≡CSiMe$_3$)(CO)$_2$.[68] Poly(pyrazole) borate co-ligated germyl-, stannyl- and plumbyl-substituted alkylidynes appear unknown at present.

8. Metal-Substituted Alkylidynes (μ-Carbido Complexes)

Good yields of Tp*Mo{≡CFe(CO)$_2$Cp}(CO)$_2$ result from the reaction of Lalor's chlorocarbyne with K[CpFe(CO)$_2$] (Scheme 26). The ^{13}C{^{1}H} NMR spectrum shows the bridging carbon at δ_C 381, which is near the range expected for C_α vinylidene shifts and below those typical of alkylidyne complexes, suggesting that a vinylidene resonance form in which both metals have formal double bonds to the carbido bridge may contribute, at least in part, to the bonding. The crystal structure determination revealed a near-linear bridge with an Fe–C–Mo angle of 172.2(5)° and a short Mo–C separation that is consistent with a normal Mo≡C triple bond.[71]

D. *Metals other than Cr, Mo and W*

The vast majority of poly(pyrazolyl)borate alkylidyne chemistry is concerned with the Group 6 metal triad Cr, Mo and W, however, rare examples of Group 7 and 8 complexes have also been isolated.

1. Group 7 Alkylidynes

Schrock and co-workers have described the synthesis of a rhenium neopentylidyne complex containing the tris(pyrazolyl)borate ligand from Re(≡C^{t}Bu)(CH$_2$^{t}Bu)$_3$(OTf).[90] Treatment of this complex with excess pyridine afforded the colourless, six-coordinate rhenium neopentylidene/neopentylidyne complex Re(≡C^{t}Bu)(OTf)(CH$_2$^{t}Bu)(=CH^{t}Bu)(py)$_2$], which was subsequently subject to ligand substitution by Na[Tp], producing the thermally stable, 18-electron complex TpRe(≡C^{t}Bu)(CH$_2$^{t}Bu)(=CH^{t}Bu) (Scheme 27).[91] Rhenium is one of the three

Scheme 26. Synthesis and canonical forms of Tp*Mo{≡CFe(CO)$_2$Cp}(CO)$_2$. (i) K[CpFe(CO)$_2$].

Scheme 27. Synthesis of TpRe(≡C^{t}Bu)(CH$_2$^{t}Bu)(=CH^{t}Bu) (R = ^{t}Bu).

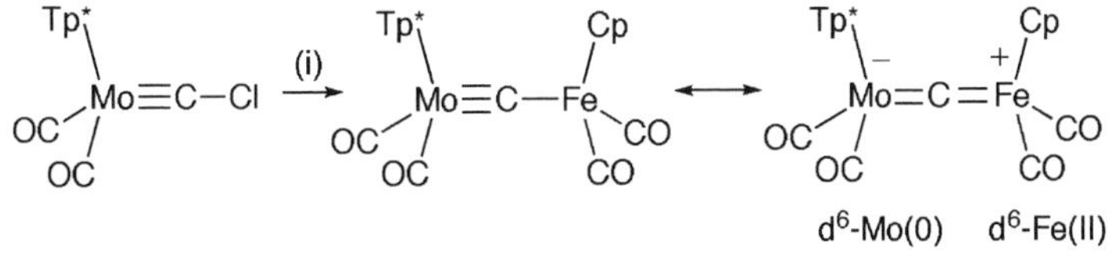

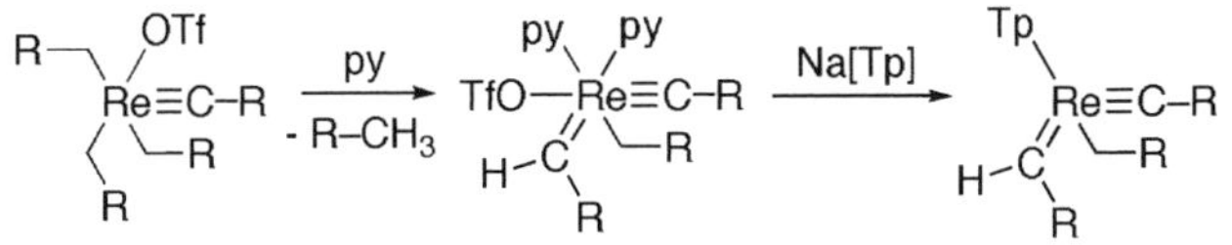

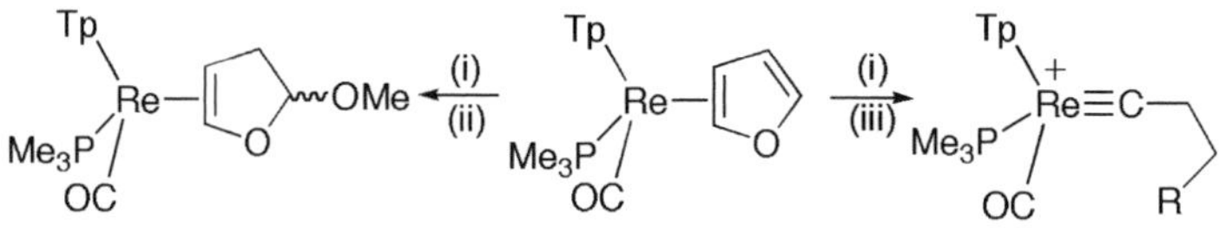

SCHEME 28. Reagents: (i) HOTf, MeOH, (ii) CH_2Cl_2, −40 °C, and (iii) THF, 25 °C (R = CH=O, $CH(OMe)_2$).

metals (along with molybdenum and tungsten) that are active in classic olefin metathesis systems; however, this alkylidyne complex showed no metathesis activity, due in part to an 18-electron configuration, a LUMO which is not metal-based,[92,93] and the low lability of the multidentate Tp ligand, which does not allow for a vacant coordination site for initial olefin binding prior to metallacyclobutane formation.[91]

The synthetic route to the rhenium(V) complex $[Tp^*Re(\equiv CC_6H_2Me_3\text{-}2,4,6)(CO)_2][OTf]$, like the Group 6 benzylidyne analogues, involves ligand substitution from the labile tetracarbonyl species $[Re(\equiv CC_6H_2Me_3\text{-}2,4,6)Cl(CO)_4][OTf]$ with K[Tp*].[94] However, in contrast to the related Group 6 systems in which the calculated energy and composition of the near frontier orbitals assigns the HOMO to a metal d orbital that is non-bonding with respect to the benzylidyne ligand, the π-system of the aryl substituent of $[Tp^*Re(\equiv CC_6H_2Me_3\text{-}2,4,6)(CO)_2][OTf]$ is in conjugation with the metal–carbon triple bond.[95,96]

In dichloromethane solution, low-temperature, acid-promoted addition of methanol to the furan complex $TpRe(\eta^2\text{-}C_4H_3O)(CO)(PMe_3)$ generates two diastereomers of the complex TpRe(4,5-η^2-2β-methoxy-2,3-dihydrofuran)(CO)(PMe_3).[97] However, if the reaction is performed at ambient temperature in tetrahydrofuran solution, rather than formation of the 2-methoxy-2,3-dihydrofuran ligand, oxidation of the metal by triflic acid yields the ring-opened alkylidyne complexes $[TpRe(\equiv CCH_2CH_2R)(CO)(PMe_3)]^+$ (R = CH=O, $CH(OMe)_2$) (Scheme 28).[97] These and other arene dearomatisation transformations are detailed in the subsequent chapter.

2. Group 8 Alkylidynes

Prior to the synthesis of the rhenium neopentylidyne complexes discussed above, osmium analogues were prepared using a similar synthetic strategy.[92] Treatment of the salt $[PPh_4]_2[OsO_2Cl_4]$ with excess dineopentylzinc afforded the dioxo complex $Os(=O)_2(CH_2{}^tBu)_2$, which underwent an alkylidene/oxo exchange reaction when treated with two equivalents of the tantalum neopentylidene complex $Ta(=CH^tBu)(CH_2{}^tBu)_3$. α-Hydrogen abstraction from the resulting bis(alkylidene) complex $Os(=CH^tBu)_2(CH_2{}^tBu)_2$ with pyridinium triflate (in the presence of excess pyridine) provided neopentane and the alkylidyne complex $Os(\equiv C^tBu)(CH_2{}^tBu)_2(OTf)(py)_2$. Subsequent treatment with Na[Tp] affords the crystallographically characterised complex $TpOs(\equiv C^tBu)(CH_2{}^tBu)_2$ (Scheme 29).[92,93] The proposed intermediate $Os(=CH^tBu)(CH_2{}^tBu)_3(OTf)$ appears similar to the stable d^0 rhenium complex $Re(\equiv C^tBu)(CH_2{}^tBu)_3(OTf)$; however, in the latter only a single α-proton is available for abstraction and two metal–ligand π-bonds are available to stabilise the metal in the high oxidation state.[91] As for the rhenium neopentylidyne complex $TpRe(\equiv C^tBu)(CH_2{}^tBu)(=CH^tBu)$, the metal–carbon multiple bond in the 18-electron osmium complex is remarkably unreactive towards olefin metathesis.

The *in situ* formation of tris(pyrazolyl)borate–ruthenium carbyne complexes resulting from electrophilic addition at C_β of ruthenium vinylidene or allenylidene precursors has recently been reported.[98,99] Kirchner and co-workers described the reversible formation of the novel ruthenium vinyl carbyne complexes $[TpRu(\equiv CCH=CR'_2)Cl(PR_3)]^+$ ($PR_3 = PPh_2{}^iPr$, $R' = Ph$, Fc; $PR_3 = PPh_3$, $R' = Ph$) in dichloromethane solution *via* protonation of the neutral ruthenium allenylidene precursor $TpRu(=C=C=CR'_2)Cl(PR_3)$ with triflic acid.[99] Similarly, formation of the ruthenium alkylidyne salts $[TpRu(\equiv CCH_2R)Cl(PPh_3)][BF_4]$ ($R = {}^tBu$, ${}^nBu$, Ph) has been observed from the low-temperature treatment of the vinylidene complexes $TpRu(=C=CHR)Cl(PPh_3)$ with excess $HBF_4 \cdot Et_2O$ (Scheme 30). Pentamethylcyclopentadienyl-supported systems ($R = {}^tBu$ and ${}^nBu$) exhibit increased stability over the corresponding tris(pyrazolyl)borate analogues, being isolable in the solid state.[98] Furthermore, the alkyl carbyne substituents prove to be more stable than the corresponding benzyl carbyne derivatives, showing stability up to –20 °C (cf. –70 °C for R = Ph).

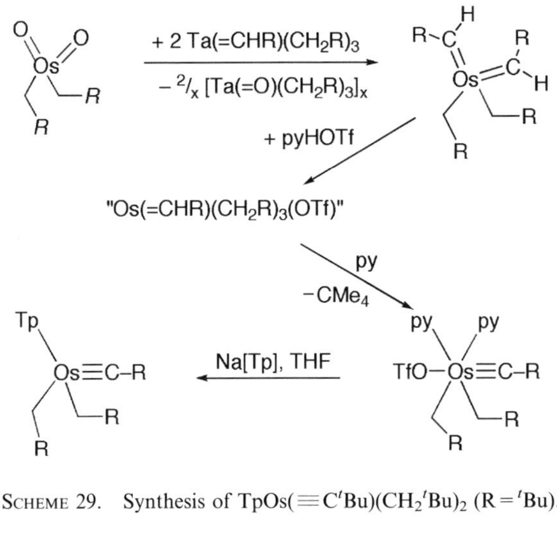

SCHEME 29. Synthesis of $TpOs(\equiv C^tBu)(CH_2{}^tBu)_2$ ($R = {}^tBu$).

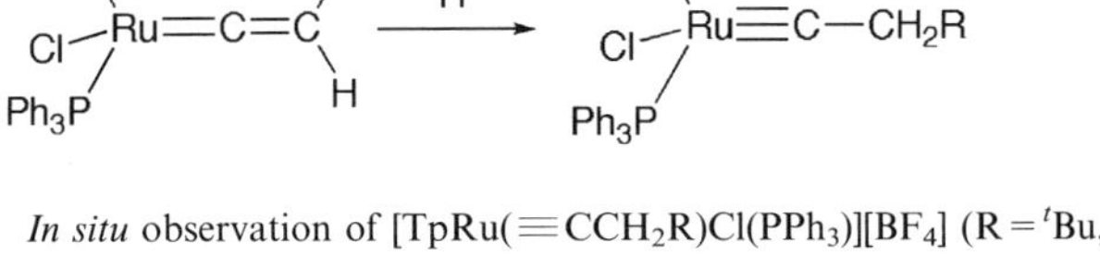

SCHEME 30. *In situ* observation of $[TpRu(\equiv CCH_2R)Cl(PPh_3)][BF_4]$ ($R = {}^tBu$, ${}^nBu$, Ph).

E. *Alkylidyne Complexes Ligated by Poly(pyrazolyl)methanes*

The pyrazolylalkane ligands, $HC(pz)_3$, $MeC(pz)_3$ and $H_2C(pz)_2$ (tris(pyrazolyl)-methane, tris(pyrazolyl)ethane and bis(pyrazolyl)methane, respectively), shown in Chart 3 are neutral analogues of the corresponding pyrazolylborate versions and display remarkably similar ligand properties.

The preparative routes for obtaining Group 6 alkylidyne complexes co-ligated by the poly(pyrazolyl)alkane ligands parallel those previously described for Tp^x, involving ligand substitution from halotetracarbonyl precursors $M(\equiv CR)Br(CO)_4$. However, alkylidyne complexes of the neutral pyrazolylalkane ligands differ from the complexes bearing pyrazolylborate co-ligands by being positively charged, and consequently a non-coordinating anion source (in this case $TlBF_4$) must also be included (Scheme 31).[100] The cationic thiocarbyne complex $\{HC(pz)_3\}[W(\equiv CSMe)(CO)_2]^+$ has also been prepared by alkylation of the nucleophilic sulfur atom in the thiocarbonyl complex $\{HC(pz)_3\}W(CO)_2(CS)$.[101]

The spectroscopic properties and the reactivity towards low-valent metal fragments of these cationic alkylidyne salts mirror those of the pyrazolylborate analogues, with one notable exception being that pyrazolylmethane ligands exhibit a (protic) reactivity at the bridgehead carbon, allowing for modification of the ligand within the alkylidyne complex (Section IV.E), a phenomenon which has not been similarly documented for $Tp^xL_nM(\equiv CR)$ complexes (hydridic B–H bond).

The anionic, *N,N,O*-chelating ligands, bis(3,5-dimethylpyrazolyl)acetate, $[HC(pz^*)_2(CO_2)]^-$,[61] and bis(3,5-dimethylpyrazolyl)(2-phenoxy)methane, $[HC(pz^*)_2(C_6H_4O\text{-}2)]^-$,[102] are unusual asymmetrical analogues of the Tp ligand that have also been employed to prepare alkylidyne complexes of the type $LW(\equiv CR)(CO)_2$ ($L = HC(pz^*)_2(C_6H_4O\text{-}2)$, $R = Me$, Ph, $C_6H_4Me\text{-}2$, $C_6H_3Me\text{-}2\text{-}OMe\text{-}4$, $C_6H_2Me_2\text{-}4,5\text{-}OMe\text{-}2$, $C_6H_3OMe\text{-}4$, $C_6H_2Me_3\text{-}2,4,6$, $C_6H_4Me\text{-}Cr(CO)_3$; $L = HC(pz^*)_2(CO_2)$, $R = C\equiv CPh$, $C\equiv CSiMe_3$).

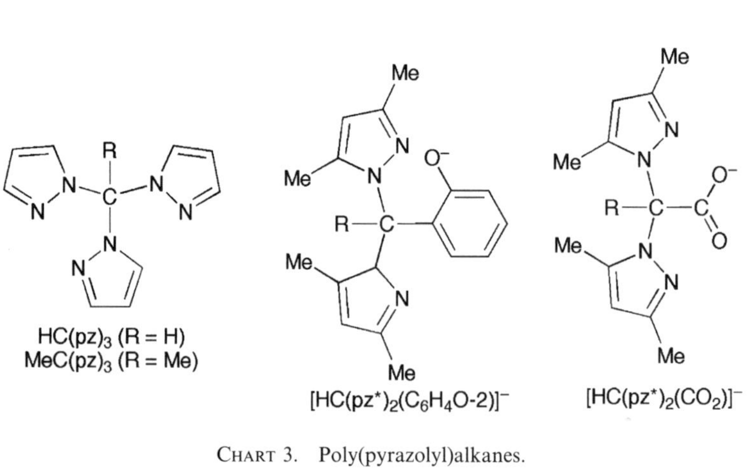

$HC(pz)_3$ (R = H)
$MeC(pz)_3$ (R = Me)

$[HC(pz^*)_2(C_6H_4O\text{-}2)]^-$

$[HC(pz^*)_2(CO_2)]^-$

CHART 3. Poly(pyrazolyl)alkanes.

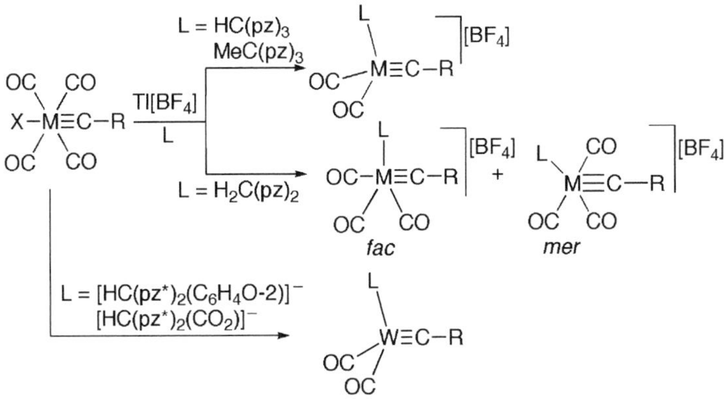

SCHEME 31. Synthesis of $[LM(\equiv CR)(CO)_n]^{x+}$ (M = Mo or W; R = alkyl, aryl or alkynyl; $x = 1$, L = $HC(pz)_3$, $MeC(pz)_3$, $n = 2$; $x = 1$, L = $H_2C(pz)_2$, $n = 3$; $x = 0$, L = $HC(pz^*)_2(CO_2)$, $HC(pz^*)_2(C_6H_4O\text{-}2)$, $n = 2$).

III
STRUCTURE AND SPECTROSCOPY

A. $^{13}C\{^1H\}$ NMR

$^{13}C\{^1H\}$ NMR resonances of the alkylidyne carbon nucleus and (where applicable) the metal carbonyls are given in Table I. Chemical shifts for the Group 6 alkylidyne carbons, C_α, range from δ_C 183.2 in $Tp^*W(\equiv CI)(CO)_2$[68] to δ_C 381 for $Tp^*Mo\{\equiv CFe(CO)_2Cp\}(CO)_2$[71] and are highly dependent on the electron-donating ability of the alkylidyne substituent. For example, for $Tp^*W(\equiv CR)(CO)_2$ the alkylidyne resonance moves from high to low field with changes in R in the order Cl (205.6)[36] > OPh (219.3)[70] > PPh_3^+ (242.1)[70] > C≡CPh (247.3)[61] ≈ NEtMe (249.0)[84] > Ph (277.9)[78] > Me (289)[67] > $P{=}C(NMe_2)_2$ (318.3).[87] The alkylidyne resonances are, however, rather insensitive to changes in the tripodal ligand. For the series $Tp^xW(\equiv CC_6H_4Me\text{-}4)(CO)_n$, the C_α resonance moves from high to low field with $Tp^x = Tp^*$ (279.6)[43] > Bp ($n = 3$) (284.5)[46] ≈ Tp (284.8)[39,40] > pzTp (286.4).[39,40] The greater electron-donating ability of Tp^* should, in principle, lead to stronger metal–alkylidyne back-bonding and a more deshielded position for C_α; however, as noted, variations are small and unlikely to be usefully diagnostic. For L = Cp and Cp* these resonances appear at δ_C 300.1[103] and 301.3,[190] respectively.

The chromium alkylidyne carbon and carbonyl resonances appear at lower field than the analogous molybdenum or tungsten compounds, which is consistent with the $^{13}C\{^1H\}$ shielding trend observed for the Group 6 metal triad. The resonances are also somewhat sensitive to the oxidation state of the metal with the $Tp^xM(\equiv CR)X_2$ complexes having C_α typically 20 ppm lower field than the analogous $Tp^xM(\equiv CR)(CO)_2$ complexes. There is, however, an exception to this rule. For $TpW(\equiv CSMe)(SMe)_2$ the alkylidyne carbon resonance is virtually unaffected by the increase in oxidation state of the tungsten with C_α appearing at δ_C 268.3,[19]

TABLE I

$^{13}C\{^{1}H\}$ NMR DATA FOR THE ALKYLIDYNE CARBON AND CO-LIGAND CARBONYL(S)[a]

Compound	δ M≡CR (ppm) (J, Hz)	δ M(CO) (ppm) (J, Hz)	Ref.
Alkyl, aryl, alkynyl, vinyl			
$[Cr(\equiv CC_6H_4Me\text{-}4)(CO)_2Tp]$	310.0	231.6	41
$[Mo(\equiv CMe)(CO)_2Tp^*]$	304.0	225.0	62
$[Mo(\equiv CMe)(CO)(NCCD_3)Tp^*]$	291.9	248.6	62[g]
$[Mo(\equiv CMe)(CO)(PMe_3)Tp^*]$	288.1 ($^2J_{CP}$ 20)	243.0 ($^2J_{CP}$ 10)	62
$[Mo(\equiv CEt)(CO)_2Tp^*]$	310.6	225.6	62
$[Mo(\equiv CPr)(CO)_2Tp^*]$	310.5	225.8	62
$[Mo(\equiv CBu)(CO)_2Tp]$	315.9	225.9	145
$[Mo(\equiv CBu)(CO)_2Tp^*]$	314.4	226.1	62
$[Mo(\equiv CBu)(CO)\{P(OMe)_3\}Tp]$	304.2 ($^2J_{CP}$ 31)	240.7 ($^2J_{CP}$ 15)	145
$[Mo(\equiv CPh)(CO)_2Tp^*]$	288.8	226.2	62,73
	288.1	225.9	73[b]
$[Mo(\equiv CC_6H_4Me\text{-}4)(CO)_2Tp]$	293.1	224.9	41,42
$[Mo(\equiv CC_6H_4Me\text{-}4)(CO)_2Tp^*]$	288.9	225.8	44[b]
$[Mo(\equiv CC_6H_4Me\text{-}4)(CO)(PPh_3)Tp]$	NR	NR	136
$[Mo(\equiv CC_6H_4OMe\text{-}4)(CO)_2Tp]$	294.2	226.0	48[b]
$[Mo(\equiv CC_4H_3S\text{-}2)(CO)_2Tp]$	276.8	225.9	48[b]
$[Mo(\equiv CC\equiv C^tBu)(CO)_2Tp]$	259.8	226.0	52
$[Mo(\equiv CC\equiv C^tBu)(CO)_2Tp^*]$	256.7	226.6	52
$[Mo\{\equiv C(CMe)_2Ph\}(OMe)(NHC_6H_3{}^iPr_2\text{-}2,6)Tp]$	301.3[k]/ 298.7[l]		60[h]
$[Mo\{\equiv C(CN)Ph_2\}(CO)_2Tp^*]$	269.47	224.20	189[b]
$[Mo(\equiv CCN)(CO)_2Tp^*]$	229.3	226.1	36[f]
$[W(\equiv CLi)(CO)_2Tp^*]$	556	232.0 ($^1J_{CW}$ 197)	68[e]
$[W(\equiv CH)(CO)_2Tp^*]$	280.6 ($^1J_{CW}$ 192)	224.5 ($^1J_{CW}$ 169)	68,89,120
$[W(\equiv CMe)(CO)_3Bp]$	293.9 ($^1J_{CW}$ 189)	222.3 ($^1J_{CW}$ 168)	46
$[W(\equiv CMe)(CO)_2Tp]$	295.2 ($^1J_{CW}$ 189)	223.3 ($^1J_{CW}$ 169)	39,40,42
$[W(\equiv CMe)(CO)_2Tp^*]$	289	223 ($^1J_{CW}$ 168)	62,68,158[b]
$[W(\equiv CMe)(SMe)_2Tp]$	303.9		19
$[W(\equiv CCH_2SiMe_3)(CO)_2Tp^*]$	295.1	225.1 ($^1J_{CW}$ 170)	66[b]
$[W(\equiv CEt)(CO)_2Tp^*]$	297	223 ($^1J_{CW}$ 168)	67,158[b]
$[W(\equiv CCH_2CH_2Ph)(CO)_2Tp^*]$	293	223 ($^1J_{CW}$ 168)	67,158[b]
$[W(\equiv CCH_2Ph)(CO)_2Tp^*]$	287.4 ($^1J_{CW}$ 189)	223.4 ($^1J_{CW}$ 166)	121
$[W\{\equiv CCH_2CH(OH)Ph\}(CO)_2Tp^*]$	289 ($^1J_{CW}$ 187)	223.5, 223.4 ($^1J_{CW}$ 166)	67[b]

$[W\{\equiv CCH_2CH(OH)(CH=CHMe)\}(CO)_2Tp^*]$	292 ($^1J_{CW}$ 188)	224.2, 224.1 ($^1J_{CW}$ 166)	67
$[W\{\equiv CCH_2C(OH)PhMe\}(CO)_2Tp^*]$	291 ($^1J_{CW}$ 187)	224.3, 224.2 ($^1J_{CW}$ 166)	67[b]
$[W\{\equiv CCH_2C(O)Ph\}(CO)_2Tp^*]$	273	223	67[b]
$[W(\equiv CCH_2CH=CH_2)(CO)_2Tp^*]$	288.0 ($^1J_{CW}$ 191)	223.0 ($^1J_{CW}$ 170)	121
$[W(\equiv C^iPr)(CO)_2Tp^*]$	301	224 ($^1J_{CW}$ 168)	67,158[b]
$[W(\equiv CCHMeSiMe_3)(CO)_2Tp^*]$	301.4	225.7, 224.8 ($^1J_{CW}$ 140)	66[b]
$[W(\equiv CCHMePh)(CO)_2Tp^*]$	293.4 ($^1J_{CW}$ 189)	223.9, 223.7 ($^1J_{CW}$ 166, 166)	121
$[W(\equiv CCHMeCH=CH_2)(CO)_2Tp^*]$	292.9 ($^1J_{CW}$ 188)	225.5, 224.4 ($^1J_{CW}$ 166, 166)	121[c]
$[W(\equiv CCHPhOH)(CO)_2Tp^*]$	284.3 ($^1J_{CW}$ 191)	224.2, 224.1 ($^1J_{CW}$ 164, 164)	68
$[W(\equiv C^tBu)Cl_2Tp^*]$	335.3 ($^1J_{CW}$ 212)		56[c,i]
$[W(\equiv C^tBu)Cl(NH_2)Tp]$	303.5		59[c]
$[W(\equiv C^tBu)Cl(NHPh)Tp]$	306.8		59[c]
$[W(\equiv C^tBu)Cl(NHPh)Tp^*]$	310.2		59[c]
$[W(\equiv C^tBu)Cl(NHC_6H_4Br\text{-}4)Tp]$	297.2		59[c]
$[W(\equiv C^tBu)Cl\{NHC_6H_3(CF_3)_2\text{-}3,5\}Tp]$	NR		59[c]
$[W(\equiv C^tBu)(NHPh)_2Tp]$	295.4		59[c]
$[W(\equiv CCMe_2CH=CH_2)(CO)_2Tp^*]$	295.6	224.7 ($^1J_{CW}$ 166)	121[c]
$[W(\equiv CCPh_2OH)(CO)_2Tp^*]$	284.4 ($^1J_{CW}$ 193)	224.3 ($^1J_{CW}$ 165)	68
$[W(\equiv CC(=O)Ph\}(CO)_2Tp^*]$	277.3 ($^1J_{CW}$ 188)	225.5 ($^1J_{CW}$ 161)	68
$[W(\equiv CCH=CHPh)(CO)_2Tp^*]$	*E*: 278 ($^1J_{CW}$ 184)	225 ($^1J_{CW}$ 166)	67[b]
$[W(\equiv CCH=CHMe)(CO)_2Tp^*]$	*E*: 280.2 ($^1J_{CW}$ 183)	223.0 ($^1J_{CW}$ 170)	121
	Z: 281.2 ($^1J_{CW}$ 183)	225.6 ($^1J_{CW}$ 166)	
$[W(\equiv CCMe=CHMe)(CO)_2Tp^*]$	*E*: 285.2	225.0	121[c]
	Z: 284.5	225.9	
$[W(\equiv CCPh=CHOPh)(CO)_2Tp^*]$	276.5[k]	224.6	65
$[W(\equiv CCPh=CHOC_6H_4Me\text{-}4)(CO)_2Tp^*]$	276.9[k] ($^1J_{CW}$ 190)	224.6 ($^1J_{CW}$ 167)	65
$[W(\equiv CC\equiv CSiMe_3)(CO)_2Tp]$	253.4 ($^1J_{CW}$ 200)	226.5 ($^1J_{CW}$ 164)	61[f]
$[W(\equiv CC\equiv CSiMe_3)(CO)_2Tp^*]$	247.3 ($^1J_{CW}$ 198)	226.1 ($^1J_{CW}$ 162)	61[b]
$[W(\equiv CC\equiv C^tBu)(CO)_2Tp]$	253.5 ($^1J_{CW}$ 202)	224.8 ($^1J_{CW}$ 168)	52
$[W(\equiv CC\equiv CPh)(CO)_2Tp]$	251.4	226.7 ($^1J_{CW}$ 164)	61[f]
$[W(\equiv CC\equiv CPh)(CO)_2Tp^*]$	247.3	227.3 ($^1J_{CW}$ 160)	61[f]
$[W(\equiv CPh)(CO)_2Tp^*]$	277.9	224.2	78
$[W(\equiv CPh)Br_2Tp]$	329.8		130,134
$[W(\equiv CPh)Br_2Tp^*]$	327.4 ($^1J_{CW}$ 208)		57,58
$[W(\equiv CC_6H_4Me\text{-}4)(CO)_3Bp]$	284.5 ($^1J_{CW}$ 189)	224.6 ($^1J_{CW}$ 167)	46
$[W(\equiv CC_6H_4Me\text{-}4)(CO)_2Tp]$	284.8 ($^1J_{CW}$ 189)	224.9 ($^1J_{CW}$ 167)	39,40
$[W(\equiv CC_6H_4Me\text{-}4)(CO)_2Tp^*]$	279.6	224.2 ($^1J_{CW}$ 168)	43
$[W(\equiv CC_6H_4Me\text{-}4)(CO)_2Tp^{Ph}]$	282.3	221.5	43

TABLE I
(CONTINUED)

Compound	δ M$\equiv$CR (ppm) (J, Hz)	δ M(CO) (ppm) (J, Hz)	Ref.
$[W(\equiv CC_6H_4Me\text{-}4)(CO)_2pzTp]$	286.4 ($^1J_{CW}$ 190)	224.7 ($^1J_{CW}$ 170)	39,40
$[W(\equiv CC_6H_4Me\text{-}4)(CO)_2(PPh_3)Bp]$	285.4	232.1, 226.5	46
$[W(\equiv CC_6H_4Me\text{-}4)(CO)_2(PMe_3)Bp]$	298.3	228.9, 224.5	46
$[W(\equiv CC_6H_4Me\text{-}4)(CO)(PMe_3)Tp]$	272.0	248.6	183
$[W(\equiv CC_6H_4Me\text{-}4)(CO)(PPh_3)Tp]$	275.4 ($^2J_{CP}$ 12.5)	246.2	50[b]
$[W(\equiv CC_6H_4OMe\text{-}2)(CO)_2Tp]$	280.5	225.1	43
$[W(\equiv CC_6H_3Me_2\text{-}2,6)(CO)_2(pic)Bp]$[o]	*fac*[k]: 294.5	225.3	49[b]
	mer[l]: 284.5	227.5, 221.5	
$[W(\equiv CC_6H_3Me_2\text{-}2,6)(CO)_2(CN^tBu)Bp]$	285.6	222.1, 211.2	49[b]
$[W(\equiv CC_6H_3Me_2\text{-}2,6)(CO)_2(CNC_6H_3Me_2\text{-}2,6)Bp]$	287.5	221.7, 210.6	49[b]
mer-$[W(\equiv CC_6H_3Me_2\text{-}2,6)(CO)_2(PMe_2Ph)Bp]$	285.1 ($^2J_{CP}$ 8.9)	227.0 ($^2J_{CP}$ 3.6), 213.4 ($^2J_{CP}$ 53.5)	49[b]
$[W(\equiv CC_6H_3Me_2\text{-}2,6)(CO)_2Tp]$	288.6	226.8	43
$[W(\equiv CC_6H_3Me_2\text{-}2,6)Cl_2Tp]$	326.5		107[b]
$[W(\equiv CC_6H_3Me_2\text{-}2,6)Br_2Tp]$	335.5		107[b]
$[W(\equiv CC_6H_2Me_3\text{-}2,4,6)(CO)_2(pic)Bp]$	*fac*[k]: 295.0 ($^1J_{CW}$ 200)	225.5 ($^1J_{CW}$ 169)	49[b]
	mer[l]: 284.9 ($^1J_{CW}$ 194)	227.6, 221.9 ($^1J_{CW}$ 169, 169)	
mer-$[W(\equiv CC_6H_2Me_3\text{-}2,4,6)(CO)_2(PMe_2Ph)Bp]$	284.4 ($^2J_{CP}$ 8.6)	227.1 ($^2J_{CP}$ 3.6), 213.4 ($^2J_{CP}$ 53.1)	49[b]
c,t,c-$[W(\equiv CC_6H_2Me_3\text{-}2,4,6)(CO)(PMe_2Ph)_2Bp]$	278.2 ($^2J_{CP}$ 11)	249.9 ($^2J_{CP}$ 5.3)	49[b]
$[W(\equiv CC_6H_2Me_3\text{-}2,4,6)(CO)_2Tp]$	288.5 ($^1J_{CW}$ 193)	226.3 ($^1J_{CW}$ 165)	107[b]
$[Re(\equiv C^tBu)(CH_2{}^tBu)(CH^tBu)Tp]$	289.3		91[c]
$[Re\{\equiv CCH_2CH_2C(O)H\}(CO)(PMe_3)Tp]^+$	308.2 ($^2J_{CP}$ 12.2)	206.3	97[g]
$[Re\{\equiv CCH_2CH_2C(OMe)_2H\}(CO)(PMe_3)Tp]^+$	310.5 ($^2J_{CP}$ 9.8)	206.3	97[g]
$[Re(\equiv CC_6H_2Me_3\text{-}2,4,6)(CO)_2Tp^*]^+$	NR	192.8	94[b]
$[Ru(\equiv CCH_2{}^tBu)Cl(PPh_3)Tp]^+$	355.2 ($^2J_{CP}$ 16.5)		98[m]
$[Ru(\equiv CCH_2{}^nBu)Cl(PPh_3)Tp]^+$	351.2 ($^2J_{CP}$ 17.5)		98[m]
$[Ru(\equiv CCH_2Ph)Cl(PPh_3)Tp]^+$	341.4 ($^2J_{CP}$ 17.8)		98[m]
$[Ru(\equiv CCH{=}CPh_2)Cl(P^iPrPh_2)Tp]^+$	324.4 ($^2J_{CP}$ 15.2)		99
$[Ru(\equiv CCH{=}CFc_2)Cl(P^iPrPh_2)Tp]^+$	328.5 ($^2J_{CP}$ 18.0)		99
$[Ru(\equiv CCH{=}CPh_2)Cl(PPh_3)Tp]^+$	326.4 ($^2J_{CP}$ 14.2)		99
$[Os(\equiv C^tBu)(CH_2{}^tBu)_2Tp]$	280.1		92,93[c]

Silyl			
$[Mo(\equiv CSiMe_2Ph)(CO)_2Tp^*]$	360.4	227.8	89
$[W(\equiv CSiMe_3)(CO)_2Tp^*]$	344.6 ($^1J_{CW}$ 160)	225.9 ($^1J_{CW}$ 154)	68
$[W(\equiv CSiMe_2Ph)(CO)_2Tp^*]$	339.0 ($^1J_{CW}$ 160)	226.1 ($^1J_{CW}$ 173)	89
Aryloxy			
$[Mo(\equiv COPh)(CO)_2Tp^*]$	218.0	225.4	70
$[Mo(\equiv COC_6H_4Me\text{-}4)(CO)_2Tp^*]$	218.8	225.5	70
$[W(\equiv COMe)(CO)_2Tp^*]$	228.2 ($^1J_{CW}$ 235)	222.4 ($^1J_{CW}$ 164)	79
$[W(\equiv COPh)(CO)_2Tp^*]$	219.3 ($^1J_{CW}$ 242)	222.9 ($^1J_{CW}$ 164)	70
$[W(\equiv COC_6H_4Me\text{-}4)(CO)_2Tp^*]$	220.2 ($^1J_{CW}$ 245)	223.1 ($^1J_{CW}$ 164)	70
$[W(\equiv COC_6H_4OMe\text{-}4)(CO)_2Tp^*]$	220.5 ($^1J_{CW}$ 244)	222.9 ($^1J_{CW}$ 164)	70
$[W(\equiv COC_6H_4NO_2\text{-}4)(CO)_2Tp^*]$	214.5 ($^1J_{CW}$ 247)	222.5 ($^1J_{CW}$ 161)	78
$[W(\equiv COC_6H_4Cl\text{-}4)(CO)_2Tp^*]$	218.0	222.7	78
Amino			
$[Cr(\equiv CN^iPr_2)(CO)_2Tp^*]$	256.4	239.2	81,82[c]
$[Cr(\equiv CN^iPr_2)(CO)(CN^tBu)Tp^*]$	254.6	242.8	81,82[c]
$[Mo(\equiv CNH^tBu)(CO)_2Tp^*]$	251.6	228.2	83
$[Mo(\equiv CNMe_2)(CO)_2Tp^*]$	253.7	228.7	83
$[Mo(\equiv CN^iPr_2)(CO)_2Tp]$	260.2	230.2	180[b]
$[Mo(\equiv CNMe^tBu)(CO)_2Tp^*]$	253.5	229.7	83
$[Mo(\equiv CNMePh)(CO)_2Tp^*]$	244.5	228.2	83
$[W(\equiv CNEtMe)(CO)_2Tp^*]$	249.0 ($^1J_{CW}$ 208)	225.5 ($^1J_{CW}$ 167)	84[j]
$[W(\equiv CNEtMe)Br_2Tp^*]$	267.1 ($^1J_{CW}$ 241)		85
$[W(\equiv CNEtMe)I_2Tp^*]$	265.5		85
$[W(\equiv CNEtMe)(CNEt)_2Tp^*]$	243.1 ($^1J_{CW}$ 232)		85[c]
$[W(\equiv CNEtMe)(\equiv CNEt_2)(CNEt)Tp^*]$	260.7		58
$[W(\equiv CNEt_2)_2(CNEt)Tp^*]$	261.5		85
$[W(\equiv CNEt_2)(CO)_2Tp]$	254.6	225.8	75[b]
$[W(\equiv CNEt_2)(CO)_2Tp^*]$	248.6 ($^1J_{CW}$ 208)	225.6 ($^1J_{CW}$ 169)	84[j]
$[W(\equiv CNEt_2)(CNEt)_2Tp^*]$	242.6 ($^1J_{CW}$ 231)		85[c]
$[W(\equiv CNEt_2)Br_2Tp^*]$	266.5 ($^1J_{CW}$ 242)		85
$[W(\equiv CNEt_2)I_2Tp^*]$	263.8		85
Chalco			
$[Mo(\equiv CSMe)(CO)_2Tp^*]$	267.6	226.8	73[b]
$[Mo(\equiv CSePh)(CO)_2Tp^*]$	261.8	226.0	189[b]
$[Mo(\equiv CTeMe)(CO)_2Tp^*]$	266.2	225.6	73[b]

TABLE I
(CONTINUED)

Compound	δ M$\equiv$CR (ppm) (J, Hz)	δ M(CO) (ppm) (J, Hz)	Ref.
$[W(\equiv CSMe)(CO)_2Tp]$	264.4	224.7	23,77[b]
$[W(\equiv CSMe)(SMe)_2Tp]$	268.3		19
$[W(\equiv CSMe)(SMe)(SPh)Tp]$	271.1		19
$[W(\equiv CSMe)(SMe)(SC_6H_4Me\text{-}4)Tp]$	270.8		19
Halo			
$[Mo(\equiv CCl)(CO)_2Tp^{iPr,4Br}]$	211.6	222.4	36[b]
$[Mo(\equiv CCl)(CO)_2Tp^*]$	208.7	224.0	36,37[b]
$[Mo(\equiv CCl)(CO)_2Tp^{3Me}]$	208.2	224.3	36[b]
$[Mo(\equiv CCl)(CO)_2Tp^{Me_2,4Cl}]$	245.01	223.1	36[b]
$[Mo(\equiv CCl)(CO)_2pzTp]$	213.5	223.6	36[b]
$[Mo(\equiv CBr)(CO)_2Tp^*]$	202.5	223.9	36[b]
$[Mo(\equiv CBr)(CO)_2Tp^{3Me}]$	201.9	224.0	36[b]
$[W(\equiv CCl)(CO)_2Tp^*]$	205.6	222.0	36[b]
$[W(\equiv CBr)(CO)_2Tp^*]$	197.97	221.88	36[b]
$[W(\equiv CI)(CO)_2Tp^*]$	183.2	223.3	68
Phosphonio			
$[W(\equiv CPCy_3)(CO)_2Tp^*][PF_6]$	251.1 ($^1J_{CW}$ 163) ($^1J_{CP}$ 15)	226.1 ($^1J_{CW}$ 163)	70
$[W(\equiv CPMe_2Ph)(CO)_2Tp^*][PF_6]$	245.9 ($^1J_{CW}$ 207)	223.0 ($^1J_{CW}$ 155)	70
$[W(\equiv CPPh_3)(CO)_2Tp^*][PF_6]$	242.1 ($^1J_{CW}$ 212)	224.5 ($^1J_{CW}$ 158)	70
Phospha- and arsaalkenyl			
$[Mo\{\equiv CP{=}C(NMe_2)_2\}(CO)_2Tp^*]$	337.5 ($^1J_{CP}$ 111.3)	229.5	87
$[Mo\{\equiv CP{=}C(NEt_2)_2\}(CO)_2Tp^*]$	338.6 ($^1J_{CP}$ 116)	230.7	87
$[Mo\{\equiv CPMeC(NMe_2)_2\}(CO)_2Tp^*][OTf]$	287.7 ($^1J_{CP}$ 94)	226.9	110
$[Mo\{\equiv CPMeC(NEt_2)_2\}(CO)_2Tp^*][OTf]$	286.7 ($^1J_{CP}$ 97)	227.2	110
$[Mo\{\equiv CP({=}O)_2C(NMe_2)_2\}(CO)_2Tp^*]$	312.6 ($^1J_{CP}$ 30)	226.5	169
$[Mo\{\equiv CP({=}O)_2C(NEt_2)_2\}(CO)_2Tp^*]$	314.4 ($^1J_{CP}$ 28)	227.3	169
$[Mo\{\equiv CAs{=}C(NMe_2)_2\}(CO)_2Tp^*]$	349.7	229.1	88
$[Mo\{\equiv CAsMeC(NMe_2)_2\}(CO)_2Tp^*][OTf]$	301.9	225.7, 226.1	36[m]
$[W\{\equiv CP{=}C(NMe_2)_2\}(CO)_2Tp^*]$	318.3 ($^1J_{CP}$ 101.2)	228.1	87
$[W\{\equiv CP{=}C(NEt_2)_2\}(CO)_2Tp^*]$	319.0 ($^1J_{CP}$ 108.3)	229.4	87
$[W\{\equiv CPMeC(NMe_2)_2\}(CO)_2Tp^*][OTf]$	273.0 ($^1J_{CP}$ 83.0)	225.5	110
$[W\{\equiv CPHC(NEt_2)_2\}(CO)_2Tp^*][BF_4]$	261.8 ($^1J_{CP}$ 80.9)	225.2	110

$[W\{\equiv CPMeC(NEt_2)_2\}(CO)_2Tp^*][OTf]$	271.9 ($^1J_{CP}$ 85.6)	225.7	110
$[W\{\equiv CP(=O)_2C(NMe_2)_2\}(CO)_2Tp^*]$	298.7 ($^1J_{CP}$ 46.2)	225.9	169
$[W\{\equiv CP(=O)_2C(NEt_2)_2\}(CO)_2Tp^*]$	299.9 ($^1J_{CP}$ 54.0)	226.5	169
$[W\{\equiv CAs=C(NMe_2)_2\}(CO)_2Tp^*]$	329.1	227.6 ($^1J_{CW}$ 183.3)	88
$[W\{\equiv CAsMeC(NMe_2)_2\}(CO)_2Tp^*][OTf]$	283.2	225.3, 225.1	88
Polymetallic			
$[CrMo\{\mu\text{-}\eta^6{:}\sigma\text{-}CC_6H_4(OMe\text{-}2)\}(CO)_5Tp]$	278.1	225.7, 225.5	53[b]
$[MoMn(\mu\text{-}\sigma{:}\eta^5\text{-}CC_5H_4)(CO)_5Tp]$	279.1	224.8	54[b]
$[Mo_2Fe(\mu\text{-}\sigma,\sigma'{:}\eta^5\text{-}CC_5H_4)_2(CO)_4Tp_2]$	293.2	225.4	55[b]
$[Mo\{\equiv CFe(CO)_2Cp\}(CO)_2Tp^*]$	381		71
$[Mo\{\equiv CP(AuCl)_2C(NMe_2)_2\}(CO)_2Tp^*]$	280.9	226.1	109
$[W\{\equiv CP(AuCl)_2C(NMe_2)_2\}(CO)_2Tp^*]$	273.0	225.1	109
$[\{Mo(\equiv CAs)(CO)_2Tp^*\}_3]$	310.6, 304.8	226.8, 226.4, 225.8	109
$[\{W(\equiv CAs)(CO)_2Tp^*\}_3]$	296.0, 291.5	225.4, 224.5	109
$[Tp^*(CO)_2Mo\equiv CCH_2C\equiv Mo(CO)_2Tp^*]$	284.7	225.1	159[b]
$[Tp^*(CO)_2W\equiv CCH_2C\equiv Mo(CO)_2Tp^*]$	Mo: 287.5; W: 273.5	Mo: 225.0; W: 223.2	159[b]
$K[Tp^*(CO)_2W\equiv CCHC\equiv Mo(CO)_2Tp^*]$	Mo: 313.7; W: 298.1	Mo: 229.6; W: 231.4 ($^1J_{CW}$ 172)	159[c]
$[Tp^*(CO)_2W\equiv CCHMeC\equiv Mo(CO)_2Tp^*]$	Mo: 294.6; W: 281.0	226.6, 225.3, 224.5, 223.3	159
$[Tp^*(CO)_2W\equiv CCMe_2C\equiv Mo(CO)_2Tp^*]$	Mo: 298; W: 285 ($^1J_{CW}$ 195)	Mo: 226; W: 224 ($^1J_{CW}$ 170)	159[b]
$[Tp^*(CO)_2Mo\equiv CC\equiv CMo(=O)_2Tp^*]$	252.8	228.1	159[b]
$[Tp^*(CO)_2W\equiv CC\equiv CMo(=O)_2Tp^*]$	246	226 ($^1J_{CW}$ 160)	159[b]
$[Tp^*(CO)_2Mo\equiv CC\equiv CW(CO)_2Tp^*]$	253	228	159[b]
$[Tp^*(CO)_2W\equiv CC(=O)C\equiv Mo(CO)_2Tp^*]$	282.3, 274.3	Mo: 227; W: 226 ($^1J_{CW}$ 162)	159[b]
$[Tp^*(CO)_2Mo\equiv CCH_2CH_2C\equiv Mo(CO)_2Tp^*]$	300.7	225.1	158[b]
$[Tp^*(CO)_2W\equiv CCH_2CH_2C\equiv W(CO)_2Tp^*]$	289.9 ($^1J_{CW}$ 180)	223.2 ($^1J_{CW}$ 160)	158[b]
$[Tp^*(CO)_2W\equiv CCHMeCHMeC\equiv W(CO)_2Tp^*]$	Isomer 1: 298.0	223.9, 223.8	158[b]
	Isomer 2: 297.7	224.2, 223.8	
$[Tp^*(CO)_2W\equiv CCH(CH_2Ph)CH(CH_2Ph)C\equiv W(CO)_2Tp^*]$	Isomer 1: 297.6	226.3, 223.1	158[b]
	Isomer 2: 297.6	226.3, 223.1	
$[Tp^*(CO)_2W\equiv CCMe_2CMe_2C\equiv W(CO)_2Tp^*]$	303.4	224.9 ($^1J_{CW}$ 170)	158[b]
$[Tp^*(CO)_2Mo\equiv CCH=CHC\equiv Mo(CO)_2Tp^*]$	285.0	226.8	158[b]
$[Tp^*(CO)_2W\equiv CCH=CHC\equiv W(CO)_2Tp^*]$	276.4	225.1	158[b]
$[Tp^*(CO)_2W\equiv CCMe=CMeC\equiv W(CO)_2Tp^*]$	283.2	226.0	158[b]
$[Tp^*(CO)_2W\equiv CC(CH_2Ph)=C(CH_2Ph)C\equiv W(CO)_2Tp^*]$	283.2	226.5	158[b]
$[Tp^*(CO)_2Mo\equiv CC\equiv CC\equiv Mo(CO)_2Tp^*]$	248.9	230.1	158[b]
$[Tp^*(CO)_2W\equiv CC\equiv CC\equiv W(CO)_2Tp^*]$	243.7	228.7	158[b]
$[Tp^*(CO)_2W(\equiv CCH_2CH=CHC\equiv W(CO)_2Tp^*]$ (*E*)	WC_2H_2: 284.7; WC_2H: 278.6	224.2 ($^1J_{CW}$ 165), 223.4 ($^1J_{CW}$ 166)	121

TABLE I
(CONTINUED)

Compound	δ M≡CR (ppm) (J, Hz)	δ M(CO) (ppm) (J, Hz)	Ref.
$[Tp^*(CO)_2W(\equiv CCH_2CH{=}CHC\equiv W(CO)_2Tp^*]$ (*Z*)	WC_2H_2: 287.0; WC_2H: 279.0	225.5, 223.6	121
$K[Tp^*(CO)_2W\equiv C(CH)_3C\equiv W(CO)_2Tp^*]$	306.1 ($^1J_{CW}$ 173)	229.6 ($^1J_{CW}$ 172)	121[e,n]
$[Tp(CO)_2W\equiv CC\equiv CC\equiv CC\equiv W(CO)_2Tp]$	NR	NR	167
$[Tp^*(CO)_2W\equiv CC\equiv CC\equiv CC\equiv W(CO)_2Tp^*]$	242.8	227.4	167
$[Tp^*(CO)_2Mo\equiv CCH_2W(\eta^2\text{-}PhC\equiv CPh)(CO)Tp]$	325.2	226.3, 226.2	161[b]
$[Tp^*(CO)_2W\equiv CCH_2W(\eta^2\text{-}PhC\equiv CPh)(CO)Tp]$	312.3	224.6, 224.5	161[b]
$K[Tp^*(CO)_2W\equiv CCH{=}W(\eta^2\text{-}PhC\equiv CPh)(CO)Tp]$	311.1	233.4, 229.7	161[c]
$[Tp^*(CO)_2Mo\equiv CC\equiv W(\eta^2\text{-}PhC\equiv CPh)(CO)Tp]$	W: 342.7 ($^1J_{CW}$ 184); Mo: 277.0 ($^2J_{CW}$ 43)	W: 218.6 ($^1J_{CW}$ 130); Mo: 234.1, 231.9	161[b]
$[Tp^*(CO)_2W\equiv CC\equiv W(\eta^2\text{-}PhC\equiv CPh)(CO)Tp]$	Tp: 348 ($^1J_{CW}$ 185); Tp*: 274 ($^1J_{CW}$ 171)	Tp: 219 ($^1J_{CW}$ 131); Tp*: 233, 231 ($^1J_{CW}$ 165, 165)	161[b]
$[Co_2\{\mu\text{-}^tBuC_2C\equiv Mo(CO)_2Tp^*\}(CO)_6]$	269.1	227.4	52[b]
$[Co_2\{\mu\text{-}^tBuC_2C\equiv W(CO)_2Tp\}(CO)_6]$	263.2	224.7 ($^1J_{CW}$ 165)	52
$[Co_2\{\mu\text{-}^tBuC_2C\equiv W(CO)_2Tp\}(\mu\text{-dppm})(CO)_4]$	278.5	227.4 ($^1J_{CW}$ 169)	52[b]
$[Mo_2\{\mu\text{-}^tBuC_2C\equiv W(CO)_2Tp\}(CO)_4Cp_2]$	272.9	233.1, 232.6, 228.0, 226.6, 226.2, 225.1 (4 CoCO, 2 WCO)	52[h]
$[Rh\{C\equiv CC\equiv W(CO)_2Tp\}(CO)(PPh_3)_2]$	258.3	227.3	162
$[Rh\{C\equiv CC\equiv W(CO)_2Tp\}I_2(CO)(PPh_3)_2]$	256.7	226.2	162[b]
$[RhFe_2(\mu\text{-}C_3W(CO)_2Tp)(CO)_{10}(PPh_3)]$	254.6	224.4	162[b]
$[Ir\{C\equiv CC\equiv W(CO)_2Tp\}_2H(CO)(PPh_3)_2]$	NR	NR	164
mer-$[Ru\{C\equiv CC\equiv W(CO)_2Tp\}H(CO)(PPh_3)_3]$	262.6	227.9	163
mer-$[Ru\{C\equiv CC\equiv W(CO)_2Tp^*\}H(CO)(PPh_3)_3]$	255.7	228.0	163
c,c,t-$[Ru\{C\equiv CC\equiv W(CO)_2Tp\}H(CO)_2(PPh_3)_2]$	260.6	227.2	163
c,c,t-$[Ru\{C\equiv CC\equiv W(CO)_2Tp^*\}H(CO)_2(PPh_3)_2]$	255.6	227.1 ($^1J_{CW}$ 168)	163
trans-$[Ru\{C\equiv CC\equiv W(CO)_2Tp\}H(CO)(Hpz^*)(PPh_3)_2]$	258.6	227.5	163
trans-$[Ru\{C\equiv CC\equiv W(CO)_2Tp^*\}H(CO)(Hpz^*)(PPh_3)_2]$	NR	227.2	163
c,c,t-$[Ru\{C\equiv CC\equiv W(CO)_2Tp\}H(CO)(CNMes)(PPh_3)_2]$	261.4	227.7 ($^1J_{WC}$ 171.1)	163
$[Ru\{C_3W(CO)_2Tp\}\{HgC_3W(CO)_2Tp\}(CO)_2(PPh_3)_2]$	NR	NR	168
$[Ru\{C_3W(CO)_2Tp^*\}\{HgC_3W(CO)_2Tp^*\}(CO)_2(PPh_3)_2]$	253.9, 253.6	227.8, 227.2	168[c]
$[Ru\{\eta^2\text{-}Tp(CO)_2WC_6W(CO)_2Tp\}(CO)_2(PPh_3)_2]$	$WC(C_2Ru)$: 300.3; $WC_3(C_2Ru)$: 253.4	230.1, 229.0	168[c]
c,c,t-$[Ru\{C_3W(CO)_2Tp^*\}_2(CO)_2(PPh_3)_2]$	254.8	226.9	168
$[Ru\{C(C_3W(CO)_2Tp){=}CHC\equiv W(CO)_2Tp\}Cl(CO)(PPh_3)_2]$	WC_2H: 297.9; WC_3: 251.7	227.7, 226.9	187[c]
$[Ru\{C(C_3W(CO)_2Tp^*){=}CHC\equiv W(CO)_2Tp^*\}Cl(CO)(PPh_3)_2]$	WC_2H: 287.9 ; WC_2: 241.6	227.9, 227.0	187[c]
$[Ru\{C(C_3W(CO)_2Tp){=}CHC\equiv W(CO)_2Tp\}Cl(CO)_2(PPh_3)_2]$	WC_2H: 287.9; WC_2: 241.6	227.7, 226.8	187[c]

$[Ru\{C(C_3W(CO)_2Tp^*\}Cl(CO)_2(PPh_3)_2]$	WC_2H: 283.6; WC_2: 244.9	228.6, 227.4	187[c]
$[W(\equiv CC\equiv CAuPPh_3)(CO)_2Tp]$	255.8 ($^1J_{WC}$ 196)	226.4 ($^1J_{WC}$ 168)	165
$[W(\equiv CC\equiv CAuPPh_3)(CO)_2Tp^*]$	251.7	226.4 ($^1J_{WC}$ 164)	165
$[(Ph_3P)_2N][Au\{C\equiv CC\equiv W(CO)_2Tp^*\}_2]$	256.6	227.3	165[b]
$[Hg\{C\equiv CC\equiv W(CO)_2Tp\}_2]$	253.6	225.8	166[d]
$[Hg\{C\equiv CC\equiv W(CO)_2Tp^*\}_2]$	246.7	225.6 ($^1J_{WC}$ 171)	166[b]

Bp, bis(pyrazolyl)borate; Tp, tris(pyrazolyl)borate; Tp^{Ph}, tris(3-phenylpyrazolyl)borate; Tp*, tris(3,5-dimethylpyrazolyl)borate; Tp^{3Me}, tris(3,4,5-trimethylpyrazolyl)borate; $Tp^{Me_2,4Cl}$, tris(4-chloro-3,5-dimethylpyrazolyl)borate; pzTp, tetrakis(pyrazolyl)borate; dppm, bis(diphenylphosphino)methane; pic, 4-picoline; fc, ferrocenyl; *c*, *cis*; *t*, *trans;* NR, not reported.

[a] $^{13}C\{^1H\}$ NMR data were obtained in CD_2Cl_2 solution at room temperature unless otherwise indicated.

[b] $CDCl_3$.

[c] C_6D_6.

[d] DMSO-d_6.

[e] THF-d_6.

[f] Acetone-d_6.

[g] CD_3CN.

[h] −40 °C.

[i] 22 °C.

[j] 20 °C.

[k] Major isomer.

[l] Minor isomer.

[m] −70 °C.

[n] −20 °C.

[o] For Bp complexes *fac* and *mer* prefixes refer to the orientation of the bidentate Bp and alkylidyne fragments.

which is similar to the chemical shift reported for the 18-electron complex TpW(≡CSMe)$(CO)_2$ (δ_C 264.4),[23] and suggests that the former is an electron-rich alkylidyne by virtue of the π-basic thiolate coordination.

The tungsten complexes show ^{13}C–^{183}W coupling constants in the range 160–250 Hz for C_α that are diagnostic of the high s character of the M–C triple bond. The corresponding one-bond coupling constants ($^1J_{CW}$) for the carbonyl carbons are typically smaller by 30–40 Hz, however, a notable exception is provided by the silyl alkylidyne complexes in which the measured $^1J_{CW}$ values for C_α and CO are comparable. The alkylidyne complexes LW(≡C–C≡C^tBu)$(CO)_2$ (L = Cp, Tp) clearly demonstrate the decrease in s-orbital contribution and subsequent decrease in $^1J_{CW}$ with increasing donor strength of the tripodal ligand, i.e., for the cyclopentadienyl complex (L = Cp), the measured one-bond coupling constants are $^1J_{CW}$ = 226 and 196 Hz for C_α and CO, respectively, vs. 202 and 168 Hz for the stronger donor Tp.[52] Table I is relatively bereft of measured $^1J_{CW}$ coupling constants due in general to the low intensity of the resonance for the quaternary carbon, such that spectra are not usually measured with sufficient signal-to-noise ratio to observe the tungsten satellites. Thus, although this information is in principle accessible, it is not generally deemed sufficiently informative as to warrant extended acquisition times, given that for all examples the hybridisation for both the sp^3d^2-tungsten and sp-carbon are constant. With interligand angles close to 90°, complexes of the form Tp^xW(≡CR)(PR'_3)(CO) show $^2J_{CP}$ values in the range 5–30 Hz. For BpW(≡CR)(PMe_2Ph)$(CO)_2$ (R = $C_6H_2Me_3$-2,4,6, $C_6H_3Me_2$-2,6), a significantly larger $^2J_{CP}$ coupling constant is measured for the *trans*-disposed carbonyl than for that which is *cis* ($^2J_{CP\ trans}$ = 53 Hz vs. $^2J_{CP\ cis}$ = 3.6 Hz).[49]

In many cases for octahedral organometallic complexes, rotation of the entire poly(pyrazol) ligand on the NMR timescale leads to all pyrazolyl rings being chemically equivalent. However, if fluxional behaviour is not occurring (as is generally the case for Tp^xM(≡CR)$(CO)_2$ alkylidyne complexes), the pattern exhibited by the pyrazolyl rings in the ^1H and $^{13}C\{^1H\}$ NMR spectra provides insight into the symmetry of the molecule in solution. For C_s symmetrically substituted alkylidyne complexes Tp^xM(≡CR)$(CO)_2$, a symmetry plane exists which contains the alkylidyne–metal bond and one pyrazole ring and bisects the two terminal metal carbonyls, resulting in a 2:1 out-of-plane:in-plane pattern for the coordinated pyrazolyl groups. A 1:1:1 pattern indicates that the molecule is chiral (C_1 symmetry). The lack of observable fluxional processes presumably reflects a barrier originating from the disparity in *trans* influences of alkylidyne and carbonyl ligands. In the case of TmW(≡CR)$(CO)_2$ (Tm = hydrotris(methimazolyl)borate), fluxionality has been observed to chemically equilibrate the methimazolyl (mt) environments;[104] however, given that the coalescence temperature (T_C) is very strongly dependent on the nature of the (remote) alkylidyne substituents (C_6H_4Me-4, 90 °C vs. N^iPr_2, 15 °C), it has been argued that equilibration proceeds *via* dissociation of one mt arm to provide a stereochemically non-rigid five-coordinate species, rather than a more conventional Bailar twist. This is consistent with both the superlative *trans* effect of alkylidynes (especially aminomethylidynes) and the geometric flexibility of Tm coordination.[104]

B. *Crystallography*

Up to March 2006 the molecular structures of about 40 alkylidyne complexes ligated by poly(pyrazolyl)borates had been reported, and structural data for the crystallographically characterised alkylidyne–metal complexes are compiled in Table II. Of these, Tp- and Tp*-ligated complexes represent more than 80%, with one third of the total known structures being bi- and polymetallic complexes. Structural features of alkylidyne complexes in general have been discussed elsewhere[7] and, accordingly, an attempt to restrict this discussion to features which are unique or characteristic of the scorpionate alkylidyne complexes has been made.

Structural features of the scorpionate complexes include a short $M{\equiv}C$ bond and an $M{\equiv}C$–R angle which departs little from linearity, as found for other alkylidyne complexes.[6] The $M{\equiv}C$ alkylidyne separations are predominantly influenced by the nature of the alkylidyne substituent rather than the tridentate co-ligand. For example, for $Tp^xMo({\equiv}CC_4H_3S\text{-}2)(CO)_2$ ($Tp^x = Tp$, Tp*) in which all other factors are constant, the $Mo{\equiv}C_\alpha$ separation for $Tp^x = Tp$ (1.809(4)Å)[105] is statistically identical to that for $Tp^x = Tp^*$ (1.810(3) Å).[105] Similarly, for $LW({\equiv}CC_6H_4Me\text{-}4)(CO)_2$, the W–$C_\alpha$ separation does not significantly differ for L = Cp, 1.82(2) Å,[106] and L = pzTp, 1.82(1) Å.[39,40]

The complexes display distorted octahedral geometry, with the N–M–N angles being contracted by approximately 10° from the idealised 90° expected for octahedral geometry due to the geometrical constraints of the scorpionate ligands.[107] For $Tp^xMo({\equiv}CC_4H_3S\text{-}2)(CO)_2$ ($Tp^x = Tp$, Tp*), the slightly expanded sum of the N–Mo–N chelate angles for Tp* compared with those of the Tp analogue (247° vs. 244°) is a modest expression of the steric bulk and relative donor ability of the pyrazolylborate ligands, with M–N and N–N bond lengths that are slightly elongated for the Tp* complex (δ M–N: 0.015 Å, δ N–N: 0.015 Å).

The occupation of three facial positions by the nitrogen atoms of the tripodal ligand forces the remaining substituents to be mutually *cis*. The pyrazolyl-N–metal separations are highly sensitive to the π-acidity of *trans*-disposed ligands. Typically, the pyrazolyl-N–metal bond *trans* to the alkylidyne carbon is appreciably lengthened with respect to the two remaining pyrazolyl groups due to the strong *trans* influence of the alkylidyne. In the crystal structure of $TpMo({\equiv}CCMe_2Ph)(OMe)(NHC_6H_3{}^iPr_2\text{-}2,6)$, the N–Mo bond lengths are consistent with the decreasing *trans* influence of the ligands in the order alkylidyne > amido > alkoxide.[60]

$M(pz)_2BH_2$ metallacycles in Bp alkylidyne complexes typically adopt a shallow boat conformation. Due to the bidentate nature of the scorpionate, the alkylidyne fragment can be oriented meridionally or facially with respect to the chelated pyrazolyl arms. The alkylidyne fragment is typically oriented *trans* to the strongest π-donor (or weakest π-acid)[49] and as such the favoured geometry depends upon the relative π-donor abilities of the remaining ligands.

The structure determination of the chloromethylidyne complex $pzTpMo({\equiv}CCl)(CO)_2$[37] revealed an anomalously long Mo–C bond length (average: 1.894(10) Å). In the related complex $Tp^{Me_3}Mo({\equiv}CCl)(CO)_2$ this distance is normal at

TABLE II

STRUCTURAL DATA FOR CRYSTALLOGRAPHICALLY CHARACTERISED ALKYLIDYNE–METAL COMPLEXES

Compound[a]	M≡C (Å)	M–C–R (°)	M–N *trans* to M≡CR (Å)	Other M–N separations (Å)	Ref.
$[Mo(\equiv CCMe_2Ph)(OMe)(NHC_6H_3\ {}^iPr_2\text{-}2,6)Tp]$	1.765(4)	177.2(4)	2.386(4)	2.265(3), 2.235(4)	60
$[Mo(\equiv CC_4H_3S\text{-}2)(CO)_2Tp]$	1.809(4)	172.7(3)	2.287(3)	2.217(2), 2.217(2)	105
$[Mo(\equiv CC_4H_3S\text{-}2)(CO)_2Tp^*]$	1.810(3)	168.5(4)	2.311(3)	2.235(3), 2.226(3)	105
$[Mo(\equiv CC_6H_4Me\text{-}4)(CO)_2Tp^*]$	1.804(4)	163.1(3)	2.306(3)	2.218(3), 2.212(3)	44
$[Mo(\equiv CSC_6H_4NO_2\text{-}4)(CO)_2Tp^*]$	1.801(4)	179.5(2)	2.290(4)	2.2142(3), 2.218(3)	73
$[Mo(\equiv CCl)(CO)_2Tp^{3Me}]$	1.798(5)	165.7(4)	2.268(4)	2.224(5), 2.234(4)	108
$[Mo(\equiv CCl)(CO)_2pzTp]$	1.894(9)		2.235(7)	2.231, 2.199	37
	1.90(1)		2.250(7)	2.217, 2.191	
	1.89(1)		2.263(7)	2.212, 2.199	
$[W(\equiv C^tBu)(NHPh)_2Tp]$	1.789(5)	166.5(4)	2.356(4)	2.239(3)	59
$[W(\equiv C^tBu)Cl(NHPh)Tp^*]$	1.77(1)	174.7(8)	2.393(8)	2.169(7), 2.223(8)	59
$[W(\equiv CPh)Br_2Tp^*]$	1.783(9)	168.0(8)	2.338(7)	2.144(7), 2.134(9)	58
$[W(\equiv CC_6H_4Me\text{-}4)(CO)_2pzTp]$	1.82(1)		2.284(6)	2.219(6), 2.186(5)	39,40
fac-$[W(\equiv CC_6H_3Me_2\text{-}2,6)(CO)_2(pic)Bp]$	1.810(6)	178.0(5)		2.201(6), 2.204(6)	49
mer-$[W(\equiv CC_6H_3Me_2\text{-}2,6)(CO)_2(PPhMe_2)Bp]$	1.825(4)	173.5(3)	2.292(3)	2.233(3)	49
$[W(\equiv CC_6H_3Me_2\text{-}2,6)Cl_2Tp]$	1.80(1)	177(1)	2.36(1)	2.138, 2.166	107
$[W(\equiv CC_6H_2Me_3\text{-}2,4,6)(CO)_2Tp]$	1.826(5)	173.9(4)	2.298(4)	2.214, 2.212	107
$[W\{\equiv CC(=O)Ph\}(CO)_2Tp^*]$	1.831(5)	168.7(4)	2.295(4)	2.2197(4), 2.222(4)	68
$[W(\equiv CCPh=CHOC_6H_4Me\text{-}4)(CO)_2Tp^*]$	1.85(1)	168(1)	2.28(1)	2.22(1), 2.22(1)	65
$[W(\equiv CC\equiv CSiMe_3)(CO)_2Tp]$	1.844(6)	176.4(5)	2.268(4)	2.202(4), 2.191(4)	166
$[W(\equiv CNEt_2)Br_2Tp^*]$	1.763(8)	174.5(5)	2.340(5)	2.131(3)	85
$[W(\equiv CPMe_2Ph)(CO)_2Tp^*][PF_6]$	1.821(9)	168.1(5)	2.291(7)	2.227(6), 2.179(6)	70
$[W\{\equiv CP=C(Net_2)_2\}(CO)_2Tp^*]$	1.838(6)	167.9(4)	2.298(5)	2.228(5), 2.231(5)	87
$[W\{\equiv CPMeC(NEt_2)_2\}(CO)_2Tp^*][OTf]$	1.840(8)	162.1(2)	2.281(3)	2.206(3), 2.208(3)	110
$[W\{\equiv CP(=O)_2C(NEt_2)_2\}(CO)_2Tp^*]$	1.813(3)	166.8(2)	2.326(2)	2.227(2), 2.219(2)	169

$[W\{\equiv CAs{=}C(NMe_2)_2\}(CO)_2Tp^*]$	1.825(9)	165.2(4)	2.304(6)	2.222(5), 2.231(6)	88
$[W(\equiv COMe)(CO)_2Tp^*]$	1.86(1)	177(1)	2.30(1)	2.22(1), 2.22(1)	79
$[W(\equiv CSMe)(SMe)_2Tp]$	1.788(7)	171.6(5)	2.288 (5)	2.222(5), 2,196(6)	19
$[Re(\equiv CC_6H_2Me_3\text{-}2,4,6)(CO)_2Tp^*][OTf]$	1.786(7)	178.8(6)	2.226(6)	2.158(6), 2.142(5)	94
$[Os(\equiv C^tBu)(CH_2\,^tBu)_2Tp]$	1.73(2)	161(2)	2.30(2)	2.20(2), 2.18(2)	92,93
Polymetallic					
$[Mo\{\equiv CFe(CO)_2Cp\}(CO)_2Tp^*]$	1.819 (6)	172.2(5)			71
$[Mo\{\equiv CP(AuCl)_2C(NMe_2)_2\}(CO)_2Tp^*]$	1.819(2)	160.6(1)	2.291(2)	2.216(2), 2.207(2)	109
$[Tp^*(CO)_2W{\equiv}CCMe{=}CMeC{\equiv}W(CO)_2Tp^*]$	1.81(2)	170(1)	2.30(1)	2.21(1), 2.221(1)	158
$[Tp^*(CO)_2W{\equiv}CC{\equiv}CC{\equiv}CC{\equiv}W(CO)_2Tp^*]$	1.862(3)	175.6(3)	2.260(3)	2.218(3), 2.190(3)	167
$[Tp^*(CO)_2W(\equiv CCH_2)W(\eta^2\text{-}PhC{\equiv}CPh)(CO)Tp]$	1.86(1)	167.5(1)	2.28(1)	2.24(1), 2.23(1)	161
$[Ru\{C{\equiv}CC{\equiv}W(CO)_2Tp^*\}H(Hpz^*)(CO)(PPh_3)_2]$	1.848(5)	171.2(4)	2.300(4)	2.230(4), 2.206(4)	163
c,c,t-$[Ru\{C{\equiv}CC{\equiv}W(CO)_2Tp^*\}_2(CO)_2(PPh_3)_2]$	1.87(1)/1.84(1)	176.9(9)/173.2(9)	2.290(5)/2.279(7)	2.238(5), 2.236(6)/2.224(6), 2.217(6)	168
$[Ru\{C(C{\equiv}CC{\equiv}[W]){=}CHC{\equiv}[W]\}Cl(CO)_2(PPh_3)_2]$ ($[W] = W(CO)_2Tp$)	1.842(3) WC_3/1.832(3), WC_2H	173.4(3)/175.9(2)	2.268(3)/2.287(3)	2.209(3), 2.212(3)/2.215(3), 2.223(3)	187
$[Rh\{C{\equiv}CC{\equiv}W(CO)_2Tp\}I_2(CO)(PPh_3)_2]$	1.816(7)	174.2(6)	2.289(6)	2.204(6), 2.213(6)	162
$[Fe_2Rh\{\mu\text{-}C_3W(CO)_2Tp\}(CO)_{10}(PPh_3)]$	1.826(6)	170.7(4)	2.274(4)	2.204(4), 2.217(5)	162
$[Ir\{C{\equiv}CC{\equiv}W(CO)_2Tp\}_2H(CO)(PPh_3)_2]$	1.85(2)/1.86(1)	175.2(8)/175.3(8)	2.273(8)/2.268(7)	2.205(9), 2.204(8)/2.216(7), 2.198(8)	164
$[W(\equiv CC{\equiv}CAuPPh_3)(CO)_2Tp^*]$	1.834(6)	177.4(4)	2.304(4)	2.203(4), 2.206(4)	165
$[Bu_4N][Au\{C{\equiv}CC{\equiv}W(CO)_2Tp^*\}_2]$	1.82(3)/1.82(2)	173(1)/170(1)	2.28(1)/2.31(1)	2.20(1), 2.22(1)/2.23(1), 2.221(1)	165
$[Hg\{C{\equiv}CC{\equiv}W(CO)_2Tp\}_2]$	1.849(5)	174.9(4)	2.267(4)	2.210(4), 2.202(4)	166

[a] Bp, bis(pyrazolyl)borate; Tp, tris(pyrazolyl)borate; Tp^*, tris(3,5-dimethylpyrazolyl)borate; Tp^{3Me}, tris(3,4,5-trimethylpyrazolyl)borate; pzTp, tetrakis(pyrazolyl)borate; dppm, bisdiphenylphosphinomethane; pic, 4-picoline.

1.798(5) Å,[108] which suggests that the discrepancy in the former is an artefact of residual disordering of the carbonyl and chloromethylidyne ligands rather than the result of π-donation from the chloroalkylidyne substituent leading to a strongly contributing $L_nMo^{(-)}{=}C{=}Cl^{(+)}$ resonance form, as was previously suggested.[7]

The M–C–R angles for the complexes are within the range 179.5(2)° for $Tp^*Mo(\equiv CSC_6H_4NO_2\text{-}4)(CO)_2$[73] to 160.56(11)° for the somewhat exotic $Tp^*Mo\{\equiv CP(AuCl)_2C(NMe_2)_2\}(CO)_2$,[109] showing greater deviation than what is typically expected for terminal alkylidynes. There is, however, no discernible pattern driving this distortion. The deviation from linearity for the bond angle at C_α has been suggested to arise from both steric interactions between the substituent and bulky Tp^x ligand[70] and crystal packing forces.[108] Support for the latter argument is provided by the structure of $Tp^*Mo(\equiv CSC_6H_4NO_2\text{-}4)(CO)_2$, which shows a near linear angle across C_α (179.5(2)°),[73] despite the large alkylidyne substituent and bulky Tp* ligand.

C. *Other Spectroscopic Methods*

In the infrared spectrum, bands due to the poly(pyrazole) ligands can be identified near 1560 and 2500 cm^{-1}, arising from the pyrazolyl ring and B–H stretching vibrations, respectively. The B–H stretching vibrations are remarkably insensitive to the nature of the alkylidyne complex, appearing (when reported) over a relatively narrow range from $\nu_{BH} = 2462\,cm^{-1}$ in $TpW(\equiv CC_6H_2Me_3\text{-}2,4,6)(CO)_2$[107] to 2560 cm^{-1} in $[Tp^*W\{\equiv CP(Me)C(NEt_2)_2\}(CO)_2][OTf]$,[110] as might be expected given that the B–H bond is protected (especially in Tp* complexes) from external influences (solid-state effects) and remote from variations in the electronic nature of the alkylidyne substituent.

The characteristic infrared absorptions for known poly(pyrazolyl)borate-ligated alkylidyne–metal complexes are collected in Table III. Complexes of the type *cis*-$L_4M(CO)_2$ display two ν_{CO} absorptions of comparable intensity,[61] with the higher frequency absorption assigned to the symmetric A_1 mode and the lower frequency absorption to the antisymmetric B_1 mode. Values of ν_{CO} and k_{CO} for the tungsten analogues are lower than those for the related molybdenum examples, as is generally observed for isostructural pairs of carbonyl complexes of 4d and 5d metals.[111] From the k_{CO} values for the series $LW(\equiv CC_6H_4Me\text{-}4)(CO)_2$, it can be seen that the donor properties of the tripodal ligands increase in the order $L = Cp < Cp^* \approx pzTp \approx Tp < Tp^{Ph} < Tp^*$. When the pyrazole-containing ligands are replaced by Cp, there is a significant shift towards higher energy for the carbonyl absorptions due to decreased electron density at the metal centre and subsequent reduction in π-back-bonding with the metal carbonyls (Table IV).

Table V shows that the position of the *cis*-$W(CO)_2$ infrared stretch is highly sensitive to the π-acidity of the alkylidyne group. Conjugation of the M≡C triple bond with π-donating alkylidyne substituents (Scheme 32) leads to a decrease in the π-acidity of the M≡C bond in line with the heteroallenic (heterovinylidene) canonical form[7] and a subsequent increase in available electron density for the π-acidic carbonyls, which leads to a decrease in the frequency of ν_{CO}.

TABLE III
CHARACTERISTIC INFRARED ABSORPTIONS FOR Tp^{x}-LIGATED ALKYLIDYNE COMPLEXES[a]

Compound	ν_{CO} (cm^{-1})	k_{CO} ($N\,m^{-1}$)	Ref.
Alkyl, aryl, vinyl			
$[Cr(\equiv CC_6H_4Me\text{-}4)(CO)_2Tp]$	1997, 1913[d]	15.44	41
$[Cr(\equiv CC_6H_4Me\text{-}4)(CO)_2Tp^*]$	1987, 1909[c]	15.33	47
$[Mo(\equiv CH)(CO)_2Tp^*]$	2001, 1913[i]	15.47	89
$[Mo(\equiv CMe)(CO)_2Tp^*]$	1982, 1889[b]	15.14	62
$[Mo(\equiv CMe)(CO)(NCCD_3)Tp^*]$	1840[e]		62
$[Mo(\equiv CMe)(CO)(PMe_3)Tp^*]$	1841[b]		62
$[Mo(\equiv CEt)(CO)_2Tp^*]$	1980, 1885[b]	15.09	62
$[Mo(\equiv CPr)(CO)_2Tp^*]$	1977, 1882[b]	15.04	62
$[Mo(\equiv CBu)(CO)_2Tp]$	1992, 1902	15.32	145
$[Mo(\equiv CBu)(CO)_2Tp^*]$	1980, 1885[b]	15.09	62
$[Mo(\equiv CBu)(CO)\{P(OMe)_3\}Tp]$	1889		145
$[Mo(\equiv C^tBu)(CO)\{P(OMe)_3\}_2Bp]$	1912[j]		135
$[Mo(\equiv CPh)(CO)_2Tp^*]$	1979, 1890[b]	15.12	62,73
	1991, 1909[c]	15.36	
$[Mo(\equiv CC_6H_4Me\text{-}4)(CO)_2Tp]$	1998, 1921[d]	15.51	41,42
	1955, 1920[l]	15.16	
$[Mo(\equiv CC_6H_4Me\text{-}4)(CO)_2Tp^*]$	1982, 1899[i]	15.21	44
$[Mo(\equiv CC_6H_4OMe\text{-}4)(CO)_2Tp]$	1991, 1906	15.34	48
$[Mo(\equiv CC_6H_4OMe\text{-}2)(CO)\{P(OMe)_3\}_2Bp]$	1919[j]		135
$[Mo(\equiv CC_4H_3S\text{-}2)(CO)_2Tp]$	1996, 1913	15.43	48
$[Mo(\equiv CC\equiv C^tBu)(CO)_2Tp]$	2001, 1919	15.52	52
$[Mo(\equiv CC\equiv C^tBu)(CO)_2Tp^*]$	1993, 1908	15.37	52
$[Mo(\equiv CC\equiv N)(CO)_2Tp^*]$	2026, 1951[c]	15.97	36
$[Mo\{\equiv C(CN)(Ph)_2\}(CO)_2Tp^*]$	2000, 1920[c]	15.52	189
$[Mo\{\equiv C(CN)(CO_2Et)(HgCl)\}(CO)_2Tp^*]$	2011, 1940[b]	15.76	69
$[Mo\{\equiv C(CN)(CO_2Et)(Cu)\}(CO)_2Tp^*]$	1989, 1898, 1885[h]		69
$[Mo\{\equiv C(CN)(CO_2Et)(N_2Ar)\}(CO)_2Tp^*]$	2015, 1940[b]	15.80	69
$[Mo\{\equiv C(CN)_2(Cu)\}(CO)_2Tp^*]$	1975, 1892[h]	15.10	69
$[W(\equiv CLi)(CO)_2Tp^*]$	1916, 1819[n]	14.09	68
$[W(\equiv CH)(CO)_2Tp^*]$	1992, 1903[k]	15.32	120
	1986, 1891[i]	15.18	89
$[W(\equiv CMe)(CO)_3Bp]$	1976, 1881	15.03	46
$[W(\equiv CMe)(CO)_2Tp]$	1983, 1899[k]	15.22	39,40
	1995, 1910[l]	15.40	42
$[W(\equiv CMe)(CO)_2Tp^*]$	1968, 1867[b]	14.86	67,158
$[W(\equiv CCH_2SiMe_3)(CO)_2Tp^*]$	1968, 1876[k]	14.93	66
$[W(\equiv CEt)(CO)_2Tp^*]$	1960, 1862[b]	14.76	67,158
$[W(\equiv CCH_2\,{}^tBu)(CO)_2Tp^*]$	1969, 1875[k]	14.93	66
$[W(\equiv CCH_2CH_2Ph)(CO)_2Tp^*]$	1963, 1864[b]	14.80	67,158
$[W(\equiv CCH_2Ph)(CO)_2Tp^*]$	1958, 1862[f]	14.74	121
$[W\{\equiv CCH_2CH(OH)(Ph)\}(CO)_2Tp^*]$	1967, 1866[b]	14.84	67
$[W\{\equiv CCH_2CH(OH)(CH=CHMe)\}(CO)_2Tp^*]$	1971, 1872[b]	14.92	67
$[W\{\equiv CCH_2C(OH)(Ph)(Me)\}(CO)_2Tp^*]$	1969, 1873[b]	14.91	67
$[W\{\equiv CCH_2C(O)Ph\}(CO)_2Tp^*]$	1971, 1876[b]	14.95	67
$[W(\equiv CCH_2CH=CH_2)(CO)_2Tp^*]$	1966, 1865[b]	14.83	121
$[W(\equiv C^iPr)(CO)_2Tp^*]$	1964, 1865[b]	14.81	67,158
$[W(\equiv CCHMeSiMe_3)(CO)_2Tp^*]$	1964, 1867[k]	14.83	66
$[W(\equiv CCHMePh)(CO)_2Tp^*]$	1965, 1868[b]	14.84	121
$[W(\equiv CCHMeCH=CH_2)(CO)_2Tp^*]$	1969, 1876[b]	14.93	121
$[W(\equiv CCHPhOH)(CO)_2Tp^*]$	1976, 1880[b]	15.02	68
$[W(\equiv CCMe_2Ph)(CO)_2Tp^*]$	1965, 1868[b]	14.84	121
$[W\{\equiv CCMe_2CH=CH_2\}(CO)_2Tp^*]$	1965, 1872[b]	14.87	121
$[W(\equiv CCPh_2OH)(CO)_2Tp^*]$	1974, 1876[f]	14.97	68
$[W\{\equiv CC(=O)Ph\}(CO)_2Tp^*]$	1999, 1918[b]	15.50	68

TABLE III
(CONTINUED)

Compound	ν_{CO} (cm^{-1})	k_{CO} ($N m^{-1}$)	Ref.
[W(≡CCH=CHPh)(CO)$_2$Tp*]	1969, 1867[b]	14.87	67
[W(≡CCH=CHMe)(CO)$_2$Tp*]	1962, 1866[b]	14.80	121
[W(≡CCMe=CHMe)(CO)$_2$Tp*]	1965, 1862[b]	14.80	121
[W{≡CCPh=CH(OPh)}(CO)$_2$Tp*]	1953, 1868[b]	14.75	65
[W{≡CCPh=CH(OC$_6$H$_4$Me-4)}(CO)$_2$Tp*]	1953, 1870[b]	14.76	65
[W(≡CC≡CSiMe$_3$)(CO)$_2$Tp]	1991, 1906[i]	15.34	61
[W(≡CC≡CSiMe$_3$)(CO)$_2$Tp*]	1982, 1896[i]	15.19	61
[W(≡CC≡C^{t}Bu)(CO)$_2$Tp]	1985, 1896	15.21	52
[W(≡CC≡CPh)(CO)$_2$Tp]	1988, 1904[i]	15.30	61
[W(≡CC≡CPh)(CO)$_2$Tp*]	1979, 1893[b]	15.14	61
[W(≡CC$_4$H$_3$O-2)(CO)$_2$Tp]	1987, 1905	15.30	42
[W(≡CPh)(CO)$_2$Tp]	1990, 1915[k]	15.40	42
[W(≡CPh)(CO)$_2$Tp*]	1969, 1876[b]	14.93	78
[W(≡CC$_6$H$_4$Me-4)(CO)$_3$Bp]	1980, 1888	15.11	46
[W(≡CC$_6$H$_4$Me-4)(CO)$_2$Tp]	1986, 1903	15.28	39,40
[W(≡CC$_6$H$_4$Me-4)(CO)$_2$Tp*]	1974, 1888[k]	15.06	43
[W(≡CC$_6$H$_4$Me-4)(CO)$_2$Tp^{Ph}]	1982, 1897[k]	15.20	43
[W(≡CC$_6$H$_4$Me-4)(CO)$_2$pzTp]	1986, 1903	15.28	39,40
[W(≡CC$_6$H$_4$Me-4)(CO)$_2$(PPh$_3$)Bp]	1991, 1905	15.33	46
[W(≡CC$_6$H$_4$Me-4)(CO)$_2$(PMe$_3$)Bp]	1989, 1899	15.27	46
[W(≡CC$_6$H$_4$Me-4)(CO)(PMe$_3$)Tp]	1866[j]		183
[W(≡CC$_6$H$_4$Me-4)(CO)(PPh$_3$)Tp]	1860		50
[W(≡CC$_6$H$_4$OMe-2)(CO)$_2$Tp]	1976, 1886	15.07	43
[W(≡CC$_6$H$_3$Me$_2$-2,6)(CO)$_2$(pic)Bp]	1978, 1886	15.08	49
[W(≡CC$_6$H$_3$Me$_2$-2,6)(CO)$_2$(CN^{t}Bu)Bp]	1998, 1918	15.49	49
[W(≡CC$_6$H$_3$Me$_2$-2,6)(CO)$_2$(CNC$_6$H$_3$Me$_2$-2,6)Bp]	1998, 1923	15.53	49
mer-[W(≡CC$_6$H$_3$Me$_2$-2,6)(CO)$_2$(PMe$_2$Ph)Bp]	1992, 1903	15.32	49
[W(≡CC$_6$H$_3$Me$_2$-2,6)(CO)$_2$Tp]	1984, 1901[k]	15.24	43
[W(≡CC$_6$H$_2$Me$_3$-2,4,6)(CO)$_2$(pic)Bp]	1975, 1888	15.07	49
mer-[W(≡CC$_6$H$_2$Me$_3$-2,4,6)(CO)$_2$(PMe$_2$Ph)Bp]	1993, 1905	15.35	49
mer-[W(≡CC$_6$H$_2$Me$_3$-2,4,6)(CO)(PMe$_2$Ph)$_2$Bp]	1860		49
[W(≡CC$_6$H$_2$Me$_3$-2,4,6)(CO)$_2$Tp]	1975, 1888	15.07	107
	1971, 1874[h]	14.93	
[Re(≡CC$_6$H$_2$Me$_3$-2,4,6)(CO)$_2$Tp*][OTf]	2076, 2010[b]	16.86	94
[Re{≡CCH$_2$CH$_2$C(=O)H}(CO)(PMe$_3$)Tp]$^+$	1996[m]		97
[Re{≡CCH$_2$CH$_2$C(OMe)$_2$H}(CO)(PMe$_3$)Tp]$^+$	1991[m]		97
Silyl			
[Mo(≡CSiMe$_2$Ph)(CO)$_2$Tp*]	1997, 1911	15.43	89
[W(≡CSiMe$_3$)(CO)$_2$Tp*]	1976, 1884[f]	15.05	68
[W(≡CSiMe$_2$Ph)(CO)$_2$Tp*]	1982, 1889	15.14	89
Aryloxy			
[Mo(≡COPh)(CO)$_2$Tp*]	1982, 1889[e]	15.14	70
[Mo(≡COC$_6$H$_4$Me-4)(CO)$_2$Tp*]	1980, 1887[e]	15.11	70
[W(≡COMe)(CO)$_2$Tp*]	1958, 1862[i]	14.74	79
[W(≡COPh)(CO)$_2$Tp*]	1967, 1870[e]	14.87	70
[W(≡COC$_6$H$_4$Me-4)(CO)$_2$Tp*]	1966, 1870[e]	14.86	70
[W{≡COC$_6$H$_4$OMe-4)(CO)$_2$Tp*]	1965, 1866[e]	14.83	70
[W{≡COC$_6$H$_4$NO$_2$-4)(CO)$_2$Tp*]	1977, 1877[i]	15.01	78
[W{≡COC$_6$H$_4$Cl-4)(CO)$_2$Tp*]	1969, 1864[b]	14.84	78
Amino			
[Cr(≡CN^{i}Pr$_2$)(CO)$_2$Tp*]	1953, 1856	14.66	81,82
	1956, 1863[j]	14.73	
[Cr(≡CN^{i}Pr$_2$)(CO)(CN^{t}Bu)Tp*]	1826, 1842[j]		81,82

Table III
(Continued)

Compound	ν_{CO} (cm^{-1})	k_{CO} ($N\,m^{-1}$)	Ref.
[Mo(≡CNH^{t}Bu)(CO)$_2$Tp*]	1948, 1852[i]	14.59	83
[Mo(≡CNMe$_2$)(CO)$_2$Tp*]	1939, 1884[b]	14.76	83
[Mo(≡CN^{i}Pr$_2$)(CO)$_2$Tp]	1937, 1835[h]	14.37	180
	1951, 1852	14.61	
[Mo(≡CNMe^{t}Bu)(CO)$_2$Tp*]	1945, 1848[b]	14.53	83
[Mo(≡CNMePh)(CO)$_2$Tp*]	1963, 1856[b]	14.74	83
[W(≡CNH$_2$)(CO)$_2$Tp]	1950, 1855	14.63	75
[W(≡CNHMe)(CO)$_2$Tp]	1943, 1837	14.44	75
[W(≡CNHEt)(CO)$_2$Tp]	1943, 1841	14.47	75
[W(≡CNHCH$_2$CH$_2$OH)(CO)$_2$Tp]	1942, 1844	14.48	75
[W(≡CNH^{i}Pr)(CO)$_2$Tp]	1943, 1843	14.48	75
[W(≡CNH^{t}Bu)(CO)$_2$Tp]	1945, 1840	14.47	75
[W(≡CNHC$_6$H$_4$Me-4)(CO)$_2$Tp]	1956, 1860	14.71	75
[W(≡CNMe$_2$)(CO)$_2$Tp]	1941, 1837	14.42	75
[W(≡CNEtMe)(CO)$_2$Tp*]	1936, 1833	14.35	84
[W(≡CNEt$_2$)(CO)$_2$Tp]	1938, 1831	14.35	75
	1930, 1830	14.28	42
[W(≡CNEt$_2$)(CO)$_2$Tp*]	1936, 1833	14.35	84
Chalco			
[Et$_4$N][Mo(≡CS)(CO)$_2$Tp*]	1886, 1794[b]	13.68	73
[Et$_4$N][Mo(≡CSe)(CO)$_2$Tp*]	1913, 1824[b]	14.11	73
[Et$_4$N][Mo(≡CTe)(CO)$_2$Tp*]	1924, 1840[b]	14.31	73
[Mo(≡CSMe)(CO)$_2$Tp*]	1987, 1904[m]	15.29	73
[Mo(≡CSePh)(CO)$_2$Tp*]	1988, 1914[c]	15.38	189
[Mo(≡CTeMe)(CO)$_2$Tp*]	1992, 1911[m]	15.39	73
[W(≡CSMe)(CO)$_2$Tp]	1979, 1893[g]	15.14	23,77
	1973, 1885	15.03	
[W(≡CSEt)(CO)$_2$Tp]	1979, 1892[g]	15.14	23
[W{≡CSC$_6$H$_4$(NO$_2$)$_2$-2,4}(CO)$_2$Tp]	1999, 1914[g]	15.47	23
Halo			
[Mo(≡CCl)(CO)$_2$Tp^{Me}]	2006, 1925[c]	15.61	36
[Mo(≡CCl)(CO)$_2$Tp^{4Tol}]	2003, 1925	15.58	36
[Mo(≡CCl)(CO)$_2$Tp^{i}Pr,4Br]	2002, 1922	15.55	36
[Mo(≡CCl)(CO)$_2$Tp*]	2005, 1921[c]	15.57	36,37
[Mo(≡CCl)(CO)$_2$Tp^{3Me}]	2005, 1920[c]	15.56	36
[Mo(≡CCl)(CO)$_2$Tp^{Me_2,4Cl}]	2010, 1929[c]	15.67	36
[Mo(≡CCl)(CO)$_2$pzTp]	2010, 1929[c]	15.67	36
[Mo(≡CBr)(CO)$_2$Tp*]	2008, 1924[c]	15.62	36
[Mo(≡CBr)(CO)$_2$Tp^{3Me}]	2007, 1923[c]	15.60	36
[Mo(≡CI)(CO)$_2$Tp*]	2009, 1927[c]	15.65	36
[W(≡CCl)(CO)$_2$Tp*]	1991, 1902[c]	15.31	36
[W(≡CBr)(CO)$_2$Tp*]	1994, 1905[c]	15.36	36
[W(≡CI)(CO)$_2$Tp*]	1992, 1907[c]	15.35	68
Phosphonio			
[W(≡CPEt$_3$)(CO)$_2$Tp*]$^+$	2020, 1935	15.80	72
[W(≡CPCy$_3$)(CO)$_2$Tp*][PF$_6$]	2020, 1933[e]	15.78	70
[W(≡CPMe$_2$Ph)(CO)$_2$Tp*][PF$_6$]	2022, 1934[e]	15.81	70
[W(≡CPPh$_3$)(CO)$_2$Tp*][PF$_6$]	2026, 1940[e]	15.89	70
Phospha- and arsaalkenyl			
[Mo{≡CP=C(NMe$_2$)$_2$}(CO)$_2$Tp*]	1946, 1864[b]	14.66	87
[Mo{≡CP=C(NEt$_2$)$_2$}(CO)$_2$Tp*]	1943, 1859[b]	15.60	87
[Mo{≡CPMeC(NMe$_2$)$_2$}(CO)$_2$Tp*][OTf]	1997, 1915[b]	15.46	110

Table III
(Continued)

Compound	ν_{CO} (cm^{-1})	k_{CO} (N m^{-1})	Ref.
[Mo{≡CPMeC(NEt$_2$)$_2$}(CO)$_2$Tp*][OTf]	1994, 1903[b]	15.34	110
[Mo{≡CP(=O)$_2$C(NMe$_2$)$_2$}(CO)$_2$Tp*]	2005, 1918[b]	15.54	169
[Mo{≡CP(=O)$_2$C(NEt$_2$)$_2$}(CO)$_2$Tp*]	2002, 1923[b]	15.56	169
[Mo{≡CAs=C(NMe$_2$)$_2$}(CO)$_2$Tp*]	1947, 1863[b]	14.66	88
[Mo{≡CAsMeC(NMe$_2$)$_2$}(CO)$_2$Tp*][OTf]	1997, 1911[b]	15.43	88
[W{≡CP=C(NMe$_2$)$_2$}(CO)$_2$Tp*]	1935, 1848[b]	14.46	87
[W{≡CP=C(NEt$_2$)$_2$}(CO)$_2$Tp*]	1933, 1845[b]	14.42	87
[W{≡CPMeC(NMe$_2$)$_2$}(CO)$_2$Tp*][OTf]	1982, 1893[b]	15.17	110
[W(≡CPHC(NEt$_2$)$_2$}(CO)$_2$Tp*][BF$_4$]	1986, 1897[b]	15.23	110
[W{≡CPMeC(NEt$_2$)$_2$}(CO)$_2$Tp*][OTf]	1981, 1884[b]	15.09	110
[W{≡CP(=O)$_2$C(NMe$_2$)$_2$}(CO)$_2$Tp*]	1987, 1893[b]	15.21	169
[W{≡CP(=O)$_2$C(NEt$_2$)$_2$}(CO)$_2$Tp*]	1989, 1894[b]	15.23	169
[W{≡CAs=C(NMe$_2$)$_2$}(CO)$_2$Tp*]	1936, 1848[b]	14.46	88
[W{≡CAsMeC(NMe$_2$)$_2$}(CO)$_2$Tp*][OTf]	1981, 1888[b]	15.12	88
Polymetallic			
[CrMo(μ-η^6:σ-CC$_6$H$_4$OMe-2)(CO)$_5$Tp]	2011, 1966, 1929, 1900[j]		53
[MoMn(μ-σ:η^5-CC$_5$H$_4$)(CO)$_5$Tp]	2027, 1996, 1947, 1914		54
[Mo$_2$Fe(μ-σ,σ':η^5-CC$_5$H$_4$)$_2$(CO)$_4$Tp$_2$]	1987, 1904	15.29	55
[Tp*(CO)$_2$Mo{≡CFe(CO)$_2$Cp}]	1947, 1865	14.68	71
[Tp*(CO)$_2$Mo{≡CP(AuCl)$_2$C(NMe$_2$)$_2$}]	2002, 1921[b]	15.54	109
[Tp*(CO)$_2$W{≡CP(AuCl)$_2$C(NMe$_2$)$_2$}]	1985, 1896[b]	15.21	109
[*cyclo*-As$_3${C≡Mo(CO)$_2$Tp*}$_3$]	1987, 1909[b]	15.33	109
[*cyclo*-As$_3${C≡W(CO)$_2$Tp*}$_3$]	1972, 1887[b]	15.04	109
[Tp*(CO)$_2$Mo≡CCH$_2$C≡Mo(CO)$_2$Tp*]	2008, 1985, 1910, 1900[b]		159
[Tp*(CO)$_2$W≡CCH$_2$C≡Mo(CO)$_2$Tp*]	1986, 1973, 1894, 1878[b]		159
[Tp*(CO)$_2$W≡CCHMeC≡Mo(CO)$_2$Tp*]	1982, 1968, 1886, 1874[b]		159
[Tp*(CO)$_2$W≡CCMe$_2$C≡Mo(CO)$_2$Tp*]	1985, 1970, 1893, 1874[b]		159
[Tp*(CO)$_2$Mo≡CC≡CMo(=O)$_2$Tp*]	1983, 1897[b]	15.21	159
[Tp*(CO)$_2$W≡CC≡CMo(=O)$_2$Tp*]	1971, 1877[b]	14.96	159
[Tp*(CO)$_2$Mo≡CC≡CW(CO)$_2$Tp*]	1985, 1904[b]	15.28	159
[Tp*(CO)$_2$W≡CC(=O)C≡Mo(CO)$_2$Tp*]	2016, 1991, 1928, 1906[b]		159
[Tp*(CO)$_2$Mo≡CCH$_2$CH$_2$C≡Mo(CO)$_2$Tp*]	1976, 1879[b]	15.01	158
[Tp*(CO)$_2$W≡CCH$_2$CH$_2$C≡W(CO)$_2$Tp*]	1963, 1867[b]	14.81	158
[Tp*(CO)$_2$W≡CCHMeCHMeC≡W(CO)$_2$Tp*]	1969, 1866[b]	14.81	158
[Tp*(CO)$_2$W≡CCHBzCHBzC≡W(CO)$_2$Tp*]	1964, 1869[b]	14.84	158
[Tp*(CO)$_2$W≡CCMe$_2$CMe$_2$C≡W(CO)$_2$Tp*]	1963, 1864[b]	14.80	158
[Tp*(CO)$_2$Mo≡CCH=CHC≡Mo(CO)$_2$Tp*]	1991, 1973, 1890[b]		158
[Tp*(CO)$_2$W≡CCH=CHC≡W(CO)$_2$Tp*]	1977, 1956, 1869[b]		158
[Tp*(CO)$_2$W≡CCMe=CMeC≡W(CO)$_2$Tp*]	1957, 1869[b]	14.79	158
[Tp*(CO)$_2$W≡CCBz=CBzC≡W(CO)$_2$Tp*]	1963, 1875[b]	14.88	158
[Tp*(CO)$_2$Mo≡CC≡CC≡Mo(CO)$_2$Tp*]	2004, 1977, 1901[b]		158
[Tp*(CO)$_2$W≡CC≡CC≡W(CO)$_2$Tp*]	1987, 1958, 1878[b]		158
[Tp*(CO)$_2$W≡CCH$_2$CH=CHC≡W(CO)$_2$Tp*]	1962, 1870[f]	14.83	121
[Tp*(CO)$_2$W≡CCH$_2$CH=CHC≡W(CO)$_2$Tp*]	1962, 1870[f]	14.83	121
K[Tp*(CO)$_2$W≡CCHCHCHC≡W(CO)$_2$Tp*]	NR		121
[Tp(CO)$_2$W≡CC≡CC≡CC≡W(CO)$_2$Tp]	1982, 1915, 1890		167
[Tp*(CO)$_2$W≡CC≡CC≡CC≡W(CO)$_2$Tp*]	1993, 1975, 1901		167

TABLE III
(CONTINUED)

Compound	ν_{CO} (cm^{-1})	k_{CO} ($N m^{-1}$)	Ref.
[Tp*(CO)$_2$Mo≡CCH$_2$W(η^2-PhC≡CPh)(CO)Tp]	1943, 1921, 1855[b]		161
[Tp*(CO)$_2$W≡CCH$_2$W(η^2-PhC≡CPh)(CO)Tp]	1933, 1918, 1842[b]		161
[Tp*(CO)$_2$Mo≡CC≡W(η^2-PhC≡CPh)(CO)Tp]	1999, 1950, 1871[b]		161
[Tp*(CO)$_2$W≡CC≡W(η^2-PhC≡CPh)(CO)Tp]	1993, 1935, 1857[b]		161
[Co$_2${μ-^{t}BuC$_2$C≡Mo(CO)$_2$Tp*}(CO)$_6$]	2090, 2055, 2032, 1975, 1898		52
[Co$_2${μ-^{t}BuC$_2$C≡W(CO)$_2$Tp}(CO)$_6$]	2089, 2053, 2029, 1978, 1902[d]		52
[Co$_2${μ-^{t}BuC$_2$C≡W(CO)$_2$Tp}(μ-dppm)(CO)$_4$]	2023, 1997, 1971, 1955, 1873		52
[Mo$_2${μ-^{t}BuC$_2$C≡W(CO)$_2$Tp}(CO)$_4$Cp$_2$]	2001, 1954, 1932, 1874, 1840		52
[Rh{C≡CC≡W(CO)$_2$Tp}(CO)(PPh$_3$)$_2$]	1979, 1872	14.98	162
[Rh{C≡CC≡W(CO)$_2$Tp}I$_2$(CO)(PPh$_3$)$_2$]	1965, 1882	14.95	162
[Fe$_2$Rh(μ-C$_3$W(CO)$_2$Tp)(CO)$_{10}$(PPh$_3$)]	1982, 1899	15.21	162
[Ir{C≡CC≡W(CO)$_2$Tp}$_2$H(CO)(PPh$_3$)$_2$]	1948, 1873	14.75	164
[Ru(C≡CC≡[W])(HgC≡CC≡[W])(CO)$_2$(PPh$_3$)$_2$] ([W] = W(CO)$_2$Tp*)	1985, 1970, 1879, 1865		168
[Ru{η^2-Tp(CO)$_2$WC$_6$W(CO)$_2$Tp}(CO)$_2$(PPh$_3$)$_2$]	1998, 1876	15.17	168
c,c,t-[Ru{C≡CC≡W(CO)$_2$Tp*}$_2$(CO)$_2$(PPh$_3$)$_2$]	2003, 1867	15.14	168
[Ru{C(C≡CC≡[W])=CHC≡[W]}Cl(CO)(PPh$_3$)$_2$] ([W] = W(CO)$_2$Tp)	1982, 1958, 1896, 1874		187
[Ru{C(C≡CC≡[W])=CHC≡[W]}Cl(CO)(PPh$_3$)$_2$] ([W] = W(CO)$_2$Tp*)	1979, 1954, 1893, 1866		187
[Ru{C(C≡CC≡[W])=CHC≡[W]}Cl(CO)$_2$(PPh$_3$)$_2$] ([W] = W(CO)$_2$Tp)	1958, 1877	14.85	187
[Ru{C(C≡CC≡[W])=CHC≡[W]}Cl(CO)(PPh$_3$)$_2$] ([W] = W(CO)$_2$Tp*)	1952, 1874	14.78	187
[W(≡CC≡CAuPPh$_3$)(CO)$_2$Tp]	1975, 1887	15.07	165
[W(≡CC≡CAuPPh$_3$)(CO)$_2$Tp*]	1967, 1878	14.93	165
[(Ph$_3$P)$_2$N][Au{C≡CC≡W(CO)$_2$Tp*}$_2$]	1953, 1868	14.75	165
[Hg{C≡CC≡W(CO)$_2$Tp}$_2$]	1992, 1905	15.34	166
[Hg{C≡CC≡W(CO)$_2$Tp*}$_2$]	1983, 1895	15.19	166

Tp^{Ph}, hydrotris(3-phenylpyrazolyl)borate; Tp^{4Tol}, hydrotris(4-tolylpyrazolyl)borate; Bz, CH$_2$Ph; dppm, bis(diphenylphosphino)methane; pic, 4-picoline.
[a]Data were obtained in CH$_2$Cl$_2$ solution at room temperature unless otherwise indicated.
[b]KBr.
[c]Cyclohexane.
[d]Light petroleum.
[e]Acetonitrile.
[f]Neat.
[g]CS$_2$.
[h]Nujol.
[i]THF.
[j]Ether.
[k]Hexane.
[l]Pentane.
[m]Unspecified.
[n]CaF$_2$.

TABLE IV

IR ABSORPTIONS FOR ALKYLIDYNE–METAL COMPLEXES WITH A RANGE OF TRIPODAL LIGANDS $[LW(\equiv CC_6H_4R\text{-}4)(CO)_2]^{x}$[a]

L	R	x	ν_{CO} (cm^{-1})	k_{CO} (N m^{-1})	Ref.
κ^3-CpCo(PO$_3$Me$_2$)$_3$	Me	0	1961, 1859	14.74	112
η^5-C$_2$B$_9$H$_9$Me$_2$	Me	1–	1956, 1874	14.82	113
κ^3-HB(mt)$_3$	Me	0	1967, 1875	14.91	104
η^5-C$_2$B$_9$H$_{11}$	Me	1–	1965, 1880	14.93	113
κ^3-HC(pz*)$_2$(C$_6$H$_4$O-2)	H	0	1890, 1958[b]	14.95	102
κ^3-Me$_3$[9]aneN$_3$	H	1+	1975, 1879[c]	15.00	114
κ^3-HB(pz*)$_3$	Me	0	1974, 1888[d]	15.06	43
κ^3-HB(pzPh)$_3$	Me	0	1982, 1897[d]	15.20	43
κ^3-HC(py)$_3^+$	H	1+	1988, 1894[c]	15.22	115
κ^3-P(py)$_3$	H	1+	1984, 1899[c]	15.23	115
η-C$_5$H$_5$	Me	0	1982, 1902	15.24	103
κ^3-(C$_6$F$_5$)AuC(pz)$_3$	Me	0	1985, 1899	15.24	116
κ^3-B(pz)$_4$	Me	0	1986, 1903	15.28	39,40
κ^3-HB(pz)$_3$	Me	0	1986, 1903	15.28	39,40
κ^3-(F$_3$B)C(pz)$_3$	Me	0	1988, 1902	15.28	117
η-C$_5$Me$_5$	Me	0	1981, 1910[d]	15.29	190
κ^3-HC(pz)$_3$	Me	1+	1995, 1912	15.42	100
η^6-C$_2$B$_{10}$H$_{10}$Me$_2$	Me	1–	1990, 1930	15.52	118
κ^3-[9]aneS$_3$	H	1+	2007, 1925[c]	15.62	115
κ^3-MeP(CH$_2$Ph$_2$)$_3$	H	1+	1999, 1934[c]	15.62	115
κ^3-PhP(C$_2$H$_4$PPh$_2$)$_2$	Me	1+	2005, 1941	15.72	119

η-C$_5$H$_5$, Cp; η-C$_5$Me$_5$, Cp*; κ^3-HB(pz)$_3$, Tp; κ^3-B(pz)$_4$, pzTp; κ^3-HB(pzPh)$_3$, Tp^{Ph}; κ^3-HB(pz*)$_3$, Tp*; κ^3-HB(mt)$_3$, Tm.

[a]Unless otherwise indicated, measurements in CH_2Cl_2 solution.

[b]$CHCl_3$.

[c]KBr.

[d]Hexane.

In line with this phenomenon for π-donating alkylidyne substituents, in the alkylidynes with the strongly donating amino groups, the ν_{CO} absorptions appear at particularly low frequencies. For cationic alkylidynes, ν_{CO} values typically appear at higher frequency. In principle, the integrated intensities of the two ν_{CO} absorptions can be used to calculate the angle between the two carbonyl ligands. However, because the approximations involved limit the accuracy of this method to $\pm 5°$ (coupled with the 'octahedral enforcer' character of the Tp and Tp* ligands), this exercise is not generally worthwhile. However, in a qualitative sense, visual inspection of the intensity profile may be useful in interpreting reactions that occur across the M≡C bond such that steric factors accompanying increased coordination at the metal can lead to a reversal of the relative intensities. For example, the sequential addition of selenium to the alkylidyne complex TpMo(≡CR)(CO)$_2$ (R = C$_4$H$_3$S-2) provides TpMo(η^2-SeCR)(CO)$_2$ and then TpMo(κ^2-Se$_2$CR)(CO)$_2$ with the relative intensity of the high-frequency absorption progressively increasing.[122]

The k_{CO} force constants for a range of Group 6 poly(pyrazolyl)borate *cis*-dicarbonyl complexes Tp^{x}M(L′)(CO)$_2$ with three-valence electron ligands (L′) that are isoelectronic with 'CR' are tabulated in Table VI and exhibit the previously

TABLE V

INFRARED ABSORPTIONS FOR ALKYLIDYNE–METAL COMPLEXES $Tp^*W(\equiv CR)(CO)_2$

Compound[a]	ν_{CO} (cm^{-1})	k_{CO} ($N m^{-1}$)	Ref.
$[W](\equiv CPMe_2Ph)^+$	2022, 1934[b]	15.81	70
$[W]\{\equiv CC(=O)Ph\}$	1999, 1918	15.50	68
$[W](\equiv CH)$	1992, 1903[c]	15.32	120
$[W](\equiv CCl)$	1991, 1902[d]	15.31	36
$[W](\equiv CC\equiv CPh)$	1979, 1893	15.14	61
$[W](\equiv CSiMe_3)$	1976, 1884[e]	15.05	68
$[W](\equiv CPh)$	1969, 1876	14.93	78
$[W](\equiv COPh)$	1967, 1870[b]	14.87	70
$[W](\equiv CMe)$	1968, 1867	14.86	67
$[W](\equiv CCMe=CHMe)$	1965, 1862	14.80	121
$[W]\{\equiv CAs=C(NMe_2)_2\}$	1936, 1848	14.46	88
$[W](\equiv CNEt_2)$	1936, 1833[f]	14.35	84

[a] $[W]=Tp^*W(CO)_2$. Unless otherwise noted, data are reported from KBr.
[b] CH_3CN.
[c] Hexane.
[d] Cyclohexane.
[e] Ether.
[f] CH_2Cl_2.

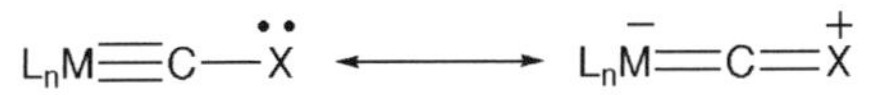

$$L_nM\equiv C-\ddot{X} \longleftrightarrow L_n\overset{-}{M}=C=\overset{+}{X}$$

SCHEME 32. Delocalisation of electron density from π-basic heteroatom alkylidyne substituents (X = Cl, NR_2, OR, SR, SeR, TeR).

observed trend of Cr > Mo > W, whereas for the same metal, k_{CO} decreases in the order L = NO > NS > NNR > CR > C_3H_5, mirroring the trend in π-acidity for L′. A curious feature is that though CS is recognised to be a stronger net acceptor than CO, the reverse appears to be true for NO and NS in this system.

IV
REACTIVITY

Reactivity modes of the poly(pyrazolyl)borate alkylidyne complexes follow a number of recognised routes for transition metal complexes containing metal–carbon triple bonds, including ligand substitution or redox reactions at the transition metal centre, insertion of a molecule into the metal–carbon triple bond, and electrophilic or nucleophilic attack at the alkylidyne carbon, C_α. Cationic alkylidyne complexes generally react with nucleophiles at the alkylidyne carbon, whereas neutral alkylidyne complexes can react at either the metal centre or the alkylidyne carbon. Substantive work has been devoted to neutral and cationic alkylidyne complexes bearing heteroatom substituents. Differences between the chemistry of the various Tp^x complexes have previously been rationalised largely on the basis of steric effects.[126,127]

TABLE VI

CHARACTERISTIC INFRARED ABSORPTIONS FOR A RANGE OF GROUP 6 Tp^x COMPLEXES $Tp^{(x)}M(L')(CO)_2$[a]

L′	Tp^x	M	ν_{CO} (cm^{-1})	k_{CO} ($N m^{-1}$)	Ref.
η^1-NO	Tp	Cr	2038, 1951	16.07	123
η^1-NO	pzTp	Cr	2039, 1951	16.08	123
η^1-NO	Tp*	Cr	2029, 1941	15.92	123
η^1-CC_6H_4Me-4	Tp	Cr	1997, 1913[d]	15.44	41
η^1-CC_6H_4Me-4	Tp*	Cr	1987, 1909	15.33	47
η^1-NO	Tp	Mo	2025, 1933[b]	15.82	123
η^1-NO	pzTp	Mo	2024, 1937	15.85	123
η^1-NO	Tp*	Mo	2016, 1925	15.69	124
			1997, 1905[c]	15.38	123,124
η^1-NS	Tp*	Mo	2003, 1924[c]	15.57	124
η^1-NNPh	pzTp	Mo	1996, 1913	15.43	123
η^1-NNPh	Tp	Mo	1994, 1904[b]	15.39	123
η^1-CC_6H_4Me-4	Tp	Mo	1998, 1921[d]	15.51	41,42
			1955, 1920[e]	15.16	
η^1-CC_6H_4Me-4	Tp*	Mo	1982, 1899[f]	15.21	44
η^3-C_3H_5	Tp	Mo	1958, 1874	14.83	125
η^3-C_3H_5	pzTp	Mo	1958, 1875	14.84	125
η^1-NO	Tp	W	2010, 1910[b]	15.52	123
η^1-NO	pzTp	W	2010, 1917	15.58	123
η^1-NO	Tp*	W	1993, 1884[c]	15.38	123
			2001, 1905	15.41	124
η^1-NS	Tp*	W	1988, 1902[c]	15.28	124
η^1-CC_6H_4Me-4	Tp	W	1986, 1903[b]	15.28	39,40
η^1-NNPh	pzTp	W	1984, 1896	15.20	123
η^1-NNPh	Tp	W	1983, 1894	15.18	123
η^1-CC_6H_4Me-4	Tp*	W	1974, 1888[g]	15.06	43
η^3-C_3H_5	Tp	W	1949, 1862	14.67	125
η^3-C_3H_5	pzTp	W	1949, 1862	14.67	125

[a]Data were obtained in cyclohexane solution unless otherwise indicated.
[b]CH_2Cl_2.
[c]KBr.
[d]Light petroleum.
[e]Pentane.
[f]THF.
[g]Hexane.

A. *The Metal Centre*

1. Ligand Substitution and Metal Oxidation

The complex $Tp^*Mo(\equiv CMe)(CO)_2$ undergoes photochemical carbonyl substitution in acetonitrile or in acetone with excess PMe_3 producing $Tp^*Mo(\equiv CMe)(L')(CO)$ (L′ = NCMe, PMe_3), in which there is no evidence for any ketenyl formation.[62]

The coordinatively labile picoline ligand in $BpW(\equiv CR)(pic)(CO)_2$ (R = $C_6H_2Me_3$-2,4,6, $C_6H_3Me_2$-2,6) can be replaced by isonitriles (in the absence of light) and phosphines to give moderate yields of $BpW(\equiv CR)(L')(CO)_2$ (L′ = CN^tBu, $CNC_6H_3Me_2$-2,6, PMe_2Ph),[49] in which the alkylidyne ligand is *trans* to a pyrazolyl arm and the ligand L′ is *trans* to a carbonyl ligand (i.e., *mer* arrangement of Bp/alkylidyne, as confirmed by the structural characterisation of L′ = PMe_2Ph, R = $C_6H_3Me_2$-2,6).[49] The analogous tricarbonyl *p*-toluidyne complex BpW(≡CR)$(CO)_3$ reacts with PPh_3 and PMe_3 similarly *via* carbonyl substitution, again with no evidence for ketenyl formation.[46] The reaction of either BpW (≡CR)(PMe_2Ph)$(CO)_2$ or BpW(≡$CC_6H_2Me_3$-2,4,6)(pic)$(CO)_2$ with excess PMe_2Ph provide the *trans*-bis(phosphine) complex BpW(≡$CC_6H_2Me_3$-2,4,6) $(PMe_2Ph)_2$(CO).[49] Formation from the monophosphine complex under mild conditions is suggested to proceed through a ketenyl intermediate followed by CO extrusion, as has been demonstrated in the synthesis of CpW(≡CC_6H_4Me-4) (PMe_3)(CO) from CpW(≡CC_6H_4Me-4)$(CO)_2$.[128,129] Notably, the reactions of TpW(≡CR)$(CO)_2$ (R = C_6H_4Me-4,[46] SMe[76]) with PMe_3 or PEt_3 (respectively) stop at the η^2-ketenyl stage, providing TpW(η^2-*C,C*-OCCR)(PR_3)(CO). The lowered steric demand and denticity of the bis(pyrazolyl)borate ligand allow its complexes to participate in ligand substitution reactions not typically observed for the bulkier Tp^x analogues.

The coordinatively saturated dicarbonyl carbyne complexes are generally inert to ligand substitution reactions without photochemical or thermal activation. However, oxidative carbonyl ligand substitution reactions occur upon treating the complexes Tp^xM(≡CR)$(CO)_2$ with X_2 (X = Cl, Br, I) at low temperature, yielding the six-coordinate dihalocarbyne complexes Tp^xM(≡CR)X_2 (Tp^x = Tp, M = W, R = Ph, X = Br).[130] Aminocarbynes behave similarly (Tp^x = Tp*, M = W, R = NR_2, X = Br, I).[85] In contrast, oxidative decarbonylations of Cp aminocarbynes with $PhICl_2$, Br_2 or I_2 afford the seven-coordinate, 18-electron aminocarbyne complexes Cp^xM(≡CNR_2)X_2(CO) (Cp^x = Cp, Cp*; X = Cl, Br, I; M = Mo, W).[131–133] This difference in reactivity can be ascribed to the greater steric bulk of the Tp^x ligand destabilising such seven-coordinate species.

Reaction of the complex TpW(≡$CC_6H_3Me_2$-2,6)$(CO)_2$ (R = $C_6H_3Me_2$-2,6) under high dilution conditions with thionyl chloride provides the dichloro alkylidyne complex TpW(≡$CC_6H_3Me_2$-2,6)Cl_2 (Scheme 33), which can also be obtained from the reaction of TpW(≡$CC_6H_3Me_2$-2,6)$(CO)_2$ with two equivalents of the non-electrophilic chlorinating reagent [Cp_2Fe]Cl.[107] Reaction with the molybdenum analogues does not proceed similarly. More sterically modest alkylidyne substituents (R = C_6H_4Me-4) yield only unreacted starting material and the complex TpWCl$(CO)_3$ in which the alkylidyne ligand has been cleaved.[107] The corresponding cyclopentadienyl complex CpW(≡$CC_6H_3Me_2$-2,6)$(CO)_2$ is much more reactive towards thionyl chloride, providing the novel metallacyclic alkylidene complex CpW{κ^2-*C,S*=CRSCRS}Cl_3 (R = $C_6H_3Me_2$-2,6) containing two xylyl groups, requiring a bimolecular dimerisation step in the mechanism.[107]

The dihaloaminocarbynes Tp*W(≡CNREt)X_2 (R = Et, Me) can undergo reductive dehalogenation by Na/Hg in the presence of CNEt to give the electron-rich aminocarbyne complexes Tp*W(≡CNREt)$(CNEt)_2$ (Scheme 23).[85]

Reaction of TpW(≡C^tBu)Cl_2 with the appropriate metal amide M[NHR] (M = Li, K, Na) yields the amido alkylidyne complexes TpW(≡C^tBu) Cl(NHR) (R = H, Ph, C_6H_4Br-4, $C_6H_3(CF_3)_2$-3,5) in moderate yields (Scheme 34). The lack of symmetry in these alkylidynes renders all of the Tp ring protons inequivalent.[59]

In the presence of excess LiNHPh, both chloride ligands are substituted and the complex TpW(≡C^{t}Bu)(NHPh)$_2$ can be isolated, in which the protons of the pyrazolyl rings appear in a 2:1 ratio, indicating the presence of a mirror plane in the molecule.[59]

The reaction of Tp*W(≡CR)X$_2$ (R = ^{t}Bu, X = Cl; R = Ph, X = Br)[57] with neutral alumina or aniline produces the compounds TpW(=CHR)(=E)X (R = ^{t}Bu, E = O, X = Cl;[56] R = Ph, X = Br, E = O, NPh) (Scheme 35). The corresponding Tp-dihalo complex, TpW(≡CPh)Br$_2$, reacts with primary amines and water in a similar manner to yield alkylidene imido and oxo complexes TpW(=CHPh)(=E)Br (E = O, N^{t}Bu, N-1-adamantyl, NC$_6$H$_3$Me$_2$-2,6) in 55–98% yield.[134] For Tp*W(≡CPh)Br$_2$, the oxocarbene species is not isolable.[58] Instead, further reaction results in formation of the dioxo-benzyl species Tp*W(CH$_2$Ph)(=O)$_2$, perhaps due to a weaker halide–metal interaction.[57] The crystallographically characterised oxo-carbene complex Tp*W(=CHR)(=O)Cl (R = ^{t}Bu) can be used as a ring-opening metathesis (ROMP) catalyst for cyclo-octene and norbornene, but only when activated with AlCl$_3$.[56]

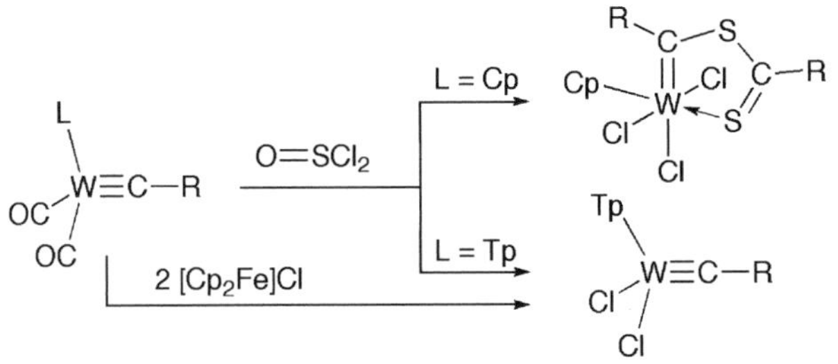

SCHEME 33. Reactions of tungsten alkylidynes with thionyl chloride.

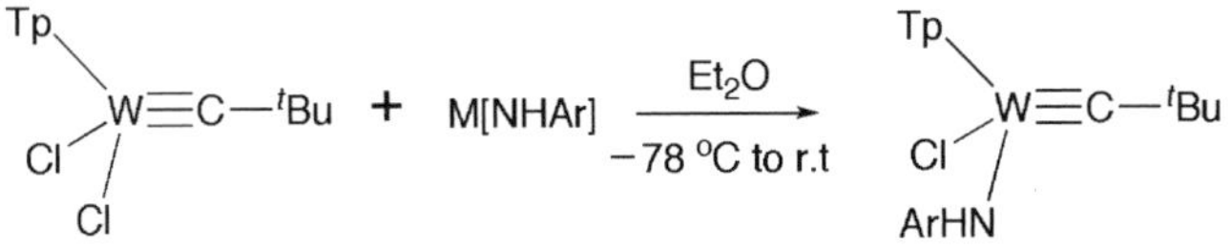

SCHEME 34. Synthesis of tungsten amido/alkylidyne complexes (M = Li, Na, K).

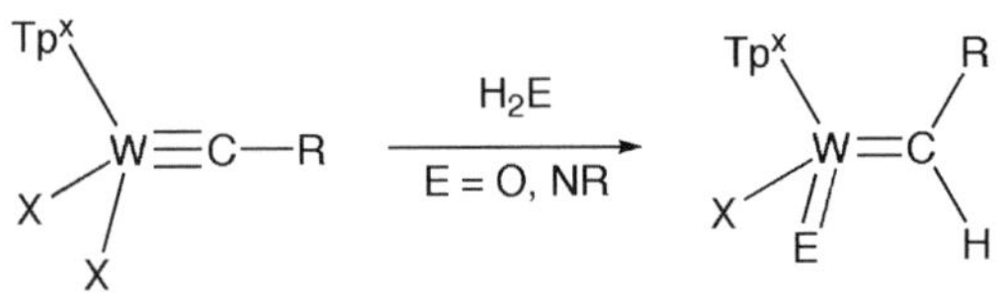

SCHEME 35. Synthesis of tungsten imido or oxo alkylidene complexes (Tp^x = Tp, Tp*; X = Cl, Br; R = ^{t}Bu, Ph; E = NR, O).

B. *The Metal–Carbon Multiple Bond*

1. Cycloaddition

The formal [2 + 2] cycloadditions of metal–carbon and carbon–carbon multiple bonds that are central to the accepted olefin/alkyne metathesis (Chauvin–Herisson) mechanism (Scheme 36) have been extended to include the reactions of metal–carbon multiple bonded compounds with molecules containing unsaturated P$\equiv$C linkages. Treatment of BpMo($\equiv$CC$_6$H$_4$OMe-2)(CO){P(OMe)$_3$}$_2$ with a stoichiometric amount of ^{t}BuC$\equiv$P under ambient conditions gave BpMo($\equiv$C^{t}Bu)(CO){P(OMe)$_3$}$_2$ in 84% yield *via* an alkylidyne exchange reaction, which is consistent with the intermediacy of a 1-phospha-3-molybdacyclobutadiene (Scheme 37), although an alternative metallaphosphatetrahedrane may not be excluded. The relative instability of the aryl-substituted phosphaalkyne P$\equiv$CC$_6$H$_4$OMe-2 is proposed to drive the equilibrium reaction to completion.[51,135]

Reactions of TpMo($\equiv$CR)(CO)(L′) (R = C$_6$H$_4$Me-4; L′ = CO, PPh$_3$) with CS$_2$ ultimately provide the metallacyclic thioketene complex TpMo{η^2-κ-(C,S),σ-κ(S')-S=C=CRC(=O)S}(=O),[136] incorporating the elements of CS$_2$, one carbonyl ligand and the alkylidyne unit into the metallacycle. The first step of the suggested mechanism, as shown in Scheme 38, involves cycloaddition of Mo$\equiv$C and C=S bonds.

A similar initial cycloaddition step is proposed in the reaction of the molybdenum alkylidynes TpMo($\equiv$CC$_4$H$_3$S-2)(CO)$_2$ with the heterocumulenic mesityl isoselenocyanate (SeCNC$_6$H$_2$Me$_3$-2,4,6) to provide a mixture of products including the

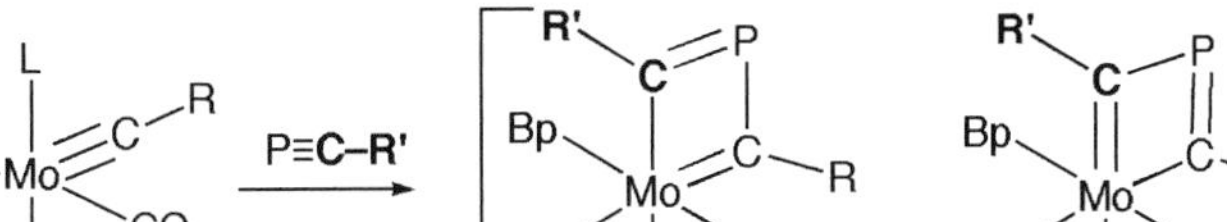

SCHEME 36. Chauvin–Herisson mechanism for alkene metathesis.

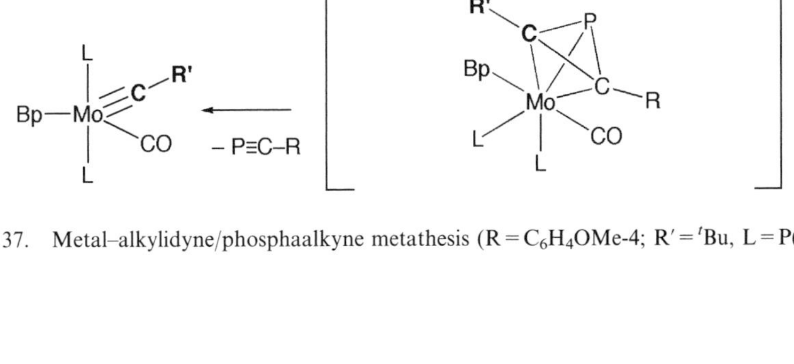

SCHEME 37. Metal–alkylidyne/phosphaalkyne metathesis (R = C$_6$H$_4$OMe-4; R′ = ^{t}Bu, L = P(OMe)$_3$).

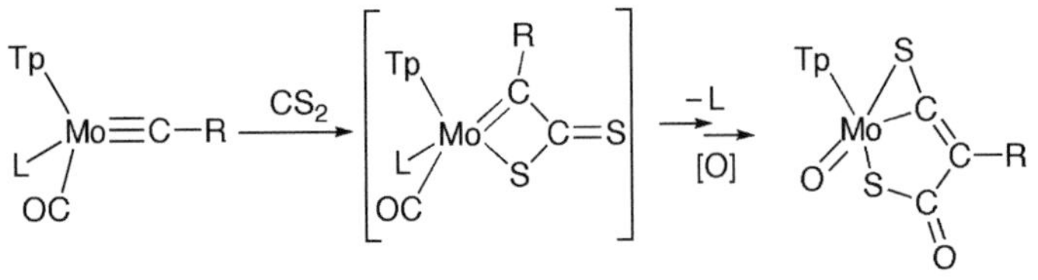

SCHEME 38. Proposed metal–alkylidyne/carbon disulfide cycloaddition step ($L = CO, PPh_3$).

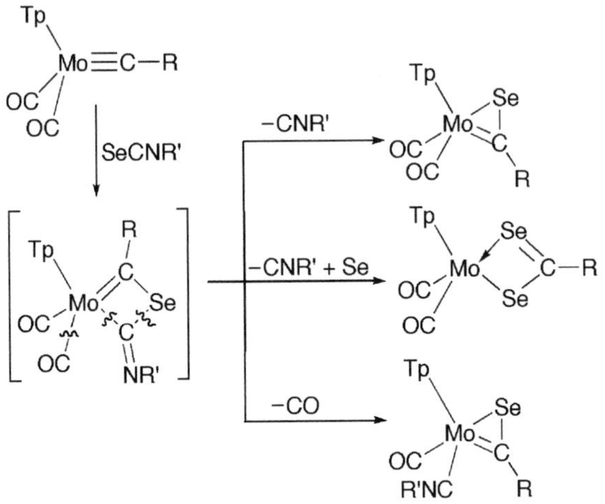

SCHEME 39. Proposed metallaselenacyclobuten-imine intermediate ($R = C_4H_3S\text{-}2$, $R' = C_6H_2Me_3\text{-}2,4,6$).

mononuclear selenoaroyl complexes $TpMo(\eta^2\text{-}SeCC_4H_3S\text{-}2)(CO)_2$ and TpMo (η^2-$SeCC_4H_3S$-2)(CO)($CNC_6H_2Me_3$-2,4,6), which may also be obtained directly from elemental Se in the presence of a catalytic amount of mesityl isocyanide (Scheme 39). Cycloaddition is proposed to produce a common metallacyclobuten-imine intermediate, from which isonitrile extrusion competes with carbonyl dissociation, followed by metallacycle collapse.[122]

2. Ketenyl Formation

The chemistry of ketenyls has been reviewed, with particular emphasis on their origins from carbyne–carbonyl coupling reactions.[11] In electron-rich carbyne complexes, the carbyne ligand is not susceptible to nucleophilic attack, rather nucleophiles cause carbonylation of the carbyne to give η^1- or η^2-ketenyl compounds (Scheme 40). Nucleophile-induced alkylidyne–carbonyl coupling is typically carried out under thermal or photolytic conditions and is facilitated by increasing the electron density of the metal centre, i.e., increasing π-back-bonding to the carbonyl ligand.

The electron-rich thiocarbyne complex $TpW(\equiv CSMe)(CO)_2$ reacts with the strongly basic phosphine PEt_3 at 40 °C to give the η^2-ketenyl derivative TpW

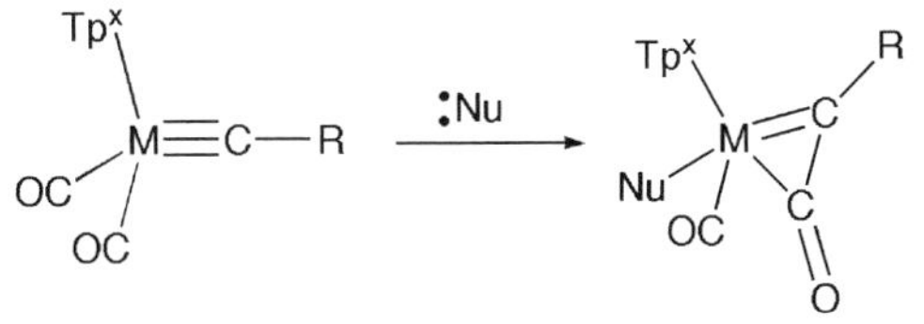

SCHEME 40. Nucleophile-induced alkylidyne–carbonyl coupling.

(η^2-*C*,*C*′-OCCSMe)(CO)(PEt_3),[76] indirectly coupling a carbonyl and thiocarbonyl ligand (the latter from which the thiocarbyne ligand was derived by electrophilic addition). This coupling reaction takes place under more forcing conditions than those for analogous Cp complexes, probably due to the greater steric hindrance provided by the Tp ligand at the metal centre.[76]

Similarly, TpW(≡CR)$(CO)_2$ (R = C_6H_4Me-4) reacts with PMe_3 to give an η^2-ketenyl complex TpW(η^2-*C*,*C*′-OCCR)(CO)(PMe_3).[46] Coupling reactions are sensitive to the steric and electronic properties of the system and an analogous product is not formed with the bulkier and less nucleophilic PPh_3 under thermal reaction conditions. Instead, TpW(η^2-*C*,*C*′-OCCR)(CO)(PPh_3) is prepared by reaction of the photochemically generated all-*trans*-W(≡CR)Br$(CO)_2(PPh_3)_2$, with late introduction of the Tp nucleophile under thermal conditions resulting in displacement of a phosphine ligand and trapping of the resultant ketenyl product.[50] The corresponding complex TpMo(η^2-*C*,*C*′-OCCR)(CO)(PPh_3) is not available by this route. Nucleophilicity at the oxygen atom of TpW(η^2-*C*,*C*′-OCCR)(CO)(PPh_3) allows subsequent treatment of the ketenyl with Cl_2PPh_3 to provide the chloroalkyne complex TpW(η^2-ClC≡CR)Cl(CO).[50] These transformations are summarised in Scheme 41.

The thermally inaccessible ketenyl complexes TpM(η^2-*C*,*C*′-OCCR)(CO)(PR'_3) (R = C_6H_4Me-4; M = Mo, W; PR'_3 = $P(OMe)_3$, PMe_2Ph, PPh_3) can be prepared *via* photochemical induction of carbonyl–alkylidyne coupling in the presence of a trapping phosphine (PR'_3). Subsequent reaction with Woollins' oxygen/selenium or Lawesson's oxygen/sulfur exchange reagent provides seleno- or thioketenyl complexes TpM(η^2-*C*,*C*′-ECCR)(CO)(PR'_3) (E = Se,[137] S,[138] respectively), one example of which (M = W, PR'_3 = PMe_2Ph, E = S) is also obtained from the reaction of TpW(η^2-ClC≡CR)Cl(CO)[50] with NaSH in the presence of PMe_2Ph.[138] In solution, the molybdenum ketenyl complexes thermally revert to either TpMo(≡CR)$(CO)_2$ or TpMo(≡CR)(CO)(PR'_3), depending on the donor properties of the phosphine.[138]

Under mild photolytic conditions, the reaction of W(≡CR)Br$(CO)_2(CNR')_2$ (R = C_6H_4Me-4, $C_6H_3Me_2$-2,6, R′ = ^{t}Bu, $C_6H_3Me_2$-2,6) with K[Bp], or of BpW (≡CR)$(CO)_2$(pic) with CNR′, both provide primarily the thermally unstable ketenyl complex BpW(η^2-*C*,*C*′-OCCR)(CO)$(CNR')_2$.[49] In contrast, the thermal reaction of BpW(≡CR)$(CO)_3$ with PR'_3 (R′ = Me, Ph) proceeds *via* carbonyl displacement to yield BpW(≡CR)$(CO)_2$(PR'_3), perhaps *via* a transiently formed ketenyl complex.[46]

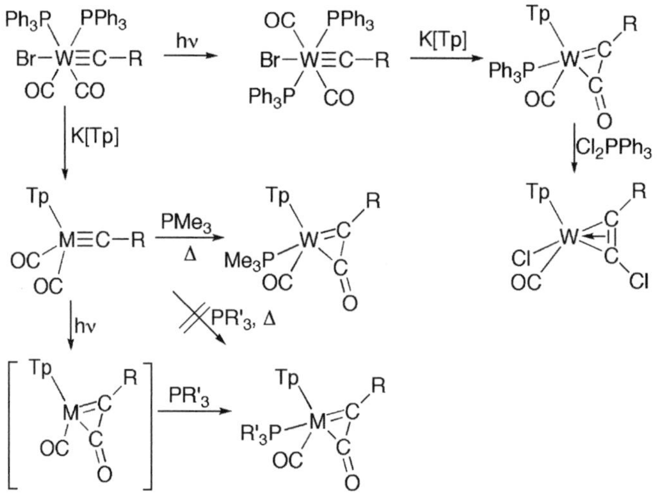

SCHEME 41. $M = W$, $PR'_3 = PPh_3$, PMe_2Ph; $M = Mo$, $PR'_3 = PPh_3$, $P(OMe)_3$; $R = C_6H_4Me\text{-}4$.

In contrast to the behaviour of the ethylidyne complex $Tp^*Mo(\equiv CMe)(CO)_2$, photoinduced carbonyl coupling occurs in acetonitrile solutions of $Tp^*Mo(\equiv CR)(CO)_2$ ($R = Ph$,[62] $C_6H_4Me\text{-}4$[44]) to generate the η^2-ketenyl complexes $Tp^*Mo(\eta^2\text{-}C,C'\text{-}OCCR)(CO)(NCMe)$. Similar reactivity is, however, not observed for the corresponding tungsten complexes $Tp^xW(\equiv CR)(CO)_2$ ($R = C_6H_4Me\text{-}4$; $Tp^x = pzTp$,[139] Tp^*[44]).

In $Tp^*Mo(\eta^2\text{-}C,C'\text{-}OCCR)(CO)(NCMe)$ ($R = Ph$,[62] $C_6H_4Me\text{-}4$[44]), the η^2-ketenyl ligand is acting as a three-electron donor (neutral convention) (**a**, Chart 4) and the nitrile a conventional two-electron σ-donor. In contrast, for the Cp^*-containing complex $Cp^*W\{\eta^1\text{-}C\text{-}C(=CO)SiPh_3\}(\eta^2\text{-}NCMe)(CO)$ produced from photoinduced carbonyl–carbyne coupling in $Cp^*W(\equiv CSiPh_3)(CO)_2$, the acetonitrile displays the unusual 'four-electron donor' mode of coordination, while an η^1-ketenyl ligand completes the tungsten centre's 18-electron configuration by serving as a one-electron ligand (**b**, Chart 4).[44] This unusual behaviour may be traced, in part, to the electronic nature of pseudo-octahedral tungsten(II) such that the d^4 centre will be most stabilised by ligand arrangements that offer a π-dative component for the one vacant t_{2g}-type orbital.

The most favourable orientation of the formally six-coordinate η^2-ketenyl complexes $Tp^xM\{\eta^2\text{-}C,C'\text{-}OCCR\}(CO)(L)$ is that in which the oxygen atom of the ketenyl lies proximal to the carbonyl and the W–C–O and ketenyl C–C–O components are co-planar in order to maximise π-overlap with the metal centre.[62]

The cationic carbyne complexes $[Tp^*W(\equiv CPMe_2R)(CO)_2]^+$ ($R = Me$, Ph) react with aryloxides of low nucleophilicity to provide η^2-ketenyl complexes $Tp^*W(\eta^2\text{-}C,C'\text{-}OCCPMe_2R)(OC_6H_4R'\text{-}4)(CO)$ ($R' = NO_2$, $R = Me$, Ph; $R' = CN$, $R = Me$, Ph; $R' = Cl$, $R = Ph$) along with the aryloxycarbynes $Tp^*W(\equiv COC_6H_4R'\text{-}4)(CO)_2$.[78] As noted above (Scheme 21), ketenyl complexes are formed simultaneously with (aryloxy)carbyne products that result from nucleophilic attack at C_α.

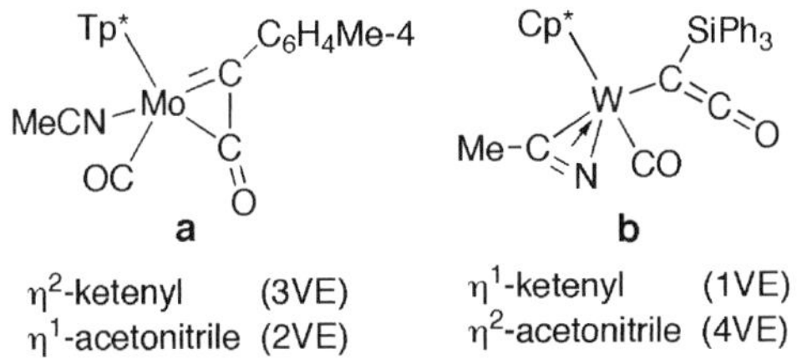

a

η^2-ketenyl (3VE)
η^1-acetonitrile (2VE)

b

η^1-ketenyl (1VE)
η^2-acetonitrile (4VE)

CHART 4. η^2- vs. η^1-ketenyl coordination with associated valence electron (VE) contributions.

The dimethylphenylphosphonio- and triphenylphosphonio-ketenyl products have similar spectroscopic characteristics. Both contain a stereogenic centre at tungsten and display 1:1:1 NMR patterns for the Tp* arms. Characteristic absorptions attributable to ν_{CO} appear at ca. 1870 and 1710 cm^{-1}, i.e., at notably low frequencies, presumably reflecting the π-dative role of the aryloxide.[78]

3. Hydroboration

Building on earlier results from Stone and co-workers for cyclopentadienyl analogues,[140,141] it was shown that hydroboration of the metal–carbon triple bond in Tp* carbyne complexes $Tp^*M(\equiv CR)(CO)_2$ with HBR'_2 leads to the novel boryl metal complexes $Tp^*M\{\eta^2\text{-}B(R')CH_2R\}(CO)_2$ (M = W, R′ = Et, R = C_6H_4Me-4 or Me; M = Mo, R′ = Et, R = C_6H_4Me-4; M = W, R′ = Ph, R = C_6H_4Me-4 or Me) (Scheme 42).[142,143]

These reaction products can be considered as comprising the addition of RBH_2 to the M–C triple bond, whereby the carbyne carbon is reduced to a methylene group and the borylene BR′ is inserted into the M–C bond. The X-ray crystal structure reveals a β-agostic C–H–M interaction between the metal atom and boryl substituent.[142] Despite the use of the dialkylborane HBR'_2, the product composition is such that it formally requires addition of dihydromonoorganoborane H_2BR' to the carbyne complex. Deuterium labelling experiments showed selective transfer of deuterium to the former carbyne carbon, with only deuterium participating in the agostic interaction.[143] This transfer of deuterium from the borane to the carbyne complex ruled out intermediates containing conventional η^2-bora-alkene ligands (i.e., without agostic interactions).[143]

In contrast, the corresponding unsubstituted Tp complexes decompose with $HBEt_2$ at 0 °C,[142] whereas the cyclopentadienyl complexes $Cp^xM(\equiv CR)(CO)_2$ (Cp^x = Cp, Cp*; R = C_6H_4Me-4) react with dialkyl boranes (9-borabicyclo[3.3.1]-nonane[140,141] or diethylborane[144]) at low temperatures to provide α-boryl-η^3-benzyl metal complexes $Cp^xM\{CH(BR'_2)R\}(CO)_2$ (M = W, $Cp^x = C_5Me_4Et$; M = Mo, Cp^x = Cp*, R′ = Et), amounting to 1,1-hydroboration of the metal–carbon triple bond at the carbyne carbon.[140,141]

The μ-‘bora-alkyne’ ditungsten complexes $Cp^x{}_2W_2\{\mu\text{-}RCB(H)CH_2R'\}(CO)_4$, obtained from the carbyne metal complexes $Cp^xW(\equiv CR)(CO)_2$ (R = C_6H_4Me-4 or Me) and $BH_3 \cdot THF$,[140,141] may be considered as the products of the starting metal carbyne and a boryl metal complex $Cp^xW(BHCH_2R')(CO)_2$, as depicted in

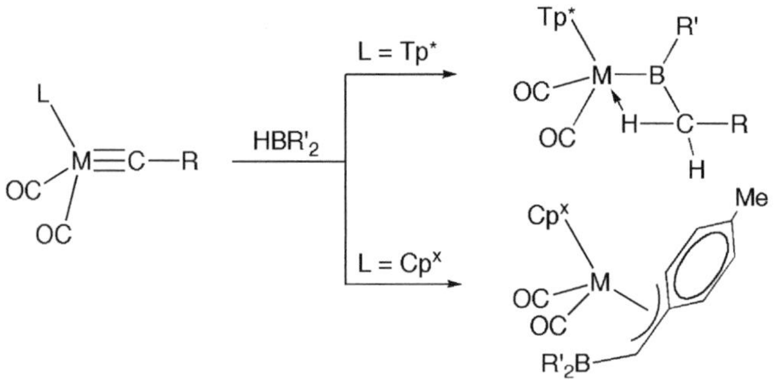

SCHEME 42. Hydroboration of alkylidyne–metal complexes ($R = C_6H_4Me$-4; M = Mo, W).

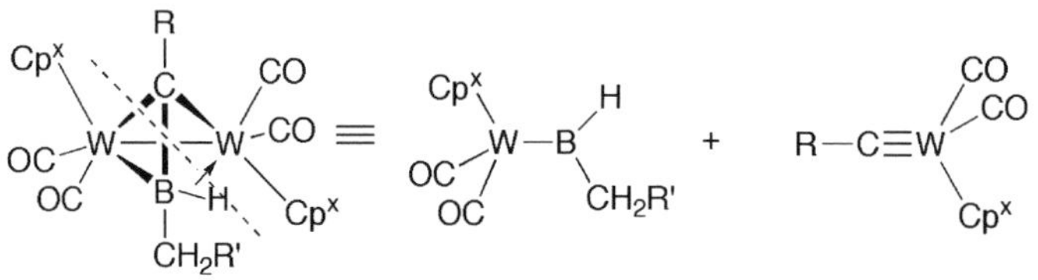

SCHEME 43. Unit composition of μ-bora-alkyne ditungsten Cp^x hydroboration complex.

Scheme 43, which is in effect comparable to the Tp* boryl metal complexes described above[143] and implies that the Cp systems are more reactive than the related Tp* systems.

4. Protonation

The product of protonation of carbyne complexes is dependent on the metal, the ancillary ligands, and the conjugate base of the acid. It is, however, not always clear that the thermodynamic site of protonation reflects the kinetic site due to the possibility of α-M–H elimination. In the presence of coordinating conjugate bases, carbyne protonation can induce further reaction to produce η^2-acyl complexes and compounds in which the original alkylidyne ligand has been lost *via* protonolysis.[145] The site of protonation (metal, alkylidyne face, or alkylidyne carbon, generating alkylidyne/hydride, face-protonated alkylidyne, and non-agostic alkylidene products, respectively) ultimately depends on the electron density at the metal centre, as depicted in Scheme 44. The Tp and Cp series of complexes display parallel protonation activity.[145]

In alkylidyne complexes that bear no π-acid co-ligands, thermodynamic protonation occurs either on the alkylidyne face or at the metal,[145] and steric considerations are invoked to explain the preference. Sub-stoichiometric quantities of H_2O or HCl catalyse the tautomerisation of TpW(≡C^{t}Bu)Cl(NHPh)[59] to the corresponding imido alkylidene complex TpW(=CH^{t}Bu)Cl(=NR). The first step in the tautomerisation is, however, suggested to involve protonation of the amide group (Scheme 45) since weaker donors (i.e., those with electron-withdrawing

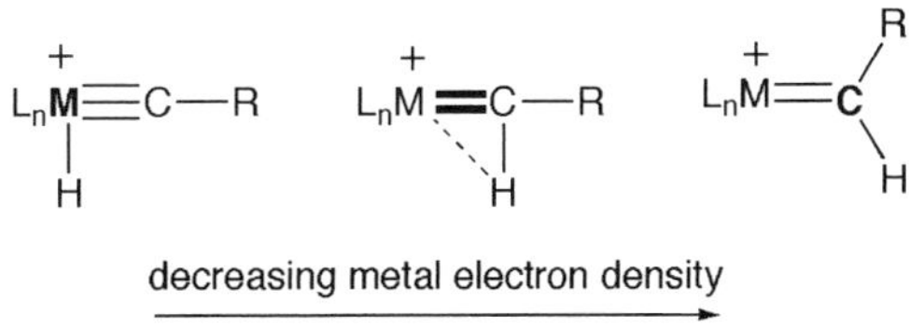

SCHEME 44. Possible sites of protonation in alkylidyne complexes.

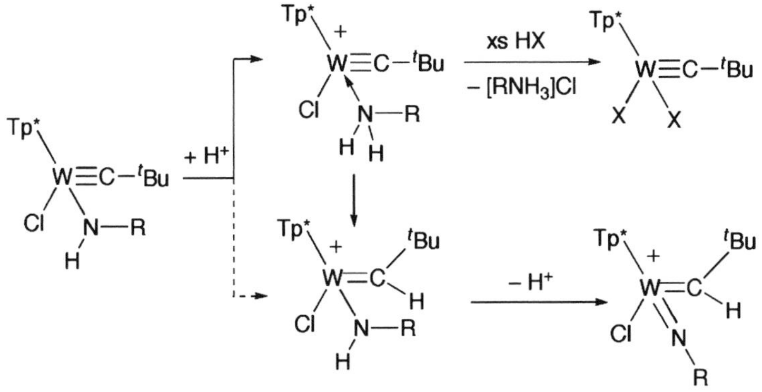

SCHEME 45. Acid-catalysed proton transfer (X = OH or Cl).

groups such as $NH_2C_6H_3(CF_3)_2$-3,5, $NH_2C_6H_4F$-4 and NH_3) dissociate from the metal under protonation conditions before proton transfer to the alkylidyne ligand can occur.[59]

In the presence of π-acidic ligands (typically prevalent for metals with higher d configurations), thermodynamic protonation shifts from the metal towards the carbyne carbon, occurring either on the alkylidyne face or at the carbyne carbon. Electronic effects appear to be the overriding factor in determining the *ultimate* site of protonation. For dicarbonyl complexes TpM(≡CR)$(CO)_2$, the electron density at the metal centre is substantially decreased by π-back-bonding. Protonation (HBF_4) of TpMo(≡C^{n}Bu)$(CO)_2$ at the carbyne carbon yields salts of the cationic carbene [TpMo(=CHBu)$(CO)_2$][BF_4]. In contrast, protonation of the Cp analogue CpMo(≡C^{n}Bu)$(CO)_2$ proceeds *via* secondary reaction of the initially formed carbene species with further starting carbyne to generate the binuclear alkyne salt [Cp_2Mo_2(μ-H)(μ-$C_2{}^n Bu_2$)$(CO)_4$][BF_4].[64,145] The poly(pyrazolyl) ligand is sufficiently sterically encumbered to prevent similar reaction of the alkylidene.

Increasing electron density at the molybdenum centre by replacement of the strong π-acid carbonyl with a phosphite ligand shifts the ultimate protonation site from the carbyne carbon to the alkylidyne face such that LMo(≡C^{n}Bu)(CO) {P(OMe)$_3$} (L = Tp, Cp) undergo protonation to yield α-agostic pentylidenes.[145]

The greater electron density of tungsten centres and the more electron-rich Tp* ligand mean that protonation of the dicarbonyl complex Tp*W(≡CH)$(CO)_2$ produces the cationic α-agostic methylidene salts [Tp*W(=CH_2)$(CO)_2$][BX_4] (X = F, $C_6H_3(CF_3)_2$-3,5).[68]

Low-temperature treatment of $Tp^*Mo\{{\equiv}CFe(CO)_2Cp\}(CO)_2$ with HBF_4 forms an agostic methyne bridge in the cationic protonation product, which serves to stabilise the unsaturated carbene.[71]

Protonation of $TpW({\equiv}CR)(CO)_2$ ($R = C_6H_4OMe$-2) with $HBF_4 \cdot Et_2O$ occurs at the carbyne carbon to give the salt $[TpW({=}CHC_6H_4OMe\text{-}2)(CO)_2][BF_4]$.[43] The related complexes ($Tp^x = Tp$, Tp^*, $R = C_6H_4Me$-4; $Tp^x = Tp$, $R = C_6H_3Me_2$-2,6; $Tp^x = Tp^{Ph}$, $R = C_6H_4Me$-4) do not form stable species under the same conditions and it is suggested that a dative interaction from the OMe-2 group on the phenyl ring stabilises the 16-electron tungsten centre.[43] Similar stabilisation has been demonstrated for the carbene complex $W\{{=}C(OMe)C_6H_4OMe\text{-}2\}(CO)_4$[146] and underpins the operational efficiency of the Grubbs–Hoveyda catalyst $RuCl_2({=}CHC_6H_4O^iPr\text{-}2)(IMesH_2)$ ($IMesH_2$ = 1,3-bis(2,4,6-trimethylphenyl)-4,5-dihydroimidazol-2-ylidene).[147]

The heterocarbyne $TpW({\equiv}CSMe)(CO)_2$ undergoes reversible protonation with the strong acids HOTf, HBF_4, and CF_3CO_2H at low temperature, giving salts of the cation $[TpW(\eta^2\text{-}CHSMe)(CO)_2]^+$ in which the carbene ligand is bonded to the metal through both the C and S atoms (Scheme 46), rather than an agostic C–H–M interaction.[74,76] Attempted protonation with HCl or HI leads to complete cleavage of the alkylidyne group, with $TpWX(CO)_3$ (X = Cl, I) being the only isolable reaction products.[76]

The methoxycarbyne complex $Tp^*W({\equiv}COMe)(CO)_2$ undergoes reversible protonation at the carbyne carbon in the presence of $[H(OEt_2)_2][BAr'_4]$ ($Ar' = C_6H_3(CF_3)_2$-3,5) to provide the α-agostic carbene salt $[Tp^*W({=}CHOMe)(CO)_2][BAr'_4]$,[79] rather than the η^2-C,O isomer which would be analogous to Angelici's thiolatocarbene.

Proton addition and subsequent phenylacetylene insertion convert (aryloxy)carbyne complexes $Tp^*W({\equiv}COAr)(CO)_2$ (Ar = Ph, C_6H_4Me-4, C_6H_4OMe-4) into η^3-vinyl carbene salts $[Tp^*W(\eta^3{=}CPhCH{=}CHOC_6H_4R\text{-}4)(CO)_2][X]$ (R = H, Me, OMe, X = BF_4; R = H, X = BAr'_4, $Ar' = C_6H_3(CF_3)_2$-3,5) (Scheme 12).[65] The intermediate agostic carbene salt $[Tp^*W({=}CHOC_6H_4OMe\text{-}4)(CO)_2][BAr'_4]$ was characterised by IR, ^1H NMR and $^{13}C\{^1H\}$ NMR spectroscopy.[65]

Protonation of $Tp^*Mo({\equiv}CNMe^tBu)(CO)_2$ in the presence of $PhC{\equiv}CH$ ultimately yields the crystallographically characterised cationic η^2-vinyliminium salt $[Tp^*Mo(\eta^2\text{-}CPh{=}CHCH{=}NMe^tBu)(CO)_2][BF_4]$ *via* alkyne insertion into the transient metal–carbene bond (Scheme 47).[83]

Proton addition to the anionic methylene carbene complex $[Tp^*W({=}CH_2)(CO)_2]^-$ (prepared *in situ via* addition of $Na[HBEt_3]$ to a tetrahydrofuran solution

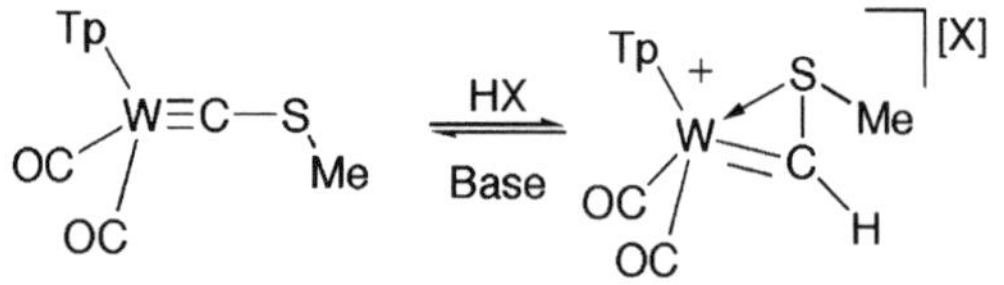

SCHEME 46. Reversible protonation of $TpW({\equiv}CSMe)(CO)_2$ (X = OTf, BF_4, O_2CCF_3; base = NaH, K_2CO_3, Et_3N).

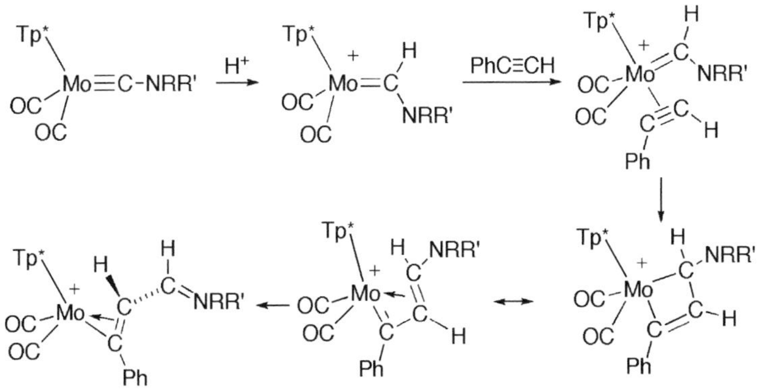

SCHEME 47. Protonation of $Tp^*Mo(\equiv CNMe^tBu)(CO)_2$ in the presence of $PhC\equiv CH$ ($R = Me$, $R' = {}^tBu$).

of $TpW(\equiv CH)(CO)_2$ at $-78\,^\circ C$)[68] results in the formation of a methyl ligand poised to insert CO. Trapping of this intermediate with phenylacetylene yields an η^1-acyl product $Tp^*W\{\eta^1\text{-}C(=O)Me\}(\eta^2\text{-}PhC\equiv CH)(CO)$. With a two-electron donor phosphine ligand, seven-coordinate methyl dicarbonyl complexes of the form $Tp^*W(CH_3)(CO)_2(PR_3)$ ($PR_3 = PMe_3$, PMe_2Ph, $PMePh_2$) are generated as the kinetic products. The seven-coordinate phosphine complexes are in equilibrium with their CO-insertion products $Tp^*W\{\eta^2\text{-}C(O)Me\}(CO)(PR_3)$.[148]

5. Reaction with Chalcogens

Group 6 alkylidyne poly(pyrazolyl)borate[46,48] and cyclopentadienyl complexes[149] $L_nM\equiv CR$ generate dichalcocarboxylates $L_nM(\kappa^2\text{-}A_2CR)$ with elemental chalcogens. The complexes $TpMo(\equiv CR)(CO)_2$, however, fail to react with tellurium and this is attributed to a kinetic phenomenon.[48] The sequential treatment of $W(\equiv CR)Br(CO)_4$ ($R = C_6H_4Me\text{-}4$) with sulfur and K[Tp] provides $TpW(\eta^2\text{-}S_2CR)(CO)_2$[150] *via* initial thioacyl or dithiocarboxylate formation and subsequent reaction with the Tp source (and, for the latter, excess sulfur).[150] Treating $TpMo(\equiv CR)(CO)_2$ ($R = C_6H_4Me\text{-}4$) with sulfur in tetrahydrofuran at reflux temperature affords $TpMo(\eta^3\text{-}S_2CR)(=O)$, containing an oxo ligand (presumably from adventitious oxygen) and a *trihapto* dithiocarboxylate ligand (Scheme 48). The oxo ligand is presumed to arise during chromatographic purification.[151]

Chalcoacyl complexes have been stabilised through coordination to polymetallic ensembles. Binuclear complexes prepared from solvent-stabilised '$Ru(CO)_2(\eta^5\text{-}7,8\text{-}C_2B_9H_{11})$' and the corresponding alkylidyne react with sulfur to generate bridging thioacyl complexes (Scheme 83).[152] This property will be elaborated upon in a later section devoted specifically to multinuclear complexes (Section IV.F). Mononuclear thioacyl complexes $TpMo(\eta^2\text{-}SCR)(CO)_2$ ($R = C_6H_4Me\text{-}4$, $C_6H_4OMe\text{-}4$, $C_4H_3S\text{-}2$) were successfully prepared through the use of 2-methylthiirane (propylene sulfide)

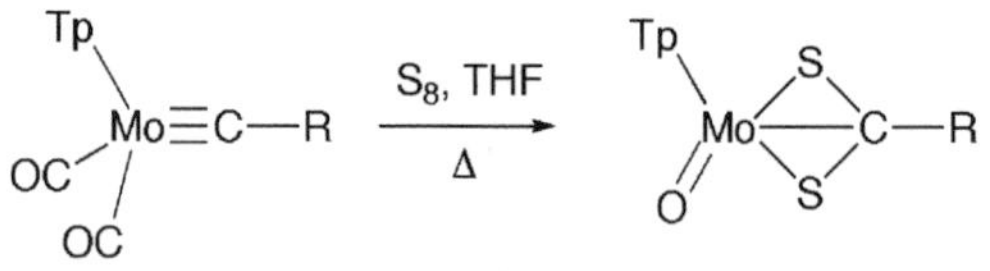

SCHEME 48. Preparation of an η^3-dithiocarboxylato complex (R = C_6H_4Me-4).

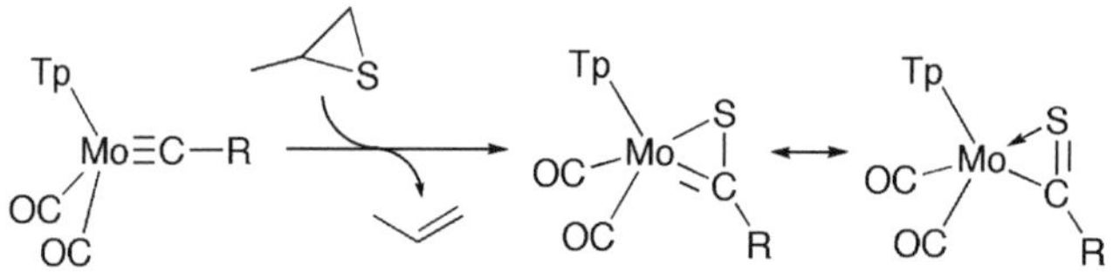

SCHEME 49. Mononuclear Group 6 chalcoacyl complexes (R = C_6H_4Me-4, C_4H_3S-2, C_6H_4OMe-4).

as a single-atom sulfur source (Scheme 49).[48,153] Only small amounts of the corresponding dithiocarboxylates TpMo(η^2-S_2CR)$(CO)_2$ were formed, which also arise from treatment of TpMo(η^2-SCR)$(CO)_2$ with further methylthiirane or elemental sulfur. This reaction fails for alkylidyne complexes of chromium, sterically congested benzylidyne complexes, and aminomethylidyne complexes. The thioacyl complexes serve as precursors for complexes bearing dithiocarboxylate, thioselenocarboxylate, thiolato-carbene and α-thioalkyl ligands, as well as thioacyl-bridged binuclear complexes.[153] With copper(II) chloride, the chalcoacyl complex TpMo(η^2-SCC_4H_3S-2)$(CO)_2$ regenerates TpMo(≡CC_4H_3S-2)$(CO)_2$ in 15% yield.[48]

The acyl formation reaction is suggested to proceed *via* initial attack at the carbyne carbon, but evidence is not conclusive.[48] Reaction of the analogous Cp complexes with cyclohexene sulfide provides exclusively the dithiocarboxylate derivatives. Presumably the second sulfur addition is considerably more rapid than for the Tp complexes due to lowered steric crowding around the molybdenum centre.[48]

C. *The Carbon Substituent*

The carbon in electron-rich transition-metal carbyne complexes is generally found to be susceptible to electrophilic attack. The SMe^+ electrophile from dimethyl(methylthio)sulfonium tetrafluoroborate adds to the carbyne carbon atom of TpM(≡CR)$(CO)_2$ (M = Mo, R = C_6H_4Me-4, C_4H_3S-2;[48] M = W, R = C_6H_4 Me-4,[154] R = SMe[19,77]) to yield the cationic η^2-thiocarbene salts [TpM(η^2-MeSCR) $(CO)_2$][BF_4] (Scheme 50). The structure of the related salt [TpW{η^2-C(H)SMe} $(CO)_2$][OTf][74] suggests that in these thiocarbenes the methyl group on the coordinated sulfur is oriented above the MCS ring toward the pyrazolyl groups and away from the metal carbonyls.

The direct conversion of carbyne complexes $TpW(\equiv CR)(CO)_2$ (R = Ph, C_6H_4Me-4) to cationic *cis*-dicarbonyl η^2-phosphinocarbene (metallaphosphacyclopropene) complexes can be achieved by the addition of chlorodiorganophosphines in the presence of sodium tetraphenylborate[155] or thallium hexafluorophosphate (Scheme 51).[156] In the absence of these sodium salts, nucleophile (Cl^-)-induced carbonyl–carbene coupling affords neutral η^3-phosphinoketene complexes.[155]

Addition of excess PMe_3 to $Tp^*W(\equiv CSMe)(CO)_2$ in dichloromethane yields the cationic bis(phosphonio)carbene complex $[Tp^*W\{=C(PMe_3)_2\}(CO)_2]^+$, isolated as the PF_6 salt by metathesis with NH_4PF_6.[72] Two non-equivalent phosphorus environments are spectroscopically evident and the molecular structure of this complex has been determined.[72] The bis(phosphonio)carbene complex is in equilibrium with the phosphoniocarbyne complex in solution and the addition of MeI readily forms the monophosphine analogue $[Tp^*W(\equiv CPMe_3)(CO)_2][PF_6]$.[72]

Nucleophiles react with carbyne complexes to promote (or trap) carbyne–carbonyl coupling products, attack the metal–carbon triple bond, or displace a substituent on the carbyne carbon. Complexes of the form $Tp^xM(\equiv CR)(CO)_2$ are coordinatively saturated and in the case of $Tp^x = Tp^*$, the metal also enjoys a substantial degree of steric protection. Accordingly, the reaction of these complexes with nucleophiles does not, in general, involve attack at the metal but rather at a co-ligand. Attempted synthesis of $Tp^*W(\equiv CMe)(CO)_2$ *via* reaction of MeLi with the chlorocarbyne complex $Tp^*W(\equiv CCl)(CO)_2$ instead proceeds *via* attack at a carbonyl carbon to provide an acyl anion (observed *in situ* by IR and ^{13}C NMR

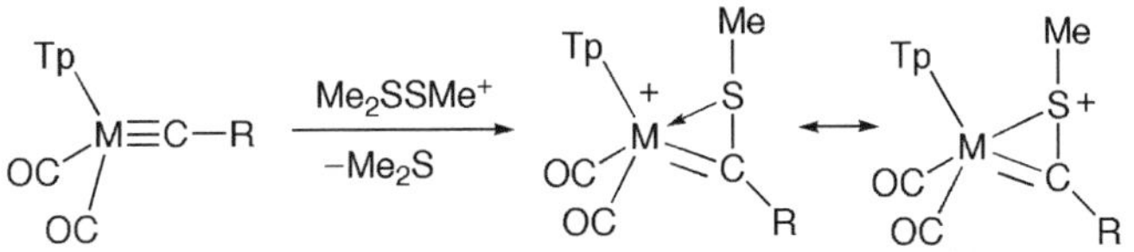

SCHEME 50. Electrophilic addition of SMe^+ by $[MeSSMe_2]BF_4$ (M = Mo, R = C_6H_4Me-4; M = W, R = SMe, C_6H_4Me-4).

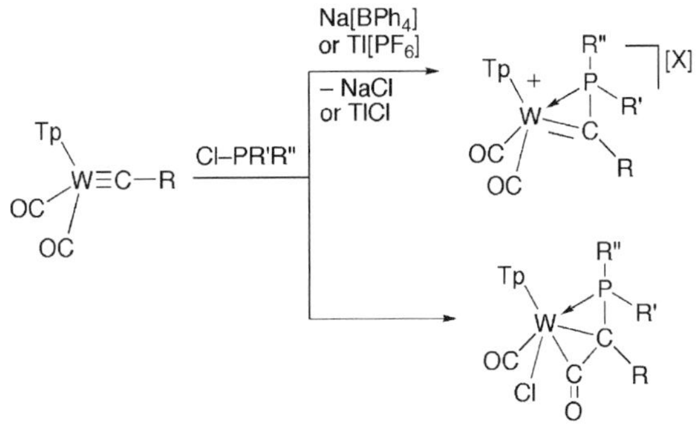

SCHEME 51. Reaction with $ClPR_2$ (R = Ph, C_6H_4Me-4; R′ = R″ = Me, Ph; R′ = Me, R″ = Cl; X = PF_6, BPh_4).

spectroscopy).[67] Treatment of $Tp^*W(\equiv CCl)(CO)_2$ with $LiMe_2Cu$ at 0 °C results in the formation of $Tp^*W(\eta^2\text{-}C(CH_3){=}CH_2)(CO)_2$ through initial chloride displacement, ensuing reaction with a second equivalent of CH_3^-, and net hydride loss. Similar behaviour was seen for the molybdenum analogue. The vinyl complexes slowly rearrange to stable η^3-allyl products in non-aromatic solvents.[67] At −22 °C, chloride displacement from $Tp^*M(\equiv CCl)(CO)_2$ (M = Mo, W) with $LiMe_2Cu$ affords the ethylidyne complexes $Tp^*M(\equiv CMe)(CO)_2$ (M = Mo, W) (Scheme 52).[67] Similarly, nucleophilic halide displacement of $Tp^*Mo(\equiv CCl)(CO)_2$ with phenyl lithium produces $Tp^*Mo(\equiv CPh)(CO)_2$ in modest yields;[73] however, wider success has been achieved with heteroatom nucleophiles (*vide supra*).

Conversion of the ethylidyne ligand in $Tp^*Mo(\equiv CMe)(CO)_2$ to a vinylidene ligand is achieved *via* deprotonation with $Na[N(SiMe_3)_2]$ in tetrahydrofuran. This vinylidene is susceptible to electrophilic attack at the β-carbon by RI (R = Et or Me) to generate elaborated alkylidyne ligands (Scheme 53).[62]

Preparation of the monomeric methylidyne complexes $Tp^*M(\equiv CH)(CO)_2$ (M = Mo, W) from $[Tp^*W(\equiv CPR_3)(CO)_2]^+$ has been achieved by hydride addition ($[Et_3BH]^-$) to provide the zwitterionic carbene complexes $Tp^*W(=CHPR_3)(CO)_2$ ($PR_3 = PMe_3$,[120] PMe_2Ph[68]), from which phosphine is abstracted as $[MePR_3]I$ *via* reaction with MeI (Scheme 54). Triphenylphosphine provides a better leaving group than PMe_2R and treatment of $[Tp^*W(\equiv CPPh_3)(CO)_2][PF_6]$ with

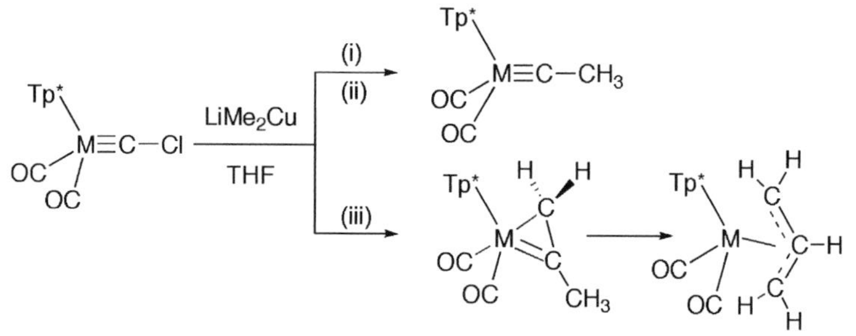

SCHEME 52. Cuprate addition to a chlorocarbyne complex. Reagents: (i) −22 °C, (ii) HCl (Et_2O), and (iii) 0–25 °C (M = Mo or W).

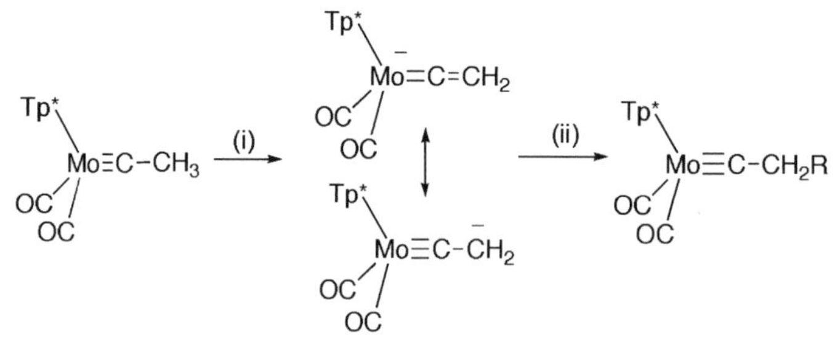

SCHEME 53. Electrophilic attack at β-carbon of vinylidene anion. Reagents: (i) $NaN(SiMe_3)_2$, −78 °C, and (ii) RI (R = Me, Et).

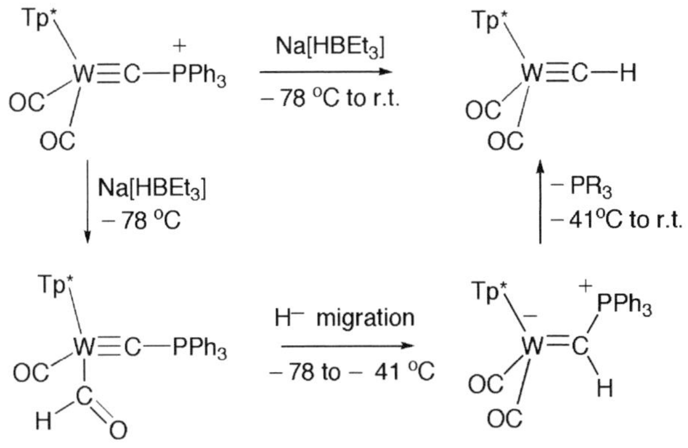

SCHEME 54. Generation of a terminal methylidyne ligand from a phosphonioalkylidyne complex.

Na[$HBEt_3$] in tetrahydrofuran forms the methylidyne complex $Tp^*W(\equiv CH)(CO)_2$ *via* the formyl and zwitterionic carbene intermediates $Tp^*W(\equiv CPPh_3)\{C(=O)H\}(CO)$ and $Tp^*W(=CHPPh_3)(CO)_2$, respectively.[68]

The tungsten alkylidynes $Tp^*W(\equiv CH)(CO)_2$ and $Tp^*W(\equiv CPh)(CO)_2$ also arise from side reactions of phosphonium alkylidyne complexes with aryloxide nucleophiles from reaction of the quaternary phosphonium salts with adventitious hydroxide to give a dissociated alkyl anion or a product resulting from a 1,2-phenyl migration. At low temperature, phenyl migration is inhibited and only the terminal dissociation product is observed. At room temperature, both phenyl and C–H alkylidyne products form in an approximate 1:1 ratio.[78]

The methylidyne complex $Tp^*W(\equiv CH)(CO)_2$ can be deprotonated with alkyllithium reagents to provide the anionic terminal carbide Li[$Tp^*W(\equiv C)(CO)_2$], characterised by a downfield resonance for C_α at δ_C 556 in the ^{13}C NMR spectrum due to substantial deshielding from the electropositive metal. The negative charge is localised primarily on the carbido carbon and addition of electrophiles (E^+) generates modified alkylidynes of the type $Tp^*W(\equiv CE)(CO)_2$ (E = Me, $SiMe_3$, I, $C(OH)Ph_2$, CH(OH)Ph and C(O)Ph).[68]

The methylidyne and above-mentioned chloromethylidyne complexes, useful in the synthesis of elaborated alkylidynes, are also employed in the synthesis of bimetallic complexes spanned by hydrocarbon and carbon-only bridges. These latter reactions are described in a later section dedicated to such polymetallic systems.

Lalor's chlorocarbynes can be utilised as a starting point for the *in situ* construction of anionic molybdenum or tungsten complexes with C-substituted vinylidene ligands $[Tp^*M(=C=CRR')(CO)_2]^-$ through use of the stabilised secondary carbanions $[CHRR']^-$ (R, R′ = CN, CO_2Et, Scheme 55).[69] Presumably, formation of the C-substituted vinylidene proceeds *via* the neutral carbyne complex which is a stronger acid than H_2CRR'. Reaction with a variety of electrophilic or oxidising reagents (HCl or PhCOCl with $[Cp_2Fe][BF_4]$, $[Ph_2I][PF_6]$ and NaOCl or with $CoCl_2$) yields neutral seven-coordinate oxametallacyclic carbene complexes, which is noteworthy given the established tendency of Group 6 Tp complexes to avoid seven-coordination.[69]

Electrophilic attack generally occurs at the β-carbon of the vinylidene ligands in $[Tp^*M(=C=CRR')(CO)_2]^-$ (R, R′ = CN, CO_2Et) to provide highly functionalised carbyne complexes $Tp^*Mo\{\equiv CC(CN)(CO_2Et)Z\}(CO)_2$ (Z = HgX, Cu, $N_2C_6H_4NMe_2$-4).[69]

The air-stable anionic carbene complexes with a metal-centred negative charge, $[Tp^*M\{=C(CN)_2\}(CO)_2]^-$ (M = Mo, W), $[Tp^{Me_2},4ClMo\{=C(CN)_2\}(CO)_2]^-$, $[Tp^{iPr,4Br}Mo\{=C(CN)_2\}(CO)_2]^-$, and $[Tp^*Mo\{=C(CN)R\}(CO)_2]^-$ [R = CMe(CN)$_2$, CMe(CN)C_6H_4Br-4, CMe(CN)C_6H_4Me-4, CMe(CN)(1-$C_{10}H_7$)], can be similarly synthesised by addition of cyanide anion to the corresponding chlorocarbyne or alkyl carbyne complexes in dimethylsulfoxide and isolated as their tetra(alkyl)ammonium salts (alkyl = Bu, Et, Scheme 56).[157] The dicyanocarbene complexes are stereochemically rigid, with the plane of the carbene ligand coinciding with the molecular mirror plane. The electron-rich carbene carbons in these dicyanocarbene complexes show some of the most highly shielded values of $\delta_{Carbene}$ so far observed, ranging from $\delta_{Carbene}$ 167.8 to 194.5. The dicyanocarbene ligands are much stronger π-acceptors than conventional carbene ligands with

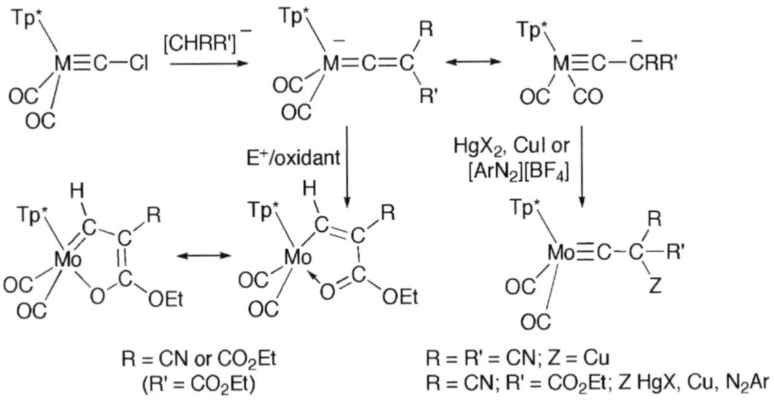

Scheme 55. Synthesis and reactivity of anionic vinylidene complexes.

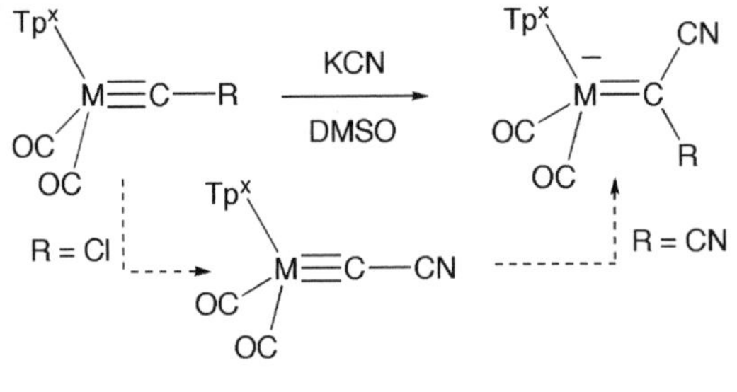

Scheme 56. Synthesis of cyanomethylidene complexes (M = Mo or W; X = Cl, R = CN; X = R = C(CN)Me(Y), Y = CN, C_6H_4Br-4, C_6H_4Me-4, 1-$C_{10}H_7$).

average carbonyl absorptions at ca. 1830 cm^{-1} due to the resonance anion-stabilising effects of the cyano-substituents.[157]

When TpM($\equiv$CX)$(CO)_2$ complexes are treated with KCN and X is a poor leaving group (R), the anionic cyano(alkyl)carbene complexes [Tp*Mo{$=$C(CN)R}$(CO)_2$] are produced (R$=$CMe$(CN)_2$, CMe(CN)C_6H_4Br-4, CMe(CN)C_6H_4Me-4, CMe(CN)(1-$C_{10}H_7$)). No reaction is observed when R$=$C(CN)Ph_2. These ligands are poorer π-acceptors than the dicyanocarbene complexes but again significantly stronger than more conventional carbene ligands that lack strongly electron-withdrawing groups.[157]

Unlike the closely related anionic vinylidene complexes [Tp*M($=$C$=$CRR′)$(CO)_2]^-$ (R, R′$=$CN, CO_2Et), the dicyanocarbene complexes did not give stable products upon reaction with electrophiles. The cyano(alkyl)carbene complexes [Tp*Mo{$=$C(CN)R}$(CO)_2]^-$ (R$=$CMe(CN)C_6H_4Br-4, CMe(CN)C_6H_4Me-4, CMe(CN)(1-$C_{10}H_7$)), however, could be doubly alkylated at the CN nitrogen to give the cationic (dialkylamino)alkyne complexes [Tp*Mo(Me_2NC$\equiv$CR)$(CO)_2]^+$ (Scheme 57).[157]

The complex [Tp*Mo{$=$C(CN)CMe$(CN)_2$}$(CO)_2]^-$ eliminates MeC$(CN)_2^-$ in the presence of excess MeI to yield purple Tp*Mo($\equiv$CC$\equiv$N)$(CO)_2$ in 74% yield, the first example of a cyanocarbyne complex and a presumed intermediate in the formation of the dicyanocarbenes (Scheme 58).[157] Infrared carbonyl frequencies are higher than for the chlorocarbyne complex, reflecting the greater inductive

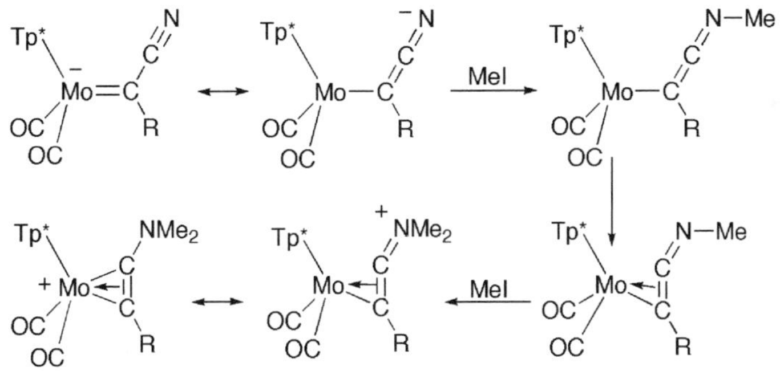

SCHEME 57. Suggested sequence of events leading to the formation of cationic dialkylaminoalkyne complexes (R$=$CMe(CN)C_6H_4Br-4, CMe(CN)(1-$C_{10}H_7$)).

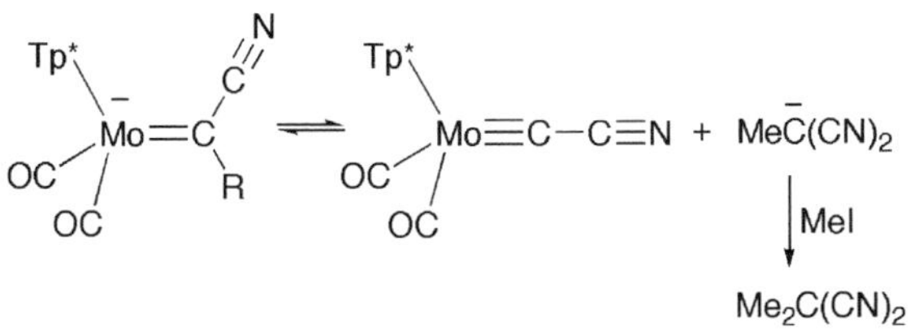

SCHEME 58. Suggested rationale for cyanocarbyne complex formation (R$=$C$(CN)_2$Me).

electron-withdrawing capacity of a cyano group compared to a chlorine atom and less efficient π-donation from the filled π-orbitals of the C≡N bond to the M≡C bond π*-orbitals than from the lone pair orbitals on chloride.[157] The cyanocarbyne complex may be considered isoelectronic with alkynylcarbynes.[52]

Templeton has reported the synthesis of a number of elaborated and bridging carbyne ligands through reactions at the carbyne substituent that *ultimately* leave the M≡C bond intact, although its formal multiplicity may vary during reaction sequences.

The methylidyne complexes $Tp^*M(\equiv CH)(CO)_2$ (M = Mo, W) are formed from fluorodesilylation of silylcarbynes $Tp^*M(\equiv CSiMe_2Ph)(CO)_2$ with $[Bu_4N]F$ in wet tetrahydrofuran solution at low temperature.[120] The yield obtained for the hydridocarbyne (30%) offers improvement over that from Templeton's previous method involving hydrido(*alkyl*phosphonio)carbene precursors.[72] In solution, spontaneous dimerisation to vinylidene-bridged complexes $Tp^*_2M_2(\mu\text{-}C{=}CH_2)(CO)_4$ occurs. The molecular structure of the tungsten derivative has been determined by X-ray crystallography, revealing a non-classical bridging vinylidene ligand that is perpendicular to the metal–metal axis but has an essentially linear $M\text{–}C_\alpha\text{–}M$ spine (Scheme 59). This unusual arrangement is noteworthy, not only for the non-classical coordination mode displayed by the vinylidene bridge, but because the complexes $Cp_2M_2(\mu\text{-}HCCH)(CO)_4$ adopt conventional dimetallatetrahedrane geometries. This dichotomy is presumably a reflection of the steric encumbrance that would result from a $Tp^*M\text{–}MTp^*$ arrangement with a direct M–M bond.

The dimeric complexes display four-band CO patterns in their infrared spectra. $^{13}C\{^1H\}$ NMR spectroscopy reveals the bridging 'carbide' carbon resonance at δ_C 346.7 (M = Mo) or 304.4 (M = W), which is consistent with a vinylidene-like quaternary centre.[120] Electrochemical studies reveal that the W and Mo vinylidene complexes are readily oxidised.[89]

In contrast to the situation with aryloxide nucleophiles, the cationic carbyne complex in the salt $[Tp^*W(\equiv CPMe_3)(CO)_2][PF_6]$ reacts with nucleophiles such as MeLi or $K[HB(O^iPr)_3]$ at the carbyne carbon to form neutral substituted phosphoniocarbene complexes, $Tp^*W\{=CMe(PMe_3)\}(CO)_2$ or $Tp^*W\{=CH(PMe_3)\}(CO)_2$, respectively.[72] Addition of excess $Na[HBEt_3]$ to neutral $Tp^*W(\equiv CH)(CO)_2$ generates the electronically similar anionic methylidene salt $Na[Tp^*W(=CH_2)(CO)_2]$, which, though not isolated, is nevertheless a valuable synthetic intermediate.[68] An ensuing reaction with PhSSPh proceeds *via* formal addition of SPh^+ to the methylidene to form the saturated tungsten product $Tp^*W(\eta^2\text{-}CH_2SPh)(CO)_2$, in which the organometallic ligand is bound to the metal

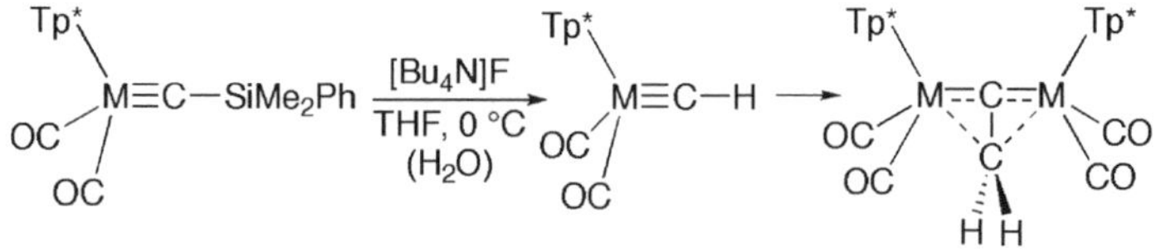

SCHEME 59. Terminal methylidyne and non-classical bridging vinylidene complexes *via* fluoride mediated desilylation of silylmethylidynes (M = Mo, W).

through both the carbon and the sulfur atoms.[68] Nucleophilic displacement of phosphine, however, occurs when $[Tp^*W(\equiv CPMe_2Ph)(CO)_2][PF_6]$ reacts with Grignard reagents RCH_2MgCl ($R = Ph$, $CH{=}CH_2$) at C_α to generate the intermediate zwitterionic carbene complexes $Tp^*W\{=C(CH_2R)(PMe_2Ph)\}(CO)_2$, which liberate phosphine to form $Tp^*W(\equiv CCH_2R)(CO)_2$.[121]

Deprotonation of the alkyl carbynes $Tp^*M(\equiv CCH_2R)(CO)_2$ with ^{n}BuLi or KO^{t}Bu at C_β forms synthetically versatile vinylidene anions $[Tp^*W(=C=CHR)(CO)_2]^-$ ($R = H$,[62] Ph, $CH{=}CH_2$,[121] Scheme 60) which add electrophiles, E^+, primarily at C_β, allowing access to elaborated alkyl carbynes $Tp^*W(\equiv CCHRE)(CO)_2$ ($R = H$, $E = Me$,[62,67] CH_2Ph;[67] $R = Ph$, $CH{=}CH_2$, $E = Me$,[121] Scheme 60). Subsequent deprotonation and alkylation (E'^+) is again possible, leading to the more highly functionalised carbynes $Tp^*W(\equiv CCMeEE')(CO)_2$.[67]

Due to resonance stabilisation, the vinylvinylidene anions $[Tp^*W(=C=CRCH{=}CH_2)(CO)_2]^-$ ($R = H$, Me) can react with electrophiles at either C_β or C_δ.[121] Alkylation at C_β is favoured but a small amount of alkylation occurs at C_δ for $R = Me$, perhaps due to steric crowding at C_β, which generates *E* and *Z* isomers of the conjugated vinyl carbyne $Tp^*W(\equiv CCMe{=}CHCH_2R)(CO)_2$.[121] By quenching a tetrahydrofuran solution of the anions $[Tp^*W(=C=CRCH{=}CH_2)(CO)_2]^-$ ($R = Me$, H), vinyl carbyne complexes of the type $Tp^*W\{\equiv CC(R){=}C(H)Me\}(CO)_2$ may be obtained selectively as a mixture of *E* and *Z* isomers.[121] Formation of the *E* isomer (in which the '$C\equiv W(CO)_2Tp^*$' unit is *trans* to the substituent on C_δ) is favoured, presumably due to unfavourable steric interactions in the *Z* isomer. Addition of NEt_3 promotes isomerisation of the allyl carbyne complex $Tp^*W(\equiv CCH_2CH{=}CH_2)(CO)_2$ to the conjugated vinyl carbyne isomer $Tp^*W(\equiv CCH{=}CHMe)(CO)_2$.[121]

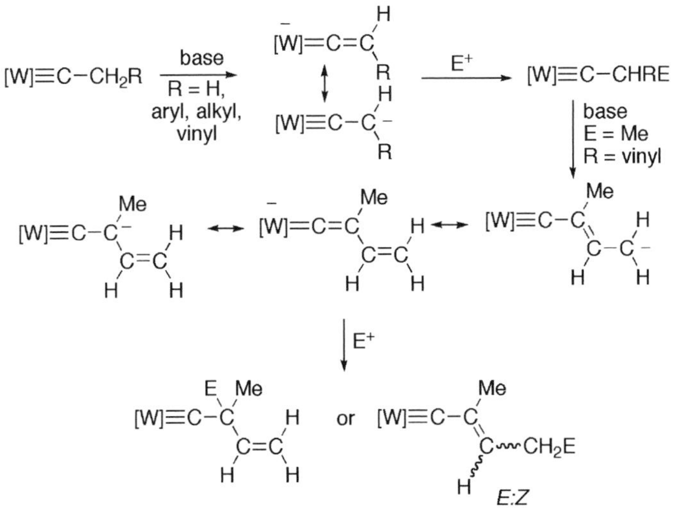

SCHEME 60. Derivatisation of carbyne complexes ($[W] = Tp^*W(CO)_2$).

Aldehydes and ketones can also be used as electrophiles in reactions with the vinylidene anion $[Tp^*M(=C=CH_2)(CO)_2]^-$. Low-temperature reaction of $[Tp^*M(=C=CH_2)(CO)_2]^-$ with $RR'C=O$ (R = Ph, ^{n}Pr, R′ = H; R = Ph, R′ = Me) followed by protonation yields $Tp^*W\{\equiv CCH_2C(OH)RR'\}(CO)_2$.[67] When R = Ph and R′ = H, one equivalent of base leads to deprotonation and hydroxide elimination to form the conjugated vinyl carbyne complex $Tp^*W(\equiv CCH=CHPh)(CO)_2$ (as the *E* isomer) in 53% yield; two equivalents of base produces a 1:1 mixture of the vinyl carbyne and the ethylidyne complex. With base, $Tp^*W\{\equiv CCH_2C(OH)PhMe\}(CO)_2$ simply regenerates the starting ethylidyne complex and ketone,[67] reminiscent of the tendency of propargylic alcohols to eliminate aldehyde or ketone under basic conditions.

The reaction of the tungsten vinylidene anion $[Tp^*W(=C=CH_2)(CO)_2]^-$ with benzoyl chloride followed by base forms the ketone-functionalised carbyne $Tp^*W\{\equiv CCH_2C(=O)Ph\}(CO)_2$ (and ethylidyne complex in approximately 1:1 ratio) that rearranges in solution to a metallafuran complex $Tp^*W(\kappa^2\text{-}O,C{=}CHCHCPhO)(CO)_2$. Spectroscopic data suggests that the dinuclear intermediate $\{Tp^*(CO)_2W\equiv CCH_2\}_2CPh(OH)$ is first formed before the addition of base produces the expected monomeric product $Tp^*W\{\equiv CCH_2C(=O)Ph\}(CO)_2$.[67]

The synthesis of the '$\equiv C(CRR')_2C\equiv$' bridged dimers (R = H, R′ = H, Me, CH_2Ph) with metal–carbon triple bond anchors is accomplished *via* oxidative dimerisation of the anionic vinylidene compounds $[Tp^*W(=C=CRR')(CO)_2]^-$.[158] Diastereomers (*meso* and d/l pairs) arise from the methyl and benzyl derivatives. Stepwise deprotonation and oxidation of the $\equiv C(CHR')_2C\equiv$ bis(carbynes) affords unsaturated bis(carbyne) complexes bridged by $\equiv CC(R')=C(R')C\equiv$ and (for R′ = H) $\equiv CC\equiv CC\equiv$ units (Scheme 61).

A crystal structure determination of $Tp^*(CO)_2W\equiv CCMe=CMeC\equiv W(CO)_2Tp^*$ reveals that the molecule lies on an inversion centre with a *trans* configuration of metal centres about the olefinic bridge.[158]

The synthetically versatile anionic vinylvinylidene anion $[Tp^*W(=C=CHCH=C_\delta H_2)(CO)_2]^-$ is capable of acting as a nucleophile towards heteroatom-substituted carbynes. No reaction occurs at C_δ with the electrophile $Tp^*W(\equiv CCl)(CO)_2$, however the phosphoniocarbyne $[Tp^*W(\equiv CPMe_2Ph)(CO)_2]^+$ adds to C_δ, forming a reactive zwitterionic phosphonium carbene intermediate that loses

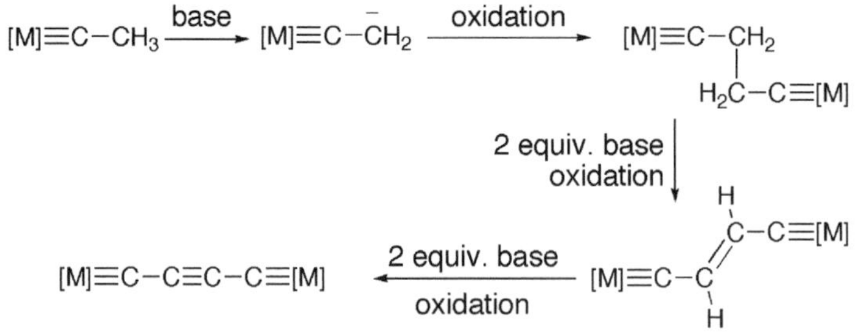

SCHEME 61. Oxidative coupling and dehydrogenation of anionic vinylidene complexes ([M] = $Tp^*M(CO)_2$; M = Mo or W).

PMe_2Ph to form the dinuclear C_5H_4-bridged bis(carbyne) compound $Tp^*(CO)_2W{\equiv}CCH_2CH{=}CHC{\equiv}W(CO)_2Tp^*$ as a 3.5:1 mixture of E and Z isomers in 56% yield (Scheme 62).[121] The reaction pathway is analogous to that suggested for the synthesis of the monomeric methylidyne complex $Tp^*W({\equiv}CH)(CO)_2$ *via* treatment of $[Tp^*W({\equiv}CPPh_3)(CO)_2][PF_6]$ with $Na[HBEt_3]$.[68] Deprotonation at C_β of the resultant C_5H_4-bridged bimetallic complex forms the deep blue, anionic $[Tp^*(CO)_2W{\equiv}C(CH)_3C{\equiv}W(CO)_2Tp^*]^-$ complex, which has a conjugated π-system between the tungsten atoms. Spectroscopic data suggest that the dinuclear anionic complex exists as a symmetrically allyl-bridged complex in solution with a '$C{\equiv}W(CO)_2Tp^*$' moiety appended to each end.[121]

The anionic vinylidene intermediate similarly acts as a nucleophile towards halocarbynes $Tp^*M({\equiv}CX)(CO)_2$ (X = Br, Cl) to produce '${\equiv}CCH_2C{\equiv}$' bridged dimers *via* halide displacement.[159] Stepwise deprotonation and alkylation of the methylene-bridged biscarbyne $Tp^*(CO)_2W{\equiv}CCH_2C{\equiv}Mo(CO)_2Tp^*$ yield, sequentially, the mono- or dimethylated products $Tp^*(CO)_2W{\equiv}CCHMeC{\equiv}Mo(CO)_2Tp^*$ and $Tp^*(CO)_2W{\equiv}CCMe_2C{\equiv}Mo(CO)_2Tp^*$, respectively.[159] Electrochemical and aerial oxidation of the doubly deprotonated anion yields the C_3-bridged complexes $Tp^*(CO)_2M{\equiv}CC{\equiv}C{-}M({=}O)_2Tp^*$ (M = Mo, W) as well as the ketone complex $Tp^*(CO)_2W{\equiv}CC({=}O)C{\equiv}Mo(CO)_2Tp^*$.[159]

The electrochemical behaviour of the bis(carbyne) complexes, $Tp^*(CO)_2M{\equiv}(C_mH_n){\equiv}M(CO)_2Tp^*$ (M = Mo, W; $m=3$, $n=2$; $m=4$, $n=0, 2, 4$), was studied.[160] These complexes exhibit two oxidation potentials between 0 and 0.5 V (vs. ferrocene). Coupling in the three-carbon saturated bridge is stronger than the saturated four-carbon bridge, as expected for a decrease in the metal–metal distance, and is comparable to that observed for the '${\equiv}CCH{=}CHC{\equiv}$' bridged complex. The difference in $E_{1/2}$ for the two oxidation waves is a function of the hydrocarbon spacer with the ${\equiv}CC{\equiv}CC{\equiv}$ spacers ($m=4$, $n=0$) exhibiting considerably larger $\Delta E_{1/2}$ values than the corresponding ${\equiv}CCH{=}CHC{\equiv}$ spacers ($m=4$, $n=2$), which were in turn more strongly coupled than the saturated analogues ($m=4$, $n=4$). The tungsten complexes are easier to oxidise and show larger $\Delta E_{1/2}$ values than their molybdenum analogues. Although the trend in

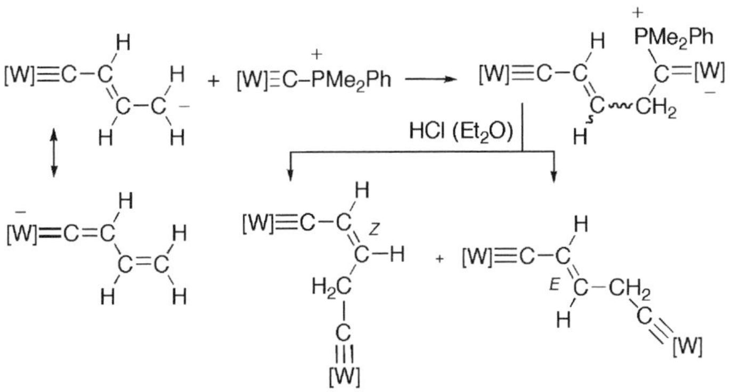

SCHEME 62. Synthesis of the neutral C_5H_4-linked dinuclear carbyne complex ([W] = $Tp^*W(CO)_2$).

electronic coupling was quantitatively as expected for the degree of unsaturation in the bridge, the absolute magnitude of the coupling, as indicated by the comproportionation constants ($K_c = 100$–104), was much smaller than that in isologous dimers where the bridge is connected to the metal centres *via* single bonds ($K_c = 108$–1012). A qualitative model based on relevant orbital occupancy as a function of metal oxidation state was developed to account for the experimental results. It is suggested that the lower coupling in the bis(carbyne) systems is due primarily to a HOMO that is oriented orthogonal to the bis(carbyne) bridge.[160]

The reaction of the anionic vinylidene $[Tp^*M(=C=CH_2)(CO)_2]^-$ with $TpWI(PhC{\equiv}CPh)(CO)$ proceeds *via* displacement of iodide and yields the dinuclear $Tp^*(CO)_2M{\equiv}CCH_2W(PhC{\equiv}CPh)(CO)Tp$ (M = Mo, W) containing an unsymmetrical CCH_2 bridge composed of a carbyne moiety and a methylene group (Scheme 63).[161]

Removal of the methylene protons and subsequent oxidation of the resulting dianion leads to net dihydrogen removal to provide the C_2-bridged bis(carbyne) dimers $Tp^*(CO)_2M{\equiv}CC{\equiv}W(PhC{\equiv}CPh)(CO)Tp$ (M = Mo, W) in good yields.[161] Conversion of the W–C single bond in the precursor complex to a triple bond in the product is accompanied by retraction of the alkyne from a four- to a two-electron donor role, providing a rare example of an alkyne–alkylidyne complex.[161]

Dewhurst *et al.* have reported the fluoride-mediated protodesilylation of silylated propargylidynes in the presence of rhodium,[162] ruthenium,[163] iridium,[164] gold[165] and mercury[166] complexes to provide a range of C_3-spanned heterobi- and trimetallic complexes.

Desilylation of the silylpropargylidynes $Tp^xW({\equiv}CC{\equiv}CSiMe_3)(CO)_2$ (Tp^x = Tp, Tp*) with moist $[Bu_4N]F$ produces *in situ* both the terminal propargylidynes $Tp^xW({\equiv}CC{\equiv}CH)(CO)_2$ and their conjugate bases $[Tp^xW({\equiv}CC{\equiv}C)(CO)_2]^-$, due to the water present in commercial tetrabutylammonium fluoride (Scheme 64). The C_3-spanned heterobimetallics $Ru\{C{\equiv}CC{\equiv}W(CO)_2Tp\}H(CO)(PPh_3)_3$[163] ($Tp^x$ = Tp, Tp*) and $Rh\{C{\equiv}CC{\equiv}W(CO)_2Tp\}(CO)(PPh_3)_2$ are produced *via* halide metathesis of the nucleophilic anion $[Tp^xW({\equiv}CC{\equiv}C)(CO)_2]^-$ with $RuHCl(CO)(PPh_3)_3$ or $RhCl(CO)(PPh_3)_2$.[162] One phosphine ligand in $Ru\{C{\equiv}CC{\equiv}W(CO)_2Tp\}H(CO)(PPh_3)_3$ is labile and readily replaced by the ligands L′ to provide $Ru\{C{\equiv}CC{\equiv}W(CO)_2Tp\}H(CO)(L')(PPh_3)_2$ (L′ = Hpz*, CO, CNR). The tetracarbonyl derivatives $Ru\{C{\equiv}CC{\equiv}W(CO)_2Tp^x\}H(CO)_2(PPh_3)_2$ are also obtained by oxidative addition of $Tp^xW({\equiv}CC{\equiv}CH)(CO)_2$ to $Ru(CO)_2(PPh_3)_3$.[163] The metal centres containing the σ-bonded '$Tp^x(CO)_2W{\equiv}CC{\equiv}C$'

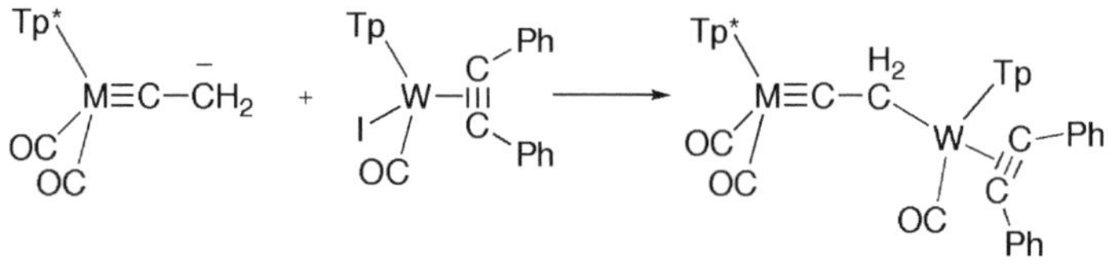

SCHEME 63. Synthesis of a CCH_2 hydrocarbon bridge between metal centres (M = Mo, W).

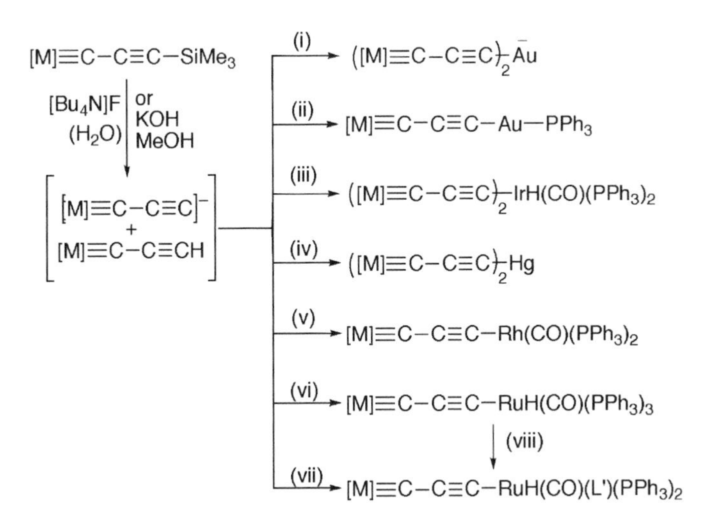

SCHEME 64. Fluoride-mediated desilylation of Tp^x silylpropargylidynes. Reagents: (i) $AuCl(SMe_2)$, (ii) $AuCl(PPh_3)$, (iii) $IrCl(CO)(PPh_3)_2$, (iv) $HgCl_2$, (v) $RhCl(CO)(PPh_3)_2$, (vi) $RuHCl(CO)(PPh_3)_3$, (vii) $Ru(CO)_2(PPh_3)_3$, L′ = CO, and (viii) L′ = Hpz*, CO, CNR. [M] = $Tp^xM(CO)_2$; Tp^x = Tp, Tp*; M = Mo, W.

fragments undergo typical ligand substitution and oxidative addition reactions, as if the group were behaving as a classical, albeit exotic, alkynyl, e.g., oxidative addition of iodine to $Rh\{C{\equiv}CC{\equiv}W(CO)_2Tp\}(CO)(PPh_3)_2$ provides the coordinatively saturated and structurally characterised rhodium complex $Rh\{C{\equiv}CC{\equiv}W(CO)_2Tp\}I_2(CO)(PPh_3)_2$.[162]

Fluoride-mediated desilylation of $Tp^xW({\equiv}CC{\equiv}CSiMe_3)(CO)_2$ in the presence of Vaska's complex $IrCl(CO)(PPh_3)_2$ or half an equivalent of mercuric chloride proceed to provide the first bis(tricarbido) trimetallic complexes $IrH\{C{\equiv}CC{\equiv}W(CO)_2Tp\}_2(CO)(PPh_3)_2$[164] and $Hg\{C{\equiv}CC{\equiv}W(CO)_2Tp\}_2$.[166]

In the presence of $AuCl(PPh_3)$, the neutral tricarbido complexes $Tp^xW({\equiv}CC{\equiv}CAuPPh_3)(CO)_2$ are obtained in addition to the formation of traces of the bis(tricarbido)aurates $[Bu_4N][Au\{C{\equiv}CC{\equiv}W(CO)_2Tp^x\}_2]$.[165] The corresponding salt $[(Ph_3P)_2N][Au\{C{\equiv}CC{\equiv}W(CO)_2Tp^*\}_2]$ may be obtained in high yield *via* propargylidyne desilylation with methanolic KOH in the presence of $AuCl(SMe_2)$ and subsequent salt metathesis with $[(Ph_3P)_2N]Cl$.[165] Spectroscopic data for the salt show that the ν_{CO} stretches are shifted to lower frequency than for the neutral complex. The crystal structure of $[(Ph_3P)_2N][Au\{C{\equiv}CC{\equiv}W(CO)_2Tp^*\}_2]$ confirms the essentially linear bis(carbido)aurate structure,[165] comparable to the isoelectronic and structurally characterised mercurial $Hg\{C{\equiv}CC{\equiv}W(CO)_2Tp^*\}_2(DMSO)_6$.

In the neutral complexes, the tungsten–carbonyl absorptions are lowered by replacement of the $SiMe_3$ substituent with the iridium(III) or gold(I) centres (ca. 40 and 15 cm^{-1}, respectively);[164,165] but replacement with Hg does not give rise to a discernible shift in the ν_{CO} values, presumably reflecting the d^{10} closed-shell configuration of mercury(II).[166] $^{13}C\{^1H\}$ Chemical shifts of the bridging carbon nuclei and crystallographic characterisation of the C_3-bridged polymetallic complexes point to a localised $M(C{\equiv}CC{\equiv}[W])_n$ ($n=1$ or 2) valence bond description along the near linear $M(C_3W)_n$ spines.[162–166]

Demercuration of the bis(tricarbido)mercurials $Hg\{C{\equiv}CC{\equiv}W(CO)_2Tp^x\}_2$ ($Tp^x = Tp$, Tp^*)[166] occurs with one equivalent of *cis*-$PtCl_2(PPh_3)_2$ or a catalytic amount of $RhCl(CO)(PPh_3)_2$, resulting in alkynyl coupling to provide the first dimetallaoctatetra-1,3,5,7-ynes $Tp^x(CO)_2W{\equiv}CC{\equiv}CC{\equiv}CC{\equiv}W(CO)_2Tp^x$ in good yields.[167] Demercuration with the rhodium complex is proposed to proceed *via* addition of a Hg–C bond followed by extrusion of elemental mercury to provide a *cis*-bis(alkynyl) rhodium(III) intermediate that reductively eliminates the diyne group.[168] The C_β and C_γ resonances are found at much higher field (δ_C: C_β 96.2, C_γ 58.6) than those of the corresponding mercurial complex (δ_C: C_β 120.5, C_γ 106.4). The alkylidyne carbon nuclei, however, are comparatively insensitive to changes in substitution at C_γ. A crystal structure confirmed the WC_6W connectivity.[167]

D. *The Heteroatom Substituent*

Neutral carbynes generally react at either the carbyne carbon atom or the metal centre. However, the phospha- and arsaalkenyl-substituted carbynes developed by Weber exhibit nucleophilicity at the heteroatoms of their carbyne substituents. Alkylation (MeOTf) occurs at the phosphorus and arsenic centres to produce the salts $[Tp^*M({\equiv}CAMe{=}C(NR_2)_2)(CO)_2][OTf]$ (A = P, As) in good yields.[88] The alkylation of the heteroatom is accompanied by strong shielding of the carbyne carbon nucleus and a significant decrease in σ-orbital contribution to the $M{\equiv}C{-}A$ bond.[88]

Protonation similarly occurs at the phosphorus centre to give $[Tp^*W\{{\equiv}CPH{=}C(NEt_2)_2\}(CO)_2]^+$, which undergoes facile rearrangement to the metallaphosphirene $[Tp^*W\{\eta^2{=}CHPC(NEt_2)_2\}(CO)_2]^+$, thereby adopting η^2-bonding to the metal centre *via* the phosphorus and carbon atoms (Scheme 65).[110] Protonation of the corresponding arsenic analogues does not, however, lead to tractable products.[88] Facile oxidation with dioxygen at the phosphorus atom cleanly affords the orange carbyne complexes $Tp^*M\{{\equiv}CP({=}O)_2C(NR_2)_2\}(CO)_2$, which are functionalised at the methylidyne with an α-carbenium phosphinate substituent.[169]

Reaction of $Tp^*M\{{\equiv}CP{=}C(NR_2)_2\}(CO)_2$ with two molar equivalents of AuCl(CO) affords the trinuclear complex $Tp^*M\{{\equiv}CP(AuCl)_2C(NR_2)_2\}(CO)_2$, in which the phosphorus atom of the carbyne substituent acts as donor to two gold centres.[109] Ligation of the two 'AuCl' fragments *via* the phosphorus atoms exerts a significant electron-withdrawal from the '$M(CO)_2$' fragment, and the resultant shift to higher energy of the ν_{CO} bands (by ca. 50–60 cm^{-1}) is attended by high field shifts for the ^{13}C nuclei of the carbyne ligand. A similar reaction with the arsenic analogue $Tp^*M\{{\equiv}CAs{=}C(NMe_2)_2\}(CO)_2$ gives rise to As=C cleavage with resultant formation of the gold carbene complex $AuCl\{C(NMe_2)_2\}$ and the functionalised cyclotriarsane $As_3\{C{\equiv}M(CO)_2Tp^*\}_3$ (Scheme 66).[109] Spectroscopic evidence is consistent with the presence of a cyclotriarsane ring in which two metal carbyne substituents are in a *cis*-disposition and the remaining one is on the

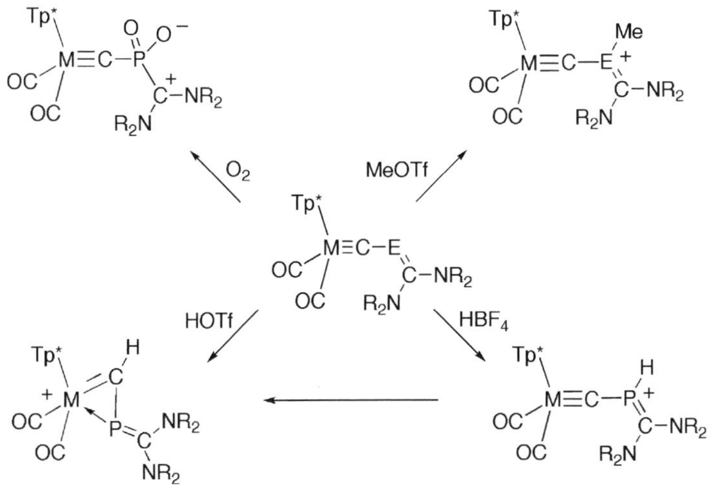

SCHEME 65. Reactivity of phospha- and arsaalkenyl alkylidyne complexes (M = Mo, W; E = As, R = Me; E = P, R = Me, Et).

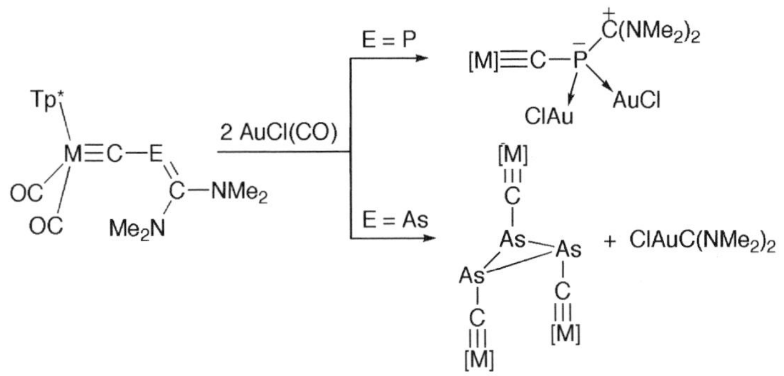

SCHEME 66. η^1-Ligation vs. pnictogen–carbon bond cleavage ([M] = $Tp^*M(CO)_2$, M = Mo, W).

opposite face of the ring. Formation of the cyclotriarsane ring decreases the σ-donor/π-acceptor capacity of the cyclotriarsane–triscarbyne ligand.[109]

The increased reactivity of the arsaalkene with respect to the phosphaalkene may be due to a decreased HOMO/LUMO energy gap in the former case and the general decrease in C–As vs. C–P bond strengths in otherwise analogous compounds.[109]

The methoxycarbyne complex $Tp^*W(\equiv COMe)(CO)_2$ also reacts at the carbyne substituent with nucleophiles attacking the methoxide Me group, delivering Me^+ and generating $[Tp^*W(CO)_3]^-$ *via* an S_N2 reaction.[79] The reactivity of the methoxycarbyne complex $Tp^*W(\equiv COMe)(CO)_2$ thus differs considerably from that of the methylthiocarbyne complex $Tp^*W(\equiv CSMe)(CO)_2$, which reacts with phosphines at the carbyne carbon to undergo nucleophilic displacement of the methylthiolate substituent to produce phosphoniocarbyne complexes.[72] The otherwise close structural similarities suggest that the difference in reactivity is due to the electronic differences between oxygen and sulfur.[79]

The bimetallic carbido complex $Tp^*Mo\{\equiv CFe(CO)_2Cp\}(CO)_2$ undergoes photolytic insertion of CS_2 into the C–Fe bond, with concomitant loss of one CO ligand from the Fe centre, to form the dithiocarboxylate-containing derivative $Tp^*Mo\{\equiv CCS_2Fe(CO)Cp\}(CO)_2$.[71] Photolytic substitution of both CO ligands at Fe forms $Tp^*Mo\{\equiv CFe(L')_2Cp\}(CO)_2$ derivatives ($L' = PMe_3$, CN^tBu). These substitutions are accompanied by a lowered frequency of the CO absorptions and the very low field carbide chemical shifts (>450 ppm) in the $^{13}C\{^1H\}$ NMR spectra.[71]

E. *The Poly(pyrazolyl)borate Ligand*

Although they are commonly employed as robust molecular scaffolds, the poly(pyrazolyl)borate ligands are not always innocent. An unusual reaction occurs during the sequential treatment of $W(\equiv CR)Br(CO)_4$ ($R = C_6H_4Me$-4) with sulfur and K[Bp] to provide $\{\kappa^3\text{-}HB(pz)_2(SCH_2R)\}W(\kappa^2\text{-}S_2CR)(CO)_2$ (**a**, Chart 5) *via* reduction of the alkylidyne–tungsten linkage.[150] The intermediacy of a thioacyl species is suggested *en route* to the bis(pyrazolyl)thiolatoborate complex and isolated examples have since followed this postulate.[48,153]

A second example involves the imido and oxo complexes $TpW(=CHPh)Br(=X)$ ($X = N^tBu$, N-1-adamantyl, $NC_6H_3Me_2$-2-6, O) prepared from dihalocarbyne complexes and primary amines or water.[134] Subsequent oxidation of these alkylidenes with bromine produces a highly reactive electron-deficient alkylidyne ligand which undergoes deprotonation and insertion into a σ-W–N bond of the tungsten tris(pyrazolyl)borate cage, affording the complexes $\{\kappa^3\text{-}HB(pz)_2(C_3H_3N_2CPh)\}WBr_2(=X)$ (**b**, Chart 5) *via* formal insertion of an alkylidyne group into one tungsten–(pyrazolyl nitrogen) bond. The $^{13}C\{^1H\}$ NMR spectra show resonances in the range δ_C 250–260 due to the former benzylidene carbons, which indicates the retention of a degree of multiple bonding with the metal centre.[134] The crystal structures of two compounds of this type (X = N-1-adamantyl and O) were determined by X-ray crystallography and display unusual geometric features due to the strained nature of the tris(pyrazolyl)borate-substituted alkylidene ligand.[130]

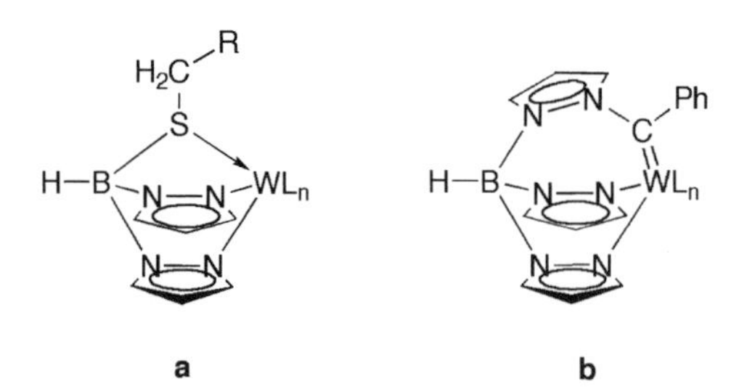

CHART 5. Examples of unusual carbyne-derived poly(pyrazolyl)borate ligands: (a) $WL_n = W(\kappa^2\text{-}S_2CR)(CO)_2$ ($R = C_6H_4Me$-4); (b) $WL_n = WBr_2(=X)$ ($X = O$, N^tBu, N-1-adamantyl, $NC_6H_3Me_2$-2,6).

The characteristic propensity for bis(pyrazolyl) and bis(methimazolyl)borates (Bm) to enter into agostic B–H–M interactions has been investigated. In the case of bis(pyrazolyl)borates, all examples of alkylidyne complexes to date do not show such an interaction. Bis(methimazolyl)borates are, however, more prone to such interactions, as has been demonstrated with the isolation of $BmMo(\equiv CC_6H_2Me_3\text{-}2,4,6)(CO)_2$ from the reaction of $Mo(\equiv CC_6H_2Me_3\text{-}2,4,6)X(CO)_2L_2$ (L = py, Hpz*; X = Br, Cl) with Na[Bm]. Attempts to prepare the corresponding tungsten derivative lead instead to the isolation of the bis(chelate) complex $(Bm)_2W(CO)$ in which one Bm ligand is bidentate-*S,S′* coordinated whereas the second adopts a tridentate κ^3-*H,S,S′* coordination mode.[170]

Treatment of the salts $[\{HC(pz)_3\}M(\equiv CR)(CO)_2][BF_4]$ (M = W, Mo; R = alkyl, aryl) with NaOEt or LiBu results in deprotonation of the bridgehead carbon of the neutral tris(pyrazolyl)methane ligand ($HC(pz)_3$). The pyrazolyl rings allow for delocalisation of the charge from the bridgehead onto the rings and subsequently onto the metal centre. Consequent reaction with $Au(C_6F_5)(tht)$ or $BF_3 \cdot Et_2O$ affords the neutral alkylidyne–metal compounds $\{(C_6F_5)AuC(pz)_3\}M(\equiv CR)(CO)_2$ (M = W, Mo, R = C_6H_4Me-4; M = W, R = Me, $C_6H_3Me_2$-2,6) or $\{(F_3B)C(pz)_3\}W(\equiv CR)(CO)_2$ (R = Me, C_6H_4Me-4), respectively, bearing poly(pyrazolyl)methane ligands which are modified at the bridgehead carbon.[116,117] $\{(F_3B)C(pz)_3\}W(\equiv CMe)(CO)_2$ may also be prepared in lower yield directly from the reaction between $W(\equiv CMe)Br(CO)_4$ and $Li[(F_3B)C(pz)_3]$ in tetrahydrofuran.[117]

F. *Bi- and Polymetallic Compounds*

Bi- and polymetallic compounds have been mentioned briefly in previous sections. Stone's recognition of the synthetic utility of the isolobal relationship between alkynes and the M≡C triple bond of carbyne metal complexes led to use of the complexes $Cp^xM(\equiv CR)(CO)_2$ (M = W, Mo, Cr; R = alkyl, aryl; Cp^x = Cp, Cp*) as precursors for the synthesis of numerous compounds containing heteronuclear and homonuclear metal–metal bonds.[171–173] The M≡C group is capable of functioning as a 'ligand' towards low-valent metal fragments, thus providing bi- and trimetallic complexes with bridging alkylidyne ligands. Typically, the low-valent metal fragments contain ligands (alkene, CO, MeCN, PR_3) that are readily displaced by the more powerful ligating properties of the M≡C groups, which typically act as μ-CR or μ_3-CR ligands. The poly(pyrazolyl)borate anions play a stabilising role in metal–alkylidyne chemistry and have allowed significant extension of the chemistry of cluster compounds containing bridging Group 6 alkylidynes, and it is this chemistry with which this section is concerned. The structures of the various products obtained are often similar to those previously reported using the Cp alkylidynes as precursors but important differences in reactivity between the poly(pyrazolyl)borate and cyclopentadienyl analogues arise. The 'TpM(≡CR)' fragment appears to serve as a superior σ-donor and weaker π-acceptor than the Cp analogue,[174] affording increased stability in the products.[40] The nature of the compounds containing heteronuclear metal–metal bonds formed from the Bp-ligated alkylidynes are generally similar to those previously prepared from the Tp analogues.[46] Angelici's thiocarbyne $TpW(\equiv CSMe)(CO)_2$ forms heteronuclear complexes similar to those formed by the analogous alkyl and aryl carbynes, however, the latter appear to act as better ligands in forming heteronuclear complexes.[175]

1. Nickel and Platinum

A characteristic reaction occurs between the platinum complex $Pt(C_2H_4)(PMe_3)_2$ and $Tp^xW(\equiv CR)(CO)_2$ (R = Me or C_6H_4Me-4, Tp^x = Tp;[39,40] R = C_6H_4Me-4, Tp^x = pzTp;[39,40] R = SMe, Tp^x = Tp,[175] Scheme 67) *via* displacement of the alkene, forming platinum–tungsten bimetallic compounds of the general formula $TpWPt(\mu\text{-}CR)(CO)_2(L')(PMe_3)$ ($L' = PMe_3$, CO).

Protonation of the bridging toluidyne complex $TpMoPt(\mu\text{-}CC_6H_4Me\text{-}4)(CO)_2\{P(OMe)_3\}_2$ with $HBF_4 \cdot Et_2O$ forms the Mo–Pt salt $[TpMoPt(\mu\text{-}\sigma{:}\eta^3\text{-}CHR)(CO)_2\{P(OMe)_3\}_2][BF_4]$ (Scheme 67), whereas the corresponding ethylidyne complex subsequently decomposes to release the carbyne (as ethane) under the same conditions due to the ability of hydrogen shifts to allow ethylidyne–ethene rearrangement.[174]

An alternative route to cationic bridged alkylidene complexes $[TpMPt(\mu\text{-}CRR')(CO)_2L_2Tp]^+$ involves the direct regioselective addition of a Pt–R′ bond (R′ = H, Me, $CH{=}CH_2$) from *trans*-$[Pt(R')L_2(acetone)][X]$ (L = PEt_3, PMe_2Ph; X = BF_4, OTf) across the M–C triple bond of the alkylidyne ligand in complexes of the type $TpM(\equiv CR)(CO)_2$ (R = Ph, 2-furyl, NEt_2, C_6H_4Me-4, Me; M = W, Mo, e.g., Scheme 68).[42]

The dinuclear products typically have a bridging CO ligand (indicated by a low-frequency ν_{CO} absorption ca. 1840 cm^{-1}) and $\mu\text{-}\eta^1,\eta^n$-alkylidene ligands so that the tungsten centres attain an 18-electron configuration. The molecular structure determination of $[TpWPt(\mu\text{-}CHNEt_2)(CO)_2(PEt_3)_2][BF_4]$ confirmed the $\mu\text{-}\eta^1,\eta^2$-coordination with the nitrogen atom of the bridging (diethylamino)methylidene ligand coordinated to the tungsten centre. In solution, the binuclear benzylidene-bridged complex exists as major and minor isomers (3:1), with only the major isomer being present in the solid state in which a PEt_3/CO ligand exchange has occurred between the two metals.[42] The bridging ethylidene complex has a

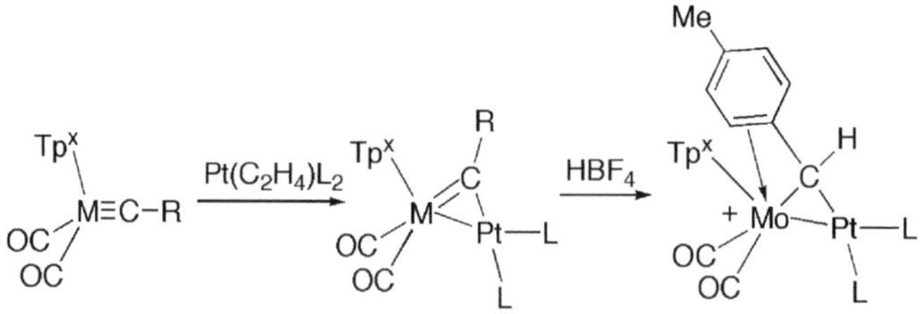

SCHEME 67. General formation of alkylidyne- and alkylidene-bridged M–Pt bimetallic complexes ($L = PMe_3$, R = Me, C_6H_4Me-4, Tp^x = Tp, pzTp).

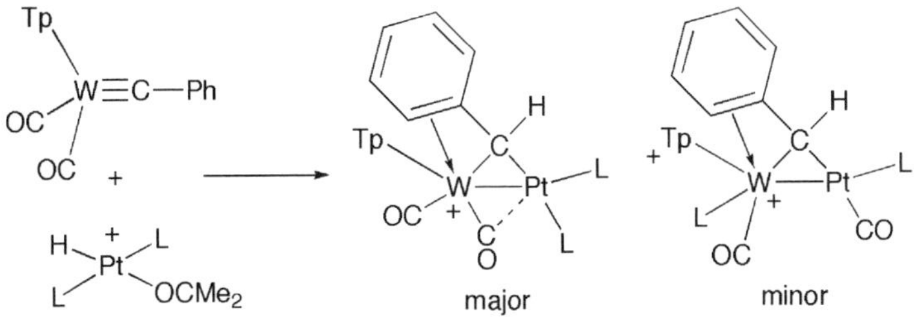

SCHEME 68. Formation of $[TpPtW(\mu\text{-}CHPh)(CO)_2(PEt_3)_2][BF_4]$ ($L = PEt_3$).

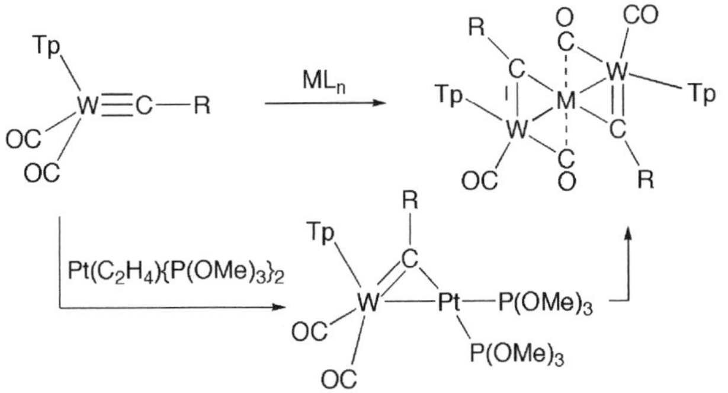

SCHEME 69. Reaction of Tp-alkylidyne complexes with unsaturated d^{10} complexes ($ML_n = Ni(cod)_2$, $Pt(C_2H_4)\{P(OMe)_3\}_2$, $Pt(C_2H_4)_3$; $R = C_6H_4Me\text{-}4$, Me, SMe).

16-electron count at the tungsten centre, and is not sufficiently stable to permit complete characterisation.[42]

The reaction of $TpM(\equiv CR)(CO)_2$ complexes with the electron-deficient platinum reagent $Pt(C_2H_4)\{P(OMe)_3\}_2$ gives extremely unstable complexes of the form $TpMPt(\mu\text{-}CR)(CO)_2\{P(OMe)_3\}_2$ (M = Cr, Mo, W, $R = C_6H_4Me\text{-}4$; M = W, R = Me) in addition to the trimetallic complexes $Tp_2M_2Pt(\mu\text{-}CR)_2(CO)_4$ (Scheme 69).[174] The trimetallic complexes presumably arise from reaction of the initially formed bimetallic species with a second equivalent of $TpM(\equiv CR)(CO)_2$, which functions as a better ligand towards platinum(0) than the phosphites.[174] From reaction of $TpW(\equiv CMe)(CO)_2$ with $Pt(C_2H_4)(PR_3)_2$ (R = Me or Et) $Pt(C_2H_4)_2(PCy_3)$ only the trimetallic complex $Tp_2W_2Pt(\mu\text{-}CMe)_2(CO)_4$ is obtained.[174]

Trimetal compounds $Tp_2M_2M'(\mu\text{-}CR)_2(CO)_4$ (M = W, R = Me, M′ = Ni or Pt;[174] M = Mo, Cr W, $R = C_6H_4Me\text{-}4$, M′ = Pt;[174] M = W, R = SMe, M′ = Pt[175]) comprising bis(μ-alkylidyne) $M'M_2C_2$-'bowtie' structures also arise by treating $TpM(\equiv CR)(CO)_2$ with the zero-valent homoleptic metal–alkene complexes $M'(alkene)_n$ (M′ = Ni, Pt; alkene = ethylene ($n = 3$) or cod ($n = 2$)) in a 2:1 ratio. The $^{13}C\{^1H\}$ NMR spectra show resonances in the range ca. δ_C 300–340, characteristic for alkylidyne carbon nuclei bridging a metal–metal bond. The Tp ligand appears to stabilise the NiW_2 species relative to the Cp analogue, being more comparable to Cp*-ligated complexes. The ferrocene-diyl–tethered bis(alkylidyne) $Tp_2Mo_2Fe(\mu\text{-}\sigma,\sigma':\eta^5\text{-}CC_5H_4)_2(CO)_4$ reacts with $Pt(cod)_2$ in a 1:1 ratio to yield the crystallographically characterised tetranuclear metal complex $Tp_2Mo_2FePt(\mu\text{-}\sigma,\sigma',\sigma'':\eta^5\text{-}CC_5H_4)_2(CO)_4$.[55]

X-ray diffraction studies of the bis(alkylidyne) 'bowtie' structures typically show that the molecules have essentially linear M–M′–M spines with the two metal–metal bonds each bridged by an alkylidyne ligand and semi-bridged by a CO group. The two alkylidyne fragments lie on the same side of the M–M′–M axis such that the compounds have approximate C_2 molecular symmetry.[55,174] The disposition of the edge-bridging alkylidyne groups is such that the two dimetallacyclopropene rings which share a common vertex (M′) are close to 90° with respect to one another.[53] For the ferrocenyl tethered tetranuclear species, this angle is expanded to 140° due to constraints imposed by the ferrocene fragment.[55]

The analogous Bp complexes $BpWPt(\mu\text{-}CR)(CO)_3(cod)$ and $Bp_2W_2M(\mu\text{-}CR)(CO)_6$ (M = Ni or Pt, R = C_6H_4Me-4; M = Pt, R = Me) have been synthesised by treating the appropriate alkylidyne–tungsten compound with $M(cod)_2$ (M = Ni or Pt, Scheme 70).[46] The reaction of $BpW(\equiv CR)(CO)_3$ with zero-valent $Pt(cod)_2$ yields the bimetallic species $BpWPt(\mu\text{-}CR)(CO)_3(cod)$ (R = C_6H_4Me-4).[46] Subsequent reaction with PMe_3 or a second equivalent of alkylidyne displaces the cod to generate $BpWPt(\mu\text{-}CR)(CO)_3(PMe_3)_2$ (which can also be prepared directly by reaction with $Pt(C_2H_4)(PMe_3)_2$) or the trimetal complex $Bp_2W_2Pt(\mu\text{-}CR)_2(CO)_6$ (which can also be obtained by treating two equivalents of the alkylidyne with $Pt(cod)_2$), respectively.[46] The trimetallic nickel complexes are similarly prepared.

The formation of *bimetallic* species from the $Pt(cod)_2$ reagent is unusual in that the analogous $Cp^xW(\equiv CR)(CO)_2$ (R = alkyl or aryl; Cp^x = Cp or Cp*)[68] and $TpW(\equiv CR)(CO)_2$[174] complexes always react to produce the trimetallic PtW_2 species, even if reactants are employed in a 1:1 ratio. Facile disproportionation of the Bp bimetallics in solution does, however, eventually provide the trimetallic W_2Pt complexes.[46]

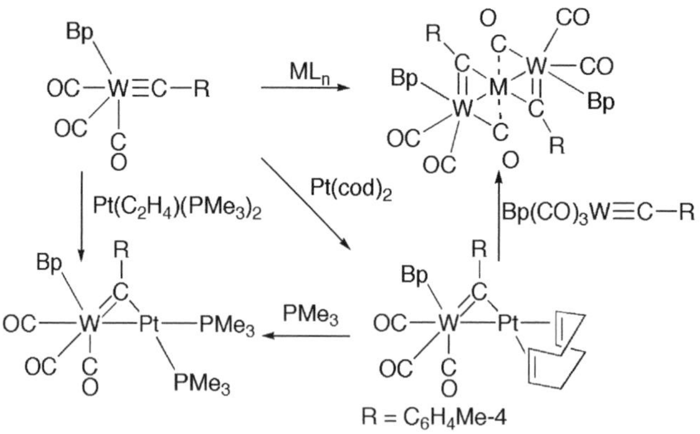

SCHEME 70. Reaction of Bp-alkylidyne complexes with unsaturated d^{10} complexes ($ML_n = Pt(cod)_2$, $Ni(cod)_2$; R = C_6H_4Me-4, Me).

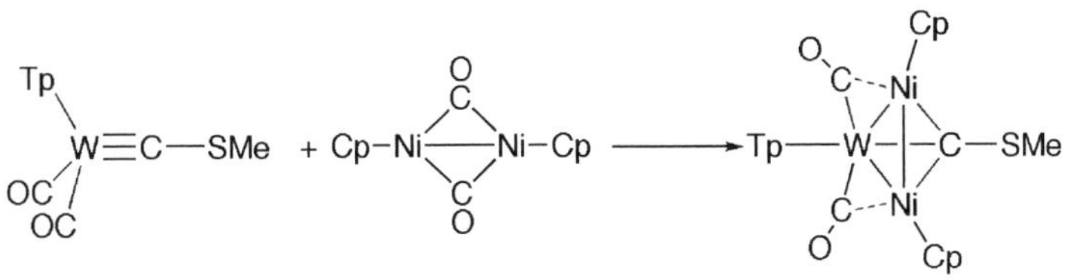

SCHEME 71. Synthesis of a tungsten dinickel cluster.

Reaction of the thiocarbyne $TpW(\equiv CSMe)(CO)_2$ with the Ni(I) dimer $Cp_2Ni_2(\mu\text{-}CO)_2$ generates a trinuclear metal complex based on a Ni_2W triangle $TpCp_2WNi_2(\mu_3\text{-}CSMe)(CO)_2$ (Scheme 71).[175] Two infrared stretches at 1875 and $1806\,cm^{-1}$ are consistent with a semi-bridging role for the tungsten carbonyls.

2. Gold

The bi- and trimetallic tungsten–gold compounds $[TpWAu(\mu\text{-}CMe)(CO)_2(PPh_3)][PF_6]$ and $[Tp_2W_2Au(\mu\text{-}CMe)_2(CO)_4][PF_6]$ are prepared by treating $TpW(\equiv CMe)(CO)_2$ with AuCl(L′) (L′ = PPh_3 or tht, respectively) in the presence of $TlPF_6$ (Scheme 72).[174,176]

The bimetallic complex displayed a $^{13}C\{^1H\}$ NMR resonance for the μ-C at δ_C 280.8, which is appreciably less deshielded than expected if the alkylidyne ligand were fully bridging the W–Au bond. The Tp ligand directs electron density to the tungsten centre,[176] increasing the stability of this complex compared with the Cp analogue which readily disproportionates in solution to give $[Cp_2W_2Au(\mu\text{-}CMe)_2(CO)_4][PF_6]$ and $[Au(PPh_3)_2][PF_6]$.[177] The neutral compound $\{(C_6F_5)AuC(pz)_3\}W(\equiv CMe)(CO)_2$ reacts with Au(X)(tht) (X = Cl or C_6F_5) to afford neutral trimetallic species $\{(C_6F_5)AuC(pz)_3\}WAu(\mu\text{-}CMe)X(CO)_2$ that retain the anionic ligand on the gold centre.[116] The μ-C resonance for X = Cl (δ_C 280.8) is again indicative of little perturbation of the $W\equiv C$ unit upon ligation of the AuX group.

In the formation of the trimetallic gold salts $[Tp_2W_2Au(\mu\text{-}CMe)_2(CO)_4][PF_6]$ from AuCl(tht), analogous to the copper, silver and gold salts $[Cp_2W_2M(\mu\text{-}CC_6H_4Me\text{-}4)_2(CO)_4][X]$ (M = Cu or Au, X = PF_6; M = Ag, X = BF_4),[178,179] the weakly coordinated tetrahydrothiophene is readily displaced by a second $TpW(\equiv CR)(CO)_2$ fragment, which acts as better donor towards Au(I) than tht (but poorer than PPh_3). The resonance for the bridging carbyne nucleus in the $^{13}C\{^1H\}$NMR spectrum occurs at δ_C 277.8, which is approximately 20 ppm upfield of its toluidyne Cp analogue (δ_C 295.8).[174] Both carbonyl ligands are terminally bound to the tungsten atoms, in contrast to the Ni and Pt complexes discussed above in which there are two strongly semi-bridging carbonyl groups present.

3. Cobalt

Octacarbonyldicobalt and the alkylidyne–molybdenum or alkylidyne–tungsten Tp^x complexes react to afford a family of species $Tp^xMCo_2(\mu_3\text{-}CR)(CO)_8$ (M = W,

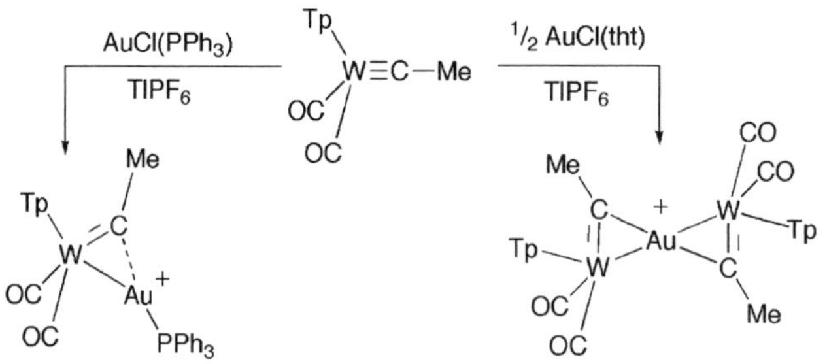

SCHEME 72. Synthesis of tungsten–gold bi- and trimetallic complexes.

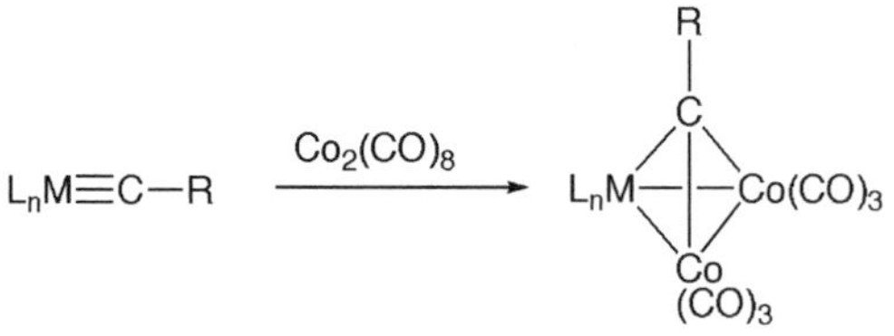

SCHEME 73. Metalladicobaltatetrahedrane synthesis ($L_nM = W(CO)_2Tp$, $W(CO)_2$ pzTp, $W(CO)_3Bp$; R = Me or C_6H_4Me-4).

R = Me or C_6H_4Me-4, $Tp^x = Tp$; R = C_6H_4Me-4, Tp^x = pzTp)[39,40] with a trimetallatetrahedrane, μ_3-$CMCo_2$ core (Scheme 73). The cobalt octacarbonyl reaction providing the analogous Bp-ligated clusters $BpWCo_2(\mu_3\text{-}CR)(CO)_9$ (R = C_6H_4Me-4)[46] proceeds in a similar manner as for the Tp analogues. $TpMo(\equiv CN^iPr_2)(CO)_2$, however, fails to react with $Co_2(CO)_8$, though it is not clear whether this is due to the steric bulk of the amino substituent, or the reduction in W–C multiple bonding that it causes.[180]

The Tp dicobalt–tungsten complexes, like their Cp analogues, exhibit rotational isomerism with respect to the '$W(CO)_2Tp$' fragment above the CCo_2 plane.[40]

For alkynyl carbynes $LM(\equiv CC\equiv CR)(CO)_2$ (L = Tp, Cp), two reactive sites are available for adduct formation with '$Co_2(CO)_6$'. Treatment of the cyclopentadienyl complexes $CpM(\equiv CC\equiv C^tBu)(CO)_2$ with $Co_2(CO)_8$ in light petroleum gives the trimetallatetrahedrane clusters $TpMCo_2(\mu_3\text{-}CC\equiv C^tBu)(CO)_8$ (M = Mo or W) in quantitative yield, structurally akin to the products obtained when R = alkyl or aryl (Scheme 74). In contrast, $Tp^xM(\equiv CC\equiv C^tBu)(CO)_2$ (Tp^xM = TpW and Tp*Mo) with $Co_2(CO)_8$ afford, respectively, the μ-alkyne trimetal compounds $Co_2\{\mu\text{-}^tBuC_2C\equiv M(CO)_2Tp^x\}(CO)_6$.[52] The more bulky hydrotris(pyrazolyl)borate ligands reduce the reactivity of the alkylidyne $M\equiv C$ bond, instead favouring reactivity at the unsaturated organic linkage. The retention of the $M\equiv C$ bond is evident in the $^{13}C\{^1H\}$ NMR spectra, with the resonance for C_α appearing ca. 15 ppm downfield of that of the precursor. Treatment of the μ-alkyne trimetal compounds with dppm produces the symmetrically phosphine bridged product *via* replacement of a CO ligand on each cobalt centre.[52]

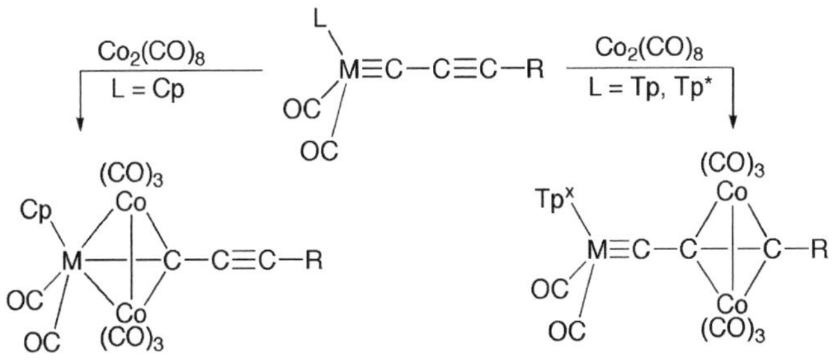

SCHEME 74. Reaction of propargylidyne complexes with $Co_2(CO)_8$ (M = Mo or W, R = ^{t}Bu).

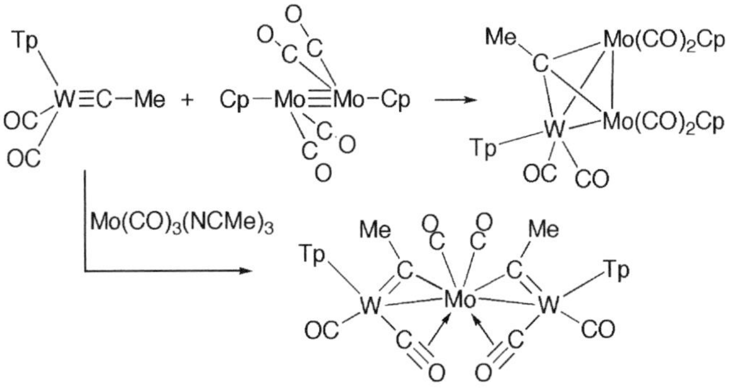

SCHEME 75. Reaction of alkynylmethylidyne complexes with $Cp_2Mo_2(CO)_4$ and $Mo(CO)_3(NCMe)_3$.

4. Molybdenum

The metal–alkylidyne complex TpW(≡CMe)$(CO)_2$ reacts with $Cp_2Mo_2(CO)_4$ or $Mo(CO)_3(NCMe)_3$ to afford the trimetal complexes $Cp_2TpMo_2W(\mu_3\text{-CMe})(CO)_6$ and $Tp_2MoW_2(\mu\text{-CMe})_2(\mu\text{-CO})_2(CO)_4$, respectively (Scheme 75).[41] In the latter complex, one of the terminal carbonyl ligands of each tungsten centre is η^2-bound to the central molybdenum atom.

For the alkynyl-substituted carbyne complex TpW(≡CC≡C^{t}Bu)$(CO)_2$, reaction with $Cp_2Mo_2(CO)_4$ gives the μ-alkyne trimetal compound Cp_2Mo_2 {μ-^{t}BuC_2C≡W$(CO)_2$Tp}$(CO)_4$ *via* addition across the organic C≡C rather than the M≡C triple bond due to steric shielding of the latter by the Tp ligand.[52] The Cp_2Mo_2{μ-^{t}BuC_2C≡W$(CO)_2$Tp}$(CO)_4$ complex is thus structurally akin to the product obtained with cobalt octacarbonyl (Scheme 74), but with the C≡C unit transversely bridging a Mo–Mo bond rather than a Co–Co linkage. Contrastingly, CpW(≡CC≡C^{t}Bu)$(CO)_2$ reacts with the dimolybdenum reagent to give the trimetallatetrahedrane cluster $Cp_3Mo_2W(\mu_3\text{-CC}{\equiv}\text{C}^t\text{Bu})(CO)_6$.[52] The $^{13}C\{^1H\}$ NMR spectrum of the complex Cp_2Mo_2{μ-^{t}BuC_2C≡W$(CO)_2$Tp}$(CO)_4$ shows a deshielded resonance at δ_C 273.1 with tungsten coupling, which unambiguously indicates that the compound possesses an intact W≡C linkage that is not directly attached to another dimetal–ligand fragment. The molecule undergoes a dynamic process in solution,[52] which is akin to the behaviour of other μ-alkyne–dimolybdenum compounds.[181,182]

5. Rhodium

The pyrazolylborate complexes $Tp^{x}W(\equiv CR)(CO)_2$ (R = Me or C_6H_4Me-4) are less reactive towards various low-valent rhodium species than their Cp analogues, requiring more forcing conditions (i.e., higher temperatures and longer reaction times) for reactions to proceed.

Cationic dimetal complexes $[TpWRh(\mu\text{-}CR)(CO)_2(L')_2][X]$ (R = Me, C_6H_4Me-4, L′ = PPh_3, X = PF_6; R = Me, L′ = ½ cod, X = BF_4) arise from treatment of $TpW(\equiv CR)(CO)_2$ with $[Rh(PPh_3)_2(cod)][PF_6]$ and $[Rh(THF)_2(cod)][BF_4]$ (the latter generated *in situ* from $Rh_2(\mu\text{-}Cl)_2(cod)_2$ and $AgBF_4$, Scheme 76).[176]

In their infrared spectra, each compound shows two bands in the CO stretching region with an absorption at approximately 1800 cm^{-1} indicating the presence of a semi-bridging carbonyl. Characteristic μ-C resonances are seen in the $^{13}C\{^1H\}$ NMR spectrum at ca. δ_C 340. The somewhat deshielded position suggests that in these compounds the W≡C bond is acting as a four-electron donor to the rhodium centre, allowing it to attain a 16- rather than 14-electron count.[176] The reaction with NaI affords the neutral, iodo-bridged complexes $TpWRh(\mu\text{-}CMe)(\mu\text{-}I)(\mu\text{-}CO)(CO)(L')_2$ (L′ = PPh_3 or ½ cod). A low-frequency infrared stretch (ca. 1760 cm^{-1}) suggests that one carbonyl ligand now fully bridges the W–Rh bond. The resonances for the μ-C nuclei at δ_C 303.7 and 294.7, respectively, are significantly less deshielded than those of the corresponding precursors, in accord with greater electronic saturation of the metal centres and a reduction in four-electron alkylidyne character.[176] The reaction of the salt $[TpWRh(\mu\text{-}CC_6H_4Me\text{-}4)(CO)_2(PPh_3)_2][PF_6]$ with hydride donors leads to decomposition (with some

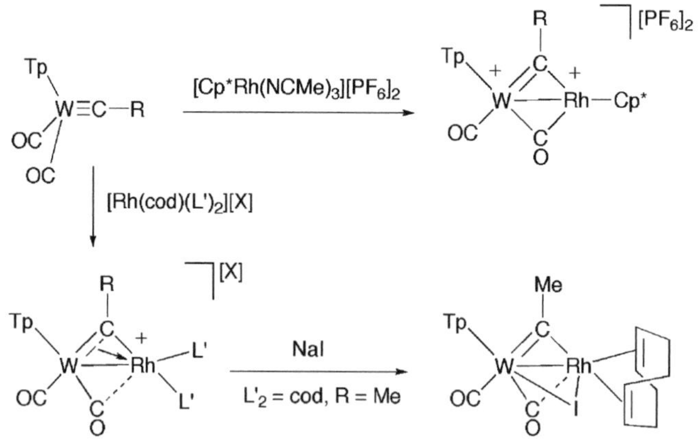

SCHEME 76. Synthesis of bimetallic tungsten–rhodium complexes with bridging alkylidyne groups (R = C_6H_4Me-4, Me; L′ = PPh_3, X = PF_6; L′ = ½ cod, X = BF_4).

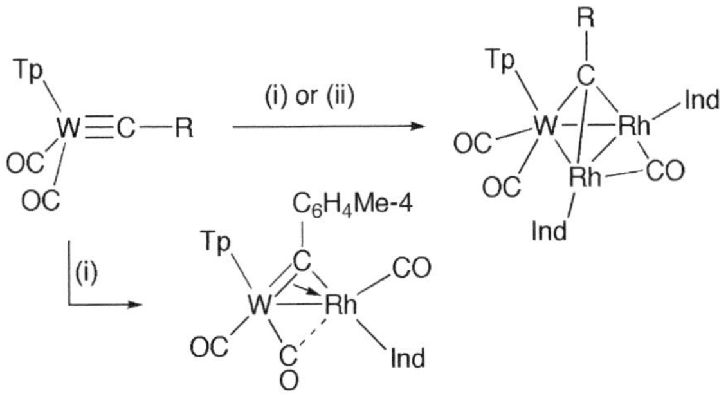

SCHEME 77. Synthesis of Rh–W complexes from Tp-alkylidyne complexes (R = C_6H_4Me-4, Me, SMe; Ind = η^5-C_9H_7, indenyl). Reagents: (i) Rh(η-C_2H_4)$_2$(η^5-C_9H_7) and (ii) Rh(CO)$_2$(η^5-C_9H_7).

evidence of hydride attack at the boron ligand), in contrast to the Cp version in which reaction with hydride produces the neutral benzylidene complex *via* attack at the bridging carbon atom. Similarly, reaction with molecular hydrogen produces an uncharacterisable product, which is, however, not akin to the dihydrido salt [CpWRh(μ-CC_6H_4Me-4)H_2(CO)$_2$(PPh_3)$_2$][PF_6] that is obtained with the Cp analogue.[176]

The reaction between [Cp*Rh(NCMe)$_3$][PF_6]$_2$ and TpW($\equiv$CMe)(CO)$_2$ gives the dicationic W–Rh complex [TpCp*WRh(μ-CMe)(μ-CO)(CO)][PF_6]$_2$ (Scheme 76).[176]

Treatment of the complexes TpW($\equiv$CR)(CO)$_2$ (R = C_6H_4Me-4, Me) with Rh (η-C_2H_4)$_2$(η^5-C_9H_7) in toluene at 60 °C gives the trimetallic complexes TpWRh$_2$ (μ_3-CR)(μ-CO)(CO)$_2$(η^5-C_9H_7)$_2$ (Scheme 77).[39,40] For the toluidyne derivative, the alkene displacement reaction provides a mixture of the bi- and trimetallic complexes TpWRh(μ-CR)(CO)$_3$(η^5-C_9H_7) and TpWRh$_2$(μ_3-CR)(μ-CO)(CO)$_2$ (η^5-C_9H_7)$_2$, respectively, the latter of which could also be prepared by heating the alkylidyne complex with Rh(CO)$_2$(η^5-C_9H_7) at 100 °C. The analogous Bp complex BpWRh$_2$(μ^3-CMe)(μ-CO)(CO)$_3$(η^5-C_9H_7)$_2$ has been synthesised by treating the appropriate alkylidyne precursor with Rh(L′)$_2$(η^5-C_9H_7) (L′ = CO or C_2H_4) in toluene at reflux.[46] A parallel reaction of the heterosubstituted carbyne TpW($\equiv$CSMe)(CO)$_2$ with two equivalents of Rh(CO)$_2$(η^5-C_9H_7) in refluxing tetrahydrofuran affords the crystalline μ_3-carbyne TpWRh$_2$(μ_3-CSMe) (μ-CO)(CO)$_2$(η^5-C_9H_7)$_2$, which has a Rh_2W triangular arrangement of metals that is akin to that of the Ni_2W cluster arising from reaction with $Cp_2Ni(CO)_2$. In the rhodium–tungsten complex, methyltriflate is capable of alkylating the sulfur of the μ_3-CSMe group to provide what might be described as a thioether stabilised carbido ligand.[175]

6. Iron

Diiron nonacarbonyl and the complexes TpM($\equiv$CR)(CO)$_2$ react to afford a family of species TpMFe(μ-CR)(CO)$_5$ (M = W, R = Me, C_6H_4Me-4;[39,183,184] M = Mo, R = C_6H_4Me-4[41]) with an unsaturated 32-valence electron count in which the M$\equiv$C bond effectively functions as a four-electron donor to iron.

Treatment of TpMo($\equiv$CR)$(CO)_2$ (R = C_6H_4Me-4) initially gives a mixture of TpMoFe(μ-CR)$(CO)_n$ (n = 5, 6),[41] but the saturated molybdenum complex readily loses CO under a nitrogen atmosphere.[41] Parallel reaction between bis(pyrazolyl) borate alkylidyne complexes and $Fe_2(CO)_9$ similarly provides the 32-valence electron dimetal species BpWFe(μ-CR)$(CO)_6$.[46] These heteronuclear bimetallic complexes have a rich further chemistry, as summarised in Scheme 78.

Infrared spectra of the unsaturated heteronuclear tris(pyrazolyl)borate M–Fe complexes typically include five carbonyl-associated bands, with a low-frequency stretch (ca. 1860 cm^{-1}) characteristic of a semi-bridging carbonyl.[183] The $^{13}C\{^1H\}$ NMR spectra show characteristic deshielded resonances for the μ-C nucleus at ca. δ_C 400, in the region typical of alkylidyne bridges serving as four-electron donors. A crystal structure determination of TpWFe(μ-CC_6H_4Me-4)$(CO)_5$ confirmed the presence of a semi-bridging carbonyl ligand and the four-electron functionality of the M$\equiv$C bond, with partially delocalised multiple bond character evident within the FeCW ring.[39] The Fe–CO ligand transoid to the π-acidic bridging alkylidyne group lies furthest from the iron atom.

The behaviour of the poly(pyrazolyl)borate alkylidyne complexes towards $Fe_2(CO)_9$ contrasts markedly with that of the Cp analogues. The complex CpW($\equiv$$CC_6H_4$Me-4)$(CO)_2$ reacts with $Fe_2(CO)_9$ to give a labile 'Fe$(CO)_3$' complex that readily affords diiron–tungsten or ditungsten–iron clusters, depending on the stoichiometry of reagents,[185] whereas the Cp* analogue, in the presence of CO, exists in equilibrium with the electronically saturated complex Cp*WFe (μ-CR)$(CO)_6$.[46] Conversion of LMFe(μ-CR)$(CO)_5$ complexes to the saturated hexacarbonyl derivatives under an atmosphere of carbon monoxide is possible, but for L = Tp the equilibrium favours the tricarbonyl species and reversion to this occurs under nitrogen purge.[39] For L = Cp, the equilibrium favours the 'Fe$(CO)_4$' adduct.[183] The enhanced stability of the 'Fe$(CO)_3$' species for L = Tp perhaps reflects the relatively greater σ-donor properties of the Tp ligand cf. Cp, which would increase electron donation by the M$\equiv$C system, thus enhancing its four-electron donor nature. Steric effects may also have a role in determining whether 'Fe$(CO)_3$' or 'Fe$(CO)_4$' adducts are formed. It is a characteristic feature that in these 32-valence electron complexes the resonances for the bridging alkylidyne carbon nuclei are approximately 80 ppm more deshielded than those of their 34-valence electron counterparts.[46] There is no evidence for addition of a CO ligand to the analogous iron–tungsten bis(pyrazolyl)borate species in solution.[46]

Treatment of the bimetallic complexes TpMFe(μ-CR)$(CO)_5$ with one equivalent of phosphine at room temperature readily gives the electronically unsaturated compounds TpWFe(μ-CR)$(CO)_4$(L′) (M = W, R = Me, R = C_6H_4Me-4, L′ = $PHPh_2$;[184] M = W, R = C_6H_4Me-4, L′ = PMe_3 or PEt_3;[183] M = Mo, R = C_6H_4Me-4, L′ = PMe_3[41]) *via* carbonyl substitution at the iron centre and ultimate retention of the four-electron donor role of the M–C bond. The unsaturated cyclopentadienyl compound CpWFe(μ-CC_6H_4Me-4)$(CO)_5$ also reacts readily with phosphines *via* initial attack at the iron centre and in this respect there is little difference in reactivity between the Cp and Tp systems.[183] The Cp complex CpWFe (μ-CC_6H_4Me-4)$(CO)_4(PMe_3)$ reacts with PMe_3 to produce the electronically saturated complex CpWFe(μ-CC_6H_4Me-4)$(CO)_4(PMe_3)_2$, which can be thermally

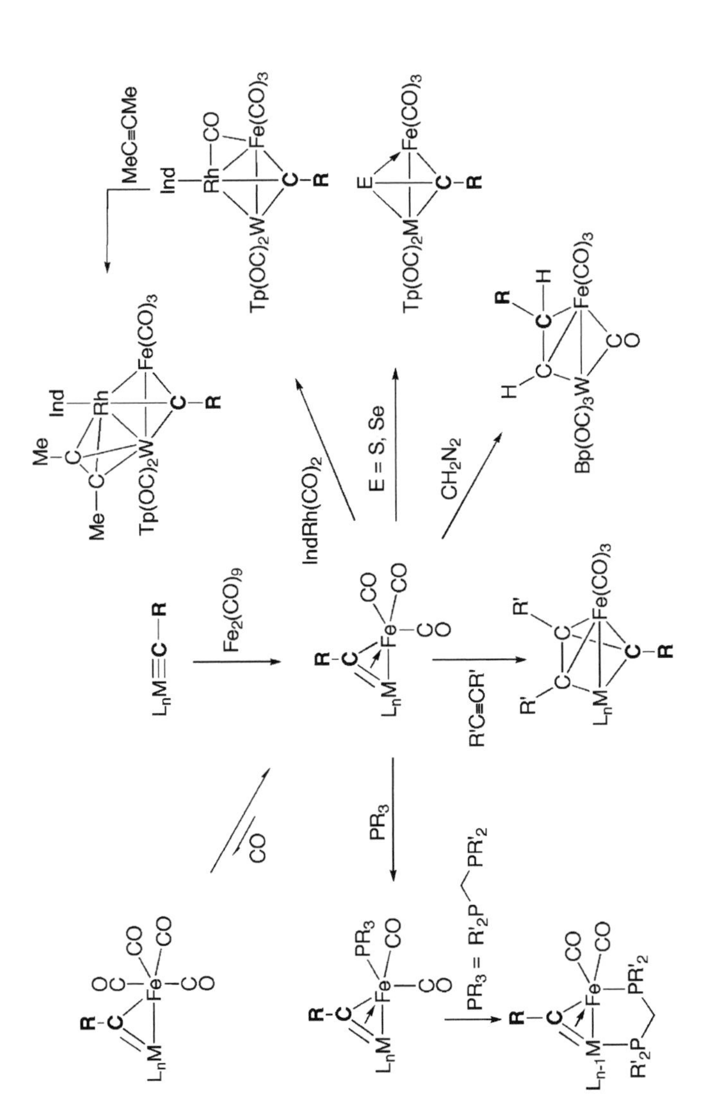

SCHEME 78. Synthesis and reactions of $Tp^xMFe(\mu\text{-}CR)L_n(CO)_3$ complexes (M = Mo, W; R = alkyl or aryl; Tp^xL_n = Tp, Bp(CO); Ind = η^5-C_9H_7).

converted to $CpWFe(\mu\text{-}CC_6H_4Me\text{-}4)(\mu\text{-}CO)(CO)_2(PMe_3)_2$ *via* release of CO, with a phosphine bound to each metal centre.[183] For Tp, excess PMe_3 gave a mixture of products including $TpW(\equiv CR)(CO)(PMe_3)$, but for the chelating phosphines (L′–L′ = dppm, dmpm), initial coordination of one phosphorus at the iron generates $TpWFe(\mu\text{-}CC_6H_4Me\text{-}4)(CO)_4(L'\text{–}L')$ and subsequent loss of CO in solution provides $TpWFe(\mu\text{-}CC_6H_4Me\text{-}4)(\mu\text{-}CO)(\mu\text{-}L'\text{–}L')(CO)_2$, in which the phosphine bridges the two metal centres (for dmpm, the non-bridging form is not isolated).[183] The unsaturated dimetal compound $BpWFe(\mu\text{-}CC_6H_4Me\text{-}4)(CO)_6$ reacts similarly with bidentate phosphines affording $BpWFe(\mu\text{-}CR)(\mu\text{-}CO)(\mu\text{-dppm})(CO)_3$.[46] *O*-Alkylation of the bridging carbonyls (ν_{CO} ca. 1700 cm^{-1}) results in the cationic complexes $[TpWFe(\mu\text{-}CR)(\mu\text{-}COMe)(\mu\text{-}L'\text{–}L')(CO)_2]^+$, which feature two distinct alkylidyne bridges.[183]

Heating $TpWFe(\mu\text{-}CC_6H_4Me\text{-}4)(CO)_4(PHPh_2)$ in toluene results in transfer of hydride from the secondary phosphine to the μ-C atom to afford the μ-alkylidene complex $TpWFe(\mu\text{-}\sigma,\eta^3\text{-}CHC_6H_4Me\text{-}4)(\mu\text{-}PPh_2)(\mu\text{-}CO)(CO)_3$, having an 18-electron configuration at both metal centres by virtue of the arene substituent that also coordinates (Scheme 79).[184] In contrast, under similar conditions, the corresponding ethylidyne complex yields the compound $TpWFe(\mu\text{-}PPh_2)(CO)_5$, presumably through decomposition of an unstable μ-CHMe analogue which eliminates ethylene and then scavenges CO.[184]

The unsaturated dimetal compounds $LWFe(\mu\text{-}CR)(CO)_n$ ($R = C_6H_4Me\text{-}4$, L = Tp, $n = 5$; L = Bp, $n = 6$) react with $R'C{\equiv}CR'$ (R′ = Me or Ph) to afford the complexes $LWFe(\mu\text{-}CRCR'CR')(CO)_n$, which may be viewed as '$Fe(CO)_3$' adducts of the metallacyclobutadienes $LW(CRCR'CR')(CO)_2$.[46] With diazomethane, $BpWFe(\mu\text{-}trans\text{-}CH{=}CHR)(\mu\text{-}CO)(CO)_6$ is produced, presumably as a result of methylene–alkylidyne coupling following hydrogen migration;[46] the Tp complex, however, does not react similary.[46]

The addition of sulfur to the bridging alkylidyne ligand in $TpMoFe(\mu\text{-}CC_6H_4Me\text{-}4)(CO)_5$ provides the heterobimetallic thioacyl complex $TpMoFe(\mu\text{-}\eta\text{-}SCR)(CO)_5$, in

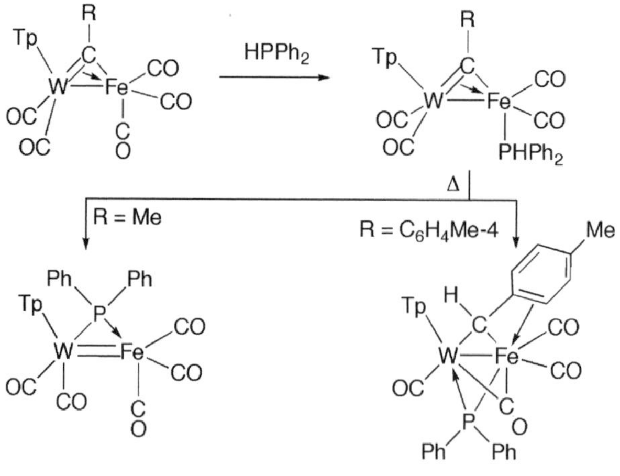

SCHEME 79. Reactivity of W–Fe bimetallic complexes containing the secondary phosphine $PHPh_2$.

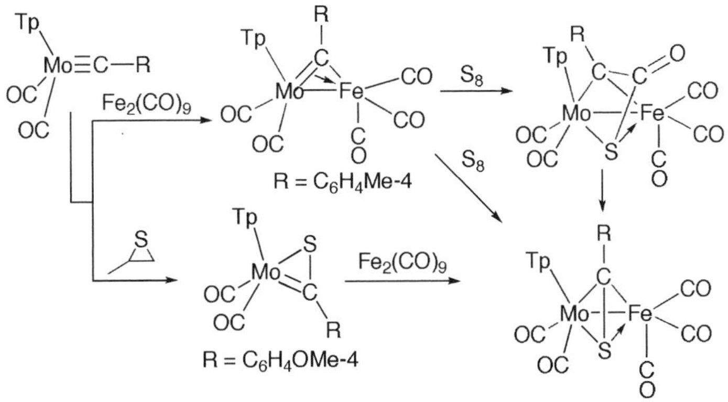

SCHEME 80. Alternative synthetic strategies for the formation of heterobimetallic thioaroyl complexes.

addition to trace amounts of a CO-insertion product which decomposes into the thioacyl complex.[41] A heterobimetallic thioaroyl complex can also be obtained from the direct reaction of TpMo(η^2-SCC_6H_4OMe-4)$(CO)_2$ with $Fe_2(CO)_9$ (Scheme 80).[48,153]

The reaction of $Fe_2(CO)_9$ and TpMo($\equiv$CN^{i}Pr$_2$)$(CO)_2$ similarly provides the thermally unstable complex TpMoFe(μ-CN^{i}Pr$_2$)$(CO)_5$, which reacts subsequently with sulfur to provide the heterobimetallic bridging thiocarbamoyl complex TpMoFe(μ-SCN^{i}Pr$_2$)$(CO)_5$.[180] Notably, the precursor aminomethylidyne complex TpMo($\equiv$CN^{i}Pr$_2$)$(CO)_2$ does not react with elemental sulfur prior to coordination to iron.

Treatment of TpFeW(μ-CR)$(CO)_5$ (R = C_6H_4Me-4, R = Me) with $Rh(CO)_2$ (η^5-C_9H_7) (*vide supra*) in diethyl ether at room temperature affords the black crystalline trimetallic complexes TpWFeRh(μ_3-CR)(μ-CO)$(CO)_5$(η^5-C_9H_7). The $^{13}C\{^1H\}$ NMR spectra of these species include four CO resonances, in contrast to the Cp analogue, which has only one. Evidently, the Tp complexes have, in this instance, a higher energy barrier to dynamic behaviour.[40] The trimetal compounds TpWFeRh(μ_3-CR)(μ-CO)$(CO)_5$(η^5-C_9H_7) react with but-2-yne in toluene at 100 °C to give the alkylidyne- and alkyne-bridged complexes TpWFeRh(μ_3-CR) (μ-MeC_2Me)$(CO)_4$(η^5-C_9H_7).[40] The crystal structure shows three essentially equal N–W distances between the Tp ligand and the tungsten, which is consistent with the reduction in bond order (and *trans* influence) of the W–C linkage.[40]

For the rhodium–propargylidyne complex Rh{C$\equiv$CC$\equiv$W$(CO)_2$Tp}(CO) $(PPh_3)_2$ (Scheme 64), shielding of the M$\equiv$C bond by the sterically cumbersome Tp ligand results in the reaction with $Fe_2(CO)_9$ proceeding *via* addition of two '$Fe(CO)_3$' units across the 'organic' C$\equiv$C bond (Scheme 81). This is similar to the situation described by Stone for the reaction of simple alkynyl-substituted carbynes with $Cp_2Mo_2(CO)_x$ (x = 4, 6) and $Co_2(CO)_8$[52] (*vide supra*), and provides Fe_2Rh {μ-$C_3W(CO)_2$Tp}$(CO)_{10}(PPh_3)$, which is a rare example of a tricarbido ligand incorporated within a cluster framework.[162]

The appearance of the tungsten alkylidyne resonance at δ_C 254.6 in the $^{13}C\{^1H\}$ NMR spectrum, together with crystallographic evidence, confirms that the W$\equiv$$C_3$

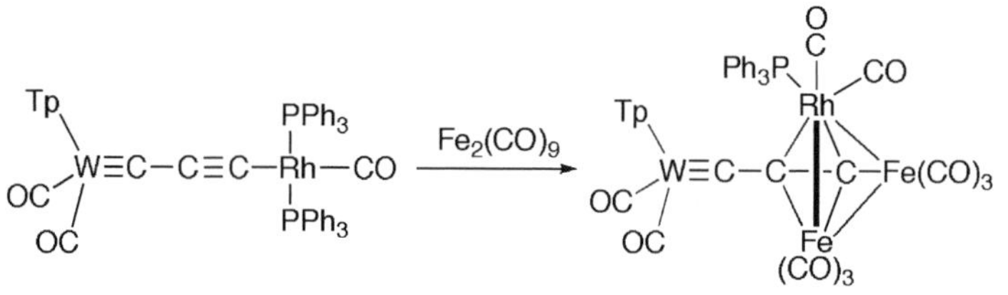

SCHEME 81. Formation of a tetrametallic $RhFe_2$ cluster from a Rh-tricarbido complex.

unit remains intact and that it straddles one Rh–Fe bond (2.552(1) Å), contracting it relative to the other Rh–Fe separation (2.721(1) Å).[162]

7. Ruthenium

The majority of complexes containing bonds between tungsten and ruthenium are structurally different from those having tungsten–iron bonds. Treatment of the compounds $TpW(\equiv CC_6H_4Me\text{-}4)(CO)_2$ with a three-fold excess of $Ru(CO)_4(C_2H_4)$, a source of the fragment '$Ru(CO)_4$', in light petroleum gives the tetranuclear metal cluster complex $TpWRu_3(\mu_3\text{-}CR)(CO)_{11}$ (Scheme 82). The cluster has a WRu_3 core which is essentially tetrahedral, with one Ru_2W face capped by a triply bridging μ-CC_6H_4Me-4 ligand. The $^{13}C\{^1H\}$ NMR spectrum shows a resonance for the asymmetrically bridging μ_3-C nuclei at δ_C 329.6, which is ca. 30 ppm more deshielded than the corresponding shift for the Cp complex due partly to the greater donor ability of the Tp-alkylidyne fragment. The formation of a tetranuclear compound reflects the ease with which carbonyl–ruthenium fragments combine to form Ru–Ru bonds.[186]

Addition of the reagent $Ru(CO)_2(THF)(\eta^5\text{-}7,8\text{-}C_2B_9H_{11})$ to the compounds $TpM(\equiv CR)(CO)_2$ (M = Mo, W; R = C_6H_4Me-4) in dichloromethane affords the bimetallic species $TpMRu(\mu\text{-}CR)(CO)_4(\eta^5\text{-}7,8\text{-}C_2B_9H_{11})$ (*vide supra*, Scheme 83).[152] The Cp analogues are very labile and have a predilection for insertion of their tolylmethylidyne ligands into an adjacent B–H bond of the carborane group. In contrast, the Tp complexes are stable in dichloromethane or toluene solution, but are readily cleaved by donor molecules including tetrahydrofuran. X-ray diffraction studies established the molecular structure, revealing long Ru–W separations (ca. 3 Å) which are asymmetrically bridged by the tolylmethylidyne groups (μ-C–W_{av} 1.92 Å vs. μ-C–Ru_{av} 2.21 Å).[152] The chemical shifts of the μ-C nuclei (ca. δ_C 290) differ little from the precursor alkylidynes and reflect the asymmetric bridging of the metal–metal bond by the tolylmethylidyne ligand.[152]

The reaction of $TpMRu(\mu\text{-}CR)(CO)_4(\eta^5\text{-}7,8\text{-}C_2B_9H_{11})$ with elemental sulfur or selenium gives the chalcoacyl compounds $TpMRu(\mu\text{-}ECR)(CO)_4(\eta^5\text{-}7,8\text{-}C_2B_9H_{11})$ and appreciable amounts of the compounds $TpM(\kappa^2\text{-}S_2CR)(CO)_2$, which had previously been obtained *via* an alternative route.[153] X-ray crystallography established the structure, displaying '$Ru(CO)_2(\eta^5\text{-}7,8\text{-}C_2B_9H_{11})$' and '$W(CO)_2Tp$' units unusually bridged by a chalcoacyl group which is η^2-coordinated to the tungsten and bound to the ruthenium centre only through the chalcogen atom, with no direct metal–metal bond.[152]

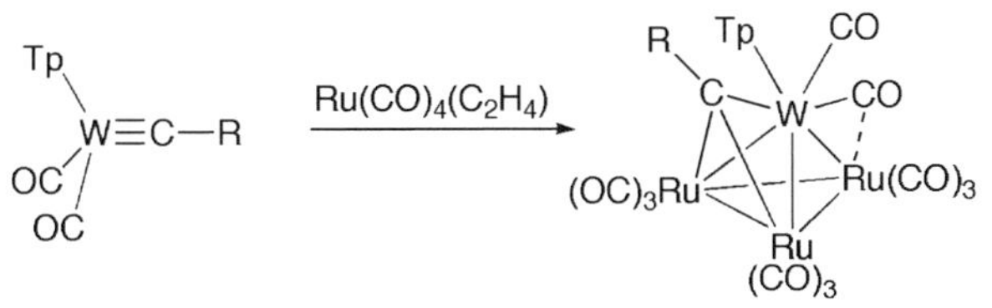

Tp
W≡C—R
OC
OC
$Ru(CO)_4(C_2H_4)$
R Tp CO
C W CO
$(OC)_3Ru$ $Ru(CO)_3$
Ru
$(CO)_3$

SCHEME 82. Formation of tetranuclear WRu_3 cluster containing a triply bridging alkylidyne ligand ($R = C_6H_4Me$-4).

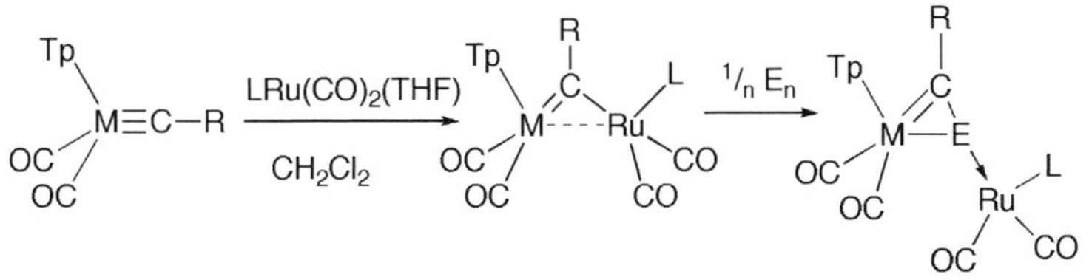

Tp
M≡C—R
OC
OC
$LRu(CO)_2(THF)$
CH_2Cl_2
Tp R
C
M Ru L
OC CO
OC CO
$^1/_n E_n$
Tp R
C
M—E
OC
OC
Ru L
OC CO

SCHEME 83. Binuclear Group 6 chalcoacyl complexes ($L = \eta^5$-7,8-$C_2B_9H_{11}$; $R = C_6H_4Me$-4; M = Mo, W; E = S, Se).

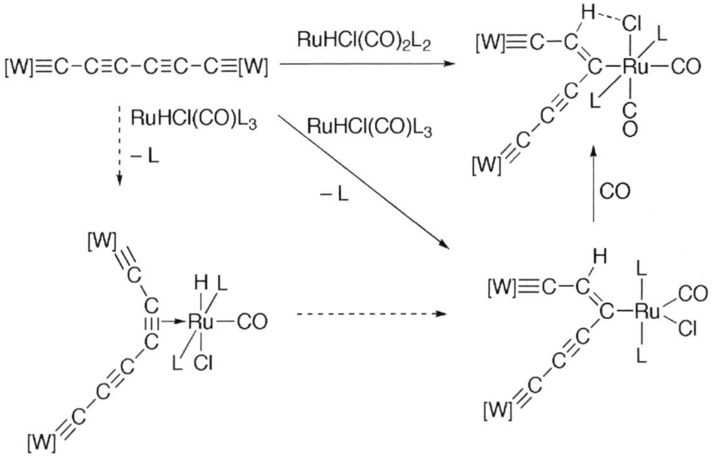

[W]≡C–C≡C–C≡C–C≡[W]
$RuHCl(CO)_2L_2$
[W]≡C—C H---Cl L
C—Ru—CO
C≡C L C
[W]≡C O
$RuHCl(CO)L_3$
– L
$RuHCl(CO)L_3$
– L
CO
[W]
C
C H L
C →Ru—CO
C L Cl
C≡C
[W]≡C
[W]≡C—C H L
C—Ru CO
C≡C Cl
[W]≡C L

SCHEME 84. Regioselective hydroruthenation of dimetallaoctatetraynes ([W] = $Tp^xW(CO)_2$, where Tp^x = Tp, Tp*; L = PPh_3).

The dimetallaoctatetra-1,3,5,7-ynes $Tp^x(CO)_2W{\equiv}CC{\equiv}CC{\equiv}CC{\equiv}W(CO)_2Tp^x$[167] ($Tp^x$ = Tp, Tp*) do not generally react at the M≡C multiple bond, instead participating in reactions at the more sterically accessible C≡C bonds of the WC_6W spine. With $RuHCl(CO)(L)(PPh_3)_2$ (L = PPh_3, CO), regioselective hydroruthenation of one C≡C bond provides the complexes $Ru\{C\{C{\equiv}CC{\equiv}W(CO)_2Tp^x\}{=}CHC{\equiv}W(CO)_2Tp^x\}Cl(CO)_n(PPh_3)_{4-n}$ (n = 1 or 2), containing hex-2-en-4-yn-3-yl-1,6-diylidyne-bridges (Scheme 84).[187]

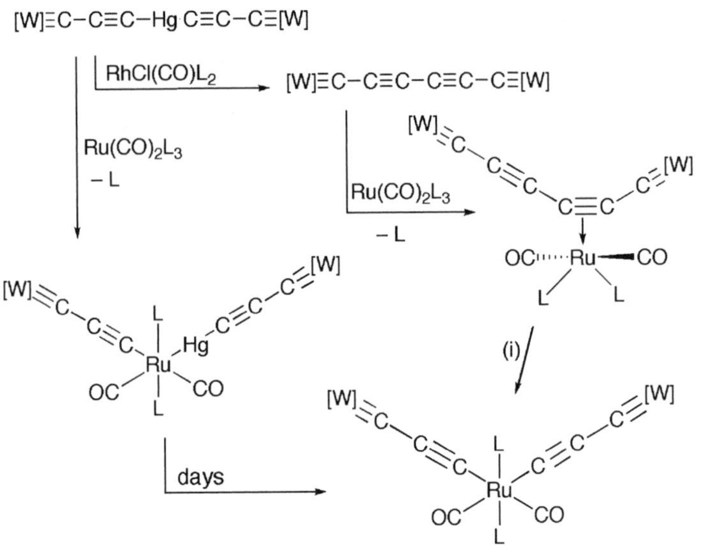

SCHEME 85. Reaction of bis(tricarbido)mercurial resulting in C–C single bond scission ([W] = $Tp^xW(CO)_2$, L = PPh_3). (i) r.t., 40 h for $Tp^x = Tp^*$; or 65 °C, 70 h for $Tp^x = Tp$.

With the zero-valent ruthenium complex $Ru(CO)_2(PPh_3)_3$, the bis(tricarbido) complexes $Ru\{C_3W(CO)_2Tp^x\}_2(CO)_2(PPh_3)_2$, which involve cleavage of the C–C single bond of the WC_6W spine, are formed *via* the adduct $Ru\{\eta^2\text{-}Tp^x(CO)_2WC_6W(CO)_2Tp^x\}(CO)_2(PPh_3)_2$. This adduct, along with $Ru\{C_3W(CO)_2Tp^x\}\{HgC_3W(CO)_2Tp^x\}(CO)_2(PPh_3)_2$, is observed as an intermediate in the reactions of $Ru(CO)_2(PPh_3)_3$ with $Hg\{C_3W(CO)_2Tp^x\}_2$ that also ultimately provide the bis(tricarbido) complex (Scheme 85).[168]

Spectroscopic data for the π-poly-yne complex $Ru\{\eta^2\text{-}Tp^x(CO)_2WC_6W(CO)_2Tp^x\}(CO)_2(PPh_3)_2$ confirm the presence of two chemically distinct '$W(CO)_2Tp^x$' groups.[168] The bis(tricarbido) complexes *cis,cis,trans*-$Ru\{C_3W(CO)_2Tp^x\}_2(CO)_2(PPh_3)_2$ display single resonances for the Ru–CO, W–CO and alkylidyne carbon nuclei. The latter complex has been crystallographically characterised.[168]

8. Zirconium and Titanium

Although a range of alkylidyne complexes $CpM(\equiv CR)(CO)_2$ react with $Cp_2M'(^nBu)_2$ or $Cp_2M'Cl_2/Mg(Hg)$ (M′ = Ti, Zr) to provide the complexes $Cp_3TiW(\mu\text{-}CC_6H_4OMe\text{-}2)(\mu\text{-}CO)(CO)$ and $Cp_3ZrM(\mu\text{-}CR)(\mu\text{-}CO)(CO)$ (M = Mo, R = C_6H_4OMe-2; M = W, R = Me, C_6H_4OMe-2, $C_6H_3Me_2$-2,4,6), the corresponding poly(pyrazolyl)borate complex $Cp_2TpZrMo(\mu\text{-}CC_6H_4OMe\text{-}2)(\mu\text{-}CO)(CO)$, generated in a similar manner and spectroscopically observable, was unstable and could not be isolated.[188]

9. Cluster Chemistry with Alkylidynes Co-Ligated by Poly(pyrazolyl)alkanes

Like the boron-based alkylidyne compounds, the tris(pyrazolyl)methane complexes have a promising, though less explored, chemistry involving polymetallic clusters.[100,116,117] Both the salts and the neutral pyrazolyl methane complexes $[LW(\equiv CR)(CO)_2]^{x+}$ (R = Me, C_6H_4Me-4; L = $HC(pz)_3$, $x = +1$; L = $\{(C_6F_5)AuC(pz)_3\}$, $x = 0$; R = Me, C_6H_4Me-4; L = $F_3BC(pz)_3$, $x = 0$) afford a variety of mixed-metal complexes upon treatment with low-valent metal–ligand fragments, the structures and properties of which reflect those obtained for the closely related Tp analogues but with a positive charge higher by one unit per ligand for L = $HC(pz)_3$.[100,116,117] In some instances, the Au–C bond of the derivatised neutral complexes are cleaved.[116]

V
CONCLUDING REMARKS

The preceding discussion indicates that a substantial amount of alkylidyne chemistry supported by TpM and Tp*M scaffolds has been developed. Early work focused on perceived and pursued similarities between these ligands and variously substituted cyclopentadienyls; indeed much of the work described herein bears out this analogy. However, distinctions soon emerged, many of which can be traced to the steric profile of the scorpionate ligands and the geometric constraints that predispose their complexes to pseudo-octahedral six-coordination, especially in the case of Tp*. From an applications point of view, the added stability offered by the Tp* and Tp scaffolds has made it possible to carry out many carbyne ligand transformations that are not yet available to Cp derivatives. This has led to a far wider range of accessible carbyne substituents including many heteroatom examples. Amongst these, halocarbynes are unique to pyrazolylborate chemistry and the synthetic potential as precursors to exotic 'C_1' ligands is enormous.

REFERENCES

(1) Fischer, E. O.; Kreis, G.; Kreiter, C. G.; Muelle, J.; Huttner, G.; Lorenz, H. *Angew. Chem.* **1973**, *85*, 618.
(2) Trofimenko, S. *J. Am. Chem. Soc.* **1966**, *88*, 1842.
(3) Gallop, M. A.; Roper, W. R. *Adv. Organomet. Chem.* **1986**, *25*, 121.
(4) Schrock, R. R. *Acc. Chem. Res.* **1986**, *19*, 342.
(5) Buhro, W. E.; Chisholm, M. H. *Adv. Organomet. Chem.* **1987**, *27*, 311.
(6) Kim, H. P.; Angelici, R. J. *Adv. Organomet. Chem.* **1987**, *27*, 51.
(7) Fischer, H.; Hofman, P.; Kreissl, F. R.; Shrock, R. R.; Schubert, U.; Weiss, K. *Carbyne Complexes*, VCH, New York, **1988**.
(8) Nugent, W. A.; Mayer, J. M. *Metal–Ligand Multiple Bonds*, Wiley, New York, **1988**.
(9) Mayr, A. *Comments Inorg. Chem.* **1990**, *10*, 227.
(10) Mayr, A.; Hoffmeister, H. *Adv. Organomet. Chem.* **1991**, *32*, 227.
(11) Mayr, A.; Bastos, C. M. *Prog. Inorg. Chem.* **1992**, *40*, 1.
(12) Engel, P. F.; Pfeffer, M. *Chem. Rev.* **1995**, *95*, 2281.
(13) Trofimenko, S. *Scorpionates – The Coordination Chemistry of Polypyrazolylborate Ligands*, Imperial College Press, London, **1999**.

(14) Curtis, M. D.; Shiu, K. B.; Butler, W. M. *Organometallics* **1983**, *2*, 1475.
(15) Curtis, M. D.; Shiu, K. B.; Butler, W. M. *J. Am. Chem. Soc.* **1986**, *108*, 1550.
(16) Trofimenko, S.; Calabrese, J. C.; Thompson, J. S. *Inorg. Chem.* **1987**, *26*, 1507.
(17) Trofimenko, S. *Chem. Rev.* **1993**, *93*, 943.
(18) Deeming, A. J.; Donovan-Mtunzi, S. *Organometallics* **1985**, *4*, 693.
(19) Doyle, R. A.; Angelici, R. J. *J. Am. Chem. Soc.* **1990**, *112*, 194.
(20) Roberts, S. A.; Young, C. G.; Kipke, C. A.; Cleland, W. E. Jr.; Yamanouchi, K.; Carducci, M. D.; Enemark, J. H. *Inorg. Chem.* **1990**, *29*, 3650.
(21) Egan, J. W. Jr.; Haggerty, B. S.; Rheingold, A. L.; Sendlinger, S. C.; Theopold, K. H. *J. Am. Chem. Soc.* **1990**, *112*, 2445.
(22) Kitajima, N.; Fujisawa, K.; Morooka, Y. *Inorg. Chem.* **1990**, *29*, 357.
(23) Greaves, W. W.; Angelici, R. J. *Inorg. Chem.* **1981**, *20*, 2983.
(24) Fischer, E. O.; Hollfelder, H.; Kreissl, F. R. *Chem. Ber.* **1979**, *112*, 2177.
(25) Fischer, E. O.; Hollfelder, H.; Friedrich, P.; Kreissl, F. R.; Huttner, G. *Angew. Chem.* **1977**, *89*, 416.
(26) Fischer, E. O.; Walz, S.; Wagner, W. R. *J. Organomet. Chem.* **1977**, *134*, C37.
(27) Fischer, H.; Fischer, E. O. *J. Organomet. Chem.* **1974**, *69*, C1.
(28) Fischer, E. O.; Schubert, U. *J. Organomet. Chem.* **1975**, *100*, 59.
(29) Himmelreich, D.; Fischer, E. O. *Z. Naturforsch.* **1982**, *37B*, 1218.
(30) Mayr, A.; McDermott, G. A.; Dorries, A. M. *Organometallics* **1985**, *4*, 608.
(31) Guggenberger, L. J.; Schrock, R. R. *J. Am. Chem. Soc.* **1975**, *97*, 2935.
(32) Wengrovius, J. H.; Sancho, J.; Schrock, R. R. *J. Am. Chem. Soc.* **1981**, *103*, 3932.
(33) Wood, C. D.; McLain, S. J.; Schrock, R. R. *J. Am. Chem. Soc.* **1979**, *101*, 3210.
(34) Chisholm, M. H.; Hoffman, D. M.; Huffman, J. C. *J. Am. Chem. Soc.* **1984**, *106*, 6806.
(35) Listemann, M. L.; Schrock, R. R. *Organometallics* **1985**, *4*, 74.
(36) Lalor, F. J.; Desmond, T. J.; Cotter, G. M.; Shanahan, C. A.; Ferguson, G.; Parvez, M.; Ruhl, B. *J. Chem. Soc., Dalton Trans.* **1995**, 1709.
(37) Desmond, T. J.; Lalor, F. J.; Ferguson, G.; Parvez, M. *J. Chem. Soc., Chem. Commun.* **1983**, 457.
(38) McDermott, G. A.; Dorries, A. M.; Mayr, A. *Organometallics* **1987**, *6*, 925.
(39) Green, M.; Howard, J. A. K.; James, A. P.; Jelfs, A. N.; De, M.; Nunn, C. M.; Stone, F. G. A. *J. Chem. Soc., Chem. Commun.* **1984**, 1623.
(40) Green, M.; Howard, J. A. K.; James, A. P.; Nunn, C. M.; Stone, F. G. A. *J. Chem. Soc., Dalton Trans.* **1986**, 187.
(41) Bermudez, M. D.; Delgado, E.; Elliott, G. P.; Ngoc Hoa Tran, H.; Mayor-Real, F.; Stone, F. G. A.; Winter, M. J. *J. Chem. Soc., Dalton Trans.* **1987**, 1235.
(42) Davis, J. H. Jr.; Lukehart, C. M.; Sacksteder, L. *Organometallics* **1987**, *6*, 50.
(43) Jeffery, J. C.; Stone, F. G. A.; Williams, G. K. *Polyhedron* **1991**, *10*, 215.
(44) Wadepohl, H.; Arnold, U.; Pritzkow, H.; Calhorda, M. J.; Veiros, L. F. *J. Organomet. Chem.* **1999**, *587*, 233.
(45) Dossett, S. J.; Hill, A. F.; Jeffery, J. C.; Marken, F.; Sherwood, P.; Stone, F. G. A. *J. Chem. Soc., Dalton Trans.* **1988**, 2453.
(46) Bermudez, M. D.; Brown, F. P. E.; Stone, F. G. A. *J. Chem. Soc., Dalton Trans.* **1988**, 1139.
(47) Cho, J.-J.; Park, J. T. *Bull. Korean Chem. Soc.* **1995**, *16*, 1130.
(48) Cook, D. J.; Hill, A. F. *Organometallics* **2003**, *22*, 3502.
(49) Hill, A. F.; Malget, J. M.; White, A. J. P.; Williams, D. J. *Eur. J. Inorg. Chem.* **2004**, 818.
(50) Hill, A. F.; Malget, J. M.; White, A. J. P.; Williams, D. J. *J. Chem. Soc., Chem. Commun.* **1996**, 721.
(51) Hill, A. F.; Howard, J. A. K.; Spaniol, T. P.; Stone, F. G. A.; Szameitat, J. *Angew. Chem.* **1989**, *101*, 213.
(52) Hart, I. J.; Hill, A. F.; Stone, F. G. A. *J. Chem. Soc., Dalton Trans.* **1989**, 2261.
(53) Fernandez, J. R.; Stone, F. G. A. *J. Chem. Soc., Dalton Trans.* **1988**, 3035.
(54) Anderson, S.; Hill, A. F. *J. Chem. Soc., Dalton Trans.* **1993**, 587.
(55) Davies, S. J.; Hill, A. F.; Pilotti, M. U.; Stone, F. G. A. *Polyhedron* **1989**, *8*, 2265.
(56) Blosch, L. L.; Abboud, K.; Boncella, J. M. *J. Am. Chem. Soc.* **1991**, *113*, 7066.

(57) Blosch, L. L.; Gamble, A. S.; Boncella, J. M. *J. Mol. Catal.* **1992**, *76*, 229.
(58) Jeffery, J. C.; McCleverty, J. A.; Mortimer, M. D.; Ward, M. D. *Polyhedron* **1994**, *13*, 353.
(59) Doufou, P.; Abboud, K. A.; Boncella, J. M. *Inorg. Chim. Acta* **2003**, *345*, 103.
(60) Vaughan, W. M.; Abboud, K. A.; Boncella, J. M. *Organometallics* **1995**, *14*, 1567.
(61) Schwenzer, B.; Schleu, J.; Burzlaff, N.; Karl, C.; Fischer, H. *J. Organomet. Chem.* **2002**, *641*, 134.
(62) Brower, D. C.; Stoll, M.; Templeton, J. L. *Organometallics* **1989**, *8*, 2786.
(63) Vaughan, W. M.; Abboud, K. A.; Boncella, J. M. *J. Organomet. Chem.* **1995**, *485*, 37.
(64) Garrett, K. E.; Sheridan, J. B.; Pourreau, D. B.; Feng, W. C.; Geoffroy, G. L.; Staley, D. L.; Rheingold, A. L. *J. Am. Chem. Soc.* **1989**, *111*, 8383.
(65) Stone, K. C.; Jamison, G. M.; White, P. S.; Templeton, J. L. T. *Organometallics* **2003**, *22*, 3083.
(66) Frohnapfel, D. S.; White, P. S.; Templeton, J. L. *Organometallics* **2000**, *19*, 1497.
(67) Woodworth, B. E.; Frohnapfel, D. S.; White, P. S.; Templeton, J. L. *Organometallics* **1998**, *17*, 1655.
(68) Enriquez, A. E.; White, P. S.; Templeton, J. L. *J. Am. Chem. Soc.* **2001**, *123*, 4992.
(69) Chaona, S.; Lalor, F. J.; Ferguson, G.; Hunt, M. M. *J. Chem. Soc., Chem. Commun.* **1988**, 1606.
(70) Jamison, G. M.; White, P. S.; Templeton, J. L. *Organometallics* **1991**, *10*, 1954.
(71) Etienne, M.; White, P. S.; Templeton, J. L. *J. Am. Chem. Soc.* **1991**, *113*, 2324.
(72) Bruce, A. E.; Gamble, A. S.; Tonker, T. L.; Templeton, J. L. *Organometallics* **1987**, *6*, 1350.
(73) Desmond, T. J.; Lalor, F. J.; Ferguson, G.; Parvez, M. *J. Chem. Soc., Chem. Commun.* **1984**, 75.
(74) Kim, H. P.; Kim, S.; Jacobson, R. A.; Angelici, R. J. *Organometallics* **1984**, *3*, 1124.
(75) Kim, H. P.; Angelici, R. J. *Organometallics* **1986**, *5*, 2489.
(76) Kim, H. P.; Kim, S.; Jacobson, R. A.; Angelici, R. J. *Organometallics* **1986**, *5*, 2481.
(77) Doyle, R. A.; Angelici, R. J. *J. Organomet. Chem.* **1989**, *375*, 73.
(78) Stone, K. C.; Jamison, G. M.; White, P. S.; Templeton, J. L. *Inorg. Chim. Acta* **2002**, *330*, 161.
(79) Stone, K. C.; White, P. S.; Templeton, J. L. *J. Organomet. Chem.* **2003**, *684*, 13.
(80) Kreissl, F. R. (Ed.). Transition Metal Carbyne Complexes, *Proceedings of the NATO Advanced Workshop on Transition Metal Carbyne Complexes, Wildbad, Kreuth, Germany, 27 September–2 October 1992. NATO ASI Ser., Ser. C*, **1993**, *392*.
(81) Filippou, A. C.; Wanninger, K.; Mehnert, C. *J. Organomet. Chem.* **1993**, *461*, 99.
(82) Filippou, A. C.; Wanninger, K.; Mehnert, C. *J. Organomet. Chem.* **1994**, *465*, C1.
(83) Gamble, A. S.; White, P. S.; Templeton, J. L. *Organometallics* **1991**, *10*, 693.
(84) Filippou, A. C.; Wagner, C.; Fischer, E. O.; Voelkl, C. *J. Organomet. Chem.* **1992**, *438*, C15.
(85) Filippou, A. C.; Hofmann, P.; Kiprof, P.; Schmidt, H. R.; Wagner, C. *J. Organomet. Chem.* **1993**, *459*, 233.
(86) Adams, R. D. *Inorg. Chem.* **1976**, *15*, 169.
(87) Weber, L.; Dembeck, G.; Boese, R.; Blaser, D. *Chem. Ber. Recl.* **1997**, *130*, 1305.
(88) Weber, L.; Dembeck, G.; Boese, R.; Blaeser, D. *Organometallics* **1999**, *18*, 4603.
(89) Jamison, G. M.; White, P. S.; Harris, D. L.; Templeton, J. L. *NATO ASI Ser., Ser. C* **1993**, *392*, 201.
(90) Ivin, K. J. *Olefin Metathesis*, Academic Press, London, **1983**.
(91) LaPointe, A. M.; Schrock, R. R. *Organometallics* **1995**, *14*, 1875.
(92) LaPointe, A. M.; Schrock, R. R. *Organometallics* **1993**, *12*, 3379.
(93) LaPointe, A. M.; Schrock, R. R.; Davis, W. M. *J. Am. Chem. Soc.* **1995**, *117*, 4802.
(94) Xue, W.-M.; Wang, Y.; Chan, M. C.-W.; Su, Z.-M.; Cheung, K.-K.; Che, C.-M. *Organometallics* **1998**, *17*, 1946.
(95) Xue, W.-M.; Chan, M. C. W.; Mak, T. C. W.; Che, C.-M. *Inorg. Chem.* **1997**, *36*, 6437.
(96) Manna, J.; Gilbert, T. M.; Dallinger, R. F.; Geib, S. J.; Hopkins, M. D. *J. Am. Chem. Soc.* **1992**, *114*, 5870.
(97) Friedman, L. A.; Harman, W. D. *J. Am. Chem. Soc.* **2001**, *123*, 8967.
(98) Beach, N. J.; Williamson, A. E.; Spivak, G. J. *J. Organomet. Chem.* **2005**, *690*, 4640.
(99) Pavlik, S.; Mereiter, K.; Puchberger, M.; Kirchner, K. *J. Organomet. Chem.* **2005**, *690*, 5497.
(100) Byers, P. K.; Stone, F. G. A. *J. Chem. Soc., Dalton Trans.* **1990**, 3499.
(101) Doyle, R. A.; Angelici, R. J. *Organometallics* **1989**, *8*, 2207.
(102) Hazra, D.; Sinha-Mahapatra, D. K.; Puranik, V. G.; Sarkar, A. *J. Organomet. Chem.* **2003**, *671*, 52.

(103) Fischer, E. O.; Lindner, T. L.; Kreissl, F. R. *J. Organomet. Chem.* **1976**, *112*, C27.
(104) Foreman, M. R. S. J.; Hill, A. F.; White, A. J. P.; Williams, D. J. *Organometallics* **2003**, *22*, 3831.
(105) Caldwell, L. M. Ph.D. Thesis, Research School of Chemistry, Australian National University, **2006**.
(106) Fischer, E. O.; Lindner, T. L.; Huttner, G.; Friedrich, P.; Kreissl, F. R.; Besenhard, J. O. *Chem. Ber.* **1977**, *110*, 3397.
(107) Anderson, S.; Cook, D. J.; Hill, A. F.; Malget, J. M.; White, A. J. P.; Williams, D. J. *Organometallics* **2004**, *23*, 2552.
(108) Desmond, T. J.; Lalor, F. J.; Ferguson, G.; Parvez, M.; Wieckowski, T. *Acta Crystallogr. C* **1990**, *C46*, 59.
(109) Weber, L.; Dembeck, G.; Loenneke, P.; Stammler, H.-G.; Neumann, B. *Organometallics* **2001**, *20*, 2288.
(110) Weber, L.; Dembeck, G.; Stammler, H.-G.; Neumann, B.; Schmidtmann, M.; Mueller, A. *Organometallics* **1998**, *17*, 5254.
(111) Kirtley, S. W. In: Wilkinson, G.; Stone, F. G. A.; Abel, E. W. (Eds.), *Comprehensive Organometallic Chemistry*, Pergamon Press, Oxford, **1982**; Vol. 3.
(112) Klaui, W.; Hamers, H. *J. Organomet. Chem.* **1988**, *345*, 287.
(113) Green, M.; Howard, J. A. K.; James, A. P.; Nunn, C. M.; Stone, F. G. A. *J. Chem. Soc., Dalton Trans.* **1987**, 61.
(114) Lee, F.-W.; Chan, M. C.-W.; Cheung, K.-K.; Che, C.-M. *J. Organomet. Chem.* **1998**, *552*, 255.
(115) Lee, F.-W.; Chan, M. C.-W.; Cheung, K.-K.; Che, C.-M. *J. Organomet. Chem.* **1998**, *563*, 191.
(116) Byers, P. K.; Carr, N.; Stone, F. G. A. *J. Chem. Soc., Dalton Trans.* **1990**, 3701.
(117) Byers, P. K.; Stone, F. G. A. *J. Chem. Soc., Dalton Trans.* **1991**, 93.
(118) Crennell, S. J.; Devore, D. D.; Henderson, S. J. B.; Howard, J. A. K.; Stone, F. G. A. *J. Chem. Soc., Dalton Trans.* **1989**, 1363.
(119) Jeffery, J. C.; Weller, A. S. *J. Organomet. Chem.* **1997**, *548*, 195.
(120) Jamison, G. M.; Bruce, A. E.; White, P. S.; Templeton, J. L. *J. Am. Chem. Soc.* **1991**, *113*, 5057.
(121) Enriquez, A. E.; Templeton, J. L. *Organometallics* **2002**, *21*, 852.
(122) Caldwell, L. M.; Hill, A. F.; Willis, A. C. *Chem. Commun.* **2005**, 2615.
(123) Trofimenko, S. *Inorg. Chem.* **1969**, *8*, 2675.
(124) Lichtenberger, D. L.; Hubbard, J. L. *Inorg. Chem.* **1984**, *23*, 2718.
(125) Trofimenko, S. *J. Am. Chem. Soc.* **1969**, *91*, 588.
(126) Desmond, T. J.; Lalor, F. J.; Ferguson, G.; Ruhl, B.; Parvez, M. *J. Chem. Soc., Chem. Commun.* **1983**, 55.
(127) Schoenberg, A. R.; Anderson, W. P. *Inorg. Chem.* **1972**, *11*, 85.
(128) Kreissl, F. R.; Friedrich, P.; Huttner, G. *Angew. Chem.* **1977**, *89*, 110.
(129) Uedelhoven, W.; Eberl, K.; Kreissl, F. R. *Chem. Ber.* **1979**, *112*, 3376.
(130) Mayr, A.; Ahn, S. *Inorg. Chim. Acta* **2000**, *300–302*, 406.
(131) Filippou, A. C.; Fischer, E. O. *J. Organomet. Chem.* **1988**, *349*, 367.
(132) Filippou, A. C.; Fischer, E. O. *J. Organomet. Chem.* **1988**, *341*, C35.
(133) Filippou, A. C.; Fischer, E. O.; Gruenleitner, W. *J. Organomet. Chem.* **1990**, *386*, 333.
(134) Ahn, S.; Mayr, A. *J. Am. Chem. Soc.* **1996**, *118*, 7408.
(135) Hill, A. F. *J. Mol. Catal.* **1991**, *65*, 85.
(136) Hill, A. F.; Malget, J. M.; White, A. J. P.; Williams, D. J. *Inorg. Chem.* **1998**, *37*, 598.
(137) Baxter, I.; Hill, A. F.; Malget, J. M.; White, A. J. P.; Williams, D. J. *Chem. Commun.* **1997**, 2049.
(138) Hill, A. F.; Malget, J. M. *Chem. Commun.* **1996**, 1177.
(139) Sheridan, J. B.; Pourreau, D. B.; Geoffroy, G. L.; Rheingold, A. L. *Organometallics* **1988**, *7*, 289.
(140) Barratt, D.; Davies, S. J.; Elliott, G. P.; Howard, J. A. K.; Lewis, D. B.; Stone, F. G. A. *J. Organomet. Chem.* **1987**, *325*, 185.
(141) Carriedo, G. A.; Elliott, G. P.; Howard, J. A. K.; Lewis, D. B.; Stone, F. G. A. *J. Chem. Soc., Chem. Commun.* **1984**, 1585.
(142) Wadepohl, H.; Arnold, U.; Pritzkow, H. *Angew. Chem., Int. Ed. Engl.* **1997**, *36*, 974.
(143) Wadepohl, H.; Arnold, U.; Kohl, U.; Pritzkow, H.; Wolf, A. *J. Chem. Soc., Dalton Trans.* **2000**, 3554.

(144) Wadepohl, H.; Elliott, G. P.; Pritzkow, H.; Stone, F. G. A.; Wolf, A. *J. Organomet. Chem.* **1994**, *482*, 243.
(145) Torraca, K. E.; Ghiviriga, I.; McElwee-White, L. *Organometallics* **1999**, *18*, 2262.
(146) Doetz, K. H.; Erben, H. G.; Staudacher, W.; Harms, K.; Mueller, G.; Riede, J. *J. Organomet. Chem.* **1988**, *355*, 177.
(147) Garber, S. B.; Kingsbury, J. S.; Gray, B. L.; Hoveyda, A. H. *J. Am. Chem. Soc.* **2000**, *122*, 8168.
(148) Stone, K. C.; Onwuzurike, A.; White, P. S.; Templeton, J. L. *Organometallics* **2004**, *23*, 4255.
(149) Gill, D. S.; Green, M.; Marsden, K.; Moore, I.; Orpen, A. G.; Stone, F. G. A.; Williams, I. D.; Woodward, P. *J. Chem. Soc., Dalton Trans.* **1984**, 1343.
(150) Hill, A. F.; Malget, J. M. *J. Chem. Soc., Dalton Trans.* **1997**, 2003.
(151) Hughes, A. K.; Malget, J. M.; Goeta, A. E. *J. Chem. Soc., Dalton Trans.* **2001**, 1927.
(152) Ellis, D. D.; Farmer, J. M.; Malget, J. M.; Mullica, D. F.; Stone, F. G. A. *Organometallics* **1998**, *17*, 5540.
(153) Cook, D. J.; Hill, A. F. *Chem. Commun.* **1997**, 955.
(154) Ostermeier, J.; Schuett, W.; Stegmair, C. M.; Ullrich, N.; Kreissl, F. R. *J. Organomet. Chem.* **1994**, *464*, 77.
(155) Kreissl, F. R.; Ostermeier, J.; Ogric, C. *Chem. Ber.* **1995**, *128*, 289.
(156) Lehotkay, T.; Wurst, K.; Jaitner, P.; Kreissl, F. R. *J. Organomet. Chem.* **1996**, *523*, 105.
(157) Lalor, F. J.; O'Neill, S. A. *J. Organomet. Chem.* **2003**, *684*, 249.
(158) Woodworth, B. E.; White, P. S.; Templeton, J. L. *J. Am. Chem. Soc.* **1997**, *119*, 828.
(159) Woodworth, B. E.; Templeton, J. L. *J. Am. Chem. Soc.* **1996**, *118*, 7418.
(160) Frohnapfel, D. S.; Woodworth, B. E.; Thorp, H. H.; Templeton, J. L. *J. Phys. Chem. A* **1998**, *102*, 5665.
(161) Woodworth, B. E.; White, P. S.; Templeton, J. L. *J. Am. Chem. Soc.* **1998**, *120*, 9028.
(162) Dewhurst, R. D.; Hill, A. F.; Willis, A. C. *Organometallics* **2004**, *23*, 5903.
(163) Dewhurst, R. D.; Hill, A. F.; Smith, M. K. *Angew. Chem., Int. Ed. Engl.* **2004**, *43*, 476.
(164) Dewhurst, R. D.; Hill, A. F.; Willis, A. C. *Organometallics* **2004**, *23*, 1646.
(165) Dewhurst, R. D.; Hill, A. F.; Smith, M. K. *Organometallics* **2005**, *24*, 5576.
(166) Dewhurst, R. D.; Hill, A. F.; Willis, A. C. *Chem. Commun.* **2004**, 2826.
(167) Dewhurst, R. D.; Hill, A. F.; Willis, A. C. *Organometallics* **2005**, *24*, 3043.
(168) Dewhurst, R. D.; Hill, A. F.; Rae, A. D.; Willis, A. C. *Organometallics* **2005**, *24*, 4703.
(169) Weber, L.; Dembeck, G.; Stammler, H. G.; Neumann, B. *Eur. J. Inorg. Chem.* **1998**, 579.
(170) Abernethy, R. J.; Hill, A. F.; Neumann, H.; Willis, A. C. *Inorg. Chim. Acta* **2005**, *358*, 1605.
(171) Stone, F. G. A. *Angew. Chem.* **1984**, *96*, 85.
(172) Stone, F. G. A. *Adv. Organomet. Chem.* **1990**, *31*, 53.
(173) Brew, S. A.; Stone, F. G. A. *Adv. Organomet. Chem.* **1993**, *35*, 135.
(174) Becke, S. H. F.; Bermudez, M. D.; Ngoc Hoa Tran, H.; Howard, J. A. K.; Johnson, O.; Stone, F. G. A. *J. Chem. Soc., Dalton Trans.* **1987**, 1229.
(175) Doyle, R. A.; Angelici, R. J.; Stone, F. G. A. *J. Organomet. Chem.* **1989**, *378*, 81.
(176) Bermudez, M. D.; Stone, F. G. A. *J. Organomet. Chem.* **1988**, *347*, 115.
(177) Carriedo, G. A.; Howard, J. A. K.; Stone, F. G. A.; Went, M. J. *J. Chem. Soc., Dalton Trans.* **1984**, 2545.
(178) Carriedo, G. A.; Howard, J. A. K.; Marsden, K.; Stone, F. G. A.; Woodward, P. *J. Chem. Soc., Dalton Trans.* **1984**, 1589.
(179) Mueller-Gliemann, M.; Hoskins, S. V.; Orpen, A. G.; Ratermann, A. L.; Stone, F. G. A. *Polyhedron* **1986**, *5*, 791.
(180) Anderson, S.; Cook, D. J.; Hill, A. F. *Organometallics* **2001**, *20*, 2468.
(181) Bailey, W. I.; Chisholm, M. H. Jr.; Cotton, F. A.; Rankel, L. A. *J. Am. Chem. Soc.* **1978**, *100*, 5764.
(182) Green, M.; Porter, S. J.; Stone, F. G. A. *J. Chem. Soc., Dalton Trans.* **1983**, 513.
(183) Green, M.; Howard, J. A. K.; James, A. P.; Jelfs, A. N. de M.; Nunn, C. M.; Stone, F. G. A. *J. Chem. Soc., Dalton Trans.* **1986**, 1697.
(184) Hoskins, S. V.; James, A. P.; Jeffery, J. C.; Stone, F. G. A. *J. Chem. Soc., Dalton Trans.* **1986**, 1709.
(185) Busetto, L.; Jeffery, J. C.; Mills, R. M.; Stone, F. G. A.; Went, M. J.; Woodward, P. *J. Chem. Soc., Dalton Trans.* **1983**, 101.
(186) Stone, F. G. A.; Williams, M. L. *J. Chem. Soc., Dalton Trans.* **1988**, 2467.

(187) Dewhurst, R. D.; Hill, A. F.; Smith, M. K. *Organometallics* **2005**, *24*, 6295.
(188) Hill, A. F.; Honig, H. D.; Stone, F. G. A. *J. Chem. Soc., Dalton Trans.* **1988**, 3031.
(189) Herrmann, W. A. (Ed.). *Synthetic Methods of Organometallic and Inorganic Chemistry*, Georg Thieme Verlag, New York, **1997**; Vol. 7, p. 189.
(190) (a) Delgado, E.; Hein, J.; Jeffery, J. C.; Ratermann, A. L.; Stone, F. G. A. *J. Organomet. Chem.* **1986**, *307*, C23. (b) Delgado, E.; Hein, J.; Jeffery, J. C.; Ratermann, A. L.; Stone, F. G. A.; Farrugia, L. J. *J. Chem. Soc., Dalton Trans.* **1987**, 1191.

Chemistry Surrounding Group 7 Complexes that Possess Poly(pyrazolyl)borate Ligands

MARTY LAIL, KARL A. PITTARD AND T. BRENT GUNNOE*

Department of Chemistry, North Carolina State University, Raleigh, NC 27695-8204, USA

I INTRODUCTION

Since the initial report of poly(pyrazolyl)borates in 1966,[1] this class of ligand has received substantial attention and has been utilized in a diverse range of applications.[2–17] The ability to synthesize dihydridobis(pyrazolyl)-, hydridotris(pyrazolyl)- and tetra(pyrazolyl)borates in combination with the flexibility of substitution at the 3, 4 and 5 positions of the pyrazolyl rings affords access to ligands with a range of properties including a variety of steric and electronic environments. This flexibility has resulted in the preparation of over 170 varieties of poly(pyrazolyl)borates.[18] In addition, the pyrazolyl rings provide potential mimics of histidine *N*-donors and have allowed poly(pyrazolyl)borate complexes to serve as models for enzyme sites. As a result of these attractive features, poly(pyrazolyl)borate ligands have supported a diversity of complexes including coordination to all elements of groups 1 through 13 as well as some lanthanides and actinides.

Herein, we present a comprehensive review that is focused on a summary of chemistry observed with Group 7 (Mn, Tc and Re) poly(pyrazolyl)borate complexes. The redox flexibility of these transition metals and the variability of poly(pyrazolyl) borate ligands have resulted in a wide range of chemistry. Oxidation states range from M(I) (d^6) to M(VII) (d^0) systems. Reactivity of these complexes ranges from highly electron-deficient and strongly oxidizing complexes (e.g., Re and Mn oxo

*Corresponding author. Tel.: +1 (919)-513-3704; Fax: +1 (919)-515-8909.
E-mail: brent_gunnoe@ncsu.edu (T.B. Gunnoe).

ADVANCES IN ORGANOMETALLIC CHEMISTRY
VOLUME 56 ISSN 0065-3055/DOI 10.1016/S0065-3055(07)56002-8

complexes) to the other extreme of electron-rich and strongly π-basic systems that can be utilized for the activation of aromatic compounds. Although portions of this chemistry have been reviewed,[19] in an effort to provide a comprehensive single source for chemistry of poly(pyrazolyl)borate Group 7 complexes, we have included topics that have been previously covered in more focused reviews. Unless otherwise noted, all tris(pyrazolyl)- and tetra(pyrazolyl)borate ligands are tridentate while bis (pyrazolyl)borates are bidentate. Throughout this chapter, the following abbreviations are incorporated: Bp = dihydrobis(pyrazolyl)borate; Bp* = dihydrobis(3,5-dimethyl-pyrazolyl)borate; Tp = hydridotris(pyrazolyl)borate; Tp* = hydridotris(3,5-dimethyl-pyrazolyl)borate; pzTp = tetra(pyrazolyl)borate. All other poly(pyrazolyl)borate abbreviations conform to the standard abbreviation format proposed by Trofimenko.[18] The review has been organized by metal oxidation state including systems with oxidation state $\geq +3$ (high oxidation state) and oxidation states $\leq +2$ (low oxidation state).

II
HIGH OXIDATION STATE COMPLEXES

A. *Rhenium Chemistry*

The parent Tp ligand has been used to support a variety of high valent rhenium systems with reactive oxo ligands including Re(VII) and Re(V) oxidation states. The coordinatively unsaturated Re(III) oxo system {TpRe(=O)}, which likely forms upon photolysis of TpRe(=O)(C_2O_4) (**1**), can be trapped by oxidants (Scheme 1).[20] For example, photolysis of **1** in the presence of O_2 yields a complex with a stoichiometry consistent with the formally Re(V) peroxo complex TpRe(=O)(O_2) (**2**), which undergoes subsequent conversion to the Re(VII) trioxo system TpRe($=O)_3$ (**3**). Isotopic labeling studies reveal that one of the two isotopically labeled oxygen atoms in $^{18}O_2$ is incorporated into the final product **3**. This provides

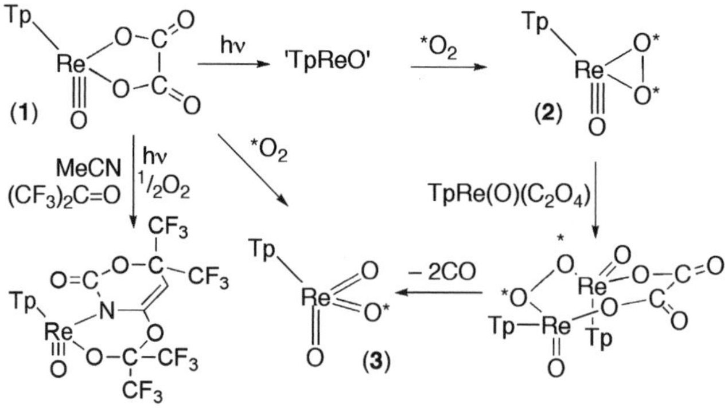

SCHEME 1. Trapping of TpRe(O) by dioxygen (*O=^{18}O) and reaction of complex **1** with perfluoroacetone in the presence of dioxygen.

evidence supporting a binuclear mechanism for the formation of **3** from the reaction of **1** and **2** (Scheme 1). Photolysis of **1** in the presence of hexafluoroacetone and NCMe yields the metallacycle TpRe(=O)($C_9HF_{12}NO_4$) (Scheme 1).[21]

Re(V) oxo complexes of the type TpRe(=O)(X)(Y) (X, Y = monoanionic ligands) form the foundation for a variety of interesting reactivity. TpRe(=O)Cl_2 (**4**) was prepared by reaction of KTp with $KReO_4$ in an acidic ethanol solution.[20] Synthesis of the terminal sulfido analog TpRe(=S)Cl_2 was accomplished independently by Tisato *et al.* upon heating **4** in the presence of B_2S_3.[22] The reduction of **4** by PPh_3 in the presence of pyridine or hex-3-yne yields O=PPh_3 and TpReCl_2(py) (py = pyridine) or TpReCl_2(EtC≡CEt), respectively. The Re(V) oxo systems TpRe(=O)(I)(Cl) (**5**) and TpRe(=O)I_2 (**6**) can be prepared by reaction of NaI with **4**.[23] Photolysis of **5** in the presence of C_6H_6 produces the phenyl complex TpRe(=O)(Cl)(Ph) (**7**),[23,24] and it has been suggested that photolysis leads to Re–I bond homolysis of **5** to produce the Re(IV) intermediate {TpRe(=O)(Cl)}, which can coordinate and activate benzene (Scheme 2). Both intermolecular and intramolecular kinetic isotope effects (KIEs) were observed upon reaction with a mixture of C_6H_6/C_6D_6 and 1,3,5-trideuteriobenzene, respectively. The intermolecular KIE was complicated and found to be dependent on the concentration of Re starting material, light intensity, percent conversion, addition of iodine and the concentration of nitrogen bases such as pyridine, acetonitrile, 2,6-lutidine or trimethylamine. The intramolecular KIE was determined to be 4.0(4) and relatively invariant on the conditions that impact the magnitude of the intermolecular KIE. These results, and other mechanistic data, were analyzed based on the multi-step mechanism shown in Scheme 2 and combination of a secondary KIE (due to coordination of benzene) and a primary KIE in the C–H(D) bond-breaking step. The C–H bond cleavage is proposed to occur by the addition of a C–H bond across the Re=O bond to form the unobserved intermediate TpRe(OH)(Cl)(Ph); hydrogen atom abstraction from the hydroxide ligand would yield the final product **7**. However, a pathway involving C–H oxidative addition to form TpRe(=O)(H)(Ph)(Cl) as an intermediate could not be discounted. This route was utilized for the preparation of the complexes TpRe(=O)(X)(Ar) (X = Cl, Br, I; Ar = 2,5-$Me_2C_6H_3$, 2,6-$Me_2C_6H_3$, *p*-C_6H_4Br,

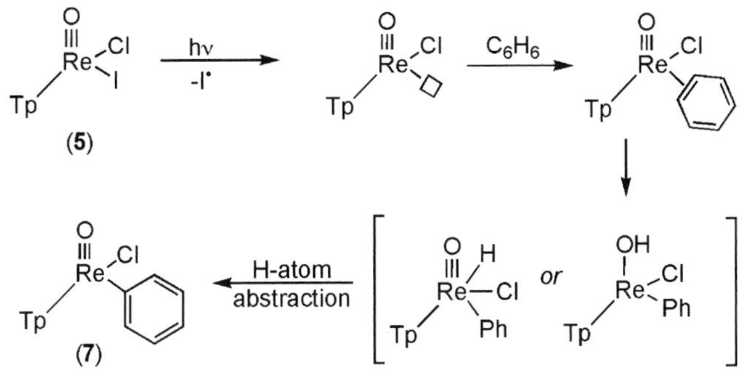

SCHEME 2. Photolysis of TpRe(=O)(I)(Cl) (**5**) in C_6H_6 to produce TpRe(=O)(Cl)(C_6H_5) (**7**).

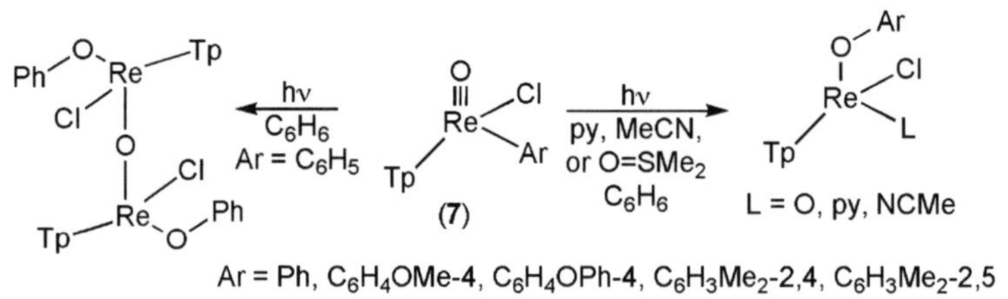

Ar = Ph, C_6H_4OMe-4, C_6H_4OPh-4, $C_6H_3Me_2$-2,4, $C_6H_3Me_2$-2,5

SCHEME 3. Conversion of Aryl-Re(V) oxo complexes to Re(III) and Re(IV) aryloxide complexes.

2,4,6-$Me_3C_6H_2Cl$, C_6H_4F, $C_6H_4(CH_3)$, *p*-$C_6H_4(OCH_3)$, 2,4-$Me_2C_6H_3$, $Cl_2C_6H_3$ and *p*-C_6H_4OPh).

Photolysis of **7** in the presence of DMSO (DMSO = dimethylsulfoxide), py or NCMe yields the phenoxide complex TpRe(OPh)(Cl)(L) (L = O, py or NCMe) (Scheme 3).[23,25] These transformations likely occur *via* an intramolecular [1,2]-shift of the phenyl ligand of **7** to the oxo terminus to produce the phenoxide functionality with subsequent coordination of the Lewis base (Scheme 3). Crossover experiments involving photolysis of TpRe(^{18}O)(C_6H_5)(Cl) and TpRe(^{16}O)(C_6D_5)(Cl) provide evidence for an intramolecular mechanism for O–C bond formation. Radical traps do not influence the reaction suggesting that Re–C bond homolysis is not involved. Analogous reactivity was observed for aryl substrates with functionality to produce *para*-substituted aryloxide ligands including TpRe(OAr)(Cl)(py) (Ar = *p*-C_6H_4OMe, *p*-C_6H_4OPh, 2,4-$Me_2C_6H_3$, 2,5-$Me_2C_6H_3$). Photolysis of **7** in the absence of DMSO, py or NCMe yields two diastereomers of the paramagnetic species $[TpRe(OPh)(Cl)]_2$ (μ-O) (Scheme 3).

Photolytic alkyl migration to a Re-oxo has also been observed. TpRe(OEt)(Cl)(py) was prepared by the photolysis of TpRe(=O)(Cl)(Et) in the presence of pyridine. In contrast to the analogous aryl migration, the formation of the Re(III) ethoxide system is thought to occur through the formation of ethyl radical. For example, photolysis of TpRe(=O)(Cl)(Et) in the presence of PhSH produces TpRe(=O)(Cl)(SPh) and ethane, which is consistent with initial bond homolysis of the Re–C bond of Re–CH_2CH_3.

In the first example of a thermally driven phenyl-to-oxo migration, the reaction of TpRe(=O)(Ph)(OTf) (OTf = trifluoromethanesulfonate) with DMSO at 25 °C produces TpRe(=O)(OPh)(OTf) (**8**) (Scheme 4).[26] A detailed mechanistic study indicates that TpRe(=O)(Ph)(OTf) initially reacts with DMSO to form [TpRe(=O)($OSMe_2$)(Ph)][OTf], which then undergoes oxygen–sulfur bond cleavage to produce the high valent Re(VII) complex [TpRe($=O)_2$(Ph)][OTf] (**9**) and free Me_2S (Scheme 4). The negative entropy of activation {$\Delta S^{\ddagger} = -20.5(25)$ eu} for phenyl-to-oxo migration is inconsistent with a reaction pathway that involves rate-determining Re–Ph bond homolysis. It has been proposed that the electrophilicity of the oxo groups in **9** is a key to the observed phenyl-to-oxo migration with the LUMO (π^*) (LUMO = lowest unoccupied molecular orbital) of the oxo group resembling the LUMO (π^*) of a M–CO fragment. Localization of the Re≡O LUMO onto the oxo ligand presumably increases the propensity for migration of a nucleophile to the oxo terminus. Complexes of the type [TpRe(=O)(R)(L)][OTf]

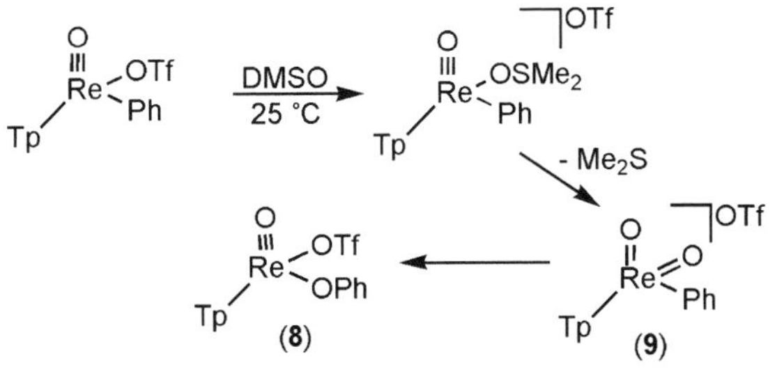

SCHEME 4. Migration of phenyl-to-oxo to yield a phenoxide ligand.

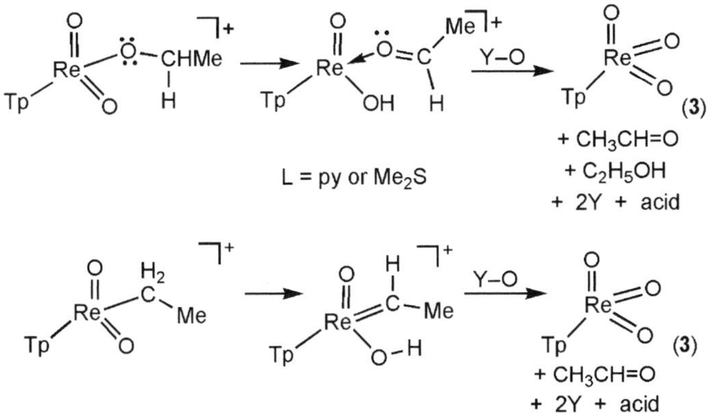

SCHEME 5. Rhenium-mediated oxidation of alkoxide and alkyl ligands.

(R = Ph, OPh; L = py, DMSO, SMe_2), TpRe(=O)(X)(X′) (X = OPh, OTf, I, Ph; X′ = OPh, Ph, $C_6H_3Me_2$-2,4, $C_6H_3Me_2$-3,4) and TpRe(=O)($O_2C_6H_4$) were also reported.

Oxidation of Re(V) alkyl complexes TpRe(=O)(OTf)R (R = Me, Et or *n*-Bu) with either pyridine-*N*-oxide or DMSO produces aldehyde, acid (HOTf or [pyH][OTf]) and pyridine or dimethylsulfide, respectively.[27] Given the observation of phenyl-to-oxo migration discussed immediately above (Scheme 4), it is appealing to consider a similar pathway for alkyl oxidation; however, it was demonstrated that the alkoxide complexes TpRe(=O)(OTf)(OR), which are prepared from TpRe(=O)OR precursors, are not intermediates in the oxidation of the alkyl complexes TpRe(=O)(OTf)R.[28] For example, reactions of TpRe(=O)(OTf)OEt (**10**) with oxygen donors such as DMSO or pyridine-*N*-oxide produce CH_3CHO, C_2H_5OH, HOTf and DMS or py. For the oxidation of TpRe(=O)(OTf)Et (**11**), the production of ethanol is not observed. These observations and additional mechanistic studies have been used to propose the pathways shown in Scheme 5. For TpRe(=O)(OTf)OEt, initial oxidation to the Re(VII) dioxo complex [TpRe($=O)_2$OEt][OTf] is followed by intramolecular *β*-hydride transfer to an oxo ligand for which the electrophilicity of the oxo ligands likely drives the hydride migration (Scheme 5, top pathway). Subsequent dissociation of acetaldehyde (observed product) and oxidation would produce $[TpRe(=O)_2OH]^+$, a strong acid that either protonates pyridine, OTf or the ethoxide ligand of TpRe(=O)(OTf)OEt to produce ethanol. For TpRe(=O)(OTf)R, oxidation to $TpRe(=O)_2R$ is likely followed by an α-hydride migration from the ligand R to an oxo ligand to produce the alkylidene TpRe(=O)(OH)(=CHR′), which in the presence of 2 equivalents of oxidant converts to $TpRe(=O)_3$ and aldehyde. During the course of these studies, a variety of complexes of the type TpRe(=O)(R)X (R = ^{n}Bu, O^{n}Bu, Me, O^{i}Pr; X = Cl or OTf) and [TpRe(=O)(L)(R)][OTf] (L = DMSO, py; R = Et, ^{n}Bu, O^{i}Pr, OEt, O^{n}Bu) were prepared and studied.[27,29] In addition, the related Tp* systems Tp*Re(=O)(OTf)Et and Tp*Re(=O)(OTf)OEt were also synthesized, and reactions that are analogous to those of the parent Tp analogs were observed (see below).[29]

Isolable transition metal complexes containing hydride and terminal oxo ligands are rare; however, Tp*Re(=O)(H)X (X = Cl, H or OTf) and TpRe(=O)(H)Cl have been synthesized, isolated and characterized.[30] Reactions of Tp*Re(=O)(H)OTf (**12**) with unsaturated substrates (e.g., ethene, propene or acetaldehyde) result in insertion of C=C or C=O bonds into the Re–H bond to yield Tp*Re(=O)(R)(OTf) (R = ethyl or propyl) or Tp*Re(=O)(OEt)(OTf) (Scheme 6). Oxidation of **12** with pyridine-*N*-oxide or DMSO produces $Tp^*Re(=O)_3$, acid and free pyridine or dimethylsulfide, respectively. A likely mechanism involves initial oxidation of **12** to produce $[Tp^*Re(=O)_2H][OTf]$ (**13**) followed by the formation of Tp*Re(=O)(OH)(OTf) (**14**) *via* a 1,2-migration of the hydride to an oxo ligand (Scheme 6). Reaction of **14** with a second equivalent of oxidant in the presence of base yields $Tp^*Re(=O)_3$ (**15**). Direct deprotonation of **13** is noted as less likely than the pathway shown in Scheme 6 due to the lack of precedent for acidity of related rhenium hydride systems.

Santos *et al.* have worked toward the preparation and study of Re(V) oxo systems.[31] A series of neutral *trans*-dioxorhenium(V) complexes of the type *trans*-(κ^2-pzTp)$Re(=O)_2L_2$ {L = py, 4-Mepy, DMAP, MeIm or L_2 = dmpe or dppe; DMAP = 4-dimethylaminopyridine; MeIm = 1-methylimidazole; dmpe = 1,2-bis(dimethylphosphino)ethane; dppe = 1,2-bis(diphenylphosphino)ethane} has been prepared by

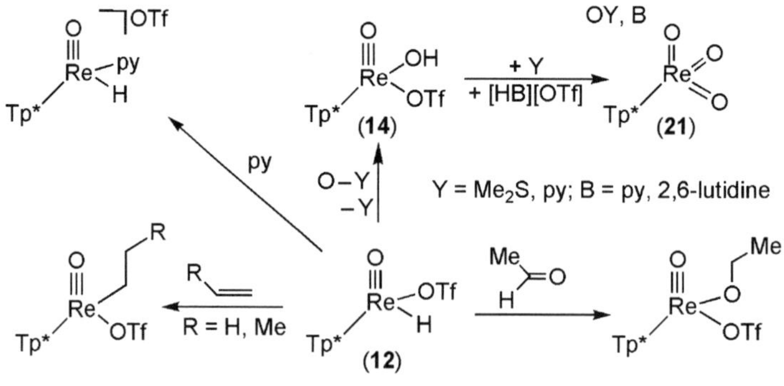

SCHEME 6. Reactivity of Tp*Re(=O)(H)(OTf) (**12**).

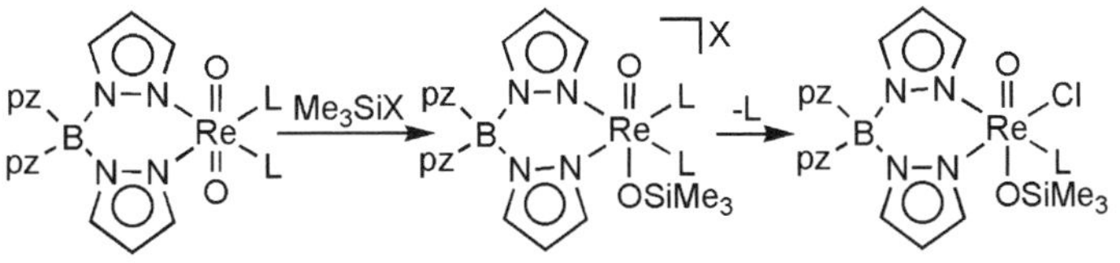

X = Cl or I; L = DMAP, MeIm, py, or L_2 = dmpe

SCHEME 7. Formation of cationic Re(V) complexes containing the κ^2-tetra(pyrazolyl)borate ligand.

reaction of the L/L_2, with (pzTp)Re(=O)$_3$ in the presence of PPh_3.[32] During these studies, *trans*-(H(pz)Bp)Re(=O)(dmpe), Re(=O){(κ^2-*N,O*)(μ-*O*)(Bpz_3)}(pz)(pzH)$_2$ and Re(=O){(κ^2-*N,O*)(μ-*O*)(Bpz_3)}(Cl)(py)$_2$ were also prepared.[32] The preparation of [(κ^2-pzTp)Re(=O)($OSiMe_3$)(DMAP)$_2$][Cl] (**16**), [(κ^2-pzTp)Re(=O)($OSiMe_3$)(MeIm)$_2$][Cl] (**17**), [(κ^2-pzTp)Re(=O)($OSiMe_3$)(dmpe)][Cl] (**18**) and [(κ^2-pzTp)Re(=O)($OSiMe_3$)(py)$_2$][X] (X = Cl or I) (**19**) are accomplished by reactions of *trans*-(κ^2-pzTp)Re(=O)$_2$L$_2$ (L = DMAP, MeIm, 1/2 dmpe or py) with Me_3SiX (Scheme 7).[31] Conversion of **16**, **17** and **19** to the neutral systems (κ^2-pzTp)Re(=O)(Cl) ($OSiMe_3$) (DMAP), (κ^2-pzTp)Re(=O)(Cl)($OSiMe_3$)(py) and (κ^2-pzTp)Re(=O)(Cl)($OSiMe_3$) (MeIm) was observed, respectively, with relative rates depending on the identity of L (MeIm > DMAP > py). This ligand substitution does not occur for complexes that possess bischelating ligands (e.g., dmpe). Reactivity of the Re(V) oxo systems *trans*-(κ^2-pzTp)Re(=O)$_2$L$_2$ with Me_3SiCl potentially provides evidence for nucleophilic character of the oxo ligands. No evidence for chloride/trimethylsiloxy metathesis was noted. This is contrary to reactivity that is observed for [(pzTp)Re(=O)(μ-O)]$_2$ (**20**) and Me_3SiCl to produce (pzTp)Re(=O)Cl_2.[33] Solid-state structures of **18**, [(κ^2-pzTp)Re(=O)($OSiMe_3$)(DMAP)$_2$][Cl] and [(κ^2-pzTp)Re(=O)($OSiMe_3$) (MeIm)$_2$][Cl] were determined by single-crystal X-ray diffraction studies.

Using the steric bulk provided by the Tp* ligand, Mayer and co-workers have accessed Re(VII) systems that are unstable using the parent Tp variant.[29] Rhenium (V) complexes of the type Tp*Re(=O)(X)(X′) (X = OH, Cl, Br, I; X′ = Cl, OTf), Tp*Re(=O)(R)(X) (R = Et, Ph, OMe, OEt, OPh, H; X = Cl, OTf), Tp*Re(=O)(R)(R′) (R = Et, Ph, OMe, H; R′ = Et, Ph, OMe, H), Tp*Re(=O)($O_2C_6H_4$), Tp*Re(=O)($S_2C_6H_4$), Tp*ReH_6 and [Tp*Re(=O)(X)(py)][OTf] (X = F, H, Et) were prepared as potential precursors to Re(VII) complexes. As anticipated, the stability of these complexes is enhanced relative to the analogous parent Tp systems. For example, Tp*Re alkyl and aryl complexes are stable indefinitely in air while the TpRe analogs decompose quickly upon exposure to air. The redox potentials of Tp*Re complexes are only 110–180 mV more negative than that of the TpRe derivatives, which indicates little electronic difference between the Tp and Tp* systems. Thus, steric effects have been proposed as the source of increased stability of the Tp* complexes.

Oxidation of the Re(V) systems Tp*Re(=O)(X)OTf to form Tp*Re(VII) systems of the type [Tp*Re(=O)$_2$X]$^+$ yields reactive species (Scheme 8). When X is a halide, it is proposed that the Re(VII) cations [Tp*Re(=O)$_2$X]$^+$ are reduced to

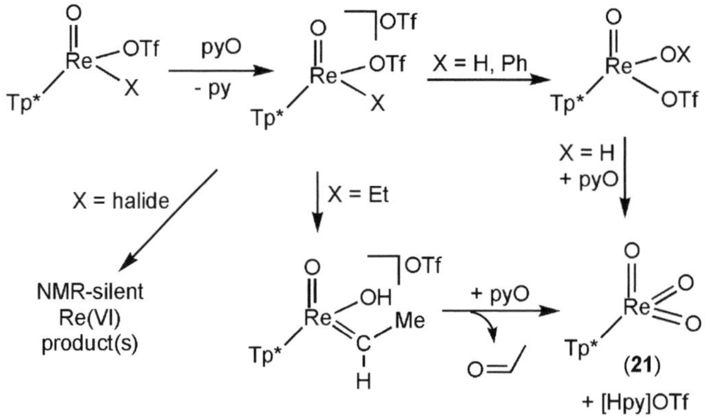

SCHEME 8. Reactivity of [Tp*Re(=O)$_2$X][OTf] (X = H, Ph, halide or alkyl).

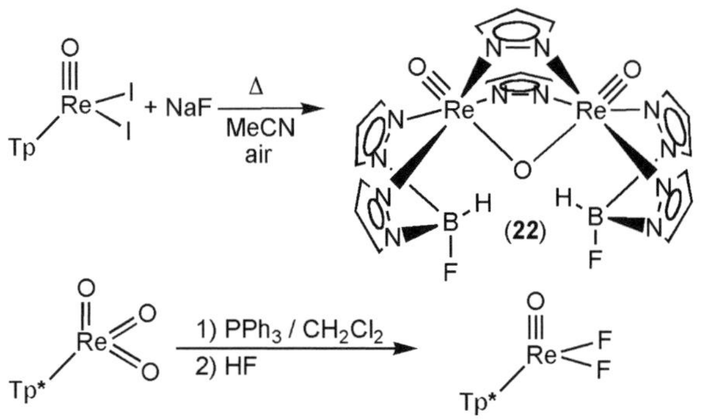

SCHEME 9. Formation of {[H(F)Bp]Re(=O)}$_2$(μ-pz)$_2$(μ-O) (**22**) and preparation of Tp*Re(=O)F$_2$.

yield paramagnetic Re(VI) (d^1) products that are NMR silent. The redox potential for [Tp*Re(=O)$_2$Cl]$^+$ of 0.93 V (vs. ferrocene) is consistent with the Re(VII) systems serving as relatively strong oxidants.[34] Proposed Re(VII) intermediates [Tp*Re(=O)$_2$R][OTf] (R = H or Ph) yield Tp*Re(=O)(OR)OTf *via* net 1,2-migration of the ligand R to an oxo ligand. When R is an ethyl group, an α-hydrogen migrates to the oxo group to form [Tp*Re(=O)(OH)(=CHMe)][OTf] as an intermediate, which subsequently converts to Tp*Re(=O)$_3$ (**21**) and acetaldehyde in the presence of additional oxidant (see above).

In order to obtain analogs of the unobservable Re(VII) intermediates [Tp*Re(=O)$_2$X][OTf] (X = halide), which are proposed to rapidly transform to the paramagnetic d^1 species, complexes of the type Tp^{x}Re(=O)X$_2$ (X = I, Cl, F or OTf; Tp^x = Tp or Tp*) were synthesized as precursors.[35] When TpRe(=O)I$_2$ is refluxed with excess NaF in acetonitrile under aerobic conditions the pyrazolyl/oxo-bridged dimer {[H(F)Bp]Re(=O)}$_2$(μ-pz)$_2$(μ-O) (**22**) is formed (Scheme 9). Complex **22** is not formed in the absence of air, and it is likely that the strong B–F bond provides a substantial driving force for its formation.

In an effort to extend radiopharmaceuticals from ^{99}Tc to ^{186}Re analogs, (pzTp)Re(=O)$_3$ (**23**) and (pzTp)Re(=O)Cl$_2$ (**24**) were synthesized.[36] The preparation of **23** is performed by reaction of Re$_2$O$_7$ with [K][pzTp], and reduction of **23** by PPh$_3$ in the presence of Me$_3$SiCl yields **24**. The conversion of **23** to **24** is reversible upon reaction of **24** with DMSO.[37] The additional pyrazolyl moiety affords increased water solubility for complexes **23** and **24** relative to Tp analogs. The reduction of **23** with PPh$_3$ in the absence of Me$_3$SiCl yields the dimeric species **20**.[33] Using **20** as a starting material, **23**, **24**, (pzTp)Re(=O)(OCH$_2$CH$_2$O) (**25**), (pzTp)Re(=O)(C$_6$H$_4$O$_2$), (pzTp)Re(=O)(OR)$_2$ (R=Me, Et, ^{i}Pr or Ph), (pzTp)Re(=O)(SPh)$_2$ and (pzTp)Re(=O)(η^2-OCONPh) were formed by reaction with DMSO, Me$_3$SiCl, ethylene glycol, pyrocathecol, alcohols, PhSH and phenyl isocyanate, respectively (Scheme 10). Complex **25** can also be prepared by reaction of **24** with HO(CH$_2$)$_2$OH. X-ray diffraction studies were carried out on (pzTp)Re(=O)(OMe)$_2$ (**26**),[38] (pzTp)Re(=O)(OPh)$_{233}$ and (pzTp)Re(=O)(SPh)$_2$.[33]

Re(V) oxo-amido complexes of the type TpRe(=O)(NRR′)Cl are formed upon reaction of **4** with several primary or secondary amines (NRR′ = NHEt, NH^{n}Pr, NH^{i}Pr, NH$_2$, NEt$_2$, NHPh, NH-p-tolyl or piperidyl) (Scheme 11).[39] The Re complexes with secondary amido ligands are more robust and less reactive than the

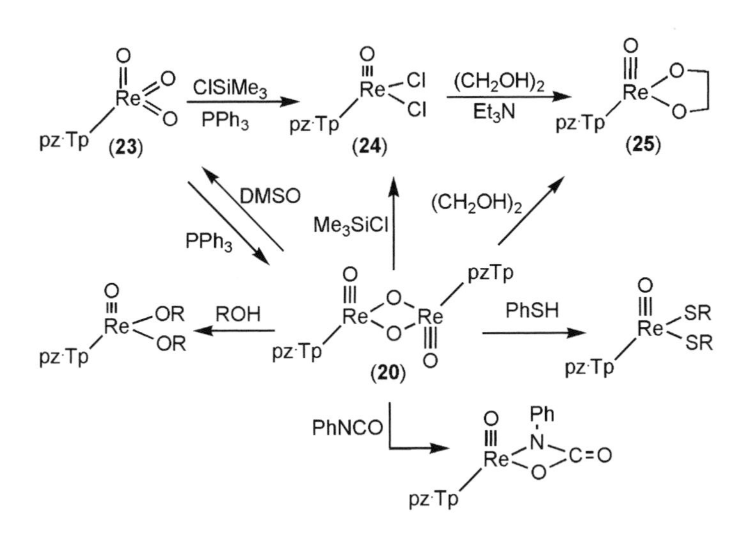

Scheme 10. Preparation and reactivity of [(pzTp)Re(=O)(μ-O)]$_2$ (**20**).

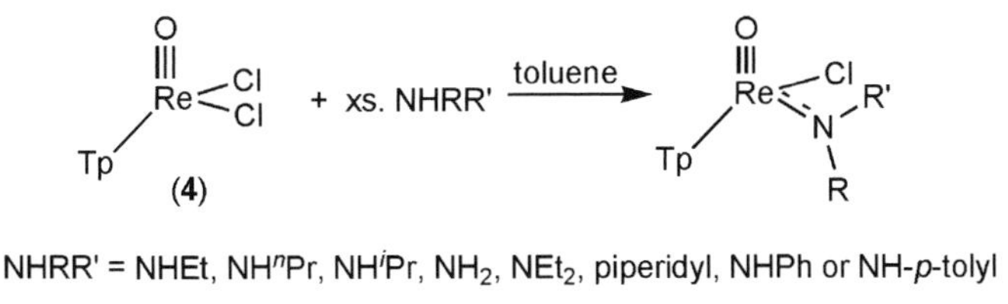

Tp (4) + xs. NHRR' —toluene→

NHRR' = NHEt, NH^{n}Pr, NH^{i}Pr, NH$_2$, NEt$_2$, piperidyl, NHPh or NH-p-tolyl

Scheme 11. Reactivity of TpRe(=O)Cl$_2$ (**4**) with amines to produce Re(V) amido complexes.

systems with primary amido ligands, the latter typically decomposing within approximately 1 h when removed from inert conditions. For all TpRe(=O)(NRR′)Cl complexes, rotation about the Re–N_{amido} bond is slow at temperatures up to 85 °C. In addition, at elevated temperatures no changes are observed in the isomer distribution, indicating that either K_{eq} is temperature independent or that the amido rotation is slow on the chemical timescale. Some Re–N multiple bond character can be construed from the solid-state structure of TpRe(=O)(NEt_2)Cl. The Re–N bond distance is 1.861(19) Å, while Re–N bond lengths for the d^6 octahedral systems Re(NHPh)(N_2)(PMe_3)$_4$ and Re(NHPh)(C_4H_6)(PMe_3)$_3$ are 2.200(14) and 2.13(4) Å, respectively.[40] Reactions of HCl with TpRe(=O)(NRR′)Cl (NRR′ = NHEt, piperidyl) produce **4** and [H_2NRR′][Cl].

Mayer *et al.* have extended their studies of high valent Re-oxo systems to isoelectronic imido analogs (Scheme 12).[39,41] Transition metal complexes containing imido ligands are of interest due to their possible role in olefin aziridinations,[42–46] metal-mediated metathesis of multiple bonds,[47–54] ammoxidation reactions,[55] C–H activation sequences[56–62] as well as catalytic hydroamination of C–C multiple bonds.[63–67] TpRe(=NTol)X_2 (X = Cl or I) complexes are synthesized by reaction of the oxo analogs TpRe(=O)X_2 with *p*-toluidine, and Re(V) complexes of the type TpRe(=NTol)R_2, TpRe(=NTol)(R)(X) and TpRe(=NTol)X_2 (R = Me, Et, Ph, *i*-Pr, *n*-Bu, OEt or SPh; X = Cl and I, OTs or OTf; Tol = *p*-tolyl) are prepared from TpRe(=NTol)X_2 (X = Cl or I). Reactions of TpRe(= A)(R)OTf (A = NTol or O) with Lewis bases result in ligand exchange with OTf to yield [TpRe(= A)(R)(L)][OTf] (L = Lewis base; see Scheme 2). For example, substitutions of OTf or OTs with pyridine for TpRe(= NTol)(Ph)(X) (X = OTf or OTs) and Tp*Re(=O)(Ph)(OTf) systems have been studied in detail.[22] Three mechanisms were proposed for the substitution of OTf/OTs with Lewis bases including: (1) dissociative exchange *via* initial loss of OTf, (2) initial dissociation of a Tp arm and (3) associative ligand exchange *via* a seven-coordinate intermediate. The relative rates of OTf/Lewis base ligand substitution are: Tp-oxo >> Tp-imido ≈ Tp*-oxo, and the rates of ligand exchange are inversely related to Re–O_{OTf} bond distances (Tp-oxo, 2.014 Å; Tp-imido, 2.077 Å; Tp*Re-oxo, 2.119 Å) as well as being inversely dependent on solvent dielectric constant. These data are inconsistent with a dissociative ligand exchange that would produce the five-coordinate ion pairs [TpRe(= A)(R)][OTf] in a rate-determining step. Access to the intermediate κ^2-TpRe(= A)(R)OTf *via* dissociation of a Tp arm has been observed upon reaction of TpRe(= NTol)(Ph)(OTf) with 1,10-phenanthroline (phen) to yield the salt

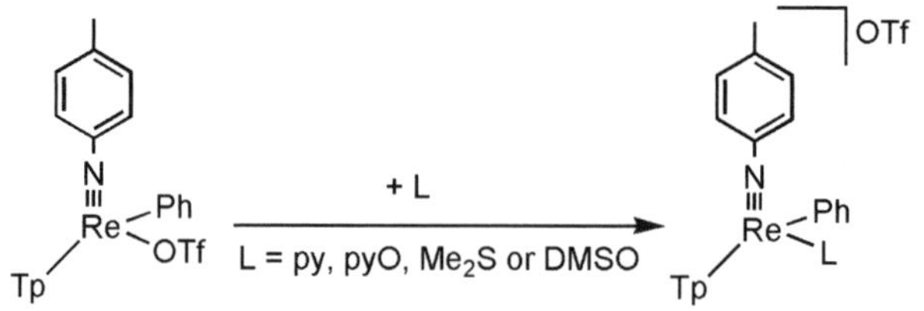

SCHEME 12. Ligand substitution reactions of TpRe(=NTol)(Ph)(OTf).

[κ^2-TpRe(= NTol)(Ph)(phen)][OTf]. Thus, the OTf/OTs–pyridine ligand exchange reactions could occur by dissociation of a Tp arm, pseudo rotation of the unsaturated five-coordinate intermediate to place the open coordinate site *cis* to the NTol/oxo ligand and coordination of the Lewis base and dissociation of OTf/OTs. However, slow degenerate exchange of NCMe of $[TpRe(=NTol)(Ph)(NCMe)]^+$ and the principle of microscopic reversibility have been used to argue against this pathway. It was concluded that the most reasonable ligand substitution pathway involves attack of a Lewis base at the metal center to form a seven-coordinate intermediate through an associative or interchange mechanism. In this scenario, the less facile (kinetically) substitution of OTf with Lewis bases for imido systems, compared with Re-oxo complexes, is likely due to the increased donating ability of imido ligands and the resulting increased energy of Re≡A π^* LUMO (A = NTol or O).

Several Re(V) diolato complexes of the type (pzTp)Re(=O)(diolato) (diolato = $OCHMeCH_2O$, OCHMeCHMeO, $OCMe_2CMe_2O$, $O(CH_2)_3O$ or $OC_6H_{10}O$) were synthesized by two different reaction pathways (Scheme 13).[38] The first pathway involves reaction of the respective diol with [K][pzTp] and Re(=O) $Cl_3(PPh_3)_2$ to produce diolato complexes in 28–54% yield. Separately, synthesis of **25** was reported by this method and confirmed by a single-crystal X-ray diffraction study.[33,68] An alternative method involving reaction of **26** with the respective diol produces the diolato system in 74–85% yield (Scheme 13).[69] Reaction of Re(=O)$Cl_3(PPh_3)_2$ and [K][pzTp] in the absence of diol produces **24**, which subsequently transforms to the paramagnetic Re(IV) species (pzTp)$ReCl_3$, whose identity was confirmed by a solid-state X-ray diffraction study. Thermolysis of **25**

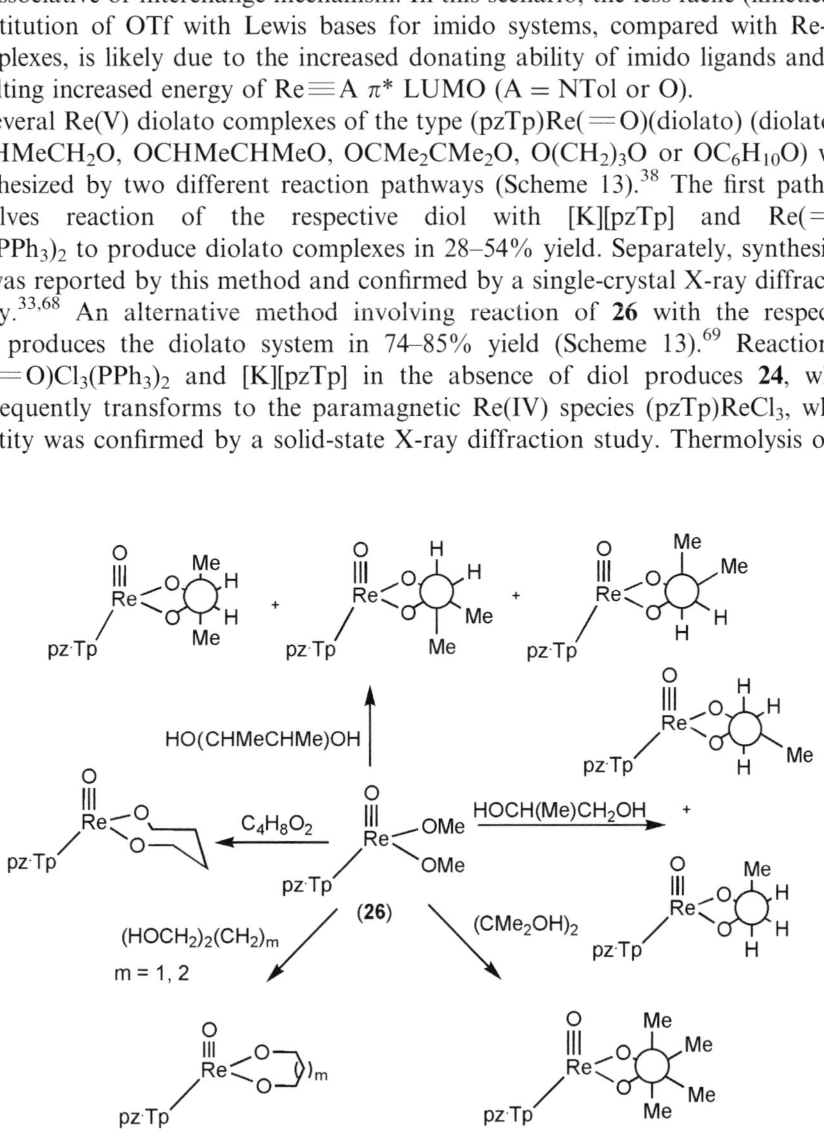

SCHEME 13. Reactivity of (pzTp)Re(=O)$(OMe)_2$ (**26**).

under reduced pressure results in the elimination of ethylene.[69] Analogous eliminations of olefins were not observed for the other diolato systems. Electrochemical experiments reveal that as the length of the diolato carbon chain is increased the Re(VI/V) oxidation potential becomes more negative. Attempts to prepare ^{99}Tc analogs of the Re(V) diolatos were unsuccessful.

The Re(V) alkoxide complexes pzTpRe(=O)(OR)$_2$ (R = Me or Et) are precursors to a variety of new Re complexes (Scheme 14).[70] For example, reactions under basic conditions with substrates that possess acidic sites allow coordination of several anionic substrates. Reaction of **26** with acetylacetone yields a mixture of (κ^2-pzTp)Re(=O)(κ^2-*O,O*-acac)(OMe) and [(κ^2-pzTp)Re(=O)(κ^2-*O,O*-acac)]$_2$ (μ-O) (acac = acetylacetonate). In addition, (κ^2-pzTp)Re(=O)(quinolin-8-olate) (OMe), (κ^2-pzTp)Re(=O)(OMe)(pz′)(pz′H), (pz′ = pz or 3,5-dimethylpyrazole), (pzTp)Re(=O)(HN(CH$_2$)$_2$NH) and (pzTp)Re(=O)(μ-nitrophenyl-*o*-diaminate) can be prepared from reaction of (pzTp)Re(=O)(OR)$_2$, 8-hydroxyquinoline and pz′H with ethylenediamine or *m*-nitrophenyl-*o*-diamine, respectively. Substitution

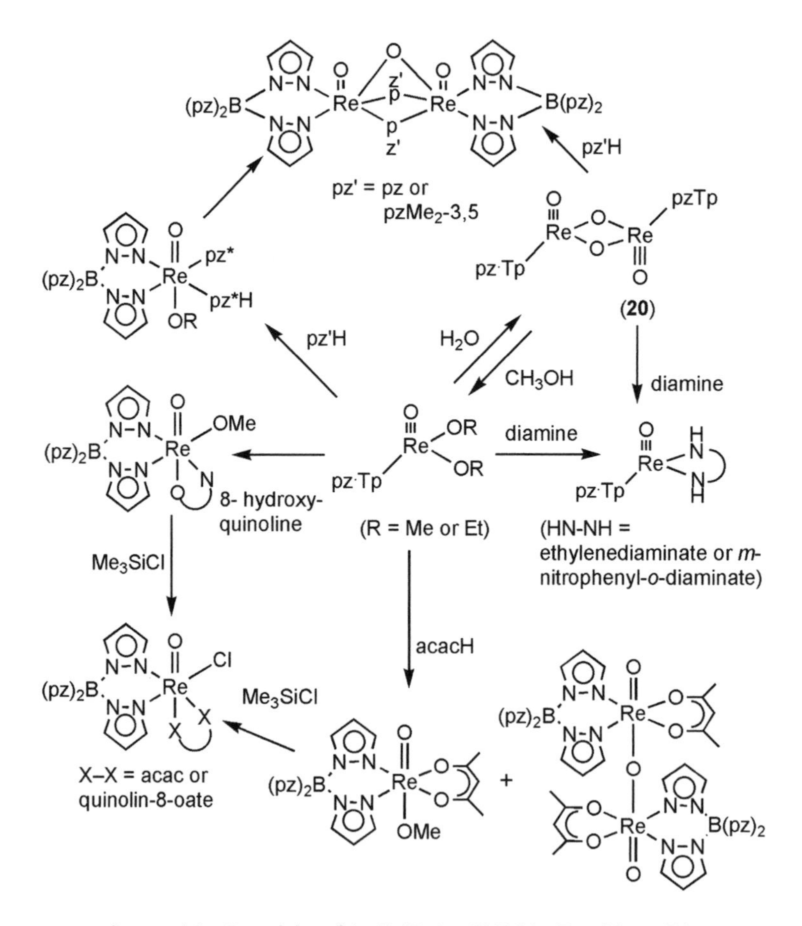

SCHEME 14. Reactivity of (pzTp)Re(=O)(OR)$_2$ (R=Me or Et).

of chloride for methoxide to produce (κ^2-pzTp)Re(=O)(κ^2-*O,O*-acac)Cl and (κ^2-pzTp)Re(=O)(quinolin-8-olate)Cl can be accomplished upon reaction of (κ^2-pzTp)Re(=O)(κ^2-*O,O*-acac)(OMe) or (κ^2-pzTp)Re(=O)(quinolin-8-olate)(OMe), respectively, with Me_3SiCl. The complex (pzTp)Re(=O)$(OMe)_2$ has been reported to reversibly react with water to produce **20** and MeOH. The combination of (pzTp)Re(=O)$(OR)_2$ and pyrazole or 3,5-dimethylpyrazole yields [(κ^2-pzTp) Re(=O)(μ-pz*)]$_2$(μ-O), which also forms upon decomposition of (κ^2-pzTp)Re(=O) (OMe)(pz*)(pz*H). X-ray diffraction studies were carried out on (κ^2-pzTp)Re(=O) (acac)(OMe), [(κ^2-pzTp)Re(=O)(acac)]$_2$(μ-O) and [(κ^2-pzTp)Re(=O)]$_2$(μ-pz)$_2$(μ-O).

At room temperature, the reaction of **26** with HSpy* (Spy* = 2-SC_5H_4N or 2-SC_5H_3N-3-$SiMe_3$) yields (κ^2-pzTp)Re(=O)(OMe)(κ^2-Spy*) (Scheme 15).[71] Upon reaction of (κ^2-pzTp)Re(=O)(OMe)(κ^2-Spy*) with Me_3SiCl, (pzTp) Re(=O)(κ^1-2-$SC_5H_4N\cdot HCl$)Cl (Spy = 2-SC_5H_4N) and **24** (Spy* = 2-SC_5H_3 N-3-$SiMe_3$) are formed. In refluxing toluene, **26** and 2-HS-C_5H_4N are converted to (pzTp)Re(=O)(κ^1-2-SC_5H_4N)$_2$. X-ray diffraction studies were conducted on complexes (pzTp)Re(=O)(OMe)(κ^2-2-SC_5H_3N-3-$SiMe_3$), (pzTp)Re(=O) (κ^1-2-SC_5H_4N)$_2$ and (pzTp)Re(=O)(κ^1-2-$SC_5H_4N\cdot HCl$)Cl.

Herrmann *et al.* have observed elimination of ethene from the Re(V) diolato complex Cp*Re(=O){$O(CH_2)_2O$} to form Cp*ReO_3,[72] and Davison *et al.* observed analogous reactivity with the Tp analog.[37] These studies were extended to complexes of the type Cp*Re(=O)(diolato) with a primary interest in understanding the mechanism of olefin extrusion,[73] and experimental data for the Cp* systems are consistent with a mechanism involving a metallaoxetane intermediate. A Hammett plot derived from a series of Cp*Re phenylethanediolates also points to metallaoxetane intermediates.[74] In order to study the influence of the ancillary ligands (i.e., substitution of Cp* with Tp*) on the cycloreversion, complexes of the type Tp*Re(=O)(diolato) (diolato = OC_2H_4O, $OCH(CH_3)CH(CH_3)O$, *cis*-OCH–(C_6H_{12})–CHO, OCH(Ph)CH_2O, OCH(Ph)CH(Ph)O, *cis*-1,2-norbornanediolato) were prepared.[75,76] Cycloreversion of these complexes yields **21** and free alkene (Scheme 16). For both Tp*Re and Cp*Re complexes, alkyl substitution on the diolato ligand does not significantly impact the rate of cycloreversion. However, the

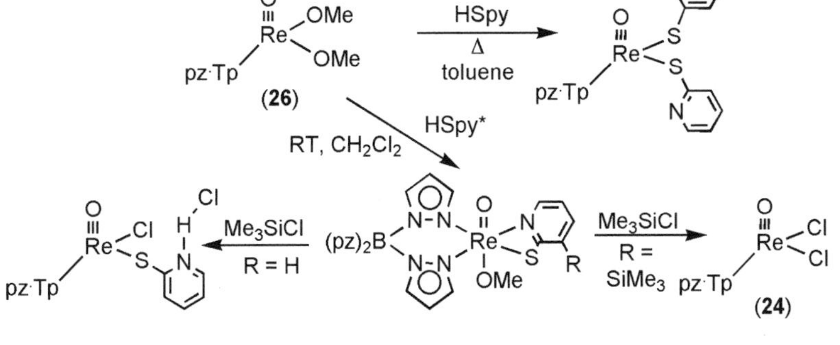

SCHEME 15. Reactivity of (pzTp)Re(=O)$(OMe)_2$ (**26**) with thiols (py = 2-pyridyl; py* = 2-SC_5H_3 N-3-$SiMe_3$).

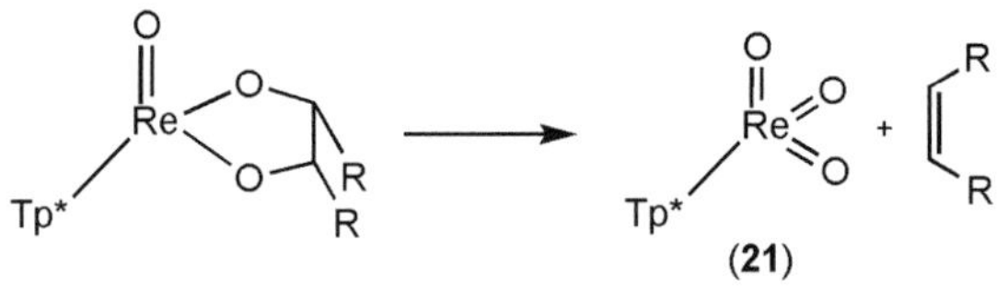

R = H, CH_3, Ph or R-R = *cis*-1,2-norbornanediolato

SCHEME 16. Cycloreversion of Tp*Re(=O)(diolato) complexes.

rate of cycloreversion is enhanced for both systems upon incorporation of a single phenyl group into the diolato backbone, yet no additional rate enhancement is observed with the incorporation of a second phenyl group. Eyring plots for the Tp* and Cp* systems gave similar $\Delta S^{\ddagger}$ values, whereas the $\Delta H^{\ddagger}$ for the Tp* systems was determined to be 2–4 kcal mol^{-1} higher than the corresponding Cp* systems. These similarities are evidence that incorporation of the Tp* ligand, compared to the Cp* ligand, results in relatively minor changes and does not substantially alter the mechanism of cycloreversion.

KIEs were probed for the cycloreversion of Tp*Re(=O){OCH(C_6H_4OMe-4) CH_2O} to **21** and 4-methoxystyrene.[77] Primary KIEs were observed based on $^{12}C/^{13}C$ with $k_{12C}/k_{13C} = 1.041(5)$ and 1.013(6) at the α- and β-positions, respectively. Secondary KIEs based on H/D substitution were observed with $k_H/k_D = 1.076(5)$ and 1.017(5) for the α- and β-positions, respectively. It is proposed that these data and the corresponding density functional theory (DFT) calculations indicate a transition state consistent with a concerted but asynchronous mechanism in which bond cleavage at the α C–O bond is more substantial than the β-position. A Hammett study on several diolato complexes was conducted revealing the dependence on rate of cycloreversions for the production of several substituted styrene products. For electron-donating groups $\rho = -0.65$ and for electron-withdrawing groups $\rho = 1.13$. The Hammett plots provide evidence for a "fluxional" transition state that is susceptible to the influence of the diolato substituents.

Re(V) complexes of the type Tp*Re(=O)($SCHRCH_2O$) (R = H, Me) (both *syn*- and *anti*-conformations are obtained when R = Me) and Tp*Re(=O) ($SCHRCH_2S$) (R = H, Ph) have been synthesized,[78] and the crystal structure of Tp*Re(=O)(SCH_2CH_2S) has been reported. Heating the dithiolate complexes to 120 °C for 1 week results in no observed reaction. Calculations on the energetics of cycloreversion of Tp/Tp*Re(=O){κ^2-X$(CH_2)_2$Y} (X = Y = O; X = O, Y = S; X = Y = S) to produce olefin and Tp/Tp*Re(=O)(X)(Y) indicate that substitution of sulfur for oxygen increases the thermodynamic barrier to cycloreversion. Stepwise incorporation of sulfur increases the thermodynamic barrier by approximately 10 kcal mol^{-1} (diolato < thiodiolato < dithiolato) for each sulfur incorporated into the ligand backbone. This is likely due to the weakened π-bonding of Re=S versus Re=O (i.e., the oxo ligand is a more proficient π-donor) and greater affinity for formation of Re=O bonds over Re=S bonds.

Of potential relevance to cycloreversion reactions of Tp*Re-diolato systems, oxygen atom transfer from epoxides to PPh_3 is catalyzed by **21** to produce

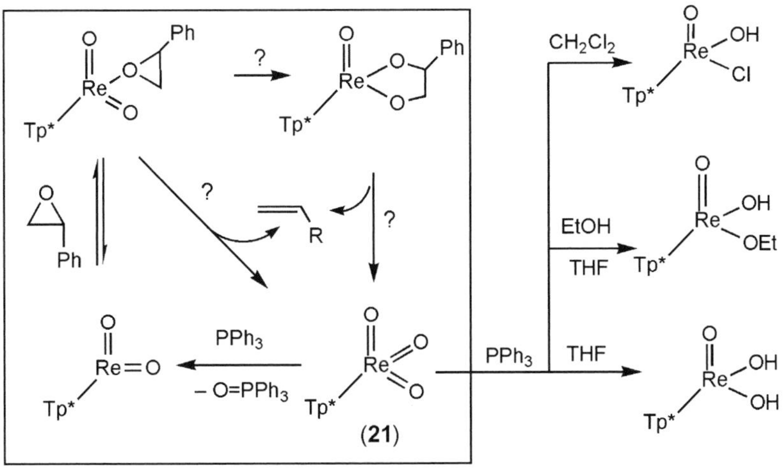

SCHEME 17. Catalytic and stoichiometric reactivity of $Tp^*Re(=O)_3$ (**21**).

$Ph_3P{=}O$ and alkene (Scheme 17).[79] Reduction of $Tp^*Re(=O)_3$ with PPh_3 in THF produces a new complex the identity of which is likely to be $Tp^*Re(=O)(OH)_2$; however, uncertainty of the exact identity of this complex was noted. The reaction of the putative intermediate $\{Tp^*ReO_2\}$ with epoxides likely occurs by one of two pathways. After coordination of the epoxide to $\{Tp^*ReO_2\}$, ring expansion would yield $Tp^*Re(=O)$(diolato). Subsequent cycloreversion would produce **21** and free alkene. Alternatively, a fragmentation pathway would directly produce **21** and alkene from the coordinated epoxide. Due to previously observed slow rates of cycloreversion, a cycloreversion mechanism to yield **21** and alkene is perhaps less likely to be operative for the majority of the epoxides. The reaction of **21** with PPh_3 in chlorinated solvents (methylene chloride) produces $Tp^*Re(=O)(Cl)(OH)$ in a reaction that possibly involves trace HCl in the methylene chloride given that independent reaction of HCl with **21** also produces $Tp^*Re(=O)(Cl)(OH)$. Reaction of **21** with ethanol in the presence of PPh_3 in THF yields $Tp^*Re(=O)(OH)(OEt)$.

To further explore the reactivity of the putative $\{Tp^*ReO_2\}$ complex, epoxides and episulfides were reacted with **21** in the presence of PPh_3.[80] Reaction of *cis*-cyclooctene oxide and **21**/PPh_3 yields a molar ratio of alkene:diolato of 24:1 after 5 days at 75 °C. The alkene:diolato ratio for the analogous reaction with *trans*-cyclooctene oxide is 0.35:1 after 6 days at 75 °C. Thus, *trans*-epoxides show an increased thermodynamic predilection toward formation of the diolato complex compared with *cis*-epoxides. The reaction of *cis*-cyclooctene episulfide with **21** and PPh_3 yields both *syn*- and *anti*-isomers of $Tp^*Re(=O)(cis\text{-}\kappa^2\text{-}S,S'\text{-}SC_8H_{14}S)$ and *cis*-cyclooctene. Reaction of the **21**, PPh_3 and *trans*-cyclooctene episulfide yields $Tp^*Re(=O)(trans\text{-}\kappa^2\text{-}S,S'\text{-}SC_8H_{14}S)$, $Tp^*Re(=O)(trans\text{-}\kappa^2\text{-}S,O\text{-}SC_8H_{14}O)$ and *trans*-cyclooctene. The reaction of complex **21**, PPh_3 with thiirane yields $Tp^*Re(=O)(\kappa^2\text{-}S,S'\text{-}SCH_2CH_2S)$. It has been proposed that the desulfurization of thiirane follows the mechanism outlined in Scheme 18.

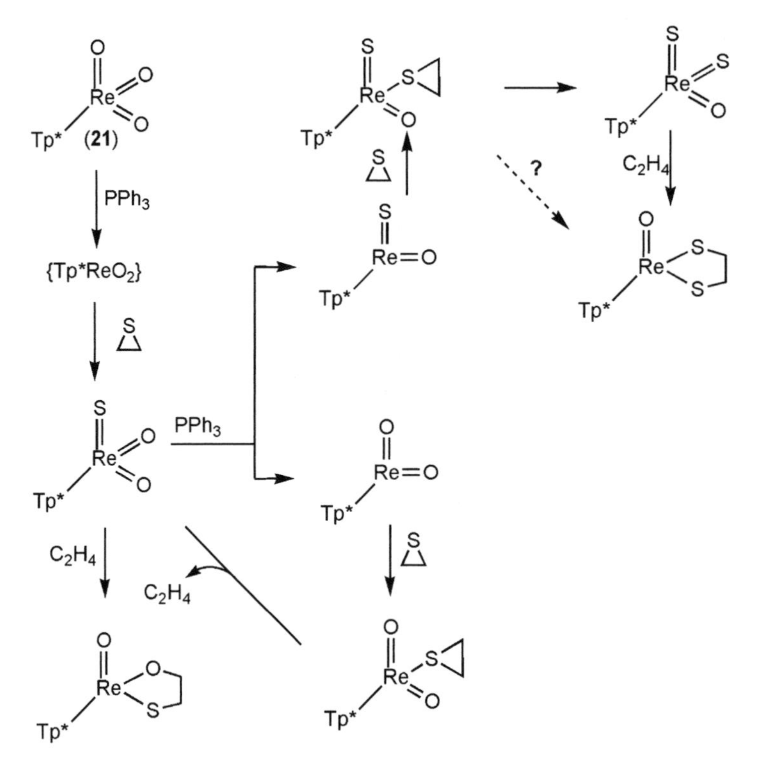

SCHEME 18. Proposed mechanism for episulfide desulfurization.

Complex **4** and TpRe(=O)Br_2 (**27**) have been prepared by reaction of **3** with PPh_3 in the presence of Me_3SiCl or Me_3SiBr (Scheme 19).[81,82] The reduction of **4** requires more forcing conditions relative to the technetium analog, consistent with the fact that reduction of rhenium complexes is typically less facile than corresponding technetium complexes. Replacement of chloride ligands with bromide does not significantly alter the redox properties of TpRe(=O)X_2 (X = Cl or Br).

TpRe(=S)Cl_2, TpRe(=O)$(SEt)_2$ and TpRe(=O)(Cl)(SEt) were synthesized by the reaction of **4** with B_2S_3, 2 equivalents each of HSEt and triethylamine and 1 equivalent of HSEt, respectively (Scheme 19).[22,83] TpRe(=S)Cl_2 is not stable in solution and reverts to TpRe(=O)Cl_2 under aerobic conditions. It was determined that the Re=S bond exhibited greater stability relative to the Tc=S bond, since TpRe(=S)Cl_2 reacts with O_2 to form **4** at a slower rate than the corresponding reaction of the Tc analog.[83]

In an effort to develop catalysts for asymmetric oxidations, TpRe(=O){(1*S*,2*S*)-κ^2-*N*,*N*-HNCHPhCHPhNH}, (S_{Re},S_{C-O},R_{C-N})-TpRe(=O)(κ^2-*N*,*O*-NMeCHMe CHPhO) and (S_{Re},S_C)-TpRe(=O)(κ^2-*N*,*O*-N$(CH_2)_3$CHCO_2) were prepared (Scheme 19).[84] Single-crystal X-ray analyses were performed for all complexes, and deviations from the octahedral paradigm of 90° L–M–L bond angles to smaller angles are proposed to originate from dπ–pπ* interactions of the pyrazolyl rings. The *O*- and *N*-ligands in these complexes, not of Tp origin, possess filled *p*-orbitals

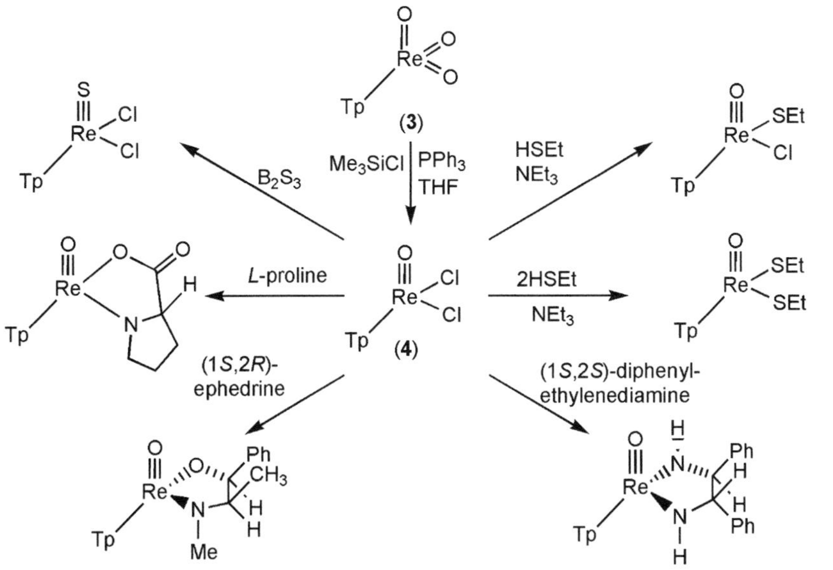

SCHEME 19. Preparation and reactivity of $TpRe(=O)Cl_2$ (**4**).

that possibly lead to distortions from 90° bond angles. Heating (S_{Re},$S_{C–O}$,$R_{C–N}$)-$TpRe(=O)(\kappa^2$-*N*,*O*-NMeCHMeCHPhO) and PMe_3 to 111 °C for 2 days did not result in reduction of the metal complex. The relatively poor oxidizing ability of those systems is likely a result of the donating ability of the chelating ligands.

[K][(κ^2-pzTp)$Re(=O)_3Me$] and **23** have been synthesized by reacting [K][pzTp] with MTO (MTO = $MeReO_3$), or $ClReO_3$ respectively.[85] With excess MTO present, no evidence for the production of [K][{$Re(CH_3)O_3$}$_2$(μ-pzTp)] was obtained.

The dihydridobis(pyrazolyl)borate chelate has been incorporated into several monomeric complexes, which provides a new route for the production of Re(V) oxo species.[86] *Trans,trans*-[$ReO_2(py)_4$][Bp_2ReO_2] and *trans,trans*-[$ReO_2(py)_4$][$Bp^*_2ReO_2$] were prepared by reaction of [$ReO_2(py)_4$)][Cl] with KBp and KBp*, respectively. Treating these products, *trans,trans*-[$ReO_2(py)_4$][Bp_2ReO_2] and *trans,trans*-[ReO_2 $(py)_4$][$Bp^*_2ReO_2$] with Me_3SiCl yields $Bp_2Re(=O)(OSiMe_3)$ and $Bp^*_2Re(=O)$ $(OSiMe_3)$, respectively. The preparation of $Bp_2Re(=O)(OSnMe_3)$ was achieved by reaction of *trans,trans*-[$ReO_2(py)_4$][Bp_2ReO_2] with Me_3SnCl. X-ray diffraction studies were performed on *trans,trans*-[$ReO_2(py)_4$][Bp_2ReO_2], $Bp_2Re(=O)(OSiMe_3)$ and $Bp^*_2Re(=O)(OSiMe_3)$. Though the crystals for $Bp^*_2Re(=O)(OSiMe_3)$ were of poor quality, the structure of the complex is consistent with assigned connectivity with the average Re–N bond distance of 2.113(5) Å consistent with a monomeric bis(pyrazolyl)borate Re-oxo complex. Variable temperature ^1H NMR studies were performed on $Bp_2Re(=O)(OSiMe_3)$ and $Bp_2Re(=O)(OSnMe_3)$. The room temperature ^1H NMR spectra of both complexes reveal broad resonances due to the BH_2 fragments. At −80 °C, the ^1H NMR spectrum of $Bp_2Re(=O)(OSnMe_3)$ reveals line broadening without splitting; however, the ^1H NMR spectrum of $Bp_2Re(=O)(OSiMe_3)$ at −80 °C shows four broad resonances due to the two BH_2

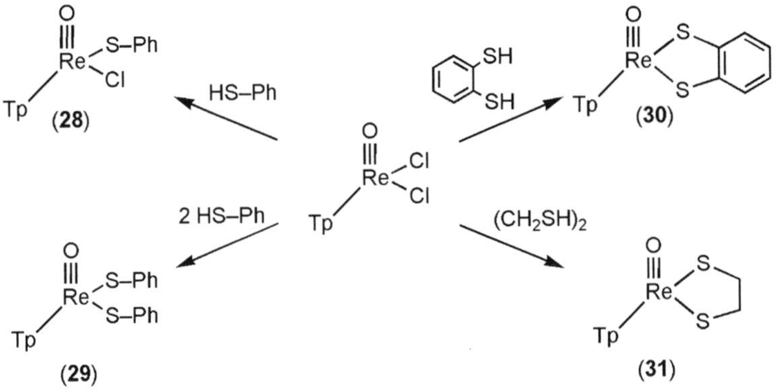

SCHEME 20. Reactivity of $TpRe(=O)Cl_2$ (**4**) with thiols.

moieties as well as splitting of the three pyrazolyl resonances into two peaks of equal intensity for each type of proton. The spectra at low temperature and fluxionality were proposed to arise from inequivalent Bp rings with a rapid "boat–boat" flip.

$TpRe(=O)(Cl)(SPh)$ (**28**), $TpRe(=O)(SPh)_2$ (**29**), $TpRe(=O)(S_2C_6H_4)$ (**30**) and $TpRe(=O)(S_2C_2H_4)$ (**31**) have been prepared upon reaction of **4** with 1 equivalent of HSC_6H_5, 2 equivalents of HSC_6H_5, 1,2-dimercaptobenzene or 1,2-dimercaptoethane, respectively (Scheme 20).[87] Single-crystal X-ray structures have been solved for **28** and **29** revealing Re=O bond distances of 1.689(5) and 1.668(5) Å, respectively, while a single-crystal X-ray diffraction study of **31** revealed a Re=O bond distance of 1.694(9) Å.[22]

The polyhydride complexes $TpReH_6$ and $TpReH_4(PPh_3)$ were prepared by Crabtree *et al.* *via* reaction of $LiAlH_4$ with $TpRe(=O)Cl_2$ and $TpReCl_2(PPh_3)$, followed by hydrolysis.[88] These complexes were proposed to be classical hydrides based on data from ^1H NMR, ^{31}P NMR and IR spectroscopy. For the hydride resonance of $TpReH_6$, it was not possible to observe a T_1 minimum. Although the observed T_1 value at 200 K ($T_1 = 63$ ms) is consistent with a non-classical system that possesses an η^2-H_2 ligand, isotopic substitution experiments suggest a classical polyhydride structure. The T_1 minimum for $TpReH_4(PPh_3)$ is 55 ms at 220 K, and the short T_1 value was attributed to "close non-bonding H···H contacts." Isotopic substitution of H with D provided strong evidence in support of a classical hydride structure for these TpRe hydride complexes.

B. *Manganese Chemistry*

Binuclear Mn complexes have been of interest due to their potential relevance to biological systems (e.g., in photosynthetic enzymes).[89] $[TpMn]_2(\mu\text{-}O)(\mu\text{-}O_2CCH_3)_2$ (**32**), $[TpMn]_2(\mu\text{-}O)(\mu\text{-}O_2CC_2H_5)_2$ (**33**), $[TpMn]_2(\mu\text{-}O)(\mu\text{-}O_2CH_2)_2$ (**34**) and $[TpMn]_2(\mu\text{-}O)(\mu\text{-}O_2CH_3)_2$ have been prepared and studied by cyclic voltammetry as well as ESR, UV–Vis, IR and Raman spectroscopy.[90] Single-crystal solid-state X-ray diffraction studies were performed for **32** · CH_3CN and **32** · $(CH_3CN)_4$. Temperature dependence of molar susceptibility and effective magnetic moment for both **32** · CH_3CN and **32** · $(CH_3CN)_4$ are consistent with weak antiferromagnetic exchange coupling between the two Mn atoms. Similarities in the electronic spectra of **32** and manganese-catalase may support the possibility of a binuclear Mn system in the enzyme.[91] X-band EPR spectroscopy was performed on the binuclear complex $[TpMn(\mu\text{-}O)(O_2CCH_3)]_2$, and spectra are consistent with an $S = 2$ state.[92]

Oxidation of water to dioxygen in biological systems that contain polymanganese sites is of interest, and a lack of detailed understanding of Mn arrangements, oxidation states and the mechanism of oxidation preclude a detailed understanding of the catalytic transformations. Attempts to prepare small molecule models of these systems have been pursued. Reaction of $MnCl_2$ with $KTp^{^iPr_2}$ in excess 3,5-$^iPr_2pz$ yields $Tp^{^iPr_2}MnCl(3,5\text{-}^iPr_2pz)$,[93] and reaction of this complex with NaOH affords $[Tp^{^iPr_2}Mn(\mu\text{-}OH)]_2$ (**35**). A single-crystal X-ray diffraction study of **35** revealed a Mn–Mn separation of 3.31 Å and statistically identical Mn–O bond distances of 2.094(4) and 2.089(5) Å for the bridging hydroxide ligands. Under anaerobic conditions, reaction of **35** with $KMnO_4$ produces $[Tp^{^iPr_2}Mn(\mu\text{-}O)]_2$ (**36**), the structure of which was confirmed by a single-crystal X-ray diffraction study that revealed Mn–O bond distances of 1.79–1.81 Å and a Mn–Mn separation of 2.70 Å. Oxidation of **35** to both **36** and **37** (Scheme 21) occurs in the presence of dioxygen.[94] The solid-state structure of **37** was determined by a single-crystal X-ray diffraction study. Isotopic labeling of the Mn–OH groups of complex **35** suggested that the bridging hydroxide groups of **35** are the source of bridging oxo in **37** and that the bridging hydroxide groups of **35** are the oxygen atom source for bridging oxo atoms of **36**. These results are consistent with the mechanism presented in Scheme 21. Reaction of **35** with O_2 produces $Tp^{^iPr_2}Mn(\mu\text{-}OH)_2(\mu\text{-}O_2)MnTp^{^iPr_2}$ (**38**), which can release H_2O_2 or H_2O to yield **36** or $Tp^{^iPr_2}Mn(\mu\text{-}O)(\mu\text{-}O_2)MnTp^{^iPr_2}$ (**39**), respectively. Conversion of **39** to $[Tp^{^iPr_2}Mn(=O)]_2(\mu\text{-}O)$ (**40**) and ultimately to **37** is proposed.

$Tp^{^iPr_2}Mn(SAr)$ (**41**) ($Ar = C_6H_4NO_2\text{-}4$) is formed upon reaction of **35** with 4-mercaptonitrobenzene.[95] Complex **41** activates dioxygen to produce **36**, **37** and ArS–SAr (Scheme 21). The proposed superoxo intermediate **42** is not observed; however, trapping of **42** with $Tp^{^iPr_2}Mn(OAc)$ yields the μ-acetato-bis(μ-oxo) complex **43**. Thus, the mechanism for activation of dioxygen by **41** is proposed to occur through initial production of **42**, followed by reaction of **42** with a second equivalent of **41** to produce $[Tp^{^iPr_2}Mn(SAr)]_2(\mu\text{-}O_2)$ (**44**). Transformation of **44** to ArS–SAr, **36**, and **37** completes the reaction sequence.

$Tp^{^iPr_2}Mn(O_2CPh)$ (**45**) and $Tp^{^iPr_2}Mn(O_2CPh)(3,5\text{-}^iPr_2pzH)$ (**46**) can be isolated after reaction of sodium benzoate with $Tp^{^iPr_2}MnCl$ or $Tp^{^iPr_2}MnCl(3,5\text{-}^iPr_2pzH)$, respectively (Scheme 22).[96] For complex **45**, analysis using infrared spectroscopy suggests bidentate coordination of the benzoate ligand while a single-crystal X-ray diffraction study of **46** revealed a monodentate benzoate ligand. Due to the interest in the role of superoxide dismutase (SOD) in the catalytic dismutation of superoxide, structural comparisons were drawn between SOD and complexes **45** and **46**. For example, the Mn-SOD carboxylate oxygen from aspartate is coordinated in a monohapto fashion similar to **46**. In addition, for complex **46** hydrogen-bonding between the N–H of the 3,5-$^iPr_2pzH$ and the unbound benzoate oxygen is observed with an H···O distance of 1.04(4) Å. Similarly, hydrogen-bonding also occurs in

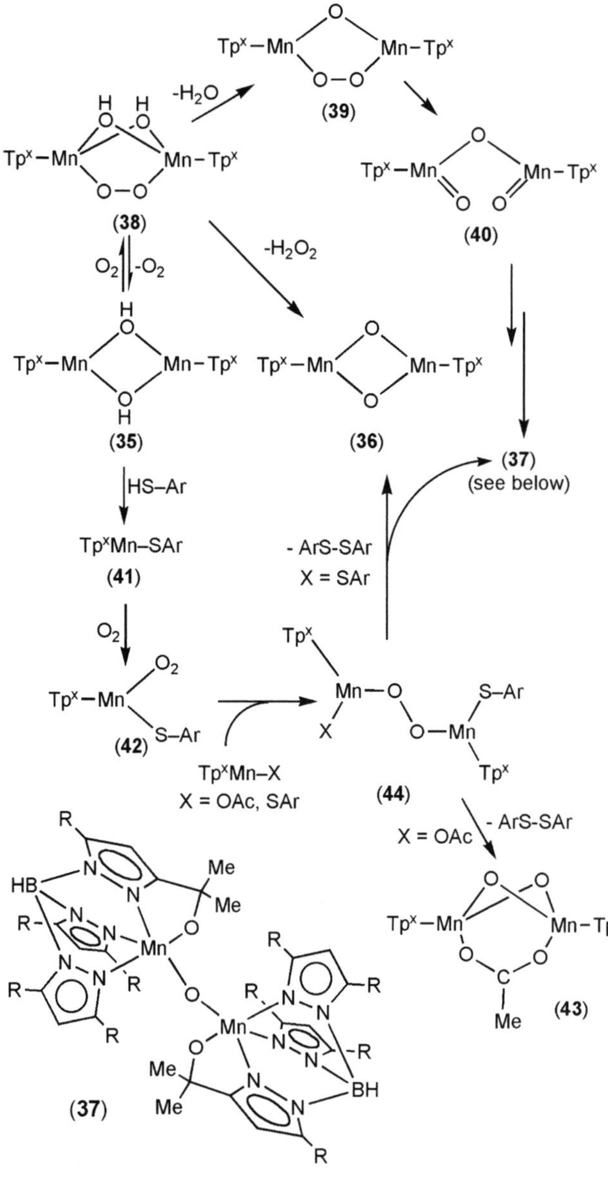

SCHEME 21. Reactivity of $[Tp^{^iPr_2}Mn(\mu\text{-}OH)]_2$ (**35**) in the presence of O_2 ($Tp^x = Tp^{^iPr_2}$, Ar$=C_6H_4NO_2$-4, R $= {}^iPr$).

Mn-SOD between the non-coordinated oxygen of aspartate and two nitrogen atoms of the peptide backbone. Studies of **45** and **46** revealed a higher activity for characteristic SOD reactivity compared to other SOD synthetic mimics. These results were attributed to the architectural similarities that exist between **45**, **46** and Mn-SOD.

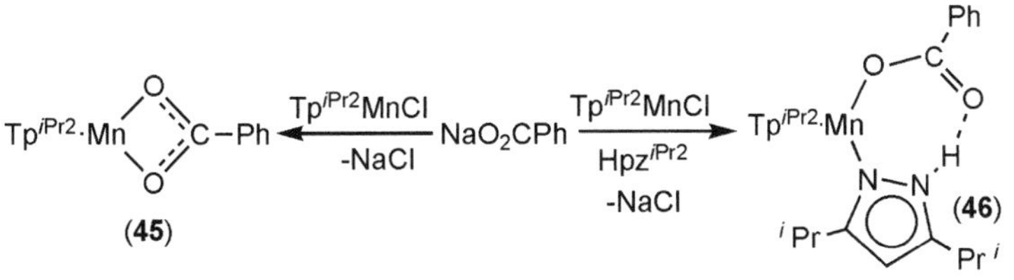

SCHEME 22. Formation of five-coordinate $Tp^{iPr_2}Mn$ systems with κ^2- and κ^1-benzoate ligands.

$$Tp^{x}_2Mn + [NO]PF_6 \longrightarrow [Tp^{x}_2Mn]PF_6 + NO$$

SCHEME 23. Single-electron oxidation of the high-spin Mn(II) complexes Tp^{x}_2Mn to yield the corresponding low-spin Mn(III) cations (Tp^x = Tp, Tp*, pzTp).

The syntheses of $[Tp_2Mn][PF_6]$, $[Tp^*_2Mn][PF_6]$ and $[(pzTp)_2Mn][PF_6]$ have been achieved by the reaction of $[NO][PF_6]$ with Tp_2Mn, Tp^*_2Mn and $pzTp_2Mn$, respectively (Scheme 23).[97] The magnetic moments of the Mn(II) and Mn(III) complexes are consistent with high-spin ($S = 5/2$) and low-spin configurations ($S = 1$), respectively. For the Mn(III) complexes, electronic spectra revealed absorptions between 36,100 and 37,600 cm^{-1}, which were assigned as charge transfer bands. A low-energy absorption that would be consistent with tetragonal distortion if high-spin Mn(III) is absent, which provides additional evidence for the low-spin state of the Mn(III) complexes.

The full characterization of the Mn(IV) complex $[Tp^*_2Mn][ClO_4]_2$, including EPR spectroscopy and electrochemical studies, has been reported.[98] An X-ray diffraction study of the octahedral complex revealed Mn–N bond distances ranging from 1.966(3) to 1.982(3) Å and N–Mn–N bond angles between 89.0(1)° and 90.7(1)°. The d^3 electron configuration was confirmed by magnetic susceptibility studies. The effective magnetic moment was measured to be $1.38\mu_B$. Cyclic voltammetry measurements yielded a reversible Mn^{IV}/Mn^{III} couple ($E_{1/2} = 1.35$ V vs. SCE) and an irreversible Mn^{III}/Mn^{II} couple ($E_{p,c} = 0.02$ V). A g-value of 3.73 was observed in the X-band EPR spectrum of $[Tp^*_2Mn][ClO_4]$ at 77 K.

C. *Technetium Chemistry*

The potential development of technetium for radiopharmaceutical applications prompted the synthesis of the Tc complexes $TpTc(=O)X_2$ (X = Cl or Br).[82,99,100] The complexes $TpTc(=O)X_2$ are obtained upon reaction of $[n\text{-}Bu_4N][Tc(=O)X_4]$ with KTp (Scheme 24). The solid-state structure of $TpTc(=O)Cl_2$ has been reported with a Tc–O bond distance of 1.656 (3) Å.[100] The reduction of $TpTc(=O)Cl_2$ with PPh_3 gives $TpTc(OPPh_3)Cl_2$ (Scheme 24),[82] and in the presence of excess phosphine $TpTc(PPh_3)Cl_2$ is produced. Ligand exchange of pyridine with PPh_3 allows the isolation of $TpTc(py)Cl_2$. More forcing conditions were required for reduction of the Re analog to form $TpRe(PPh_3)Cl_2$.

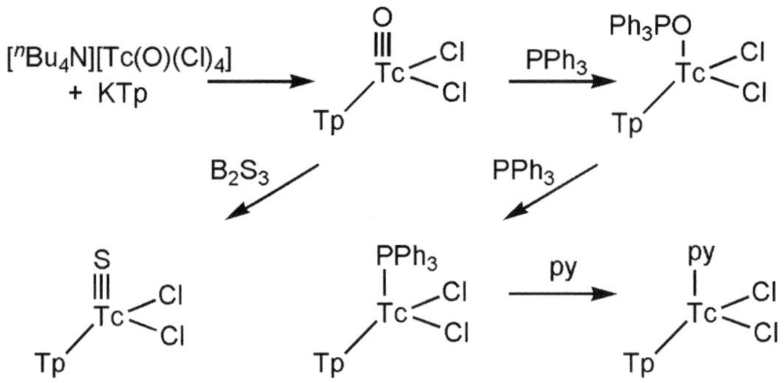

SCHEME 24. Formation and reactivity of $TpTc(=O)Cl_2$.

$TpTc(=O)Cl_2$ reacts with B_2S_3 to give the Tc(V) sulfido complex $TpTc(=S)Cl_2$ (Scheme 24),[22,83] which was the first reported Tc complex containing a terminal sulfur atom. Magnetic susceptibility data are consistent with a diamagnetic Tc(V) complex.

III
LOW OXIDATION STATE COMPLEXES

A. *Rhenium Chemistry*

$ReBr(CO)_5$ reacts with KBp to give $BpRe(CO)_3(Hpz)$. Pyrazole (Hpz) is assumed to be derived from dihydrobis(pyrazolyl)borate by a disproportionation mechanism. $BpRe(CO)_3$(pyrazole) reacts with the phosphites $P(OMe)_3$ and $EtC(CH_2O)_3P$ (4-ethyl-2,6,7-trioxa-1-phosphabicyclo[2,2,2]octane) to form the bisphosphite complexes $BpRe(CO)_2\{P(OMe)_3\}_2$ and $BpRe(CO)_2(EtC(CH_2O)_3P)_2$, respectively, in which the phosphites ligands are oriented *trans*.[101]

Two modified scorpionate complexes have been recently synthesized based on the $\{Re(CO)_3\}$ fragment. KTp^{4py} and KTp^{2py} are synthesized from 3-(4-pyridyl)pyrazole and 3-(2-pyridyl)pyrazole, respectively, and KBH_4.[102] Stirring KTp^{4py} or KTp^{2py} in the presence of $ReCl(CO)_5$ yields the Re(I) systems $Tp^{4py}Re(CO)_3$ or $Tp^{2py}Re(CO)_3$.[102] Solid-state structures of both complexes reveal C_3 molecular symmetry with geometric features similar to $TpRe(CO)_3$. For $Tp^{4py}Re(CO)_3$, the pyridyl substituents are orthogonal to the pyrazole rings. The pyridyl substituents of $Tp^{2py}Re(CO)_3$ are twisted 47.4° from the mean plane of the pyrazole ring.[102]

Facile removal of CO from $TpRe(CO)_3$ renders it a useful entry for the preparation of a diverse array of new Re(I) dicarbonyl systems.[103] For example, $TpRe(CO)_2(THF)$ is synthesized by photolysis of $TpRe(CO)_3$ in THF,[103] and the lability of the THF ligand provides access to complexes of the type $TpRe(CO)_2(L)$ ($L = PPh_3$, MeCN, pyridine or PMe_2Ph; Scheme 25).[103] For $TpRe(CO)_2(THF)$, $Tp^*Re(CO)_2(THF)$ and $Cp^*Re(CO)_2(THF)$, the displacement of THF by NCMe

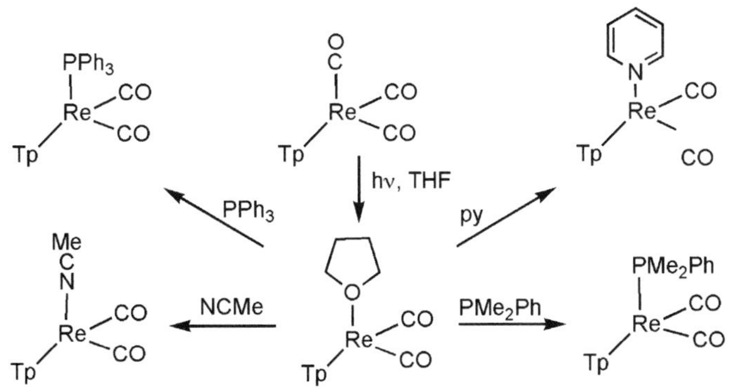

SCHEME 25. Ligand substitution reactions of $TpRe(CO)_2THF$.

was determined to proceed by a dissociative mechanism.[104] The THF/NCMe ligand exchange proceeds slowest for the Tp complex and most rapid for the Cp* complex. Consistent with a dissociative process, the rate of substitution of 2,5-dimethyltetrahydrofuran with acetonitrile for $TpRe(CO)_2$(2,5-dimethyltetrahydrofuran) is approximately 25 times more rapid than the analogous reaction with $TpRe(CO)_2$(THF). Use of $TpRe(CO)_2L$ systems as precursors to π-basic Re(I) complexes is discussed in detail below.

Starting from $[Re(CO)_4Cl]_2$, $Tp^*Re(CO)_3$ and $Re(3,5\text{-}Me_2pz)_2(CO)_3Cl$ can be prepared. The latter complex is proposed to arise from a Re catalyzed decomposition of Tp*.[105] In addition, treatment of $Tp^*Re(CO)_3$ with bromine incorporates bromide at the pyrazolyl 4-position to yield $Tp^{Me,Br,Me}Re(CO)_3$. $Tp^*Re(CO)_2$(THF) has also been reported.[106]

A survey of available IR and CV data has revealed that relative donor properties of Cp, Cp*, Tp and Tp* can vary with metal identity and oxidation state.[107] Analysis of $TpRe(CO)_3$ and $Tp^*Re(CO)_3$ shows that the two complexes are electronically similar.[108] For example, sharp peaks at 2010 and 2020 cm^{-1} are assigned as the symmetric CO stretching absorption for $TpRe(CO)_3$ and $Tp^*Re(CO)_3$, respectively, while broad absorptions at 1900 and 1890 cm^{-1} were assigned as the asymmetric ν_{CO}, respectively. Analysis of ν_{CO} for $CpRe(CO)_3$ and $TpRe(CO)_3$ suggest that, for these systems, the Tp ligand is a more effective electron donor compared to Cp.[108–110]

DFT was used to compare the energetics of oxidative addition of methane to $\{TpRe(CO)_2\}$ and $\{CpRe(CO)_2\}$ fragments to produce the Re(III) systems $TpRe(CO)_2(CH_3)H$ and $CpRe(CO)_2(CH_3)H$.[107] In addition, the tris(azo)borate ligand (Tab) was used to model the full Tp ligand. Interest in these calculations is derived from the observation of oxidative addition of C–H bonds by $\{CpRe(CO)(PMe_3)\}$ while $\{TpRe(CO)L\}$ fragments do not appear to undergo analogous reactions.[107,111–113] Consistent with the experimental observations, calculated enthalpies starting from $TpRe(CO)_2$ and $CpRe(CO)_2$ reveal that the formation of $TpRe(CO)_2(CH_3)H$ is endothermic by 6.4 kcal mol^{-1} while the formation of

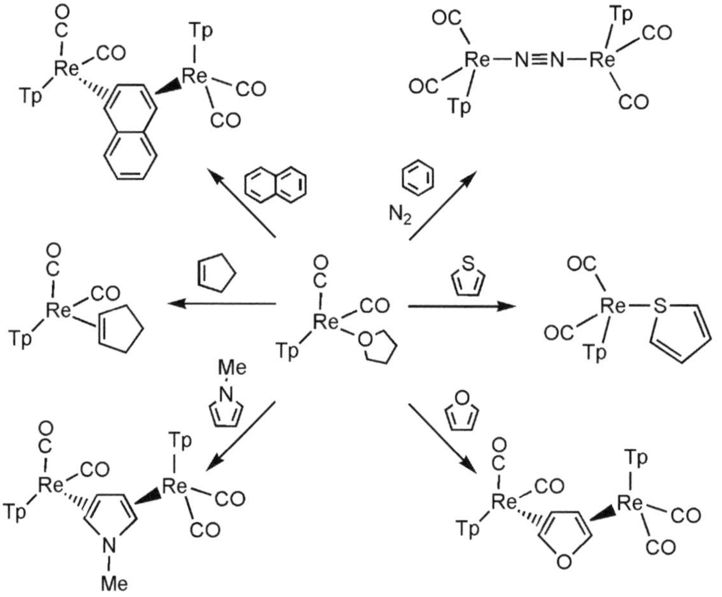

SCHEME 26. Reactions of $TpRe(CO)_2(THF)$ with aromatic substrates, cyclopentene and dinitrogen.

$CpRe(CO)_2(CH_3)H$ is exothermic by 7.9 kcal mol^{-1}. Thus, the calculated difference in energy of the seven-coordinate Re(III) systems is 14.3 kcal mol^{-1}. In contrast, the calculated difference in energy of the six-coordinate methane adducts $TpRe(CO)_2(CH_4)$ and $CpRe(CO)_2(CH_4)$ is only 2.5 kcal mol^{-1} with the CpRe system calculated to be more favorable. The free energies of activation for reductive elimination of CH_4 from $TpRe(CO)_2(CH_3)H$ and $(Tab)Re(CO)_2(CH_3)H$ are calculated to be small (0.1 and 0.6 kcal mol^{-1}, respectively), and neither seven-coordinate complex is expected to be kinetically stable. For reductive elimination of CH_4 from $CpRe(CO)_2(CH_3)H$, a free energy of activation of 6 kcal mol^{-1} was calculated. Differences in the steric profile of Tp versus Cp were suggested to account for the different predilection toward C–H oxidative addition.

In pursuit of transition-metal-based dearomatization agents, Harman *et al.* have utilized $TpRe(CO)_2(THF)$ as a precursor for dihapto-coordination of aromatic compounds (Scheme 26).[114] Reactions of $TpRe(CO)_2(THF)$ with aromatic molecules such as furan, naphthalene and *N*-methylpyrrole result in the formation of η^2-C,C-bound systems; however, isolated products are binuclear Re(I) complexes with $\eta^2:\eta^{2'}$-bridging aromatic ligands. Although isolation of monomeric η^2-aromatic species was not achieved, evidence that they form as intermediates *en route* to the thermally stable binuclear systems was obtained using ^1H NMR spectroscopy for the reaction of $TpRe(CO)_2(THF)$ with furan. These systems are rare examples of relatively stable η^2-aromatic ligands using a d^6 metal center.[19,115–117] The reaction of $TpRe(CO)_2(THF)$ with thiophene gives the *S*-bound complex $TpRe(CO)_2(\sigma\text{-}S\text{-}SC_4H_4)$. The reaction of $TpRe(CO)_2(THF)$ with benzene under an atmosphere of dinitrogen results in the formation of the dinitrogen-bridged complex

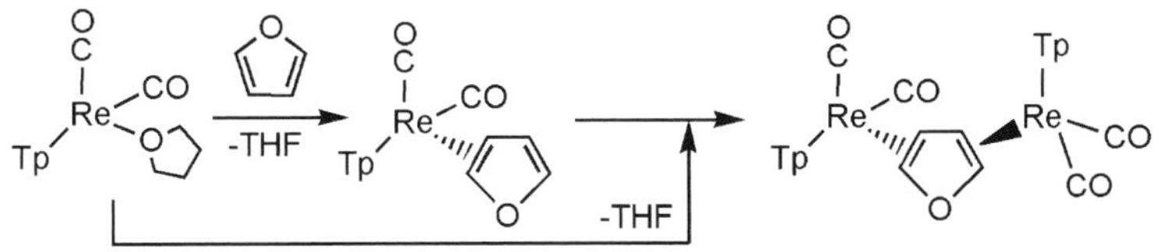

SCHEME 27. Proposed route for formation of binuclear Re complexes bridged by η^2: $\eta^{2'}$-aromatic ligands (furan used as example).

$[TpRe(CO)_2]_2(\mu\text{-}N_2)$, which is characterized by a single-crystal X-ray diffraction study. Although the fragment $\{TpRe(CO)_2\}$ binds aromatic substrates in a dihapto-coordination mode, the efficacy of these complexes for aromatic activation is limited by their dimeric nature. Although spectroscopic data suggest that the binuclear systems form *via* $TpRe(CO)_2(\eta^2$-aromatic) intermediates, the reduction of electron density at Re(I) due to π-back-bonding with two strongly π-acidic CO ligands likely renders the monomeric systems kinetically and thermally unstable (Scheme 27).

In order to access a more electron-rich Re(I) system, Harman *et al.* synthesized $TpRe(CO)(PMe_3)(\eta^2$-cyclohexene). The synthesis was reported from both heating a pressure tube containing the Re(III) system $TpReCl_2(PMe_3)$ in a benzene solution with 20 equivalents of cyclohexene, 10 psi CO and Na/Hg as well as using 20 equivalents cyclohexene, 1 atm of CO and Na.[118,119] The IR spectrum of TpRe $(CO)(PMe_3)(\eta^2$-cyclohexene) reveals an electron-rich system with $\nu_{CO} = 1796\,cm^{-1}$. Cyclic voltammetry shows a reversible oxidation for Re(II/I) at 0.23 V (versus NHE). Oxidation of $TpRe(CO)(PMe_3)(\eta^2$-cyclohexene) by AgOTf gives the Re(II) system $[TpRe(CO)(PMe_3)(\eta^2$-cyclohexene)][OTf], and heating this system produces free cyclohexene and $TpRe(CO)(PMe_3)(OTf)$. Reduction of $TpRe(CO)(PMe_3)OTf$ by ferrocene or Na/Hg in the presence of excess naphthalene, phenanthrene, cyclopentadiene or furan gives monomeric Re(I) complexes with η^2-bound substrates (Scheme 28). In contrast to σ-*S* coordination of thiophene to the $\{TpRe(CO)_2\}$ fragment (see above), for $\{TpRe(CO)(PMe_3)\}$ thiophene coordinates in a dihapto mode in equilibrium with $TpRe(CO)(PMe_3)(\sigma\text{-}S\text{-}SC_4H_4)$. TpRe(CO) $(PMe_3)(\eta^2$-naphthalene) undergoes ligand exchange with acetone ($t_{1/2} \sim 8$ h) to give $TpRe(CO)(PMe_3)(\eta^2$-acetone), which can be independently prepared upon reduction of $TpRe(CO)(PMe_3)OTf$ in acetone.

A potentially important aspect of the asymmetric $TpRe(CO)(PMe_3)(\eta^2$-aromatic) systems is the predilection toward stereoselective η^2-coordination of prochiral substrates. For heteroaromatic substrates an additional coordination mode, η^1 through the heteroatom, is also possible and is the more common binding mode for thiophenes.[120] The fragment $\{TpRe(CO)(PMe_3)\}$ favors atypical η^2-2,3-coordination of thiophene by 5.2:1 cf. σ-*S*-coordination through sulfur. The diastereoselectivity for η^2-coordination of the thiophene face is 3.2:2 with orientation of the heteroatom away from PMe_3 being slightly favored. An analogous facial preference is observed for furan though with a slightly higher diastereoselectivity of 2.1:1. Coordination of naphthalene occurs with high diastereoselectivity such that $TpRe(CO)(PMe_3)$

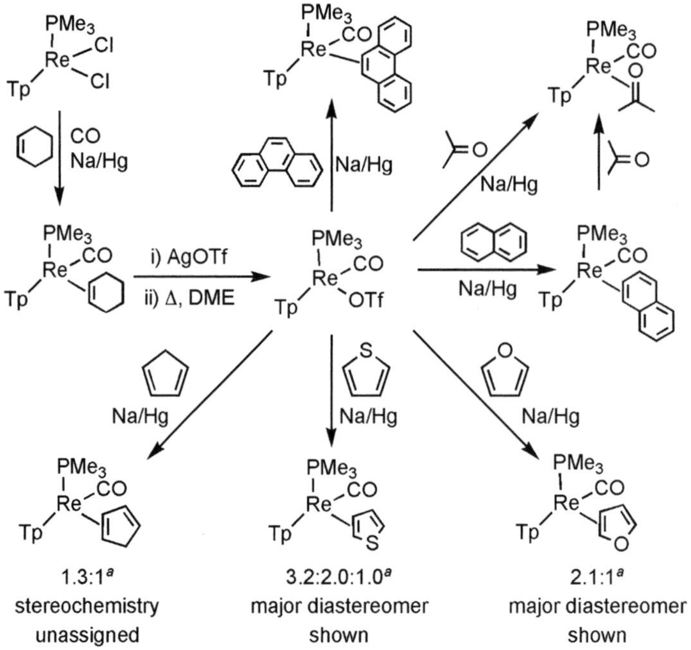

SCHEME 28. Dihapto-coordination of various ligands upon reduction of TpRe(CO)(PMe_3)OTf. [a]Diastereomeric ratio.

(η^2-naphthalene) forms as a single diastereomer with the uncomplexed ring oriented away from coordinated PMe_3. Phenanthrene coordinates dihapto across C(9) and C(10) of the internal ring.[119]

Modification of Mayer's synthesis of the Re(III) complex $TpReBr_2(py)$ gives $TpReBr_2(L)$ complexes where the identity of L includes ^{t}BuNC, NH_3, MeIm and 1-butylimidazole (BuIm) as well as the previously mentioned PMe_3.[25,82] In the presence of CO, reducing agent (e.g., sodium amalgam) and an aromatic ligand, these precursors give complexes of the type TpRe(CO)(L)(η^2-aromatic).[121–124] The range of aromatic ligands is extensive and depends on the identity of L and includes benzene, anisole, naphthalene, 2-methoxynaphthalene, 1,8-dimethylnaphthalene, 2,6-lutidine, *N*-methylpyrrole, 2-methylpyrrole, 2,5-dimethylpyrrole, thiophene, furan and 2-methylfuran. In addition, complexes of the type TpRe(CO)(L)(η^2-L′) can be synthesized from TpRe(CO)(L)(η^2-aromatic) precursors where η^2-L′ = ethylene, propene, *trans*-2-butene, 2-methylpropene, 3-methyl-3-butene, 2-methyl-2-butene, 2,3-dimethyl-2-butene, cyclopentene, cyclohexene, 1,3-cyclohexadiene, acetone, 2-butanone, 3-pentanone, 2-methyl-3-butanone, cyclohexanone, formaldehyde or acetaldehyde.[125,126] A computational study that compares various binding energies for benzene over a range of L for systems of the type TpRe(CO)(L)(η^2-benzene) as well as variation of the identity of the metal has been conducted.[127] The binding energy of the coordinated benzene was linked to the extent of charge transfer from the metal to the coordinated benzene.

In order to probe the impact of the donating ability of ancillary ligands on metal–ligand (dihapto) bonds, the η^2-ethylene complexes TpRe(CO)(L)(η^2-$H_2C{=}CH_2$) (L = ^{t}BuNC, PMe_3, MeIm, py or NH_3) were studied by infrared spectroscopy, cyclic voltammetry and dynamic NMR spectroscopy.[125] At 20 °C, all of the complexes exhibit two sharp resonances in their ^{1}H NMR spectra due to the ethylene ligand. Varying the temperature allowed the energetics of ethylene rotation to be studied. At the temperatures of the measurements, interfacial isomerization and substitution of ethylene are not operative, and rotation of ethylene is assumed to be the only dynamic process. Spin saturation transfer experiments and the Forsen–Hoffman method were used to determine the $\Delta G^{\ddagger}$ for ethylene rotation, which varied from 8.0 kcal mol^{-1} (for L = ^{t}BuNC) to 12.7 kcal mol^{-1} for L = NH_3. Comparisons of the free energies of activation for ethylene rotation with the electron density of each metal center, as determined by CO stretching frequencies or Re(II/I) reduction potential, reveal strong correlations. That is, as electron density is increased, as indicated by lower ν_{CO} values and/or more negative Re(II/I) potentials, the free energy of activation for ethylene rotation increases.

TpRe(CO)(MeIm)(η^2-benzene) also undergoes ligand exchange with carboxylic acid derivatives to give single diastereomers (−20 °C) in which the new ligand is bound η^2 through the carbonyl moiety.[128] Ethyl acetate, acetic anhydride, *N*-methylsuccinimide, *N*-acetylpyrrole and *N*-methylmaleimide (NMM) were used as substrates. For each substrate, the η^2-carbonyl oxygen is *syn* to MeIm (Scheme 29). At ambient temperatures, an interfacial migration occurs giving the corresponding diastereomer, presumably through η^1-oxygen coordination, inversion, then return to η^2-carbonyl coordination. Use of NMM provides a means for observing the binding preference for an alkene or carbonyl with the η^2-carbonyl preferred in a 2:1 molar ratio in what is likely to be a kinetic ratio. As shown in Scheme 30, reaction of TpRe(CO)(MeIm)(η^2-benzene) and NMM also results in a [4+2] cycloaddition (see below). Studies also indicate that {TpRe(CO)(MeIm)} promotes amide isomerization through dihapto-coordination of the amide carbonyl,

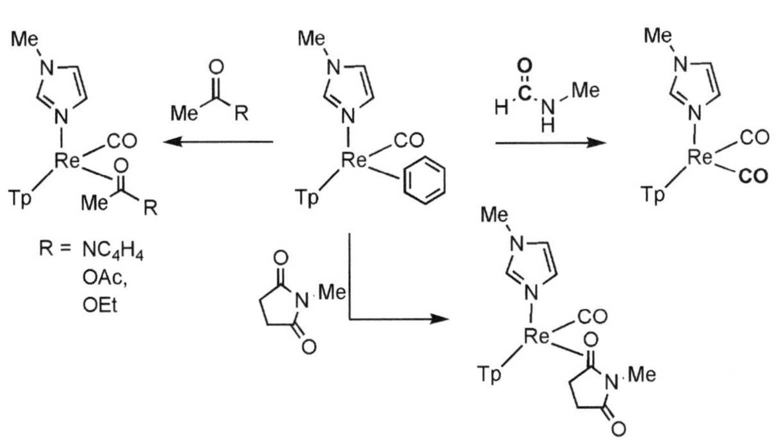

SCHEME 29. Reactions of TpRe(CO)(MeIm)(η^2-benzene) with carboxylic acid derivatives.

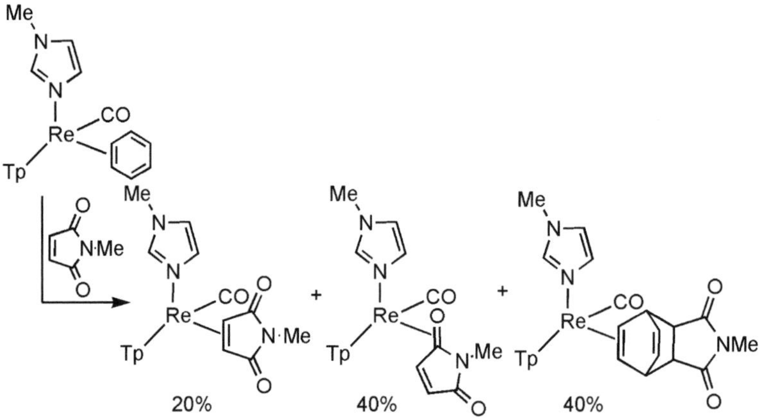

SCHEME 30. Reaction of TpRe(CO)(MeIm)(η^2-benzene) with N-methylmaleimide.

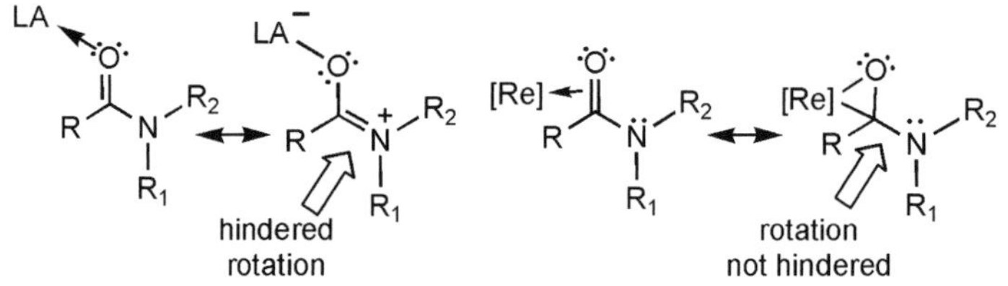

SCHEME 31. Promotion of isomerization of carboxamides by Re(I) π-base compared to interaction with σ-Lewis acid (LA = σ-Lewis acid).

which likely attenuates the electrophilicity of the carbonyl carbon and disrupts donation from the nitrogen lone pair to the carbonyl fragment. The latter interaction inhibits isomerization in the free amide as well as in the presence of a σ-Lewis acid (Scheme 31).[129]

TpRe(CO)(MeIm)(η^2-benzene) can be used as a synthetic precursor to other η^2-aromatic systems. Benzene undergoes ligand exchange with naphthalene, 2,6-lutidine, N-methylpyrrole and anisole to give the corresponding product (Scheme 32). At −20 °C, NMR spectroscopy reveals that TpRe(CO)(MeIm)(η^2-naphthalene) exists as a diastereomeric ratio of 20:1. At 20 °C, the diastereomeric ratio of TpRe(CO)(MeIm)(η^2-naphthalene) is 7:1. In contrast to TpRe(CO)(PMe_3)(η^2-naphthalene) (see above), the uncoordinated naphthalene ring of TpRe(CO)(MeIm)(η^2-naphthalene) is adjacent to the methylimidazole ligand in the major diastereomer.

Comparing the free energies of activation for intrafacial migration of TpRe(CO)(MeIm)(η^2-aromatic) reveals that $\Delta G^{\ddagger}$ is more substantial for furan and thiophene than for N-methylpyrrole, which may result from thiophene and furan forming more stable π-bound ligands or the methyl group of N-methylpyrrole interacting with the ancillary ligands on rhenium. Attempts at isolating TpRe(CO)(MeIm)(η^2-pyrrole) resulted in N–H activation and isolation of the paramagnetic σ-N-bound heterocyclic

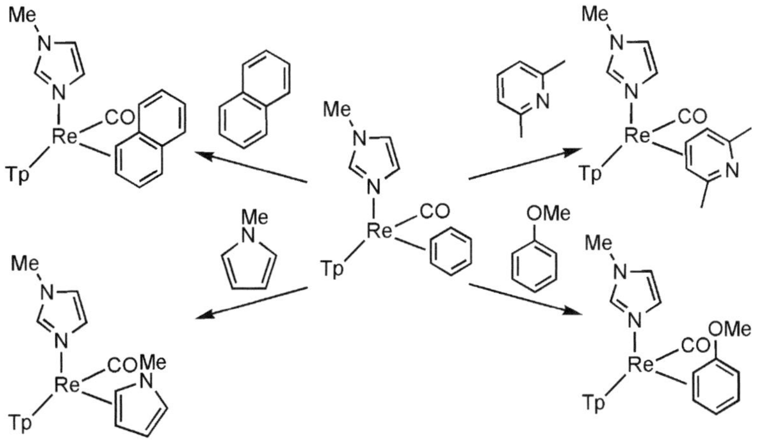

SCHEME 32. Synthesis of TpRe(CO)(MeIm)(η^2-aromatic) complexes.

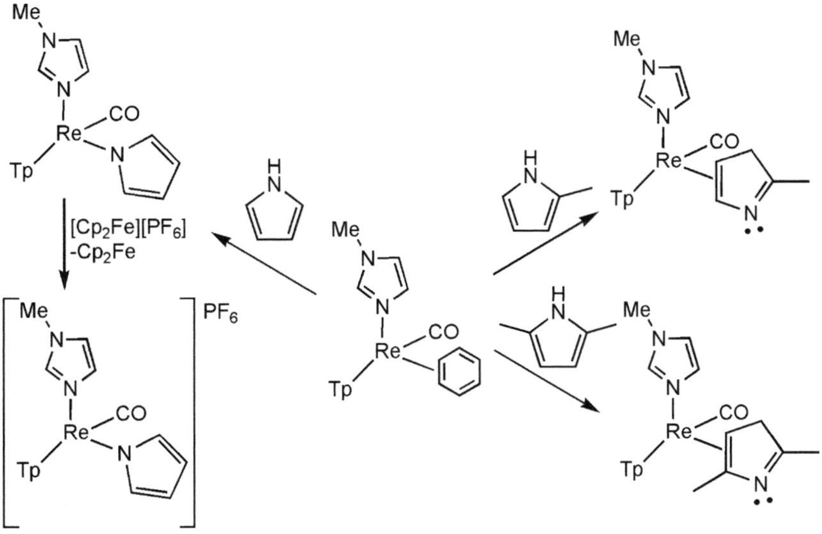

SCHEME 33. Reaction of TpRe(CO)(MeIm)(η^2-benzene) with pyrroles.

compound.[124] The Re(II) pyrrolyl is oxidized by [Cp_2Fe][PF_6] to give [TpRe(CO)(MeIm)(σ-*N*-pyrrolyl)][PF_6]. In contrast, treatment of "TpRe(CO)(MeIm)(η^2-C_6H_6)" with 2-methylpyrrole or 2,5-dimethylpyrrole yields the imine tautomers TpReCO(MeIm)(η^2-3H-2-methylpyrrole) and TpReCO(MeIm)(η^2-3H-2,5-dimethylpyrrole), respectively, *via* proton transfer from nitrogen to the three position (Scheme 33). In contrast to the *N*-methylpyrrole analog (Scheme 32), these complexes give static ^{1}H NMR spectra at room temperature.[124] TpReCO(MeIm)(η^2-3H-2-methylpyrrole) is a 2:1 mixture of diastereomers at room temperature while TpReCO(MeIm)

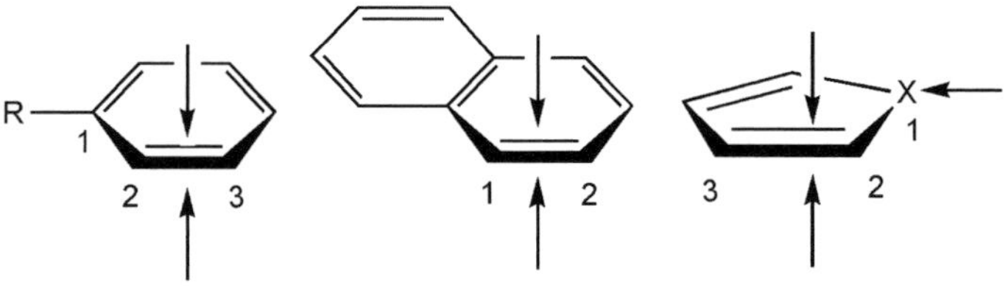

SCHEME 34. Common binding positions for some functionalized, bicyclic and heteroaromatic substrates.

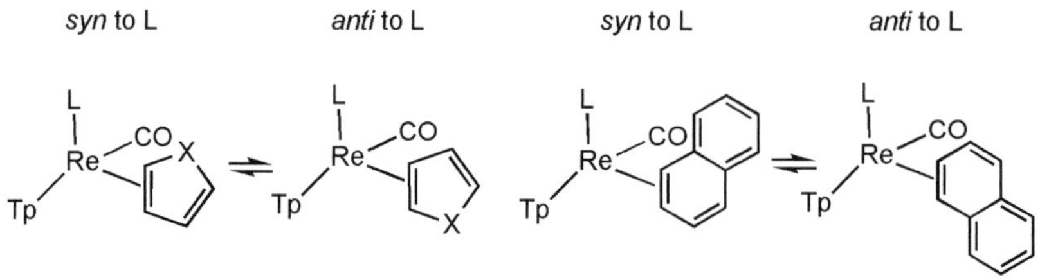

SCHEME 35. Orientation of η^2-aromatic ligands.

(η^2-3H-2,5-dimethylpyrrole) is a single diastereomer. For both systems, the favored diastereomer orients the pyrrole nitrogen *anti* to MeIm.

For heteroaromatic and substituted benzene ligands, upon coordination to "TpRe(CO)(L)" several coordination isomers are possible depending on the identity of the aromatic ligand and substitution pattern (Scheme 34). The regiochemistry of η^2-arene coordination to Re(I) is presumably governed by electronic factors that are similar to those observed for the π-base penta(ammine)osmium(II).[115] Thus, five-membered heteroaromatic substrates generally coordinate across C2/C3, naphthalene coordinates across C1/C2 and anisole is bound across C2/C3 (Scheme 34). For "TpRe(L)(CO)" systems, the orientations of the η^2-aromatic ligand *syn* to the Tp ligand are not observed. However, the orientation of the η^2-aromatic ligand with respect to L varies with both L and the η^2-aromatic and exhibits a subtle energetic balance (Scheme 35). For instance, complexes of the type TpRe(CO)(^{t}BuNC)(η^2-heteroaromatic) exhibit major isomers where the heteroatom is *anti* to ^{t}BuNC. The major isomer of TpRe(CO)(^{t}BuNC)(η^2-naphthalene) has the uncoordinated ring of naphthalene *syn* to ^{t}BuNC. In contrast, all complexes of the type TpRe(CO)(MeIm)(η^2-aromatic) have either their heteroatom, uncoordinated ring or functional group *syn* to the ligand MeIm.

Linkage isomerization of η^2-coordinated aromatic ligands can occur by dissociative migration, non-dissociative interfacial migration (face-flipping) or non-dissociative intrafacial migration (ring-walking) (Scheme 36). Mechanistic studies indicate that non-dissociative pathways are most likely for TpRe(CO)(L)(η^2-aromatic) systems. For example, linkage isomerization of TpRe(CO)(L)(η^2-aromatic) complexes in solvents such as acetone-d_6 is much faster than formation of the thermally stable complex TpRe(CO)(L)(η^2-acetone-d_6). In addition, ^{1}H NMR spin saturation transfer

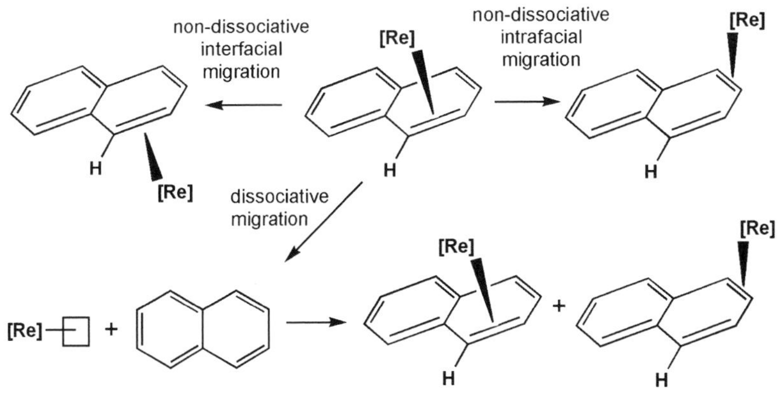

non-dissociative interfacial migration
non-dissociative intrafacial migration
dissociative migration
[Re]
H
[Re]—□ +

SCHEME 36. Stereolinkage isomerization routes {[Re] = TpRe(CO)(L)}.

experiments are inconsistent with dissociative mechanisms. The possible mechanisms of isomerization vary depending on the η^2-aromatic substituent. Migrations of aromatic heterocycles may proceed through a heteroatom-bound intermediate while migrations of arenes are void of this possibility.

For heteroaromatic ligands, four possible interfacial pathways are (i) "heteroatom-assisted" migration, (ii) σ-C–H interaction, (iii) C–H oxidative addition and (iv) inversion of metal configuration and a 90° rotation of the aromatic ligand (Scheme 37). Pathway (iv) is unlikely given the high activation barriers for inversion at the rhenium center for the closely related systems $[CpRe(NO)(PPh_3)(\eta^2\text{-}CH_2{=}CHR)]^+$.[130] In addition, α-pinene was used to resolve enantiomers of the {TpRe(CO)(MeIm)} fragment which do not undergo inversion at Re at elevated temperatures.[128] Pathways (ii) and (iii) are the extremes of a three-center interaction between rhenium, carbon and hydrogen. Thus, the two most likely interfacial pathways for heteroaromatic substrates are a heteroatom-assisted migration involving η^1-coordination through the heteroatom or a C–H σ-bond pathway with either 180° rotation around a C–H σ-bond or oxidative addition of a C–H σ-bond followed by 180° rotation around the metal-C_{ipso} bond then reductive elimination. For homocyclic arenes, the possible interfacial mechanisms are reduced due to the lack of a heteroatom (Scheme 38).

Intrafacial migration involves movement of the metal from one-coordinated double bond to an adjacent coordinated double bond while maintaining the same binding face. This can be done *via* (i) migration to a σ-C–H bond without rotation around the C–H bond followed by migration to an adjacent carbon–carbon π-bond, (ii) oxidative addition of a C–H bond to the metal center followed by reductive elimination, (iii) ring-walk *via* an η^1-arenium intermediate or (iv) ring-walk *via* an η^3-allyl intermediate. The four possible pathways are shown in Scheme 39. The reaction coordinate is relatively straightforward for simple aromatic substrates (e.g., benzene), but becomes more complicated for heteroaromatic substrates, naphthalenes and substituted benzenes due to the presence of multiple coordination sites.

For systems of the type TpRe(CO)(L)(η^2-aromatic), ^{1}H NMR spin saturation transfer and dynamic NMR experiments were used to elucidate details of

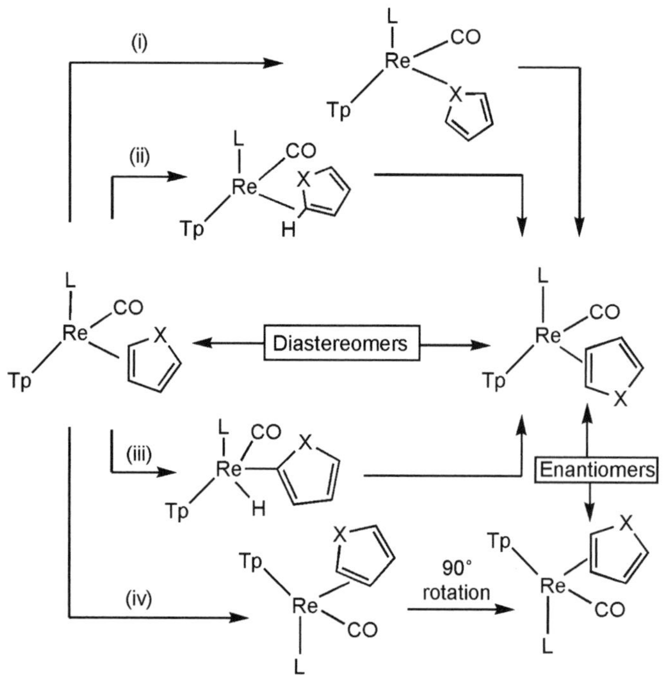

SCHEME 37. Possible pathways for interfacial migration of heteroaromatic substrates. (i) heteroatom-assisted; (ii) σ-C–H interaction; (iii) C–H oxidative addition; (iv) inversion of metal configuration.

non-dissociative isomerization and to distinguish interfacial and intrafacial conversion.[131–134] Evidence was obtained for both interfacial and intrafacial conversions with the relative rates dependent on the identity of L and the aromatic substrate. Comparing free energies of activation for interfacial migration and electronic properties ($E_{p,a}$ and ν_{CO}) between {TpRe(CO)(L)} systems reveals the effect of metal-based electron density on interfacial isomerization.[135] When L is a more potent donor, the activation energy for interfacial isomerization increases due to enhanced π-back-bonding from Re(I) to the η^2-bound substituent. The magnitude of the $\Delta G^{\ddagger}$ for isomerization of TpRe(CO)(L)(η^2-aromatic) systems (aromatic = furan, thiophene or naphthalene) is L = ^{t}BuNC < PMe$_3$ < py < MeIm. An analogous correlation exists for intrafacial migrations. In addition, a similar relation exists between the $\Delta G^{\ddagger}$ for dissociation of η^2-naphthalene ligands and the electronic properties of the metal. For example, in acetone, TpRe(CO)(L)(η^2-naphthalene) complexes convert to TpRe(CO)(L)(η^2-acetone) systems *via* dissociation of the aromatic ligand.[135] The activation energy for this conversion decreases in the order MeIm > py > ^{t}BuNC. These results suggest that increased electron density at the metal center enhances metal-to-aromatic ligand π-back-bonding and decreases the predilection toward any process that breaks the metal–aromatic dihapto-coordination.

The utilization of chiral {TpRe(CO)(L)} fragments for enantioselective dearomatization of aromatic substrates could prove a versatile and useful synthetic tool.

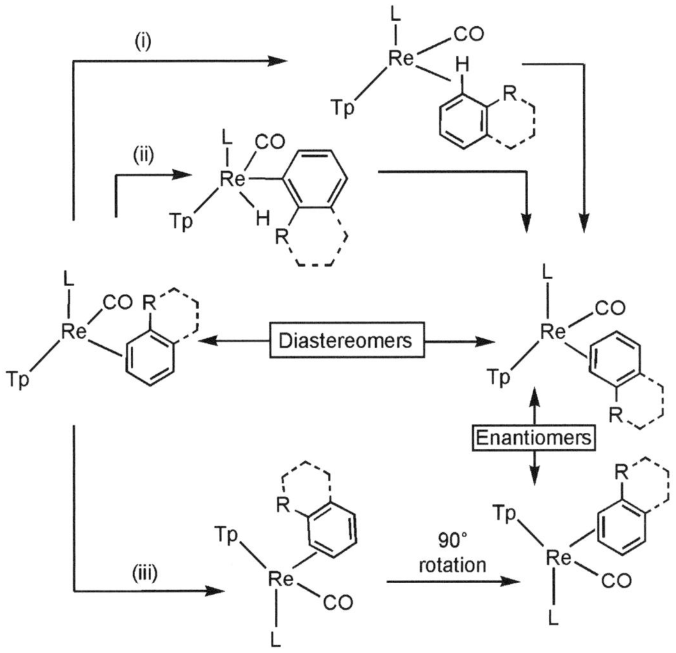

SCHEME 38. Possible pathways for interfacial migration of arenes. (i) σ-C–H interaction; (ii) C–H oxidative addition; (iii) inversion of metal configuration.

In order to access such transformations, the Re systems must be resolved into single enantiomers. This has been accomplished for TpRe(CO)(MeIm)(η^2-benzene) using (1*R*)-(+)-α-pinene and (1*S*)-(−)-β-pinene (Scheme 40).[128] When racemic TpRe(CO)(MeIm)(η^2-benzene) is stirred for 30 h with (1*R*)-(+)-α-pinene two diastereomers are formed in a 1:1 ratio. Continued stirring of this mixture for 60 h in benzene gives (*S*)-TpRe(CO)(MeIm)(η^2-(*R*)-α-pinene) and (*R*)-TpRe(CO)(MeIm)(η^2-benzene). The two compounds can be separated on silica gel to give enantiomerically resolved metal centers. Oxidative removal of pinene from (*S*)-TpRe(CO)(MeIm)(η^2-(*R*)-α-pinene) followed by reduction in benzene gives (*S*)-TpRe(CO)(MeIm)(η^2-benzene).[119,122] The enantiopure complexes are stable for days at 25 °C. The resolved benzene complexes are precursors for other enantiopure TpRe(CO)(MeIm)(η^2-aromatic) complexes. Interestingly, when (*S*)-TpRe(CO)(MeIm)(η^2-(*R*)-α-pinene) is first oxidized by AgOTf then reduced with Na/Hg in the presence of benzene followed by reaction with (1*S*)-(−)-β-pinene in THF, a 1:1 mixture of diastereomers is obtained. However, if the first step is carried out in a mixture of THF/benzene, (*S*)-TpRe(CO)(MeIm)(η^2-(*S*)-β-pinene) is obtained in 95:5 diastereomeric ratio.

Another way to obtain enantio-enriched tandem addition products from racemic coordination diastereomers is to use solid-state induced control of TpRe(CO)(L)(η^2-aromatic) stereoisomers where the aromatic substrate is prochiral.[136] This method is effective for L = MeIm, BuIm, py or PMe_3 and η^2-aromatic = anisole, 3-methylanisole

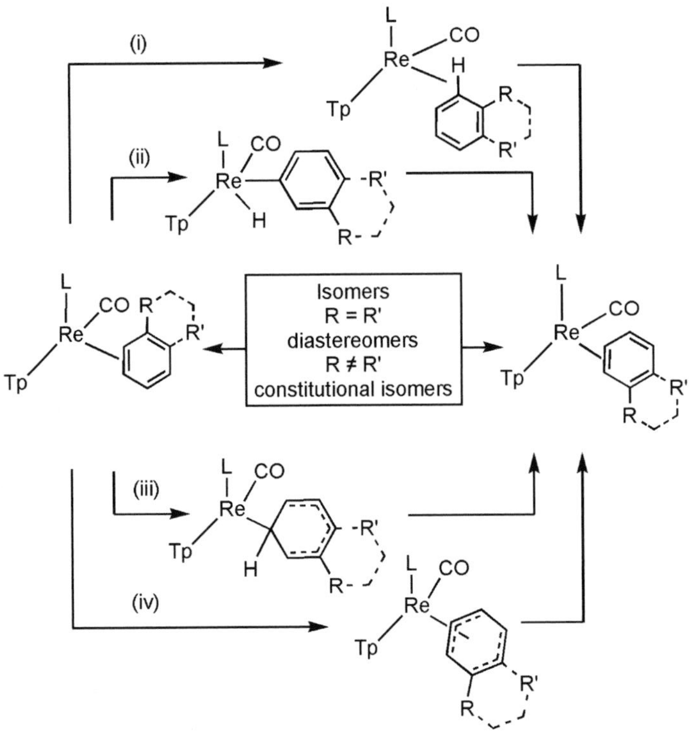

Isomers
R = R'
diastereomers
R ≠ R'
constitutional isomers

SCHEME 39. Possible routes for intrafacial migration of arenes. (i) Migration to a σ-C–H; (ii) C–H oxidative addition; (iii) σ-arenium ring-walk; (iv) η^3-allyl ring-walk.

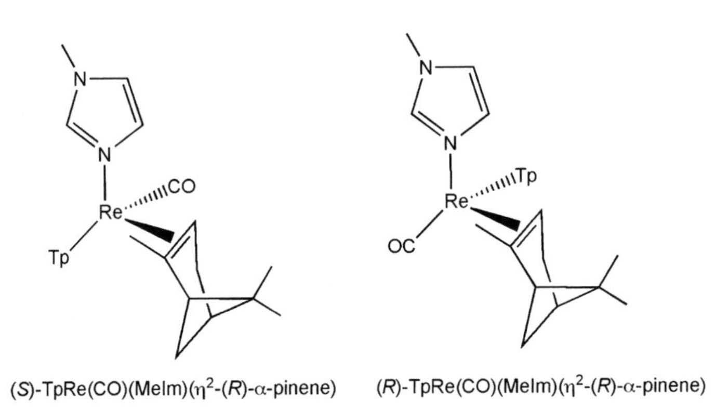

SCHEME 40. Resolution of {TpRe(CO)(MeIm)} enantiomers using (*R*)-α-pinene.

or naphthalene. In the solid-state, these complexes favor diastereomeric distributions that are quite distinct from their solution equilibria, and this feature can be used advantageously where rates of reactions with other substrates are more rapid than the rates of isomerization in solution. Diastereomerically enriched solids can be obtained *via* slow growth of crystals or by precipitation of an amorphous solid from solution using hexanes. As an example of a typical increase in diastereomeric ratio upon precipitation, in solution TpRe(CO)(MeIm)(η^2-anisole) exists as a 2:1 ratio of diastereomers while the solid has an approximately 20:1 ratio of diastereomers.

Strongly π-basic metals can stabilize arenium complexes that result from electrophilic addition to η^2-aromatic ligands.[137–143] The stabilization of arenium complexes generated from TpRe(CO)(L)(η^2-aromatic) systems was investigated for L = MeIm, BuIm, py or PMe_3 and η^2-aromatic = benzene, anisole, 3-methylanisole, 4-methylanisole, naphthalene, toluene, *o*-xylene, *m*-xylene, *N*-methylpyrrole, 2-methylpyrrole or 2,5-dimethylpyrrole.[123,124] For example, in the presence of strong acids TpRe(CO)(BuIm)(η^2-aromatic) is converted to various arenium complexes that have been characterized by NMR and/or X-ray crystallography. From η^2-benzene or η^2-toluene complexes, conversion to two arenium diastereomers is observed upon treatment with Brønsted acids such as methanesulfonic acid or diphenylammonium triflate. Protonation of TpRe(CO)(BuIm)(5,6-η^2-anisole) at the 2- and 4-position occurs with acids such as phosphoric acid to produce four isomers (Scheme 41). The ratio of protonation at C2/C4 increases with decreasing temperature and increasing acid concentration with an optimized regioselectivity of 20:1 at $-60\,^{\circ}$C with 2 equivalents of acid. Addition of CH_3OD to a diastereomer of [TpRe(CO)(BuIm){5,6-η^2-(2H-anisolium)}]$^+$ results in stereoselective H/D exchange with the proton *exo* to the metal center. With 1 equivalent acid-d_1 at $20\,^{\circ}$C, rapid intrafacial isomerization incorporates deuterium at both protonated positions while use of multiple equivalents of acid-d_1 at $20\,^{\circ}$C or 1 equivalent of acid-d_1 at $-60\,^{\circ}$C results in deuterium incorporation into a single position (dr>20:1). TpRe(CO)(L)(5,6-η^2-3-methylanisole) is significantly more basic than TpRe(CO)(L)(5,6-η^2-anisole). For example, a mixture of TpRe(CO)(BuIm)(5,6-η^2-3-methylanisole)

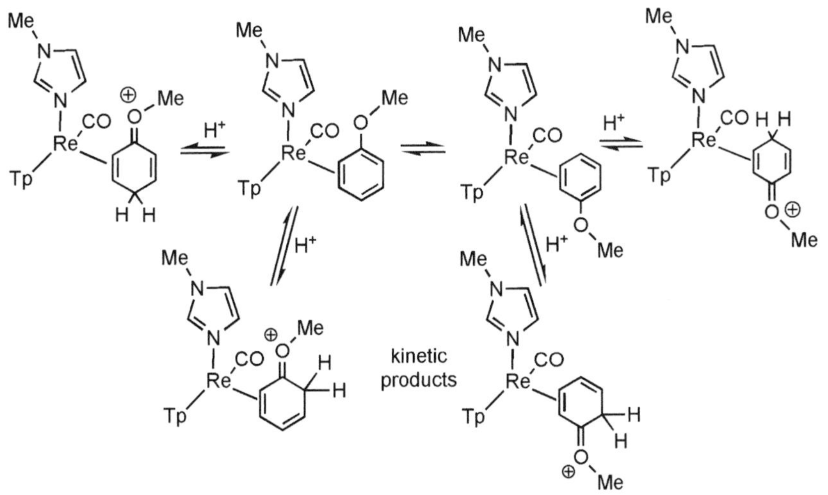

SCHEME 41. Protonation of a π-base stabilized anisole ligand.

and [TpRe(CO)(BuIm)(5,6-η^2-(2H-anisolium)]$^+$ is completely converted to the 3-methylanisolium complex and TpRe(CO)(BuIm)(5,6-η^2-anisole). TpRe(CO)(BuIm)(3,4-η^2-naphthalene) is slightly less basic than the anisole analog, being readily protonated by diphenylammonium triflate ($pK_a = 0.8$) but not phosphoric acid ($pK_a = 2.12$). Overall, differences in basicities were not extensive when varying L. For instance, naphthalene complexes of the fragments {TpRe(CO)(PMe_3)}, {TpRe(CO)(py)}, {TpRe(CO)(MeIm)}, {TpRe(CO)(BuIm)} and {TpMo(NO)(MeIm)} are protonated by diphenylammonium triflate but do not react with phosphoric acid. Protonations of TpRe(CO)(MeIm)(η^2-*N*-methylpyrrole), TpRe(CO)(MeIm)(η^2-3H-2-methylpyrrole) and TpRe(CO)(MeIm)(η^2-3H-2,5-dimethylpyrrole) were carried out with Brønsted acids such as triethylammonium triflate. TpRe(CO)(MeIm)(η^2-*N*-methylpyrrole) is protonated at the pyrrole three position while TpRe(CO)(MeIm)(η^2-3H-2-methylpyrrole) and TpRe(CO)(MeIm)(η^2-3H-2,5-dimethylpyrrole) are protonated at the pyrrole nitrogen providing η^2-3H-pyrrolium complexes. Evidence for isomerization between 3H and 2H pyrrolium complexes has been obtained by NMR spectroscopy.

Michael additions are accessed upon reaction of TpRe(CO)(L)(5,6-η^2-anisole) and substituted anisole complexes with various Michael acceptors such as methylvinylketone (MVK), 3-penten-2-one, 2-cyclopentenone, 2-cyclohexenone, NMM or methylpropenoate in the presence of a Lewis acid (e.g., $BF_3 \cdot OEt_2$; Scheme 42).[144] Addition at the *para* carbon occurs *anti* to the metal center.

In the presence of $BF_3 \cdot OEt_2$, complexes of the type TpRe(CO)(MeIm)(η^2-furan) add Michael acceptors to the coordinated furan through a 1,3-propene dipole (Scheme 43).[145,146] The 1,3-propene dipole undergoes cyclization to yield a cyclopentene ring.MVK, 3-pentene-2-one, cyclopentenone, 2,4-hexadienal, crotonaldehyde, methacrolein and methylenenorbornanone were successfully deployed. When the enantiopure complex of 2,5-dimethylfuran was combined with racemic methylenenorbornanone, a single diastereomer was isolated in 80% ee (Scheme 44).

A proposed mechanism for cyclopentannulation is shown in Scheme 45. Lewis acid catalyzed epimerization likely accounts for scrambling of the stereochemistry at C3

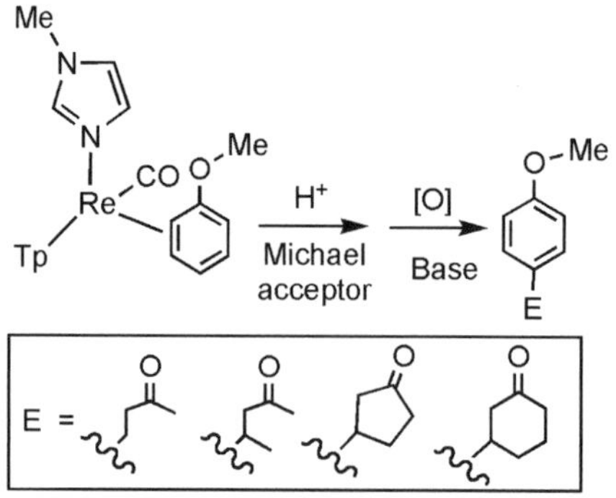

SCHEME 42. Examples of Michael addition reactions to an η^2-anisole ligand.

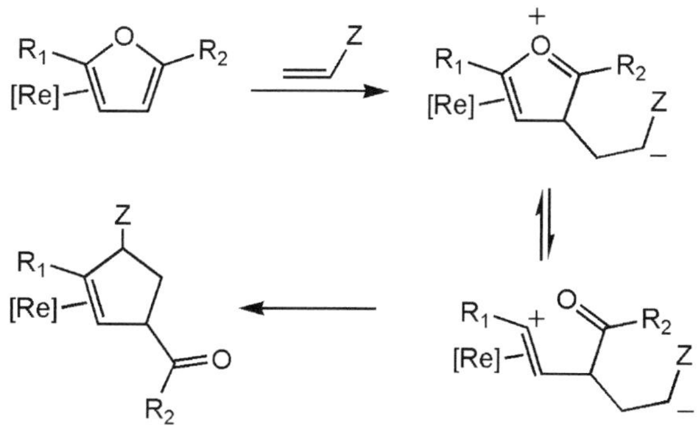

SCHEME 43. Rhenium-promoted cyclopentannulation through formation of a 1,3-propene dipole {[Re] = TpRe(CO)(MeIm)}.

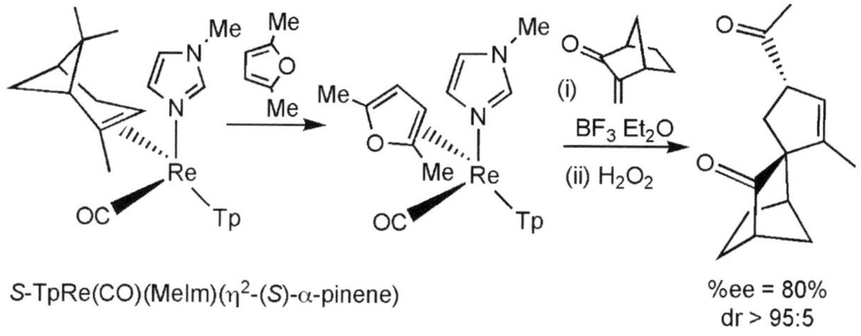

S-TpRe(CO)(MeIm)(η^2-(S)-α-pinene)

%ee = 80%
dr > 95:5

SCHEME 44. Michael addition of enantioenriched (*S*)-TpRe(CO)(MeIm)(η^2-2,5-dimethylfuran).

and C5 for TpRe(CO)(MeIm)(η^2-2-methylfuran) complexes resulting in the thermally preferred product where acetyl groups are *anti* to the metal. Epimerization does not occur with TpRe(CO)(MeIm)(η^2-2,5-dimethylfuran), which provides kinetic control of stereoselectivity. For TpRe(CO)(MeIm)(η^2-2,5-dimethylfuran), epimerization does not likely occur *via* deprotonation at C3 or C5 since the addition of CD_3OD does not result in deuterium incorporation at these sites. Attempts at synthesizing a cyclopentene functionalized with both a ketone and an ester from TpRe(CO)(MeIm)(η^2-2-methoxyfuran) results in the production of two carbene isomers.

The addition of aldehydes (e.g., acetaldehyde, benzaldehyde or 3-formylfuran) to η^2-furan or η^2-2-methylfuran complexes of the type TpRe(CO)(^{t}BuNC)(η^2-furan) gives three dihydrofuran products (Scheme 46).[147] The product **47** comes from *anti*-addition of the Lewis acid modified aldehyde to C3 of the η^2-furan followed by formation of a 1,3-propene dipole. Rotation around the former C3–C4 of furan positions the aldehydic oxygen for nucleophilic attack on the positive end of the dipole and hence the new furan with the acetyl group *syn* to the metal. This pathway

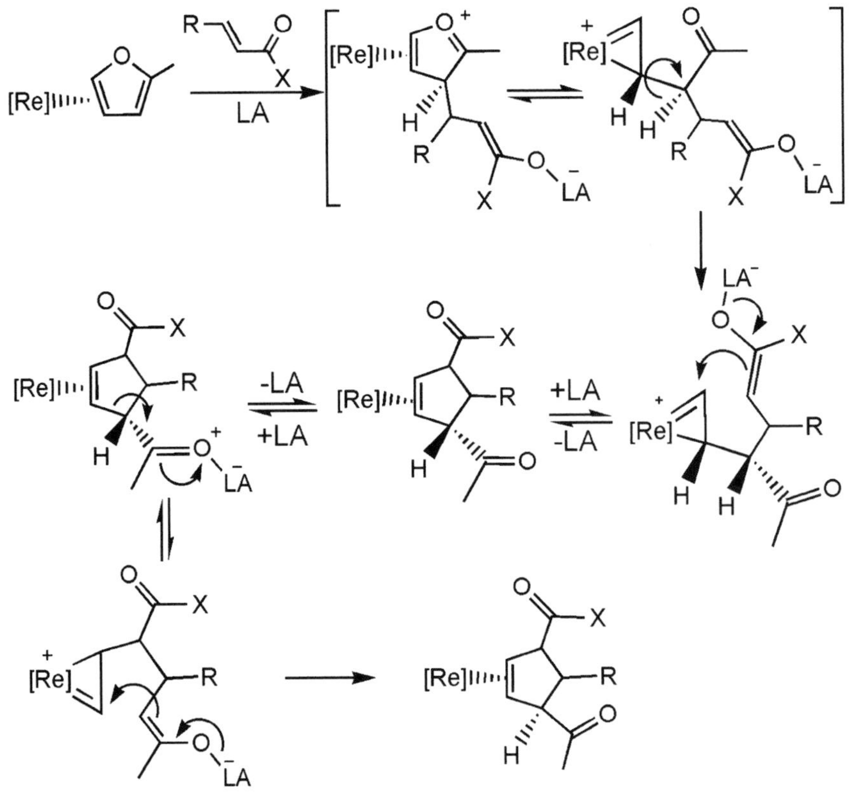

SCHEME 45. Conversion of furan to functionalized cyclopentenes mediated by Re(I) ([Re] = TpRe(CO)(MeIm), LA = Lewis acid).

occurs for both η^2-furan and η^2-2-methylfuran, originates from the least abundant coordination diastereomer where the heteroatom is *anti* to ^{t}BuNC and gives the most abundant addition product. In the case of η^2-2-methylfuran interfacial isomerization is observed at ambient conditions over a period of hours to reposition the acetyl group *anti* to the metal. The other two products originate from the thermally preferred coordination diastereomer by initial addition of aldehyde to C3 followed by addition of a second equivalent *via* nucleophilic attack on the aldehydic carbonyl carbon. The nucleophilic electrons in the oxygen–Lewis acid bond can attack the C2 position of the furan ligand giving the dioxine **48**. Alternatively, simple loss of the Lewis acid from the second equivalent of aldehyde and hydride transfer gives **49**.

Cycloaddition reactions, [3+2] or [2+2] depending on L, of TpRe(CO)(L) (η^2-furan) complexes have been reported.[148] The electrophilic alkene tetracyanoethylene (TCNE) and the alkyne dimethylacetylenedicarboxylate (DMAD) were used as dienophiles. For L = PMe_3 and ^{t}BuNC, 1,3-dipolar cycloaddition of TCNE to η^2-furan gives [3+2] products *via* a carbonyl ylide (Scheme 47). Addition of

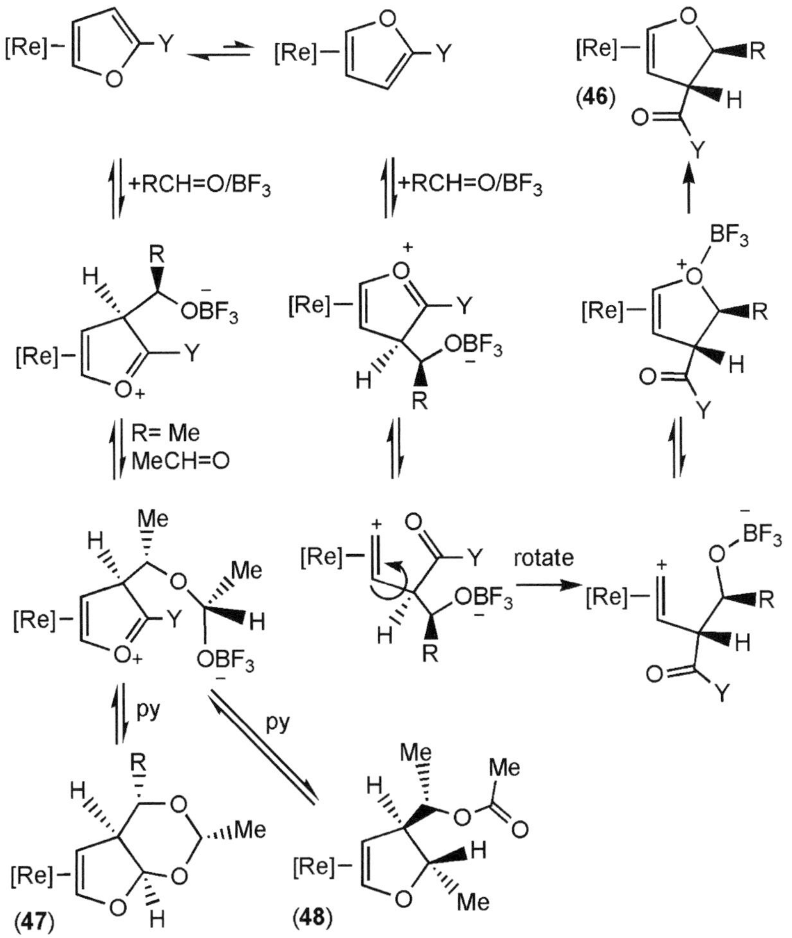

[Re]– Y ⇌ [Re]– Y
+RCH=O/BF$_3$ +RCH=O/BF$_3$
[Re]– R H (46) O Y
R H OBF$_3$ [Re]– Y O+
+ O [Re]– Y H OBF$_3$ R
BF$_3$ + O [Re]– R H O Y
R= Me MeCH=O
Me H O Me [Re]– Y H O+ OBF$_3$
[Re]– + O Y OBF$_3$ H R rotate
BF$_3$ O + [Re]– R H O Y
py py
R H O Me [Re]– O O H (47)
Me Me H O O [Re]– H O Me (48)

SCHEME 46. Addition of aldehydes to TpRe(CO)(^{t}BuNC)(η^2-furan) complexes {R = Me or MeCHO, Y = H or Me, [Re] = TpRe(CO)(^{t}BuNC)}.

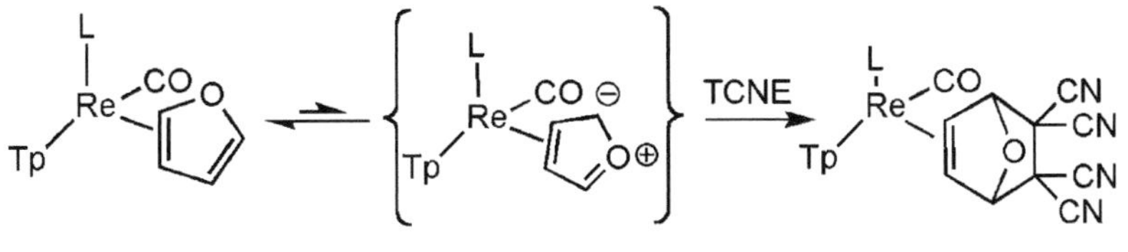

L Re CO O Tp ⇌ L Re CO ⊖ Tp O⊕ TCNE L Re CO Tp O CN CN CN CN

SCHEME 47. Rhenium-mediated [3 + 2] cycloaddition to η^2-furan ligand (L = PMe_3, CN^{t}Bu; TCNE = tetracyanoethene).

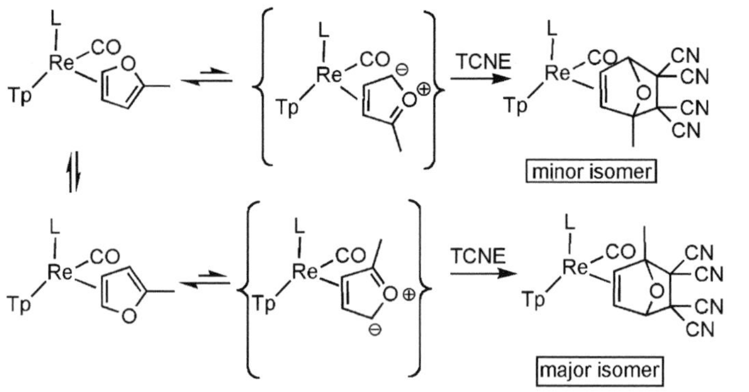

L
Re
CO
Tp
O
TCNE
CN
CN
CN
CN
minor isomer
TCNE
major isomer

SCHEME 48. Rhenium-mediated [3 + 2] cycloaddition to 2-methylfuran (L = PMe_3, CN^tBu; TCNE = tetracyanoethylene).

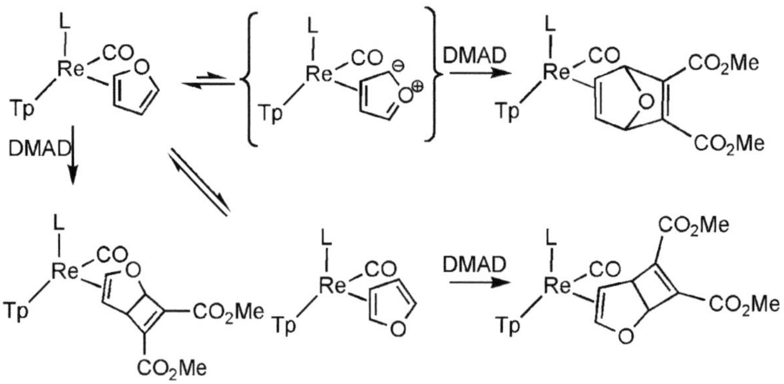

L
Re
CO
Tp
O
DMAD
CO_2Me
CO_2Me
DMAD
CO_2Me
CO_2Me
DMAD
CO_2Me
CO_2Me

SCHEME 49. Cycloaddition reactions of TpRu(CO)(MeIm)(η^2-furan) (L = PMe_3, CN^tBu; DMAD = dimethylacetylenedicarboxylate).

TCNE to η^2-2-methylfuran gives two diastereomeric products of [3 + 2] addition with a diastereomeric ratio of 20:1 (Scheme 48). When L = MeIm, TCNE oxidizes the furan complex, but with DMAD, three products arise, one from [3 + 2] addition to the carbonyl ylide and two products from Michael type [2 + 2] addition (Scheme 49). The substituted 2-methylfuran complex gives only the two products of [2 + 2] addition, which are coordination diastereomers, while 2,5-dimethylfuran gives only a single diastereomer of [2 + 2] addition.

In the presence of triflic or trifluoroacetic acid, a tandem addition occurs with TpRe(CO)(L)(η^2-furan) and methanol at −40 °C (L = ^tBuNC, PMe_3, py or MeIm).[149] Unlike the typical addition of an electrophile to free furan, which is favored at the 2-position, electrophilic addition occurs at the 3-position of the π-bound furan to give a rhenium-stabilized η^3-3H-furanium. Nucleophilic attack *anti* to the metal at the 2-position gives TpRe(CO)(L)(2-methoxy-2,3-dihydrofuran), and the products form as two diastereomers that differ in the relative orientation of the

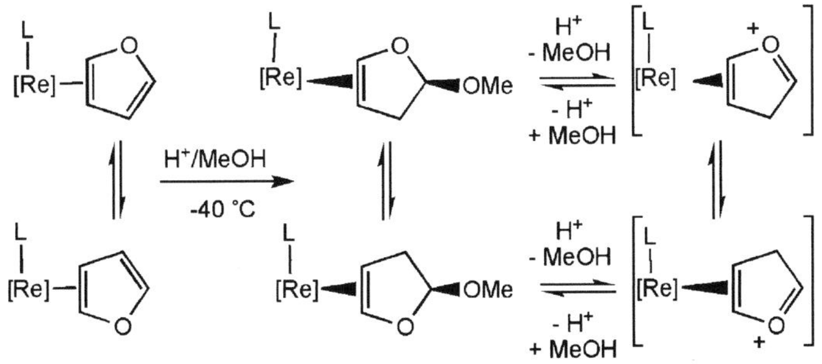

SCHEME 50. Tandem addition to TpRe(CO)(PMe$_3$)(η^2-furan) providing two diastereomers each resulting in 2,3-dihydrofuran with the ring O *syn* to PMe$_3$ ([Re] = TpRe(CO); L = ^{t}BuNC, PMe$_3$, py, or MeIm).

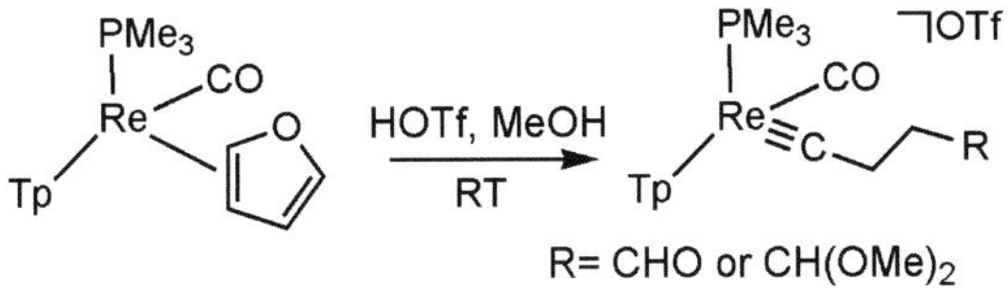

SCHEME 51. Reaction of triflic acid and methanol with TpRe(CO)(PMe$_3$)(η^2-furan) at room temperature to form a Re carbyne complex. After 20 min of reaction time R = CH(OMe)$_2$ while after 45 min R = CHO.

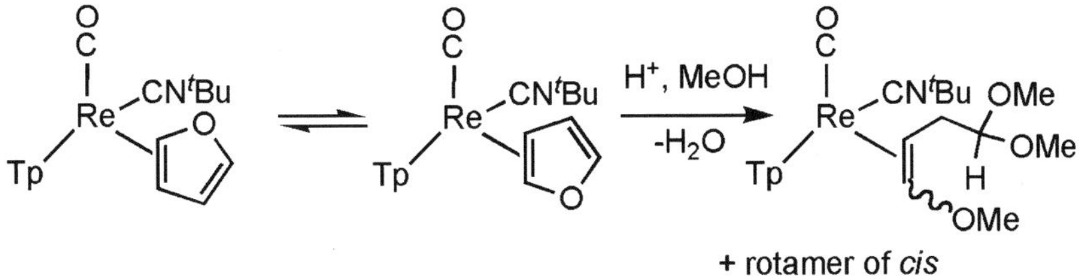

SCHEME 52. Methanol addition to TpRe(CO)(CN^{t}Bu)(η^2-furan).

oxygen and ligand "L" (Scheme 50). The ratio of diastereomers varies with the identity of L. A direct correlation exists between electron density at Re, as evaluated by IR spectroscopy and cyclic voltammetry, and the rate of the electrophile addition. Detailed mechanistic studies provide evidence that two diastereomeric products likely interconvert *via* a face-flip at the oxonium intermediate as depicted in Scheme 50. Addition of HOTf at room temperature results in oxidation of the metal center to give a Re(V) carbyne (Scheme 51). For L = ^{t}BuNC, the reaction of acid and MeOH with TpRe(CO)(L)(η^2-furan) also produces ring-opened vinyl ethers (Scheme 52).

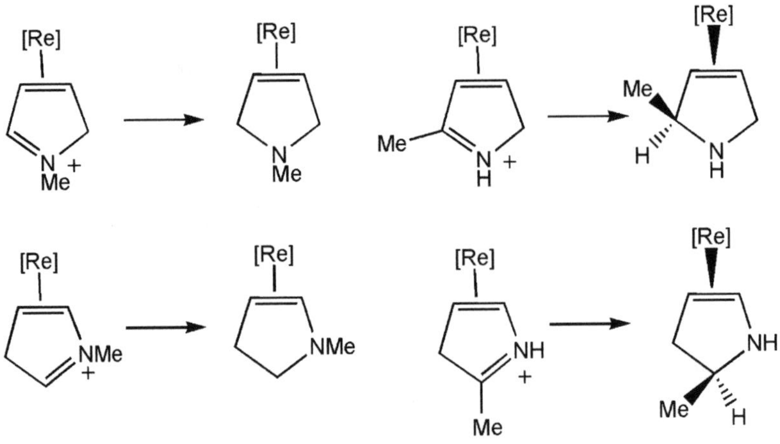

SCHEME 53. Examples of hydride addition ($NaBH_4$, H_2O) to [TpRe(CO)(MeIm)(η^2-pyrrolium)]$^+$ complexes ([Re] = [TpRe(CO)(MeIm)]).

Nucleophilic addition (e.g., hydride addition) has been observed for the 2H- and 3H-pyrrolium complexes [TpRe(CO)(MeIm)(η^2-*N*-methylpyrrolium)]$^+$, [TpRe(CO)(MeIm)(η^2-2-methylpyrrolium)]$^+$ and [TpRe(CO)(MeIm)(η^2-2,5-methylpyrrolium)]$^+$ (see above).[124] When $NaBH_4$ is combined with these complexes, a hydride is added to the iminium carbon converting them to 2,3-dihydropyrrole and 2,5-dihydropyrrole complexes (Scheme 53). The hydride addition occurs at the N=C face oriented away from Re.

TpRe(CO)(MeIm)(η^2-aromatic) complexes undergo a variety of Diels–Alder reactions as shown in Scheme 54.[150] TpRe(CO)(MeIm)(η^2-benzene) combines with NMM to give a single diastereomer in 65% yield. NOESY spectral data and an X-ray crystal structure determination confirm that the reaction occurs on the arene face opposite to metal coordination. The organic cycloadduct can be liberated from the metal using a variety of oxidizing agents/conditions (e.g., $CuBr_2$, AgOTf, $[Cp_2Fe][PF_6]$ or O_2/TFA). Demonstrating the substantial activating ability of the Re(I) π-basic fragment, the Re-promoted cycloaddition of TpRe(CO)(MeIm)(η^2-benzene) and NMM is approximately twice as fast as the cycloaddition of NMM to free 1,3-cyclohexadiene under identical reaction conditions.

TpRe(CO)(MeIm)(η^2-anisole) reacts with DMAD to form an η^2-barrelene-complex (Scheme 54). Demetalation gives the barrelene and a trisubstituted benzene that is the product of retrocycloaddition involving elimination of ethyne. TpRe(CO)(MeIm)(5,6-η^2-anisole) also reacts with NMM to give a cycloadduct with the methoxy group at a bridgehead position.[144] In addition, TpRe(CO)(MeIm)(η^2-1-methylpyrrole) undergoes a 1,3-dipolar cycloaddition reaction with dimethylfumarate to give 7-azabicyclo-[2.2.1]heptene (Scheme 54).

The stability of cationic Re(I) arenium complexes (see above) renders the arenium ligand susceptible to a second controlled transformation with a nucleophile. Thus, an electrophile can be added to a dihapto-coordinated aromatic ligand followed by a nucleophile in a stepwise tandem addition. For example, TpRe(CO)(MeIm)

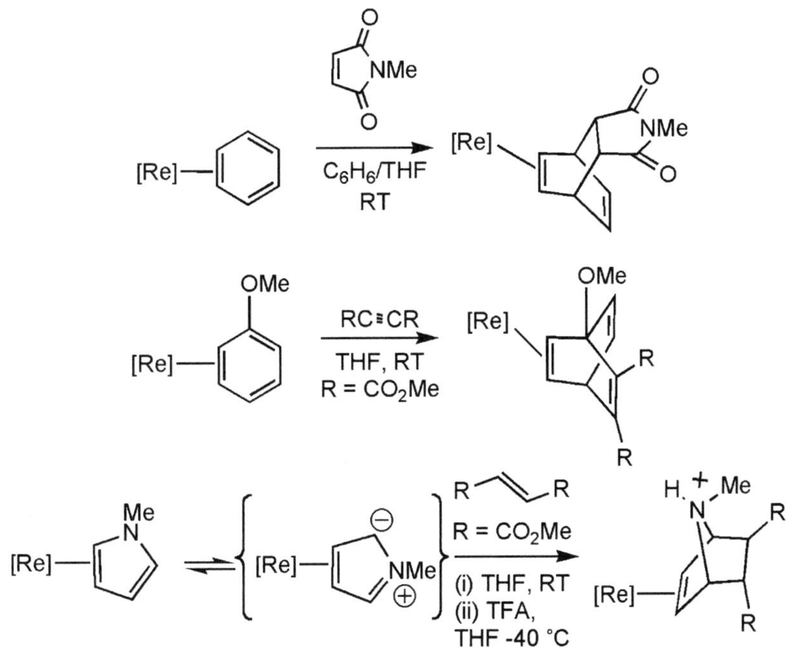

[Re] — NMe — C_6H_6/THF RT — [Re] — NMe

[Re] — OMe — RC≡CR THF, RT R = CO_2Me — [Re] — OMe — R — R

[Re] — Me N — [Re] — NMe — R = CO_2Me (i) THF, RT (ii) TFA, THF -40 °C — H + Me N — R — [Re] — R

SCHEME 54. Rhenium-promoted cycloaddition reactions of η^2-aromatic ligands ([Re] = {TpRe(CO)(MeIm)}).

(η^2-benzene) undergoes tandem 1,4-additions of electrophiles and nucleophiles.[151] The current scope of electrophiles includes dimethoxymethane (DMM) and methyl vinyl ketone. The range of nucleophiles includes 1-methoxy-2-methyl-1-(trimethylsiloxy)propene (MMTP), phenyl anion (PhLi/CuCN), 2-trimethylsiloxypropene and dimethylmalonate. The products are *cis*-1,4-substituted cyclohexadienes in yields of 46–84%. For example, an acetonitrile solution of TpRe(MeIm)(η^2-benzene) stirred in the presence of DMM and TBSOTf at $-20\,^\circ$C gives two benzenium diastereomers (Scheme 55). Addition of MMTP gives two diastereomers that result from 1,4-tandem addition.

The coordination of toluene to form TpRe(CO)(MeIm)(η^2-toluene) results in a mixture of four regio- and stereoisomers.[151] The addition of HOTf and MMTP to TpRe(CO)(MeIm)(η^2-toluene) results in a tandem addition to the coordinated toluene to produce *cis*-3-dimethylcarbomethoxymethyl-6H-hydro-1,4-cyclohexadiene in 62% isolated yield. Despite the mixture of Re isomers in the parent solution, this is the only organic product isolated from the reaction. Likely due to the additional steric bulk of the methyl group, Michael acceptors or acetals were not incorporated at C3.

TpRe(CO)(L)(η^2-naphthalene) complexes undergo tandem addition reactions with control of regioselectivity dependent upon the identity of L.[152,153] At 20 °C, TpRe(CO)(PMe_3)(η^2-naphthalene) exists as a 10:1 mixture of diastereomers with the uncoordinated ring *anti* to PMe_3 for the major diastereomer. When treated with

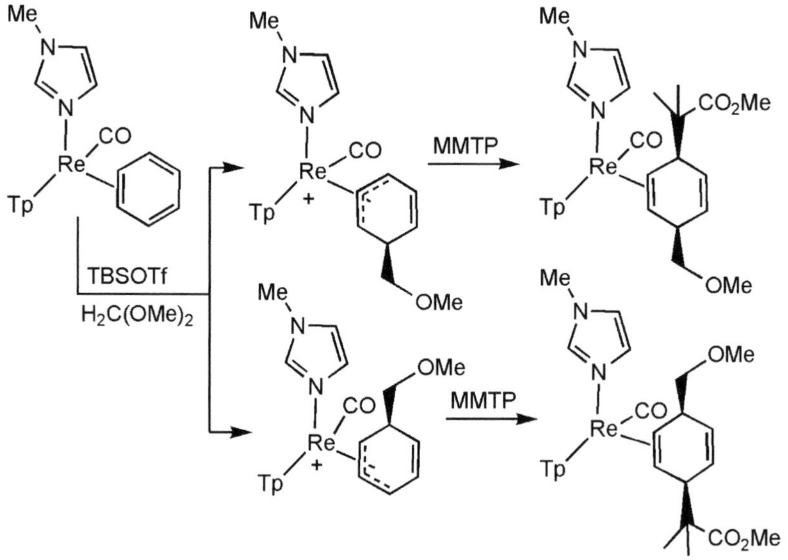

SCHEME 55. Tandem electrophile/nucleophile addition to the η^2-benzene ligand of TpRe(CO)(MeIm)(η^2-benzene) (MMTP = 1-methoxy-2-methyl-1-(trimethylsiloxy)propene).

triflic acid (HOTf) followed by MMTP, the *cis*-1,4-tandem addition product is attained with 12:1 regioselectivity over the 1,2-addition product. At 20 °C, TpRe(CO)(MeIm)(η^2-naphthalene) exists as a 1:5 mixture of diastereomers favoring alignment of the uncoordinated ring *syn* to MeIm, and addition of HOTf/MMTP to this complex results in a *cis*-1,2-tandem addition with 23:1 regioselectivity versus the 1,4-addition (Scheme 56). Among other factors, the regioselectivity seems attributable to π/π interactions between L and the uncoordinated ring of naphthalene to give the 1,2-addition. For instance, a single-crystal X-ray diffraction study of the 1,2-addition product TpRe(CO)(py)(η^2-1H-2-dimethylcarbomethoxymethyldihydronaphthalene) indicates π-stacking between pyridine and the unbound ring of naphthalene. In contrast, steric interactions between PMe_3 and the uncoordinated naphthalene ring are sufficient to favor the opposite orientation and thus 1,4-addition.

DMM and various Michael acceptors can also be used to generate TpRe(CO)(L)(η^3-1H-naphthalenium) (L = py, MeIm or PMe_3) complexes, which subsequently react with stabilized enolates, silyl ketene acetals or enols. Dimethoxymethane must be used in the presence of TBSOTf and 1,8-diazabicyclo[5.4.0]undec-7-ene. Reactions with DMM yield only the 1,4-tandem addition products, regardless of the identity of L (Scheme 57). The combination of TpRe(CO)(L)(η^2-naphthalene), dimethylmalonate and TBSOTf followed by MMTP results in 1,2-addition with incorporation of a proton as the electrophile (Scheme 58). In this reaction, 2 equivalents of MMTP are incorporated into the product, and this transformation is a rare example of activation of the unbound naphthalene ring by the π-basic Re(I) fragment. Chiral resolution using α-pinene

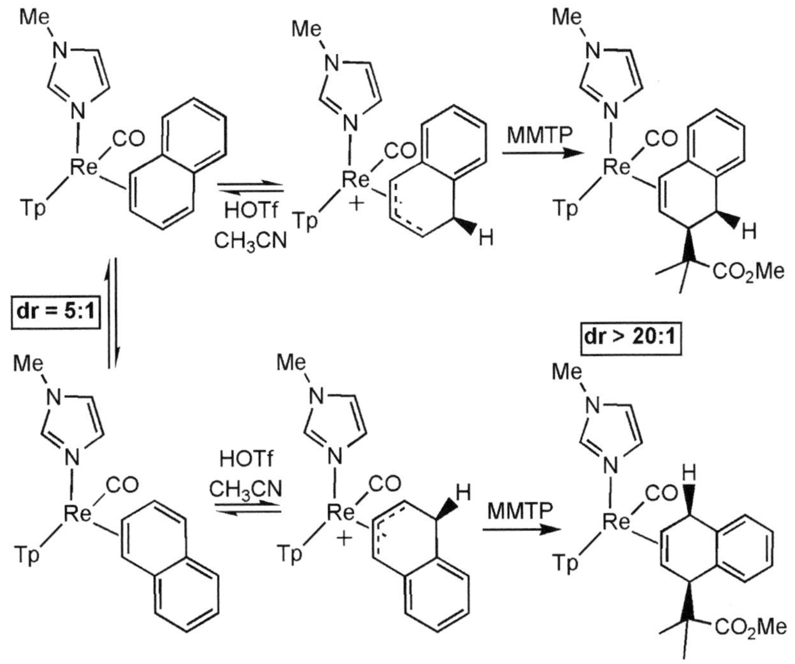

SCHEME 56. 1,2-Tandem addition of H^+/1-methoxy-2-methyl-1-(trimethylsiloxy)-propene (MMTP) to TpRe(CO)(MeIm)(η^2-naphthalene).

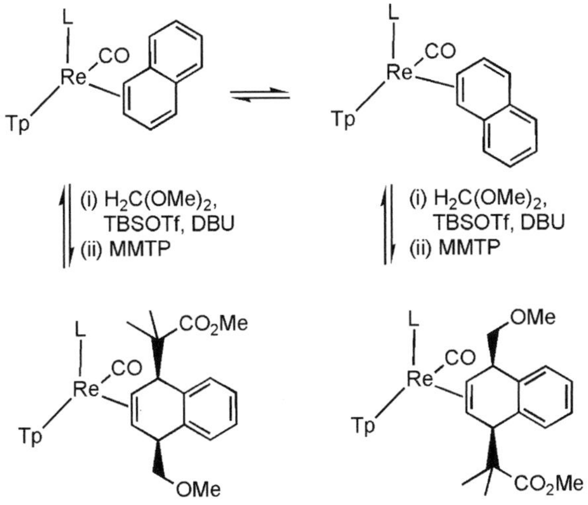

SCHEME 57. 1,4-Tandem addition of DMM/MMTP using TpRe(CO)(L)(η^2-naphthalene) (L = PMe_3, py, MeIm).

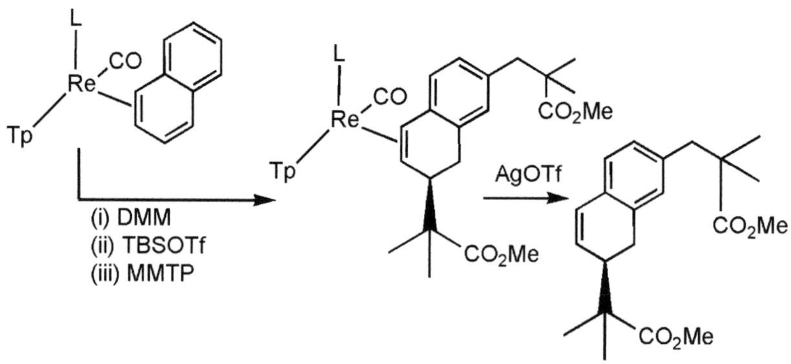

SCHEME 58. Addition to the unbound ring of naphthalene (TBS = tributylsilyl, L = py, MeIm.

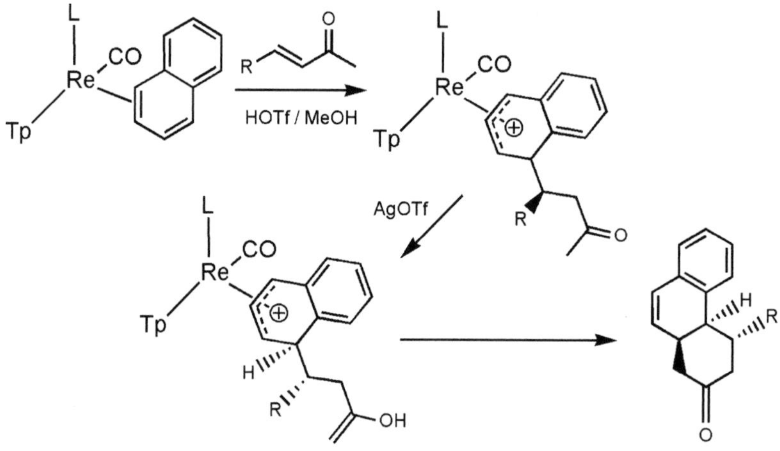

SCHEME 59. Ring-closing reaction of η^2-naphthalene with α,β–unsaturated ketones to produce phenanthrene ring systems. Use of enantiopure Re (L = py) and optimized reactions conditions provides the organic product with 96% ee.

(see above) can be employed *in situ* in combination with tandem addition of DMM and MMTP in a single pot synthesis. The free organic product is generated upon oxidative demetallation using AgOTf with 90% *ee* for the isolated organic product.

Michael acceptors such as methyl vinyl ketone and 3-penten-2-one are added to TpRe(CO)(L)(η^2-naphthalene) complexes (L = py or MeIm) to give η^3-1H-naphthalenium complexes which can undergo intramolecular nucleophilic addition to give ring-closed phenanthrene ring systems (Scheme 59).[153] For example, reaction of TpRe(CO)(L)(η^2-naphthalene) with 3-penten-2-one in the presence of triflic acid and methanol yields *S*-tetrahydro-1H-phenanthrene-2-one as a single diastereomer for L = MeIm and as an 8:1 mixture of diastereomers for L = py. Using a resolved Re complex (with (*R*)-α-pinene, see above) with L = py and optimized reaction conditions, the free organic product was produced with 96% ee.

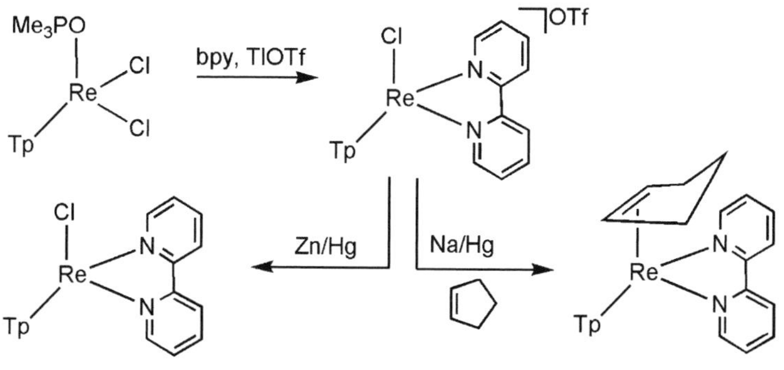

SCHEME 60. Low-valent TpRe(bpy) complexes.

Low-valent rhenium complexes with Tp and bpy (bpy = bipyridine) ligands that do not include CO in the coordination sphere have been synthesized. Complexes of the type TpRe(bpy)Y (Y = Cl or cyclopentene; bpy = 2,2′-bipyridyl) are derived from [TpRe^{III}(bpy)Cl][OTf] with the accompanying one or two electron reduction by Zn/Hg or Na/Hg amalgam, respectively (Scheme 60).[154] TpRe(bpy)(η^2-cyclopentene) is a potent π-base as evidenced by the II/I oxidation potential of −0.66 V versus NHE, 0.89 V more cathodic than TpRe(CO)(PMe_3)(η^2-cyclohexene).

B. *Manganese Chemistry*

For poly(pyrazolyl)borate complexes of low-valent Mn, when sterically unhindered pyrazole arms are employed homoleptic poly(pyrazolyl)borates can be accessed. Bp_2Mn, Bp^*_2Mn, Tp_2Mn, Tp^*_2Mn, and $(pzTp)_2Mn$ are isolated by precipitation or extraction from aqueous solutions of the potassium poly(pyrazole) borates.[155–158] Magnetic susceptibility data suggest that these complexes are tetrahedral (for Bp and Bp*) or octahedral (for Tp, Tp* or pzTp; Scheme 61).[158] A thorough account of the electrochemical, magnetic, and spectroscopic properties of these complexes has been made.[97] It was determined that ligand field strength follows the trend Tp* < Tp < pzTp, which was proposed to result from the influence of steric profile on spectroscopic properties. Redox potentials indicate that overall donating ability for these systems is ordered Tp* > Tp > pzTp.

Modified poly(pyrazolyl)borates have been recently synthesized for which addition of the proper chelating group to the pyrazolyl-position makes each scorpionate arm a bidentate chelate.[102,159] Termed podands, these ligands have been shown to coordinate in a hexadentate manner to some transition metals. Coordination of these ligands to manganese has been explored.[159] The reaction of [K][HB(3-(2-pyridyl)pz)$_3$] with $Mn(OAc)_2$ leads to formation of the unusual tetranuclear complex [Mn_4{HB(3-(2-pyridyl)pz)$_3$}$_4$] as characterized by X-ray crystallography. In this structure, each scorpionate arm coordinates to three different metal centers (each in a κ^2 mode) rendering a tetrahedral metal skeleton with four face-capping scorpionate ligands.

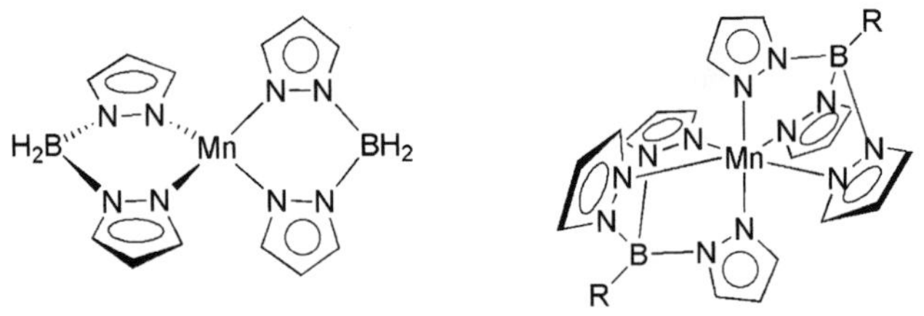

SCHEME 61. Homoleptic poly(pyrazolyl)borate complexes of Mn(II) (R = H, pz).

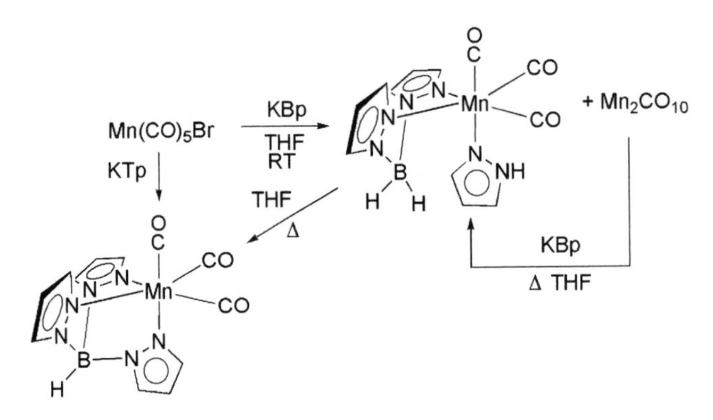

SCHEME 62. Poly(pyrazolyl)borate complexes obtained from $Mn(CO)_5Br$ (RT = room temperature).

Five-coordinate Mn(II) salts [A][Bp_2MnX] (X = Cl, A = PPh_4 or $AsPh_4$; X = NCS or N_3, A = NEt_4) have been synthesized.[160] The coordination geometry was confirmed by X-ray crystallography to be square pyramidal for [NEt_4][Bp_2MnCl]. The magnetic moment is consistent with high-spin d^5 Mn(II).

$BpMn(CO)_3$(pyrazole) is prepared from reaction of $Mn(CO)_5Br$ and KBp, accompanied by $Mn_2(CO)_{10}$ as a byproduct.[101] The dimer $Mn_2(CO)_{10}$ can be heated in THF in the presence of KBp to give $BpMn(CO)_3$(pyrazole). KBp is stable in refluxing THF, thus it may be inferred that the pyrazole ligand of $BpMn(CO)_3$ (pyrazole) is likely derived from a metal-mediated process. In addition, $BpMn(CO)_3$(pyrazole) can be heated under reflux in THF to give $TpMn(CO)_3$, which can be independently synthesized from $Mn(CO)_5Br$ and KTp (Scheme 62).

$BpMn(CO)_3$(pyrazole) reacts with various phosphines to form $BpMn(CO)_2L_2$ complexes with *cis* CO ligands and phosphines that may be either *cis* or *trans*.[101] For example 1,2-bis(diphenylphosphino)ethane (dppe) displaces 1 equivalent of CO and pyrazole giving BpMn(*cis*-$(CO)_2$)(dppe). Reaction of $BpMn(CO)_3$(pyrazole) with trimethylphosphite leads to the *cis*-dicarbonyl *trans*-bisphosphite complex $BpMn(CO)_2\{P(OMe)_3\}_2$.[101] Anionic Mn(I) salts of the type $[BpMn(CO)_3X]^-$ (X = I or Br) are formed upon reaction of $BpMn(CO)_3$(pyrazole) with tetra-propylammonium iodide or tetraethylammonium bromide.

$TpMn(CO)_3$ (**50**), (pzTp)$Mn(CO)_3$ (**51**) and Tp*$Mn(CO)_3$ (**52**) were all prepared from $Mn(CO)_5Br$, and IR spectroscopy suggests that these systems are more electron-rich than the cyclopentadienyl analogs.[161,162] The phosphine complexes

$TpMn(CO)_2L$, $(pzTp)Mn(CO)_2L$ and $Tp^*Mn(CO)_2L$ {L = PF_3, PCl_3, $P(OPh)_3$, PPh_3, $P(n\text{-}C_4H_9)_3$, $P(C_6H_{11})_3$} were prepared from **50**, **51** and **52**, respectively, using photolytic methods. For most procedures, 1 equivalent of CO is liberated from the tricarbonyl precursor to generate a stable THF adduct, which reacts with L under either photolytic or thermal conditions.[163] When excess L {L = PF_3, $P(OMe)_3$ or $P(OPh)_3$} is used $(pzTp)Mn(CO)L_2$ is obtained. For these Mn complexes, the electronic character of pzTp appears similar to that of Cp. Differences in mono- and di-substitution suggest that the size of the polypyrazolyl borate and L play a role in reactivity.[164]

Incorporation of bulky substituents at the pyrazolyl three position imparts a predilection toward formation of tetrahedral complexes.[5,17,165] For instance, $MnCl_2$ reacts with [K][HB(3-^{i}Pr,4-Brpz)$_3$] to form the four-coordinate tetrahedral complex $Tp^{^iPrBr}MnCl$.[165] $MnCl_2$ also reacts with [K][(HB(3,5-^{i}Pr$_2$pz)$_3$)] to form $Tp^{^iPr_2}Mn(Cl)$ (**53**).[93,166] In the presence of 5 equivalents of 3,5-^{i}Pr$_2$pzH, $Tp^{^iPr_2}Mn(Cl)(3,5\text{-}^iPr_2pz)$ (**54**) is formed.[93] Grinding $MnCl_2$ in an agate mortar with [Tl][(HB(3-Ph,5-Mepz)$_3$)] and an acetonitrile–water mixture yields $Tp^{Ph,Me}MnCl$ (3-Ph,5-Mepz).[167] Decomposition by hydrolysis resulting in B–N bond cleavage and coordination of a free pyrazole ring was however found to be problematic.[167]

Treatment of either **53** or **54** with NaOH affords $[Tp^{^iPr_2}Mn]_2(\mu\text{-}OH)_2$ which has been characterized by a single-crystal X-ray diffraction study.[93] The hydroxo ligands are separated from each Mn center by 2.094(4) and 2.089(5) Å. The observed Mn–Mn separation (3.31 Å) is shorter than expected but longer than other reported bis(μ-oxo) dinuclear Mn complexes ($\sim$2.7 Å).[93] Magnetic susceptibility measurements (6.88 $\mu B\,mol^{-1}$ at 298 K) are characteristic of a weak interaction between two high-spin Mn(II) ions, and X-band EPR spectroscopy (77 K) is consistent with a weakly coupled dinuclear complex. A comprehensive account of this chemistry has been provided.[2] In toluene under 1 atm of CO_2, $[Tp^{^iPr_2}Mn]_2(\mu\text{-}OH)_2$ (**35**) is converted to $[Tp^{^iPr_2}Mn]_2(\mu\text{-}CO_3)$ as verified by ^1H and ^{13}C NMR spectroscopy, FD-MS and X-ray crystallography.[166] $[Tp^{^iPr_2}Mn]_2(\mu\text{-}CO_3)$ reverts to complex **35** in a toluene solution of NaOH. In the presence of $KMnO_4$ under anaerobic conditions, complex **35** is oxidized to $[Tp^{^iPr_2}Mn]_2(\mu\text{-}O)_2$ (**36**). The bis(μ-oxo) complex can also be formed by the reaction of **35** with O_2 or H_2O_2; however, the aerobic oxidation leads to additional products (see Scheme 21).[93,94,168,169] This transformation was the first example of the oxidative conversion of a bis(μ-hydroxo) moiety to a bis(μ-oxo) moiety on a polynuclear manganese center, which is thought to be an important function in the Mn-based photosystem II oxygen-evolving center.[93]

If **35** is treated with H_2O_2 in the presence of 2 equivalents of 3,5-^{i}Pr$_2$pzH (−78 °C) the blue mononuclear Mn(III) complex $Tp^{^iPr_2}Mn(O_2)(3,5\text{-}^iPr_2pz)$ is formed with O_2 coordinated side-on and hydrogen-bonding with the pyrazole ligand.[169] Warming from −78 to −20 °C produces a brown isomer that lacks the hydrogen-bonding interaction. In the presence of 1 equivalent of 3,5-^{i}Pr$_2$pzH, **35** gives $[Tp^{^iPr_2}Mn]_2(\mu\text{-}OH)(\mu\text{-}3,5\text{-}^iPr_2pz)$ (Scheme 63). Formation of the bis(μ-pyrazolato) complex was not observed even at elevated concentrations of 3,5-^{i}Pr$_2$pzH.[169] The mononuclear complex $Tp^{^iPr_2}Mn(3,5\text{-}^iPr_2pz)$ is formed upon ion exchange of [Na][3,5-^{i}Pr$_2$pz] with $Tp^{^iPr_2}MnCl$; however, this reactive complex decomposes upon work-up and forms

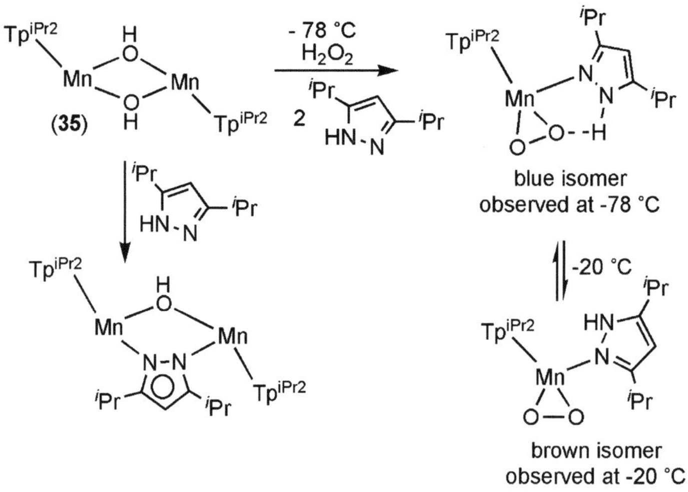

SCHEME 63. Transformations of $[Tp^{^iPr_2}Mn]_2(\mu\text{-OH})_2$ with 3,5-diisopropylpyrazole.

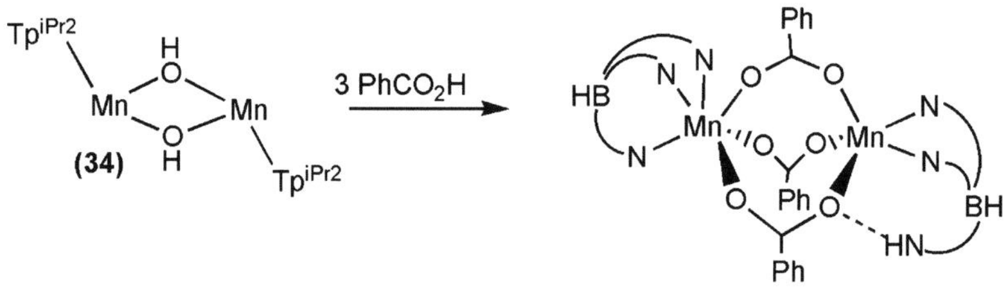

SCHEME 64. Conversion of $[Tp^{^iPr_2}Mn]_2(\mu\text{-OH})_2$ to $Tp^{^iPr_2}Mn(\mu\text{-}O_2CPh)_3Mn(HB(3,5\text{-}^{i}Pr_2pz)_2(3,5\text{-}^{i}Pr_2pzH)$ upon reaction with benzoic acid.

$[Tp^{^iPr_2}Mn]_2(\mu\text{-OH})(\mu\text{-}3,5\text{-}^iPr_2pz)$. Further hydrolysis results in the formation of $[Tp^{^iPr_2}Mn]_2(\mu\text{-OH})_2$ (**35**).

The addition of benzoic acid to $[Tp^{^iPr_2}Mn]_2(\mu\text{-OH})_2$ (**35**) results in the formation of $Tp^{^iPr_2}Mn(\mu\text{-}O_2CPh)_3Mn\{HB(3,5\text{-}^iPr_2pz)_2(3,5\text{-}^iPr_2pzH\}$ where two arms of the scorpionate are coordinated to a single Mn center and the third arm is protonated and hydrogen-bonding with a bridging benzoato ligand (Scheme 64).[170] The hydrogen-bonding interaction is apparently sufficient to inhibit hydrolysis of the B–N linkage under acidic conditions.

Transition metal–alkylperoxo complexes may be integral to catalytic oxygenation reactions and can serve as analogs to hydroperoxo complexes thought to play biologically important roles in dioxygen metabolism. Mn(II) complexes with alkylperoxo ligands have been synthesized.[171] $Tp^{^tBu,^iPr}Mn(OH)$ was prepared by reaction of $Tp^{^tBu,^iPr}Mn(OAc)$ with NaOH. The Mn(II)-hydroxide complex reacts with cumylhydroperoxide to yield $Tp^{^tBu,^iPr}Mn(OOCMe_2Ph)$ (Scheme 65). The solid-state structure of $Tp^{^tBu,^iPr}Mn(OOCMe_2Ph)$ was determined. The alkylperoxo complex was not, however, isolable when $Tp^{^iPr_2}$ was used as the scorpionate. It is

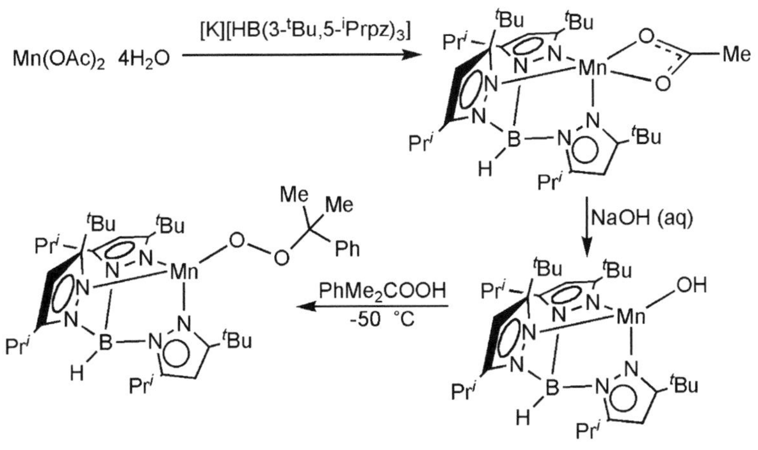

SCHEME 65. Synthesis of a scorpionate Mn(II) alkylperoxo complex *via* $Tp^{^tBu,^iPr}Mn(OAc)$ and $Tp^{^tBu,^iPr}Mn(OH)$ complexes.

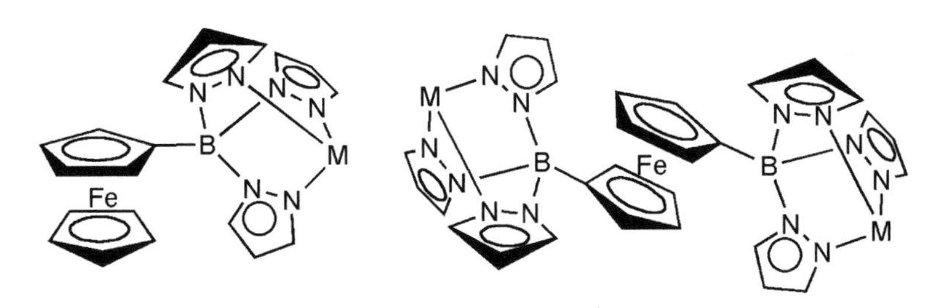

SCHEME 66. Heterodinuclear and trinuclear ferrocenyltris(1-pyrazolyl)borate (M=Li or Tl).

likely that the steric bulk of the ^{t}Bu group stabilizes the low-valent Mn complex at −20°C and hinders reduction by PPh_3, Me_2S and hydrocarbons; however, the complex decomposes at room temperature.

Poly(pyrazolyl)borate ligands have also been used with Mn(II) as metalloenzyme mimics to demonstrate stabilizing intramolecular hydrogen-bonding interactions between H_2O ligands and an ancillary backbone.[172] [K][HB((3-methyl-5-carboxyethyl)pyrazolyl)$_3$] reacts with $Mn(ClO_4)_2 \cdot 6H_2O$ in MeOH to form $[Tp^{Me,CO_2Et}Mn(OH_2)_3][ClO_4]$. A solid-state structure of this octahedral complex indicates two hydrogen-bonding interactions for each H_2O ligand. The bonding partners are two carboxyethyl carbonyl oxygens, one carboxyethyl carbonyl oxygen and one solvent water, or one carboxyethyl carbonyl oxygen and one oxygen from the perchlorate anion, depending on the specific water ligand. Along with other first row divalent analogs, this complex is a model for the vicinal oxygen chelate family of enzymes.[172]

Extensive studies have been conducted on ferrocene-based tris(1-pyrazolyl)borate ligands (Scheme 66).[173,174] One drawback to these systems is a problematic steric interaction between the ferrocenyl substituent and the hydrogen atom at the

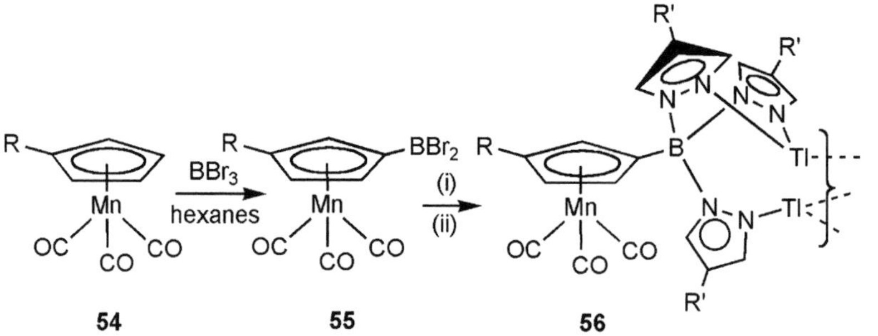

56a, R = R' = H; **56b**, R = Me, R' = H; **56c**, R = Me, R' = Br; **56d**, R = Me, R' = Ph

SCHEME 67. Synthesis of cymantrenylpoly(pyrazolyl)borates. (i) 3 4-R′ pyrazole, 2 NEt_3, toluene; (ii) TlOEt, toluene.

5-position of a pyrazole ring.[175] Cymantrenyl Mn(I) complexes have been synthesized in efforts to avoid this problem.

Cyclopentadienyl manganese tricarbonyl ("cymantrene," Cym) is brominated upon treatment with BBr_3 to yield **55** (Scheme 67).[176] Subsequent reactions with various 4-substituted pyrazoles, 4R-pyrazole $\{R = H, Br, CH_2(C_6H_{11})\}$ and 2 equivalents of NEt_3 followed by reaction with TlOEt produces the corresponding Tl(I) salt, **56a**, **56b**, **56c** and **56d** (Scheme 67). The ^1H NMR spectra of these complexes suggest hindered rotation around the B–C bond as evidenced by broad resonances due to the pyrazolyl fragments, which sharpen upon heating.[176] Solid-state crystal structures of **56a** and **56b** were solved revealing formation of polymeric chains in the solid-state of **56a** while **56b** packs as cyclic tetramers. Despite the differences in the solid-state packing, the monomeric structures of **56a** and **56b** are similar. Both bind one Tl(I) κ^2 by two pyrazole nitrogen atoms while the third pyrazole ring coordinates to a different Tl(I) ion. The source for the discrepancy in crystal packing might be the different orientation of one pyrazole ring in each monomer as evidenced by the large difference in the C–B–N–N torsion angles for the respective rings. The Tl–N bonding interactions are thought to be relatively weak as NMR solutions suggest C_3 symmetry.

Ferrocenyltris(1-pyrazolyl)borate (FcTp) has been combined with $MnCl_2$ to form a heterotrinuclear complex with one Mn and two Fe centers (Scheme 68).[177] Cyclic voltammetry reveals an irreversible Mn(III/II) oxidation at 0.37V (versus SCE). This oxidation occurs at more negative potential than the two-electron ferrocenyl oxidation (one per ferrocene), which occurs at 0.54V.

Cymantrenyltris(1-pyrazolyl)borate (CymTp) has also been incorporated into heterodimetallic and heterotrimetallic motifs.[178] [Tl][CymTp] reacts with $Mn(CO)_5Br$ to produce $(CymTp)Mn(CO)_3$ (**57**) (Scheme 69). If adventitious water is present during the synthesis, complex **58** is formed. The reaction of 2 equivalents of [Tl][CymTp] with $ZnBr_2$ produces the heterotrimetallic complex **59**. IR absorptions for complex **58** were assigned as $\nu_{CO} = 2030 cm^{-1}$ for the $Mn(CO)_3$ coordinated to the scorpionate and $2023 cm^{-1}$ for the cymantrenyl moiety reflecting

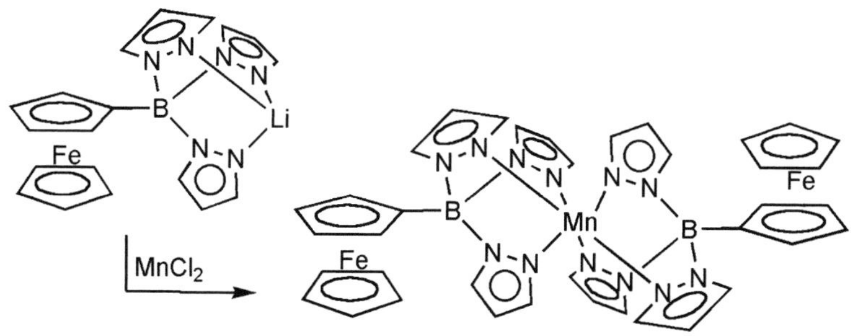

SCHEME 68. The heterotrimetallic complex $(FcTp)_2Mn$.

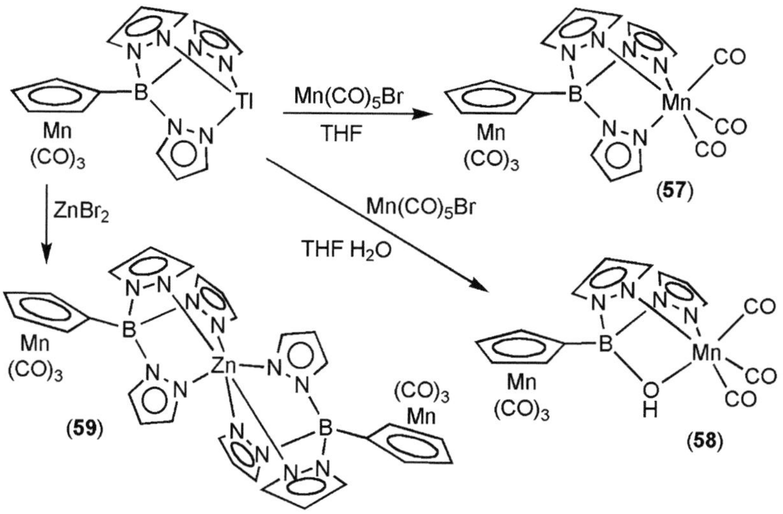

SCHEME 69. Heterodimetallic and heterotrimetallic derivatives of [Tl][CymTp].

the electronic similarity of the two environments. Solid-state structures were determined for all three complexes **57**–**59**.

Another example of a polymetallic scorpionate based on Mn(II) which possesses a metal–metal bond has been synthesized.[2,179] $Tp^{iPr_2}MnCl$ reacts with $AgPF_6$ in acetonitrile to produce $[Tp^{iPr_2}Mn(NCMe)_3][PF_6]$. The Mn(II) salt reacts with $[Co(CO)_4]^-$ to form $Tp^{iPr_2}Mn–Co(CO)_4$. This complex represents a new polymetallic species with a polarized metal–metal bond, "Werner-like" at one metal and "soft" (i.e., organometallic) at the other metal center. This complex and others with scorpionate bound metals including Ni, Co and Fe have been characterized by IR spectroscopy and X-ray crystallography. When $Tp^{iPr_2}Mn–Co(CO)_4$ is dissolved in NCMe the ion pair $[Tp^{iPr_2}Mn(NCMe)_3][Co(CO)_4]$ is formed (Scheme 70). In a toluene solution, PPh_3 irreversibly displaces a CO ligand to yield $Tp^{iPr_2}Mn–Co(CO)_3(PPh_3)$.

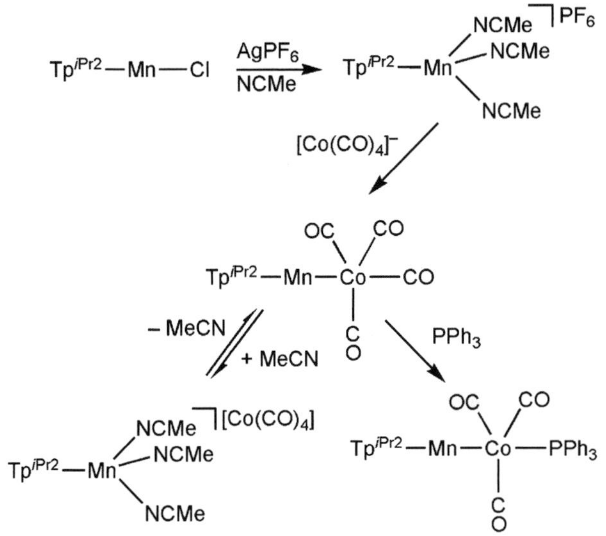

Tp^{iPr2}—Mn—Cl
$AgPF_6$
NCMe
Tp^{iPr2}—Mn(NCMe)(NCMe)(NCMe) PF_6
$[Co(CO)_4]^-$
OC CO
Tp^{iPr2}—Mn—Co—CO
C
O
– MeCN
+ MeCN
PPh_3
NCMe $[Co(CO)_4]$
Tp^{iPr2}—Mn(NCMe)(NCMe)(NCMe)
OC CO
Tp^{iPr2}—Mn—Co—PPh_3
C
O

SCHEME 70. Synthesis and reactivity of a scorpionate Mn(II) "xenophilic" complex.

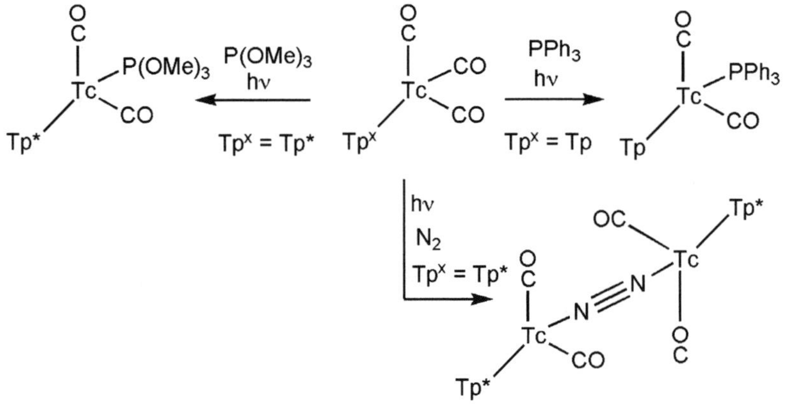

O
C
$P(OMe)_3$
Tc
CO
Tp*
$P(OMe)_3$
hν
Tp^x = Tp*
O
C
CO
Tc
CO
Tp^x
PPh_3
hν
Tp^x = Tp
O
C
PPh_3
Tc
CO
Tp
hν
N_2
Tp^x = Tp*
OC
Tp*
Tc
O
C
N≡N
Tc
CO
Tp*
O
C

SCHEME 71. Photolysis of $TpTc(CO)_3$ and $Tp^*Tc(CO)_3$.

C. *Technetium Chemistry*

Due in part to the importance of Tc in radio diagnostics, low-valent Tc scorpionates have been synthesized and investigated. $Tc(CO)_5Br$ reacts with KTp or KTp* to produce the octahedral $TpTc(CO)_3$ and $Tp^*Tc(CO)_3$ complexes, respectively, which have been characterized by X-ray crystallography.[105] Irradiation of THF solutions of either complex yields unstable $TpTc(CO)_2THF$ as well as the isolable complex $Tp^*Tc(CO)_2THF$. Photolysis of $TpTc(CO)_3$ or $Tp^*Tc(CO)_3$ in the presence of PPh_3, $P(OMe)_3$ or N_2 produces the isolable complexes $TpTc(CO)_2(PPh_3)$,

$Tp^*Tc(CO)_2\{P(OMe)_3\}$ or dinuclear $[Tp^*Tc(CO)_2]_2(\mu\text{-}N_2)$ (Scheme 71).[106] TpTc $(CO)_2(PPh_3)$ is also prepared from $Tc(CO)_3(PPh_3)_2Cl$ and KTp.[180]

IV
SUMMARY

Poly(pyrazolyl)borate ligands support numerous complexes involving Group 7 transition metals. The flexibility of these ligands has allowed access to a range of oxidation states and geometries as well as provided Mn, Tc and Re complexes with a diversity of reactivity.

References

(1) Trofimenko, S. *J. Am. Chem. Soc.* **1966**, *88*, 1842–1844.
(2) Akita, M.; Hikichi, S. *Bull. Chem. Soc. Jpn.* **2002**, *75*, 1657–1679.
(3) Trofimenko, S. *Acc. Chem. Res.* **1971**, *4*, 17–22.
(4) Trofimenko, S. *Chem. Rev.* **1972**, *72*, 497–509.
(5) Trofimenko, S. *Chem. Rev.* **1993**, *93*, 943–980.
(6) Kitajima, N.; Moro-oka, Y. *Chem. Rev.* **1994**, *94*, 737–757.
(7) Reger, D. L. *Coord. Chem. Rev.* **1996**, *147*, 571–595.
(8) Etienne, M. *Coord. Chem. Rev.* **1996**, *156*, 201–236.
(9) Janiak, C. *Coord. Chem. Rev.* **1997**, *163*, 107–216.
(10) Slugovc, C.; Schmid, R.; Kirchner, K. *Coord. Chem. Rev.* **1999**, *185–186*, 109–126.
(11) Menger, F. M. *Top. Curr. Chem.* **1986**, *136*, 1–15.
(12) Trofimenko, S. *Prog. Inorg. Chem.* **1986**, *34*, 115–210.
(13) Han, R. *J. Inorg. Biochem.* **1993**, *49*, 105–121.
(14) Byers, P. K. *Adv. Organomet. Chem.* **1992**, *34*, 1–65.
(15) Parkin, G. *Adv. Inorg. Chem.* **1995**, *42*, 291–393.
(16) Santos, I. *New J. Chem.* **1995**, *19*, 551–571.
(17) Kitajima, N.; Tolman, W. B. *Prog. Inorg. Chem.* **1995**, *43*, 419–531.
(18) Trofimenko, S. *Scorpionates: The Coordination Chemistry of Polypyrazolylborate Ligands*, Imperial College Press, London, **1999**.
(19) Keane, J. M.; Harman, W. D. *Organometallics* **2005**, *24*, 1786–1798.
(20) Brown, S. N.; Mayer, J. M. *Inorg. Chem.* **1992**, *31*, 4091–4100.
(21) Brown, S. N.; Mayer, J. M. *Inorg. Chem.* **1995**, *34*, 3560–3562.
(22) Tisato, F.; Bolzati, C.; Duatti, A.; Bandoli, G.; Refosco, F. *Inorg. Chem.* **1993**, *32*, 2042–2048.
(23) Brown, S. N.; Mayer, J. M. *J. Am. Chem. Soc.* **1994**, *116*, 2219–2220.
(24) Brown, S. N.; Myers, A. W.; Fulton, J. R.; Mayer, J. M. *Organometallics* **1998**, *17*, 3364–3374.
(25) Brown, S. N.; Mayer, J. M. *Organometallics* **1995**, *14*, 2951–2960.
(26) Brown, S. N.; Mayer, J. M. *J. Am. Chem. Soc.* **1996**, *118*, 12119–12133.
(27) DuMez, D. D.; Mayer, J. M. *J. Am. Chem. Soc.* **1996**, *118*, 12416–12423.
(28) Dumez, D. D.; Mayer, J. M. *Inorg. Chem.* **1995**, *34*, 6396–6401.
(29) Matano, Y.; Northcutt, T. O.; Brugman, J.; Bennett, B. K.; Lovell, S.; Mayer, J. M. *Organometallics* **2000**, *19*, 2781–2790.
(30) Matano, Y.; Brown, S. N.; Northcutt, T. O.; Mayer, J. M. *Organometallics* **1998**, *17*, 2939–2941.
(31) Paulo, A.; Domingos, A.; Garcia, R.; Santos, I. *Inorg. Chem.* **2000**, *39*, 5669–5674.
(32) Paulo, A.; Reddy, K. R.; Domingos, A.; Santos, I. *Inorg. Chem.* **1998**, *37*, 6807–6813.
(33) Paulo, A.; Domingos, A.; Marcalo, J.; Dematos, A. P.; Santos, I. *Inorg. Chem.* **1995**, *34*, 2113–2120.
(34) DuMez, D. D.; Mayer, J. M. *Inorg. Chem.* **1998**, *37*, 445–453.
(35) DuMez, D. D.; Northcutt, T. O.; Matano, Y.; Mayer, J. M. *Inorg. Chem.* **1999**, *38*, 3309–3312.

(36) Domingos, A.; Marcalo, J.; Paulo, A.; Dematos, A. P.; Santos, I. *Inorg. Chem.* **1993**, *32*, 5114–5118.
(37) Thomas, J. A.; Davison, A. *Inorg. Chim. Acta* **1991**, *190*, 231–235.
(38) Nunes, D.; Domingos, A.; Paulo, A.; Patricio, L.; Santos, I.; Carvalho, N. N.; Pombeiro, A. J. L. *Inorg. Chim. Acta* **1998**, *271*, 65–74.
(39) Masui, C. S.; Mayer, J. M. *Inorg. Chim. Acta* **1996**, *251*, 325–333.
(40) Chiu, K. W.; Wong, W. K.; Wilkinson, G.; Galas, A. M. R.; Hursthouse, M. B. *Polyhedron* **1982**, *1*, 31–36.
(41) McNeil, W. S.; DuMez, D. D.; Matano, Y.; Lovell, S.; Mayer, J. M. *Organometallics* **1999**, *18*, 3715–3727.
(42) Li, Z.; Quan, R. W.; Jacobsen, E. N. *J. Am. Chem. Soc.* **1995**, *117*, 5889–5890.
(43) Brandt, P.; Södergren, M. J.; Andersson, P. G.; Norrby, P.-O. *J. Am. Chem. Soc.* **2000**, *122*, 8013–8020.
(44) Müller, P.; Fruit, C. *Chem. Rev.* **2003**, *103*, 2905–2920.
(45) Evans, D. A.; Faul, M. M.; Bilodeau, M. T. *J. Org. Chem.* **1991**, *56*, 6744–6746.
(46) Díaz-Requejo, M. M.; Pérez, P. J. *J. Organomet. Chem.* **2001**, *617–618*, 110–118.
(47) Cantrell, G. K.; Meyer, T. Y. *J. Am. Chem. Soc.* **1998**, *120*, 8035–8042.
(48) Meyer, K. E.; Walsh, P. J.; Bergman, R. G. *J. Am. Chem. Soc.* **1995**, *117*, 974–985.
(49) Zuckerman, R. L.; Krska, S. W.; Bergman, R. G. *J. Am. Chem. Soc.* **2000**, *122*, 751–761.
(50) Krska, S. W.; Zuckerman, R. L.; Bergman, R. G. *J. Am. Chem. Soc.* **1998**, *120*, 11828–11829.
(51) Meyer, K. E.; Walsh, P. J.; Bergman, R. G. *J. Am. Chem. Soc.* **1994**, *116*, 2669–2670.
(52) Burland, M. C.; Pontz, T. W.; Meyer, T. Y. *Organometallics* **2002**, *21*, 1933–1941.
(53) Bruno, J. W.; Li, X. J. *Organometallics* **2000**, *19*, 4672–4674.
(54) McInnes, J. M. *Chem. Commun.* **1998**, 1669.
(55) Reddy, K. L.; Sharpless, K. B. *J. Am. Chem. Soc.* **1998**, *120*, 1207–1217.
(56) Bennett, J. L.; Wolczanski, P. T. *J. Am. Chem. Soc.* **1997**, *119*, 10696–10719.
(57) Cummins, C. C.; Baxter, S. M.; Wolczanski, P. T. *J. Am. Chem. Soc.* **1988**, *110*, 8731–8733.
(58) de With, J.; Horton, A. D. *Angew. Chem. Int. Ed.* **1993**, *32*, 903–905.
(59) Hoyt, H. M.; Michael, F. E.; Bergman, R. G. *J. Am. Chem. Soc.* **2004**, *126*, 1018–1019.
(60) Schaller, C. P.; Cummins, C. C.; Wolczanski, P. T. *J. Am. Chem. Soc.* **1996**, *118*, 591–611.
(61) Walsh, P. J.; Baranger, A. M.; Bergman, R. G. *J. Am. Chem. Soc.* **1992**, *114*, 1708–1719.
(62) Walsh, P. J.; Hollander, F. J.; Bergman, R. G. *J. Am. Chem. Soc.* **1988**, *110*, 8729–8731.
(63) Ackermann, L.; Bergman, R. G. *Org. Lett.* **2002**, *4*, 1475–1478.
(64) Baranger, A. M.; Walsh, P. J.; Bergman, R. G. *J. Am. Chem. Soc.* **1993**, *115*, 2753–2763.
(65) Johnson, J. S.; Bergman, R. G. *J. Am. Chem. Soc.* **2001**, *123*, 2923–2924.
(66) Straub, B. F.; Bergman, R. G. *Angew. Chem. Int. Ed.* **2001**, *40*, 4632–4635.
(67) Walsh, P. J.; Hollander, F. J.; Bergman, R. G. *Organometallics* **1993**, *12*, 3705–3723.
(68) Paulo, A.; Domingos, A.; Dematos, A. P.; Santos, I.; Carvalho, M. F. N. N.; Pombeiro, A. J. L. *Inorg. Chem.* **1994**, *33*, 4729–4737.
(69) Nunes, D. Graduate thesis, University of Lisbon, **1995**.
(70) Paulo, A.; Domingos, A.; Santos, I. *Inorg. Chem.* **1996**, *35*, 1798–1807.
(71) Paulo, A.; Domingos, A.; Santos, I. *J. Chem. Soc. Dalton Trans.* **1999**, 3735–3740.
(72) Herrmann, W. A.; Herdtweck, E.; Flöel, M.; Kulpe, J.; Küsthardt, U.; Okuda, J. *Polyhedron* **1987**, *6*, 1165–1182.
(73) Gable, K. P.; Phan, T. N. *J. Am. Chem. Soc.* **1994**, *116*, 833–839.
(74) Gable, K. P.; Juliette, J. J. J. *J. Am. Chem. Soc.* **1996**, *118*, 2625–2633.
(75) Gable, K. P.; AbuBaker, A.; Zientara, K.; Wainwright, A. M. *Organometallics* **1999**, *18*, 173–179.
(76) Gable, K. P.; Phan, T. N. *J. Am. Chem. Soc.* **1993**, *115*, 3036–3037.
(77) Gable, K. P.; Zhuravlev, F. A. *J. Am. Chem. Soc.* **2002**, *124*, 3970–3979.
(78) Gable, K. P.; Chuawong, P.; Yokochi, A. F. T. *Organometallics* **2002**, *21*, 929–933.
(79) Gable, K. P.; Brown, E. C. *J. Am. Chem. Soc.* **2003**, *125*, 11018–11026.
(80) Gable, K. P.; Khownium, K.; Chuawong, P. *Organometallics* **2004**, *23*, 5268–5274.
(81) Degnan, I. A.; Herrmann, W. A.; Herdtweck, E. *Chem. Ber.* **1990**, *123*, 1347–1349.
(82) Abrams, M. J.; Davison, A.; Jones, A. G. *Inorg. Chim. Acta* **1984**, *82*, 125–128.
(83) Duatti, A.; Tisato, F.; Refosco, F.; Mazzi, U.; Nicolini, M. *Inorg. Chem.* **1989**, *28*, 4564–4565.

(84) Faller, J. W.; Lavoie, A. R. *Organometallics* **2000**, *19*, 3957–3962.
(85) Drew, M. G. B.; Felix, V.; Goncalves, I. S.; Kuhn, F. E.; Lopes, A. D.; Romao, C. C. *Polyhedron* **1998**, *17*, 1091–1102.
(86) Reddy, K. R.; Domingos, A.; Paulo, A.; Santos, I. *Inorg. Chem.* **1999**, *38*, 4278–4282.
(87) Degnan, I. A.; Behm, J.; Cook, M. R.; Herrmann, W. A. *Inorg. Chem.* **1991**, *30*, 2165–2170.
(88) Hamilton, D. G.; Luo, X. L.; Crabtree, R. H. *Inorg. Chem.* **1989**, *28*, 3198–3203.
(89) Depaula, J. C.; Brudvig, G. W. *J. Am. Chem. Soc.* **1985**, *107*, 2643–2648.
(90) Sheats, J. E.; Czernuszewicz, R. S.; Dismukes, G. C.; Rheingold, A. L.; Petrouleas, V.; Stubbe, J.; Armstrong, W. H.; Beer, R. H.; Lippard, S. J. *J. Am. Chem. Soc.* **1987**, *109*, 1435–1444.
(91) Kono, Y.; Fridovich, I. *J. Biol. Chem.* **1983**, *258*, 13646–13648.
(92) Dexheimer, S. L.; Gohdes, J. W.; Chan, M. K.; Hagen, K. S.; Armstrong, W. H.; Klein, M. P. *J. Am. Chem. Soc.* **1989**, *111*, 8923–8925.
(93) Kitajima, N.; Singh, U. P.; Amagai, H.; Osawa, M.; Moro-oka, Y. *J. Am. Chem. Soc.* **1991**, *113*, 7757–7758.
(94) Kitajima, N.; Osawa, M.; Tanaka, M.; Morooka, Y. *J. Am. Chem. Soc.* **1991**, *113*, 8952–8953.
(95) Komatsuzaki, H.; Nagasu, Y.; Suzuki, K.; Shibasaki, T.; Satoh, M.; Ebina, F.; Hikichi, S.; Akita, M.; Moro-oka, Y. *J. Chem. Soc. Dalton Trans.* **1998**, 511–512.
(96) Kitajima, N.; Osawa, M.; Tamura, N.; Morooka, Y.; Hirano, T.; Hirobe, M.; Nagano, T. *Inorg. Chem.* **1993**, *32*, 1879–1880.
(97) De Alwis, D. C. L.; Schultz, F. A. *Inorg. Chem.* **2003**, *42*, 3616–3622.
(98) Chan, M. K.; Armstrong, W. H. *Inorg. Chem.* **1989**, *28*, 3777–3779.
(99) Thomas, R. W.; Davison, A.; Trop, H. S.; Deutsch, E. *Inorg. Chem.* **1980**, *19*, 2840–2842.
(100) Thomas, R. W.; Estes, G. W.; Elder, R. C.; Deutsch, E. *J. Am. Chem. Soc.* **1979**, *101*, 4581–4585.
(101) Bond, A.; Green, M. *J. Chem. Soc. A* **1971**, 682–685.
(102) Adams, H.; Batten, S. R.; Davies, G. M.; Duriska, M. B.; Jeffery, J. C.; Jensen, P.; Lu, J. Z.; Motson, G. R.; Coles, S. J.; Hursthouse, M. B.; Ward, M. D. *J. Chem. Soc. Dalton Trans.* **2005**, 1910–1923.
(103) Angaroni, M.; Ardizzoia, G. A.; D'Alfonso, G.; La Monica, G.; Masciocchi, N.; Moret, M. *J. Chem. Soc. Dalton Trans.* **1990**, 1895–1900.
(104) Bengali, A. A.; Mezick, B. K.; Hart, M. N.; Fereshteh, S. *Organometallics* **2003**, *22*, 5436–5440.
(105) Joachim, J. E.; Apostolidis, C.; Kanellakopulos, B.; Maier, R.; Marques, N.; Mayer, D.; Müller, J.; de Matos, A. P.; Nuber, B.; Rebizant, J.; Ziegler, M. L. *J. Organomet. Chem.* **1993**, *448*, 119–129.
(106) Joachim, J. E.; Apostolidis, C.; Kanellakopulos, B.; Meyer, D.; Nuber, B.; Raptis, K.; Rebizant, J.; Ziegler, M. L. *J. Organomet. Chem.* **1995**, *492*, 199–210.
(107) Bergman, R. G.; Cundari, T. R.; Gillespie, A. M.; Gunnoe, T. B.; Harman, W. D.; Klinckman, T. R.; Temple, M. D.; White, D. P. *Organometallics* **2003**, *22*, 2331–2337.
(108) Diaz, F. G.; Campos, V. M.; Klahn, O. A. G. *Vib. Spectrosc.* **1995**, *9*, 257–264.
(109) Lokshin, B. V.; Klemenkova, Z. S.; Makarov, Y. V. *Spectrochim. Acta* **1972**, *28A*, 2209–2215.
(110) Diaz, G.; Klahn, A. H. *Spectrosc. Lett.* **1990**, *23*, 87–109.
(111) Bergman, R. G.; Seidler, P. F.; Wenzel, T. T. *J. Am. Chem. Soc.* **1985**, *107*, 4358–4359.
(112) Wenzel, T. T.; Bergman, R. G. *J. Am. Chem. Soc.* **1986**, *108*, 4856–4867.
(113) Godoy, F.; Higgitt, C. L.; Klahn, A. H.; Oelckers, B.; Parsons, S.; Perutz, R. N. *J. Chem. Soc. Dalton Trans.* **1999**, 2039–2048.
(114) Gunnoe, T. B.; Sabat, M.; Harman, W. D. *J. Am. Chem. Soc.* **1998**, *120*, 8747–8754.
(115) Harman, W. D. *Chem. Rev.* **1997**, *97*, 1953–1978.
(116) Harman, W. D. *Coord. Chem. Rev.* **2004**, *248*, 853–866.
(117) Brooks, B. C.; Gunnoe, T. B.; Harman, W. D. *Coord. Chem. Rev.* **2000**, *206–207*, 3–61.
(118) Gunnoe, T. B.; Sabat, M.; Harman, W. D. *J. Am. Chem. Soc.* **1999**, *121*, 6499–6500.
(119) Gunnoe, T. B.; Sabat, M.; Harman, W. D. *Organometallics* **2000**, *19*, 728–740.
(120) Angelici, R. J. *Acc. Chem. Res.* **1988**, *21*, 387–394.
(121) Meiere, S. H.; Brooks, B. C.; Gunnoe, T. B.; Sabat, M.; Harman, W. D. *Organometallics* **2001**, *20*, 1038–1040.
(122) Meiere, S. H.; Brooks, B. C.; Gunnoe, T. B.; Carrig, E. H.; Sabat, M.; Harman, W. D. *Organometallics* **2001**, *20*, 3661–3671.

(123) Keane, J. M.; Chordia, M. D.; Mocella, C. J.; Sabat, M.; Trindle, C. O.; Harman, W. D. *J. Am. Chem. Soc.* **2004**, *126*, 6806–6815.
(124) Myers, W. H.; Welch, K. D.; Graham, P. M.; Keller, A.; Sabat, M.; Trindle, C. O.; Harman, W. D. *Organometallics* **2005**, *24*, 5267–5279.
(125) Friedman, L. A.; Meiere, S. H.; Brooks, B. C.; Harman, W. D. *Organometallics* **2001**, *20*, 1699–1702.
(126) Meiere, S. H.; Harman, W. D. *Organometallics* **2001**, *20*, 3876–3883.
(127) Harman, W. D.; Trindle, C. *J. Comput. Chem.* **2004**, *26*, 194–200.
(128) Meiere, S. H.; Valahovic, M. T.; Harman, W. D. *J. Am. Chem. Soc.* **2002**, *124*, 15099–15103.
(129) Cox, C.; Ferraris, D.; Murthy, N. N.; Lectka, T. *J. Am. Chem. Soc.* **1996**, *118*, 5332–5333.
(130) Peng, T. S.; Gladysz, J. A. *J. Am. Chem. Soc.* **1992**, *114*, 4174–4181.
(131) Faller, J. W. In: Nachod, F. C.; Zuckerman, J. J. (Eds.), *Determination of Organic Structure by Physical Methods*, New York, Academic Press, **1973**; Vol. V, p. 75.
(132) Forsen, S.; Hoffman, R. A. *J. Chem. Phys.* **1963**, *39*, 2892–2901.
(133) Forsen, S.; Hoffman, R. A. *J. Chem. Phys.* **1964**, *40*, 1189–1196.
(134) Hoffman, R. A.; Forsen, S. In: *Progress in NMR Spectroscopy*, Pergamon Press, Oxford, England, **1966**; Vol. 1, p. 15.
(135) Brooks, B. C.; Meiere, S. H.; Friedman, L. A.; Carrig, E. H.; Gunnoe, T. B.; Harman, W. D. *J. Am. Chem. Soc.* **2001**, *123*, 3541–3550.
(136) Keane, J. M.; Ding, F.; Sabat, M.; Harman, W. D. *J. Am. Chem. Soc.* **2004**, *126*, 785–789.
(137) Winemiller, W. D.; Kopach, M. E.; Harman, W. D. *J. Am. Chem. Soc.* **1997**, *119*, 2096–2102.
(138) Leong, V. S.; Cooper, N. J. *J. Am. Chem. Soc.* **1988**, *110*, 2644–2646.
(139) Thompson, R. L.; Lee, S.; Rheingold, A. L.; Cooper, N. J. *Organometallics* **1991**, *10*, 1657–1659.
(140) Veauthier, J. M.; Chow, A.; Fraenkel, G.; Geib, S. J.; Cooper, N. J. *Organometallics* **2000**, *19*, 661–671.
(141) Vigalok, A.; Shimon, L. J. W.; Milstein, D. *J. Am. Chem. Soc.* **1998**, *120*, 477–483.
(142) Vigalok, A.; Rybtchinski, B.; Shimon, L. J. W.; Ben-David, Y.; Milstein, D. *Organometallics* **1999**, *18*, 895–905.
(143) Vigalok, A.; Milstein, D. *Acc. Chem. Res.* **2001**, *34*, 798–807.
(144) Smith, P. L.; Keane, J. M.; Shankman, S. E.; Chordia, M. D.; Harman, W. D. *J. Am. Chem. Soc.* **2004**, *126*, 15543–15551.
(145) Friedman, L. A.; You, F.; Sabat, M.; Harman, W. D. *J. Am. Chem. Soc.* **2003**, *125*, 14980–14981.
(146) You, F.; Friedman, L. A.; Bassett, K. C.; Lin, Y.; Sabat, M.; Harman, W. D. *Organometallics* **2005**, *24*, 2903–2912.
(147) Schiffler, M. A.; Friedman, L. A.; Brooks, B. C.; Sabat, M.; Harman, W. D. *Organometallics* **2003**, *22*, 4966–4972.
(148) Friedman, L. A.; Sabat, M.; Harman, W. D. *J. Am. Chem. Soc.* **2002**, *124*, 7395–7404.
(149) Friedman, L. A.; Harman, W. D. *J. Am. Chem. Soc.* **2001**, *123*, 8967–8973.
(150) Chordia, M. D.; Smith, P. L.; Meiere, S. H.; Sabat, M.; Harman, W. D. *J. Am. Chem. Soc.* **2001**, *123*, 10756–10757.
(151) Ding, F.; Harman, W. D. *J. Am. Chem. Soc.* **2004**, *126*, 13752–13756.
(152) Ding, F.; Valahovic, M. T.; Keane, J. M.; Anstey, M. R.; Sabat, M.; Trindle, C. O.; Harman, W. D. *J. Org. Chem.* **2004**, *69*, 2257–2267.
(153) Valahovic, M. T.; Gunnoe, T. B.; Sabat, M.; Harman, W. D. *J. Am. Chem. Soc.* **2002**, *124*, 3309–3315.
(154) Gunnoe, T. B.; Meiere, S. H.; Sabat, M.; Harman, W. D. *Inorg. Chem.* **2000**, *39*, 6127–6130.
(155) Trofimenko, S. *J. Am. Chem. Soc.* **1967**, *89*, 3904–3905.
(156) Trofimenko, S. *J. Am. Chem. Soc.* **1967**, *89*, 3170–3177.
(157) Trofimenko, S. *Inorg. Synth.* **1970**, *12*, 99.
(158) Jesson, J. P.; Trofimenko, S.; Eaton, D. R. *J. Am. Chem. Soc.* **1967**, *89*, 3148–3158.
(159) Paul, R. L.; Amoroso, A. J.; Jones, P. L.; Couchman, S. M.; Reeves, Z. R.; Rees, L. H.; Jeffery, J. C.; McCleverty, J. A.; Ward, M. D. *J. Chem. Soc. Dalton Trans.* **1999**, 1563–1568.
(160) Di Vaira, M.; Mani, F. *J. Chem. Soc. Dalton Trans.* **1990**, 191–194.
(161) Trofimenko, S. *J. Am. Chem. Soc.* **1969**, *91*, 588–595.

(162) Macneil, J. H.; Roszak, A. W.; Baird, M. C.; Preston, K. F.; Rheingold, A. L. *Organometallics* **1993**, *12*, 4402–4412.
(163) Schoenberg, A. R.; Anderson, W. P. *Inorg. Chem.* **1972**, *11*, 85–87.
(164) Schoenberg, A. R.; Anderson, W. P. *Inorg. Chem.* **1974**, *13*, 465–469.
(165) Brunker, T. J.; Hascall, T.; Cowley, A. R.; Rees, L. H.; O'Hare, D. *Inorg. Chem.* **2001**, *40*, 3170–3176.
(166) Kitajima, N.; Hikichi, S.; Tanaka, M.; Morooka, Y. *J. Am. Chem. Soc.* **1993**, *115*, 5496–5508.
(167) Kolotilov, S. V.; Addison, A. W.; Trofimenko, S.; Dougherty, W.; Pavlishchuk, V. V. *Inorg. Chem. Commun.* **2004**, *7*, 485–488.
(168) Kitajima, N.; Komatsuzaki, H.; Hikichi, S.; Osawa, M.; Moro-oka, Y. *J. Am. Chem. Soc.* **1994**, *116*, 11596–11597.
(169) Komatsuzaki, H.; Ichikawa, S.; Hikichi, S.; Akita, M.; Moro-oka, Y. *Inorg. Chem.* **1998**, *37*, 3652–3656.
(170) Singh, U. P.; Singh, R.; Hikichi, S.; Akita, M.; Moro-oka, Y. *Inorg. Chim. Acta* **2000**, *310*, 273–278.
(171) Komatsuzaki, H.; Sakamoto, N.; Satoh, M.; Hikichi, S.; Akita, M.; Moro-oka, Y. *Inorg. Chem.* **1998**, *37*, 6554–6555.
(172) Hammes, B. S.; Carrano, M. W.; Carrano, C. J. *J. Chem. Soc. Dalton Trans.* **2001**, 1448–1451.
(173) de Biani, F. F.; Jäkle, F.; Spiegler, M.; Wagner, M.; Zanello, P. *Inorg. Chem.* **1997**, *36*, 2103–2111.
(174) Herdtweck, E.; Peters, F.; Scherer, W.; Wagner, M. *Polyhedron* **1998**, *17*, 1149–1157.
(175) Guo, S. L.; Bats, J. W.; Bolte, M.; Wagner, M. *J. Chem. Soc. Dalton Trans.* **2001**, 3572–3576.
(176) Renk, T.; Ruf, W.; Siebert, W. *J. Organomet. Chem.* **1976**, *120*, 1–25.
(177) Guo, S.; Peters, F.; de Biani, F. F.; Bats, J. W.; Herdtweck, E.; Zanello, P.; Wagner, M. *Inorg. Chem.* **2001**, *40*, 4928–4936.
(178) Ilkhechi, A. H.; Guo, S. L.; Bolte, M.; Wagner, M. *J. Chem. Soc. Dalton Trans.* **2003**, 2303–2307.
(179) Uehara, K.; Hikichi, S.; Akita, M. *Organometallics* **2001**, *20*, 5002–5004.
(180) Alberto, R.; Herrmann, A. W.; Kiprof, P.; Baumgaerter, F. *Inorg. Chem.* **1992**, *31*, 895–899.

The Organometallic Chemistry of Group 8 Tris(pyrazolyl)borate Complexes

EVA BECKER, SONJA PAVLIK AND KARL KIRCHNER*

Institute of Applied Synthetic Chemistry, Vienna University of Technology, Getreidemarkt 9, A-1060 Vienna, Austria

I
INTRODUCTION

The tris(pyrazolylborate) anion (Tp) as a ligand in transition metal complexes was introduced by Trofimenko in 1966.[1] Since then, Tp and its derivatives, such as Tp* (hydridotris(3,5-dimethylpyrazolyl)borate) and pzTp (tetrapyrazolylborate), have found increasing applications in all areas of chemistry with most of the transition metals.[2] The present review which covers the literature up to 2006 emphasizes the organometallic aspects of Tp chemistry of the group 8 metals, i.e., synthesis, reactivity, reaction mechanisms of the metal–carbon bond and other organic and inorganic functionalities present in the molecule, organometallic reagents in organic, inorganic and polymer synthesis, and catalytic processes in which an organometallic compound is the precatalyst or catalyst or in which organometallic species are intermediates. With respect to group 8 metals especially, the development of Tp ruthenium chemistry has dramatically accelerated in the last decade,[3] while that of iron and osmium still remains comparatively underdeveloped but is steadily growing.

Tp is often compared with Cp (cyclopentadienyl) and Cp* (the pentamethyl derivative) due to the same charge and number of electrons donated (anionic 6e donors) as well as the facial geometry typically adopted. Accordingly, in the case of

*Corresponding author. Tel.: +43 1 58801 15340; Fax: +43 1 58801 16399.
E-mail: kkirch@mail.zserv.tuwien.ac.at (K. Kirchner).

ADVANCES IN ORGANOMETALLIC CHEMISTRY
VOLUME 56 ISSN 0065-3055/DOI 10.1016/S0065-3055(07)56003-X

iron, ruthenium and osmium this similarity was, at least originally, a major driving force and motivation for many researchers to utilize Tp ligand systems. Despite these formal similarities, however, differences in size and electronic properties are obvious, resulting in quite different chemistry. The cone angle of parent Tp close to 180° is well above the 100° and 146° calculated for Cp and Cp*, respectively. As far as ruthenium and osmium are concerned, the steric bulk and the rigidity of the Tp ligand appear to disfavor higher coordination numbers of the metal center. In addition, there are different geometries of the orbitals involved in constructing the complexes, with Cp being a π-donor, but Tp also being a good σ-donor. The π-bonding properties of the Tp ligand come into play in the presence of appropriate, i.e, π-accepting, co-ligands such as CO. This interplay is particularly effective in the case of C_{3v} symmetry where all three π orbitals of the metal are equally participating in metal–ligand bonding as in $[TpRu(NCMe)_3]^+$. Thus, TpM (M = Ru, Os) complexes are substitutionally very inert as compared to MCp and MCp* analogues. For comparison, the MeCN exchange at 298 K is more than eight orders of magnitude slower in $[TpRu(NCMe)_3]^+$ than in the isoelectronic complex $[CpRu (NCMe)_3]^+$ being 1.2×10^{-8} and $5.6\,s^{-1}$, respectively.[4]

Accordingly, in sharp contrast to the $[TpFe]^+$ fragment, the $[TpRu]^+$ and $[TpOs]^+$ fragments are hybridized so as to strongly bias the preferential binding of three additional ligands such that an octahedral six-coordinate structure is obtained and maintained. As a consequence, the chemistry of ruthenium and osmium Tp complexes is restricted to octahedral, coordinatively saturated $18e^-$ compounds with the metal center being most commonly in the formal oxidation state +II. Processes involving coordination number increase, e.g., oxidative additions, associative substitutions, are thus rare for Tp systems within group 8. Complexes with coordination number 7 and 5 are unknown. Furthermore, both low- and high-valent TpM complexes are rare and only a few examples of M(I), M(III), M(IV) (M = Ru, Os) and M(VI) (M = Os) Tp complexes are reported. Presumably due to the smaller size of iron, a variety of tetrahedral FeTp compounds with sterically demanding Tp^x ligands have been described where iron is typically in the oxidation state +II. Only a few examples are known where iron is in the oxidation state +I and +III, while higher oxidation states remain rather speculative. It has to be noted that the organometallic chemistry of the TpFe fragment is frequently limited due to the high tendency to form highly stable homoleptic complexes of the type $FeTp_2$.

II

IRON TRIS(PYRAZOLYL)BORATE CHEMISTRY

Since Trofimenko synthesized Tp_2Fe,[5] $(pzTp)_2Fe$ and Tp^*_2Fe[6] in 1967, iron homoscorpionate compounds featuring all types of substituted Tp ligands have gained considerable interest. The vast majority of the research deals with inorganic coordination compounds and can be divided into three fields: First, highly stable octahedral Tp_2Fe analogues, which are easily prepared in quantitative yield by the reaction of Fe(II) salts with stoichiometric amounts of all types of substituted NaTp,

KTp or TlTp derivatives unless they bear exceptionally bulky substituents on the 3-position of the pyrazole.[7] Second, oxo-bridged diiron centers, which are ubiquitous in metalloenzymes such as hemerythrin, methane mono-oxygenase and rubrerythrin. Accordingly, complexes such as $Tp_2Fe_2(\mu\text{-}O)(\mu\text{-}O_2CCH_3)$ having one μ-oxo and two μ-carboxylato bridges serve as model compounds in which the Tp ligands are considered to mimic three histidine moieties.[8] Five-coordinate iron Tp complexes like $Tp^{^iPr_2}Fe(OAc)$ and their derivatives[9] have been found to act as models for non-heme metalloproteins. Not only the structural, spectroscopic and magnetic properties of all these coordination compounds have been well investigated, but also their ability to catalyze oxidation reactions in the presence of O_2.[10] Third, facially capped building blocks for clusters of the type $[Tp^{R,R'}Fe(CN)_3]^-$ (R = H, Me) have attracted considerable interest as useful synthons for the construction of single-molecule magnets and single-chain magnets to afford compounds that are tuneable in their magnetic, electrical and optical properties.[11]

Apart from these extensively investigated inorganic coordination compounds, there are only a few examples of organometallic chemistry derived from Tp^xFe species as there is a strong tendency for all kinds of organometallic intermediates to be converted into the highly stable octahedral Tp^x_2Fe compounds. Generally, there are two strategies to generate organometallic Tp^xFe complexes: (i) introducing the Tp^x ligand to complexes featuring iron–carbon bonds requires co-ligands that are strongly bound to the iron center to inhibit the formation of Tp^x_2Fe and (ii) four-coordinate Tp^RFeX complexes with very bulky substituents 'R' can easily be converted into organometallic species upon reaction with alkyl lithium or Grignard reagents.

In the case of the mixed sandwich complexes $TpFeCp^x$ ($Cp^x = C_5H_5$, C_5Me_5) all attempts to prepare TpFeCp failed, and only $FeCp_2$ and Tp_2Fe were isolated.[12,13] Brunker *et al.*[14] who successfully prepared Cp*TpFe from the reaction of $[Cp^*Fe(NCMe)_3]PF_6$ with KTp observed substantial contamination of the product with Cp^*_2Fe and Tp_2Fe formed by ligand redistribution (Scheme 1).

Iron carbonyl compounds turned out to be very promising precursors for organometallic iron Tp complexes.

King and Bond reported the first examples of such species, namely, $TpFe(CO)_2(\eta^1\text{-}CH{=}CHMe)$ and $TpFe(CO)_2(\eta^1\text{-}(C{=}O)CH{=}CHMe)$[15] as products of the reaction of $Fe(CO)_3I(\eta^3\text{-}C_3H_5)$[16] with KTp (Scheme 2) and $TpFe(CO)_2(C_3F_7)$ from $TpFe(CO)_4I(C_3F_7)$ as starting materials (Scheme 3). In both cases $FeTp_2$ was, however, the main product.

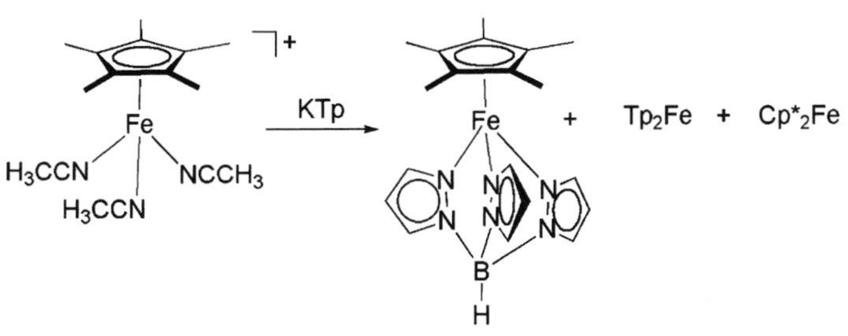

SCHEME 1.

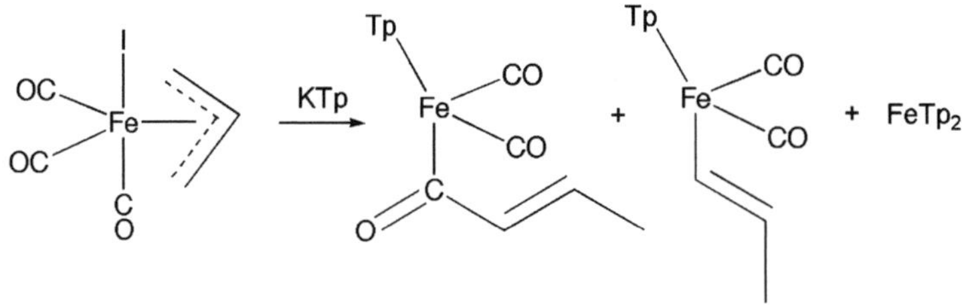

SCHEME 2.

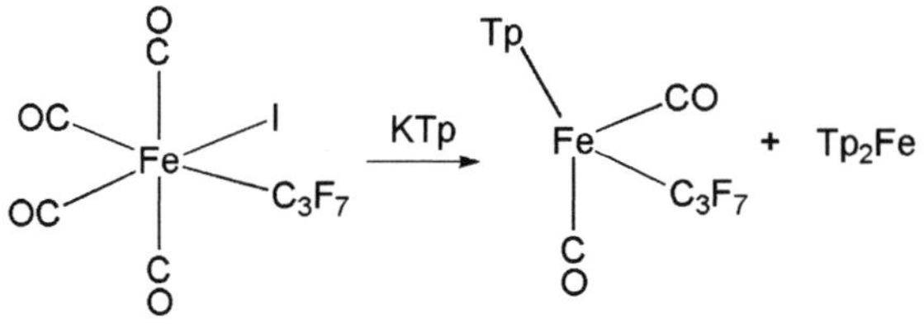

SCHEME 3.

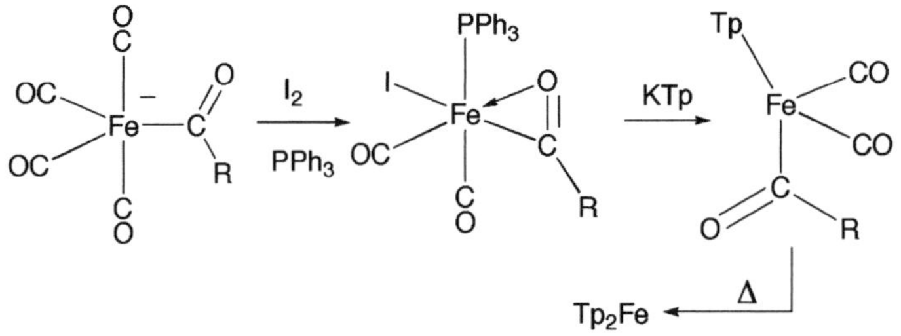

SCHEME 4. $R = C_6H_3Me_2$-2,6.

Hill *et al.* used the thermally stable aroyl precursor $Fe(\eta^2\text{-}OCC_6H_3Me_2\text{-}2,6)I(CO)_2(PPh_3)$ to prepare the monodentate aroyl complex $TpFe\{C(=O)C_6H_3Me_2\text{-}2,6\}(CO)_2$[17] by reaction with KTp in 54% yield (Scheme 4). The related acyl complex $TpFe\{(C(=O)Me\}(CO)_2$ has also been described but was obtained in only 8% yield from the reaction of KTp with $Fe_2(CO)_9$ and MeI.[18]

It should be noted that the analogous reaction of the carbamoyl complex $Fe(\eta^2\text{-}OCN^iPr_2)I(CO)_2(PPh_3)$ with KTp provides the binuclear compound $Fe_2(\mu\text{-}\sigma,\sigma'\text{-}OCN^iPr_2)(CO)_5(PPh_3)$[19] without Tp incorporation, rather than the expected carbamoyl complex $TpFe\{C(=O)N^iPr_2\}(CO)_2$.

An unusual rearrangement of the carbamoyl moiety was reported by Hill *et al.*[20] The reaction of the iron trifluoromethyl complex $Fe(\eta^2\text{-}OCN^iPr_2)(CF_3)(CO)_2(PPh_3)$ with KTp led to the formation of the novel ferraoxetene $TpFe\{\kappa^2{=}C(N^iPr_2)OCF_2\}(CO)$ in 73% yield with loss of CO and PPh_3 *via* a postulated difluorocarbene intermediate (Scheme 5).

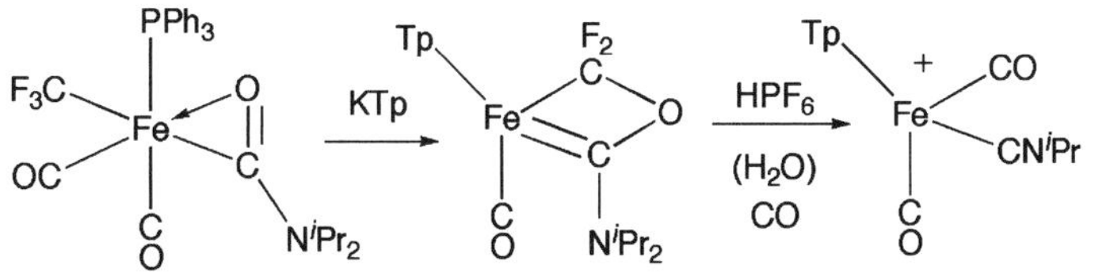

SCHEME 5.

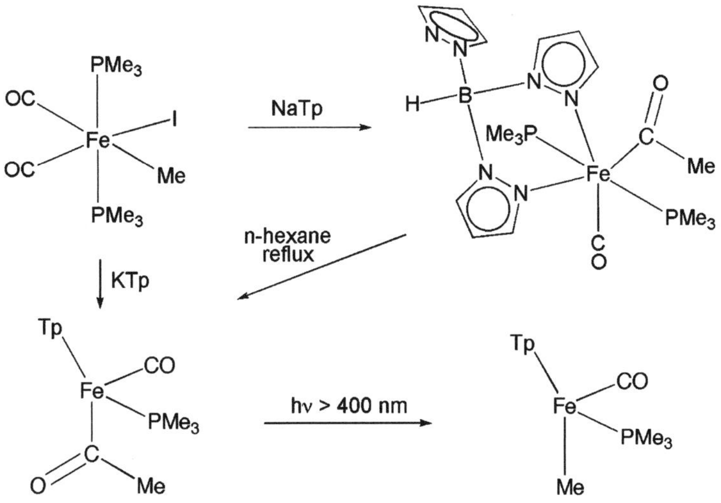

SCHEME 6.

Hydrolysis of this metallacycle with aqueous HPF_6 led to the formation of the cationic isonitrile complex $[TpFe(CN^iPr)(CO)_2]PF_6$ in 60% yield which is only obtained if the reaction is carried out under an atmosphere of carbon monoxide.

Tilset and Hamon reported that the reaction of Fe $MeI(CO)_2(PMe_3)_2$ with KTp at room temperature cleanly afforded the acetyl complex TpFe{C(=O)Me}(CO)(PMe_3) in 78% yield (Scheme 6).[21] As this compound is thermally stable in boiling *n*-hexane, they used visible light irradiation in toluene for 1 h to perform the decarbonylation reaction to isolate the methyl iron complex TpFe(Me)(CO)(PMe_3) that slowly decomposes to give $FeTp_2$. It is noteworthy that when Macchioni *et al.*[22] performed the same reaction with $FeMeI(CO)_2(PMe_3)_2$ employing NaTp instead of KTp, the isolated species was (κ^2-Tp)Fe{C(=O)Me}(CO)$(PMe_3)_2$, obtained in 80% yield due to the Na^+ vs. K^+ salt effect. This complex can be completely converted into the κ^3-Tp complex *via* loss of one phosphine upon heating in *n*-hexane under reflux for 1 h.

Parkin *et al.* chose a different approach to organometallic iron Tp^x chemistry. The reaction of $FeCl_2$ with the sterically demanding KTp^{tBu} afforded the first example of a four-coordinated iron Tp^x complex, $Tp^{tBu}FeCl$.[23] To avoid the possibility of ligand degradation *via* reactions at the B–H bond, they used the modified Li($PhTp^{tBu}$) to

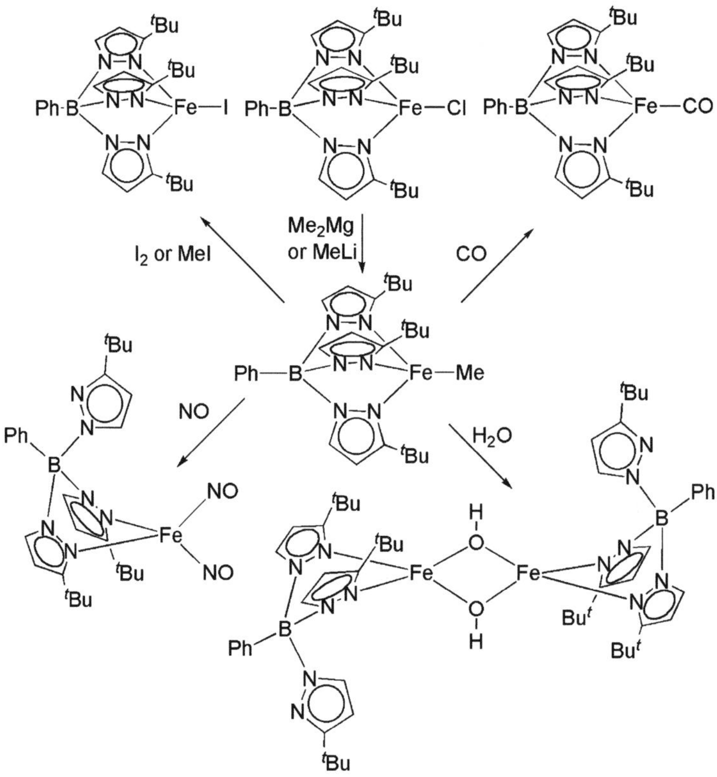

SCHEME 7.

prepare ($PhTp^{^tBu}$)FeCl as starting material for the reaction with Me_2Mg or MeLi to give the iron methyl complex ($PhTp^{^tBu}$)FeMe.[24] The four-coordinate complex ($PhTp^{^tBu}$)FeMe reacts with I_2 at room temperature or with MeI at 120 °C to yield the iodide complex ($PhTp^{^tBu}$)FeI (Scheme 7). The reaction of the methyl complex with water affords the dinuclear complex $\{(\kappa^2\text{-}PhTp^{^tBu})Fe(\mu\text{-}OH)_2\}_2$ in which one of the pyrazolyl donors is replaced as a result of the hydroxide bridges. Tridentate to bidentate conversion of the $PhTp^{^tBu}$ ligand is also observed during the reaction with NO affording the mononuclear 17-electron dinitrosyl complex (κ^2-$PhTp^{^tBu}$) $Fe(NO)_2$. The reaction of ($PhTp^{^tBu}$)FeMe with CO at 120 °C yields the mononuclear 15-electron *monocarbonyl* complex ($PhTp^{^tBu}$)FeCO that is of interest given that Fe(I) is an uncommon valence state for iron.

Akita *et al.* have employed the bulky hydridotris(3,5-diisopropylpyrazolyl)borate ligand as a 'tetrahedral enforcer'. Reactions between $Tp^{^iPr_2}$FeCl and various Grignard reagents such as ethyl magnesium bromide and benzyl magnesium bromide provide $Tp^{^iPr_2}$FeEt[25] and $Tp^{^iPr_2}FeCH_2Ph$,[26] respectively (Scheme 8). It is noteworthy that the iron ethyl complex is resistant to β-hydride elimination. Both complexes form under CO atmosphere the hexacoordinated acyl-dicarbonyl complexes $Tp^{^iPr_2}Fe\{C(=O)Et\}(CO)_2$ and $Tp^{^iPr_2}Fe\{C(=O)CH_2Ph\}(CO)_2$, respectively, due to the smaller size of $Tp^{^iPr_2}$ compared to $PhTp^{^tBu}$.

Tp^{iPr2}ǦFeǦR ← RMgBr / THF — Tp^{iPr2}ǦFeǦCl — RMgBr / CO → $Tp^{iPr2}Fe(CO)_2(C(=O)R)$

R= η^1-allyl, CH_2Ph, Et

R= Et, CH_2Ph

SCHEME 8. $Tp^{iPr_2} = HB(pz^iPr_2\text{-}3,5)_3$.

In contrast to the analogous Ni and Co complexes $Tp^{iPr_2}MCl$ that react with allyl magnesium bromide during the formation of $Tp^{iPr_2}M(\eta^3\text{-}C_3H_5)$, $Tp^{iPr_2}FeCl$ gives under the same reaction conditions the tetrahedral 14-electron η^1-allyl iron complex $Tp^{iPr_2}Fe(\eta^1\text{-}CH_2CH{=}CH_2)$.[27]

III
RUTHENIUM TRIS(PYRAZOLYL)BORATE CHEMISTRY

A. *Precursors and Useful RuTp Compounds*

The vast majority of tris(pyrazolyl)borate ruthenium complexes prepared until now contains the parent Tp ligand. In contrast to other transition metals, ruthenium compounds which bear bulky Tp ligands remain still rare and are, thus far, limited to Tp*, Tp^{iPr_2}, $Tp^{iPr,Br}$ and Tp^{Ms} as shown in Chart 1.

An extremely versatile and useful precursor providing an entry point into TpRu chemistry is TpRuCl(COD). This compound was originally prepared by Singleton *et al.*[28] several years ago by the reaction of $[Ru(COD)(NH_2NMe_2)_3H]PF_6$ with KTp. The reaction proceeds *via* the hydrido complex TpRuH(COD) which is readily converted to TpRuCl(COD) upon treatment with CCl_4 or $CHCl_3$. Since the yield of $[RuH(COD)(NH_2NMe_2)_3]PF_6$ is comparatively low (ca. 50%), the reaction is better performed with $[RuH(COD)(NH_2NMe_2)_3]BPh_4$, which is available in higher yield (81%) (Scheme 9).[29] The same methodology has been applied by Akita *et al.*[30,31] for the synthesis of the bulky complexes Tp^{iPr_2}RuH(COD) and $Tp^{iPr,Br}$RuH(COD). Spectroscopic and crystallographic analyses revealed, however, the presence of two isomeric structures, a square pyramidal structure wherein the Tp ligands are coordinated in a κ^2-(N,N') fashion (**I**) and an octahedral structure (**II**) where the Tp ligands are κ^3-(N,N',H)-coordinated with a 3-center-2-electron Ru–H–B interaction, rather than the expected κ^3-(N,N',N'') coordination (Scheme 10). Isomerization from the former structure to the latter was observed by means of ^{1}H NMR spectroscopy.

Very recently, Theopold *et al.*[32] demonstrated that the reaction of [RuH(COD) $(NH_2NMe_2)_3]BPh_4$ with $TlTp^{Ms}$ constitutes an efficient route for obtaining Tp^{Ms} RuH(COD). Interestingly, in this compound the Tp ligand is clearly coordinated in the κ^3-mode, based on both spectroscopic data and an X-ray crystallographic study. Notably, this complex could also be obtained by the reaction of RuHCl(bpzm) (COD) (bpzm = bis(pyrazolyl)methane) with $TlTp^{Ms}$.

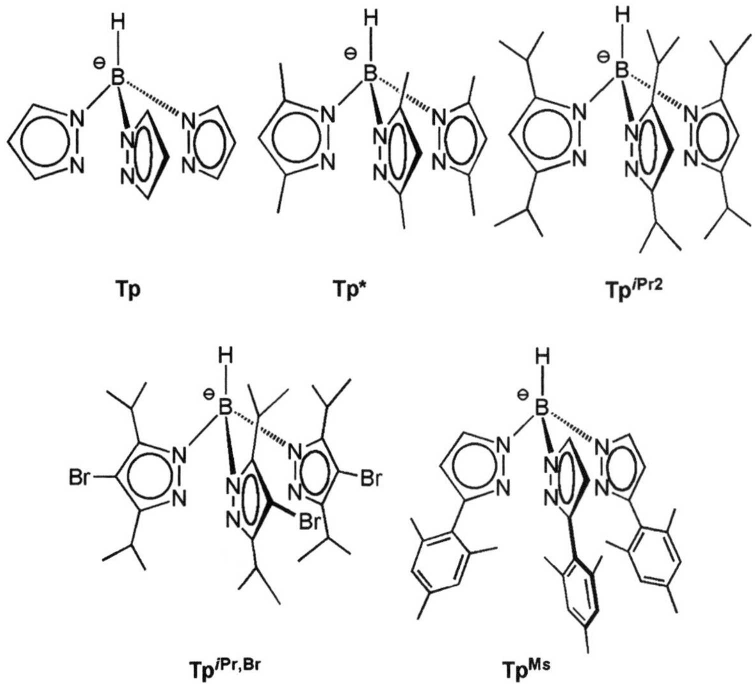

Tp **Tp*** **Tp^{iPr2}**

Tp^{iPr,Br} **Tp^{Ms}**

CHART 1. Tp^{x} ligands used in organoruthenium chemistry.

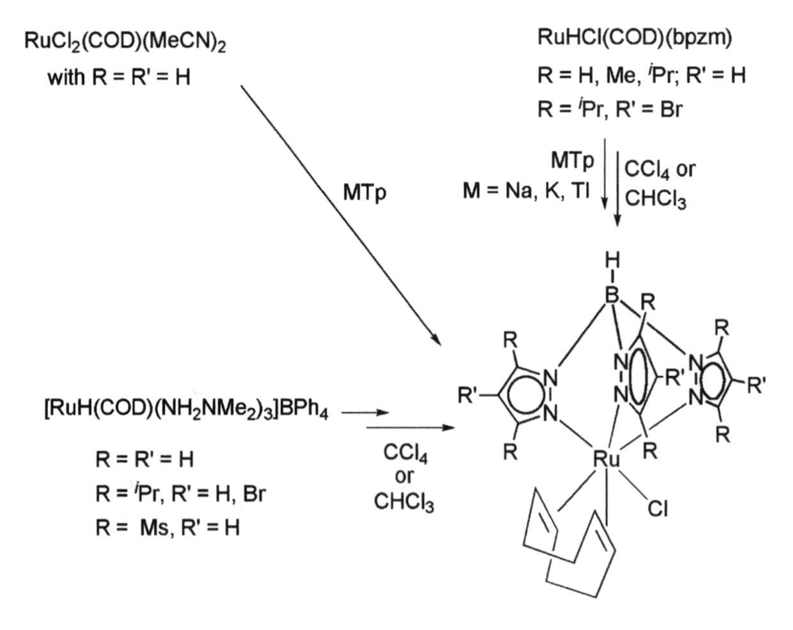

SCHEME 9.

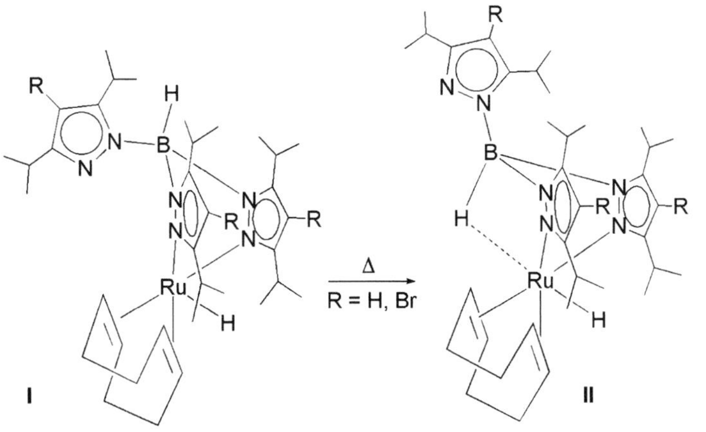

SCHEME 10.

According to Jalon *et al.*,[33] the polymer $[RuCl_2(COD)]_x$ reacts with KTp to directly provide TpRuCl(COD), while RuHCl(COD)(bpzm) reacts with KTp to yield initially TpRuH(COD), which then transforms into TpRuCl(COD) on treatment with chlorinated solvents. Chaudret *et al.* also prepared TpRuH(COD), Tp*RuH(COD) and $Tp^{^{i}Pr,Br}$RuH(COD) *via* the reactions of RuHCl(COD)(bpzm) with KTp, KTp* and $KTp^{^{i}Pr,Br}$, respectively (Scheme 9).[34–36]

An alternative high-yield approach to TpRuCl(COD) (and TpRuCl(NBD) (NBD = norbornadiene)) has been described by Wilson and Nelson[37] from the reaction of $RuCl_2(COD)(NCMe)_2$ (and $RuCl_2(NBD)(NCMe)_2$) with MTp (M = Na, K) in dichloroethane (Scheme 9).

The cyclo-octadiene ligand in TpRuCl(COD) is substitutionally inert in this complex, in contrast to the analogous complexes CpRuCl(COD) and Cp*RuCl(COD),[38] requiring vigorous conditions to be applied. Thus, in boiling DMF, TpRuCl(COD) converts with L or L_2 (e.g., mono- and bidentate *N*-, *O*- and *P*-donor ligands such as tertiary phosphines, tertiary phosphites, aminophosphines, amines, imines, DMSO, DMF) to $TpRuCl(L_2)$ and $TpRuCl(L)_2$.[29,40–43]

Substitution of the cyclo-octadiene ligand in TpRuCl(COD) is thus a convenient and general synthetic route provided the ligands L and L_2 are stable under these conditions. Some representative examples are depicted in Scheme 11 including $TpRuCl(\kappa^1\text{-}S\text{-}DMSO)_2$,[44] TpRuCl(TMEDA)[45] and TpRuCl(pn) (pn = $Me_2NCH_2CH_2PPh_2$, $Et_2NCH_2CH_2PPh_2$, $^{i}Pr_2NCH_2CH_2PPh_2$).[46] Chloride abstraction from TpRuCl(tmeda) with $NaBPh_4$ or $NaPF_6$ in the presence of L (L = acetone, DMF, CH_3CN, $HC{\equiv}CPh$, $HC{\equiv}CSiMe_3$) led to the formation of the cationic complexes $[TpRu(tmeda)(L)]^+$ (L = acetone, DMF, MeCN, =C=CHPh, $=C=CHSiMe_3$). Similarly, the coordination chemistry of complexes $[TpRu(pn)(L)]^+$ (L = H_2O, acetone, MeCN, N_2, CO, =C=CHPh) was studied. Attempts to prepare a coordinatively unsaturated compound were, however, unsuccessful. If halide abstraction is performed under a N_2 atmosphere, $[TpRu(pn)(N_2)]^+$ is obtained with the dinitrogen ligand coordinated in an end-on fashion. This was the first TpRu complex of

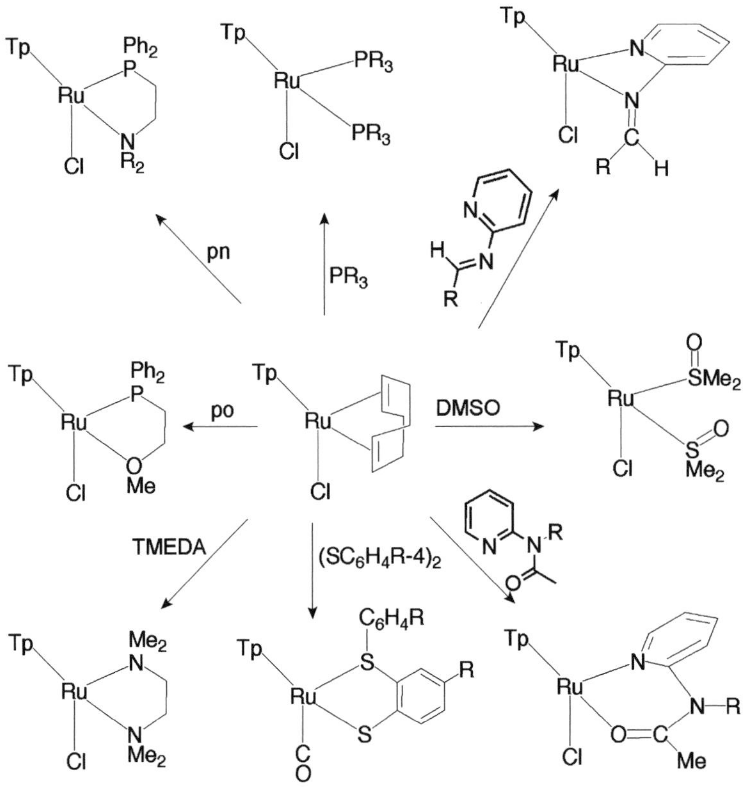

SCHEME 11. Cyclo-octadiene substitution in TpRuCl(COD). All reactions in refluxing DMF.

dinitrogen.[46] The reaction of TpRuCl(COD) with one equivalent RC_6H_4S-SC_6H_4R (R = H, Me) in boiling DMF affords the complexes TpRu(κ^2-*S*,*S*′-S-C_6H_3R-S-C_6H_4R)(CO) containing 2-arylmercapto-arylmercaptan and CO ligands (Scheme 11). The formation of this unusual product involves decomposition of the solvent DMF with release of the base $HNMe_2$, which also acts as an HCl scavenger.[47]

By heating a DMF solution of TpRu(COD)Cl under reflux in the presence of one equivalent of PR_3 (PR_3 = PPh_3, $PPh_2{}^iPr$, P^iPr_3 and PCy_3), the air-sensitive DMF intermediate TpRuCl(PR_3)(DMF) is formed (Scheme 12). The DMF molecule is easily replaced by other monodentate ligands yielding TpRuCl(PR_3)(L) (L = PR'_3, pyridine, MeCN).[48–50] The syntheses of the following complexes have been described: TpRuCl(PPh_3)(L) (L = pyridine, =C=CHPh, CO, PMe_3) and TpRuX(PPh_3)(L) (L = CH_3CN, X = Cl; L = CH_3CN, X = H). On exposure to air in the presence of CCl_4, TpRuCl(PR_3)(DMF) is readily converted to the corresponding Ru(III) complexes TpRuCl_2(PR_3).[51] These compounds are air-stable, readily accessible, permit easy variations as far as electronic and steric properties of the co-ligands PR_3 are concerned, and are reducible to a variety of TpRu(II) complexes. Treating TpRuCl(COD) with PCy_3 in boiling DMF yields TpRuCl(PCy_3)(DMF), which readily reacts with $HOCH_2R$ (R = H, Me) to give the

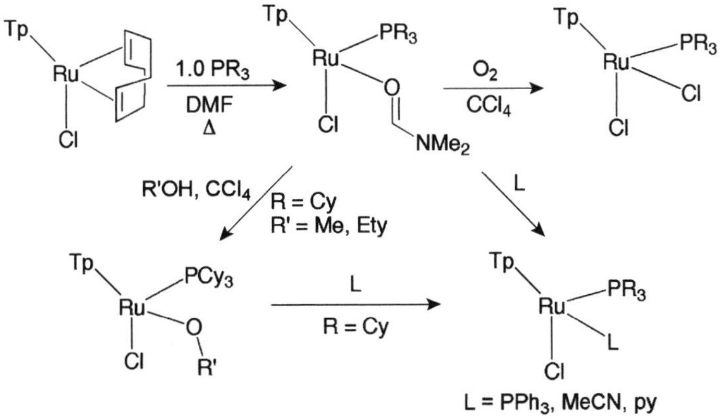

SCHEME 12.

paramagnetic complexes $TpRuCl(OCH_2R)(PCy_3)$ ($\mu_{eff} = 1.83\mu_B$ at 295 K) in moderate-to-good yield.[52] $TpRuCl(OCH_3)(PCy_3)$ has been crystallographically characterized. Both complexes react with L = CH_3CN, pyridine, CO, $P(OMe)_3$ and PMe_3 to afford the (diamagnetic) Ru(II) compounds $TpRuCl(PCy_3)(L)$ [Eq. (1)].

$$2\,TpRuCl(OEt)(PCy_3) + 2L \rightarrow 2\,TpRuCl(PCy_3)(L) + CH_3CH = O + EtOH \quad (1)$$

The chloride ligand in TpRuCl(COD) is also readily replaceable in the presence of KX (X = Br^-, I^-, CN^-) to give the corresponding TpRuX(COD) complexes.[53] Chloride abstraction by means of $AgCF_3SO_3$, $TlCF_3SO_3$ or $NaBPh_4$ in a coordinating solvent leads to the formation of cationic $[TpRu(COD)(solvent)]^+$ (solvent = DMSO, pyridine, CH_3CN, H_2O) complexes. In the non-coordinating solvent CH_2Cl_2, the aquo complex $[TpRu(COD)(H_2O)]^+$ has been formed in 96% yield with the water molecule arising from the residual water in the CH_2Cl_2 solution, despite rigorous drying.

From TpRuCl(COD), also the neutral complexes $TpRu(\eta^4\text{-diene})Cl$ (diene = butadiene, isoprene, 2,4-hexadiene) can be obtained,[54] where the diene ligand is coordinated in *S-trans* fashion as proven by X-ray crystallography. In the synthesis, TpRuCl(COD) is heated in xylene in the presence of air. The dark-green paramagnetic (i.e., Ru(III)) complex, tentatively formulated as $\{TpRuCl\}_x$, thus obtained is redissolved in MeOH. Upon addition of Zn in the presence of the diene, $TpRuCl(\eta^4\text{-diene})$ is formed (Scheme 13).

Other prominent TpRu precursors are $TpRuCl(PPh_3)_2$, $TpRuH(CO)(PPh_3)$ and $TpRuCl(CO)(PPh_3)$. The bis(phosphine) complex $TpRuCl(PPh_3)_2$ was prepared for the first time by Hill *et al.* from $RuCl_2(PPh_3)_3$ and KTp in 90% isolated yield (Scheme 14).[55] This complex is a valuable precursor for complexes of the types $TpRuCl(PPh_3)(L)$, $TpRuCl(L)_2$ and $[TpRu(PPh_3)(L)(L')]^+$, as shown, for instance, by Jia *et al.*,[56,57] Grubbs *et al.*,[58] and our group.[59] Tenorio *et al.*[60] treated TpRuCl $(PPh_3)_2$ with dippe (dippe = 1,2-bis(diisopropylphosphino)ethane) and isolated TpRuCl(dippe) in 55% yield. $TpRuCl(P^iPr_2Me)_2$ and $TpRuCl(PEt_3)_2$ were

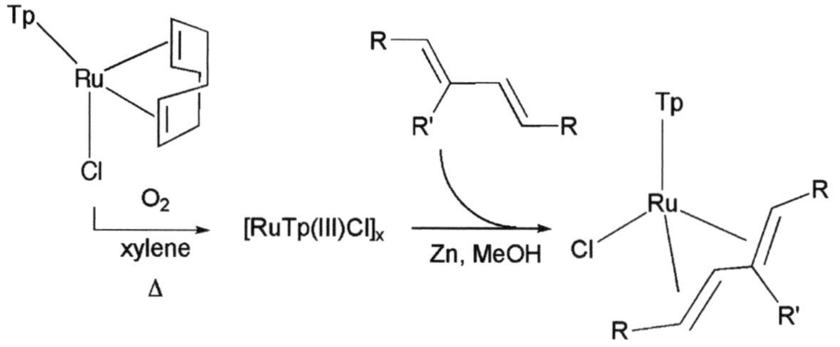

SCHEME 13.

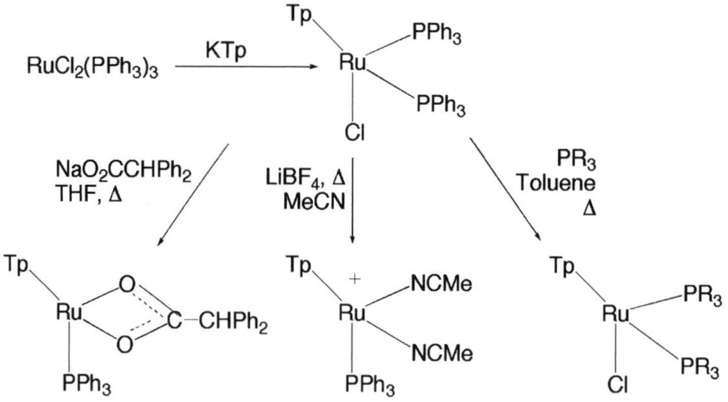

SCHEME 14.

obtained in similar fashion.[61] Phosphite complexes $TpRuCl(PPh_3)\{P(OEt)_3\}$, $TpRuCl(PPh_3)\{PPh(OEt)_2\}$ and $TpRuCl\{P(OEt)_3\}_2$ were prepared by Albertin *et al.* also following the above methodology.[62] Upon chloride abstraction in the presence of L (L = CO, CN^tBu, acetone, THF, N_2), the corresponding cationic $[TpRu(dippe)L]^+$ complexes are formed. The paramagnetic compound $[TpRu(OMe)(dippe)]^+$ has been obtained by treating $[TpRu(dippe)(H_2)]^+$ with methanol.

The reactivity of $TpRuCl(PPh_3)_2$ towards $[Me_4N][B_3H_8]$ was found to provide the ruthenatetraborane $TpRu(B_3H_8)(PPh_3)$.[63] TpRu complexes containing aliphatic amines as co-ligands are rare. Treating $TpRuCl(PPh_3)_2$ with $NaBH_4$ provides $TpRuH(PPh_3)_2$, protonation of which results in conversion to the cationic dihydrogen complex $[TpRu(H_2)(PPh_3)_2]^+$. The dihydrogen ligand is labile and is readily displaced by L = CH_3CN, H_2O and N_2. This process was found to be reversible when $[TpRu(PPh_3)_2(L)]^+$ is pressurized with H_2.[56] Similar reactions were studied by Jia using $[TpRu(H_2)(dppe)]^+$, $[TpRu(H_2)(PPh_3)(CO)]^+$ and $[TpRu(H_2)(PPh_3)(NCMe)]^+$, and comparisons with other co-ligands such as Cp and 1,4,7-triazacyclononane were undertaken.[64]

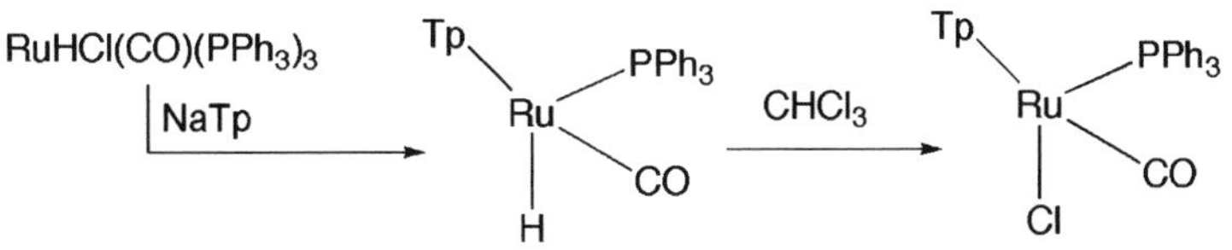

SCHEME 15.

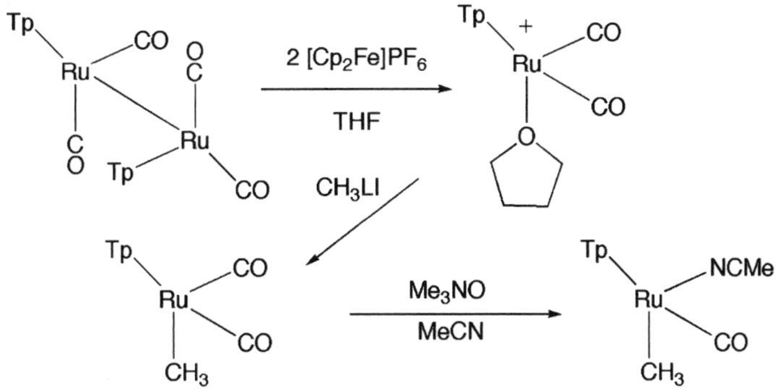

SCHEME 16.

The syntheses of $TpRuH(CO)(PPh_3)$ and $TpRuCl(CO)(PPh_3)$ were described by Sun and Simpson (Scheme 15).[65] The hydrido complex $TpRuH(CO)(PPh_3)$ was obtained in 92% isolated yield from the reaction of NaTp with $RuHCl(CO)(PPh_3)_3$. The corresponding chloride was then obtained in 90% yield by treatment of $TpRuH(CO)(PPh_3)$ with $CHCl_3$.

Finally, Gunnoe and co-workers[66] showed that $TpRu(CH_3)(CO)(NCMe)$ is a very useful starting material for a variety of remarkable stoichiometric and catalytic transformations of unsaturated organic substrates (*vide infra*). The synthesis starts from the solvent-coordinated complex $[TpRu(CO)_2(THF)]PF_6$, which is readily available by oxidation of the Singleton *et al.*'s dimeric $Tp_2Ru_2(CO)_4$[67] with two equivalents of $[Cp_2Fe]PF_6$.[68] Methylation of $[TpRu(CO)_2(THF)]PF_6$ with $LiCH_3$ yields the methyl complex $TpRu(CH_3)(CO)_2$. Subsequent treatment of this compound with Me_3NO in refluxing acetonitrile affords $TpRu(CH_3)(CO)(NCMe)$ (Scheme 16).

B. *Ruthenocene Analogues*

One of the first ruthenium compounds containing the Tp ligand was TpCpRu. This complex was synthesized by Singleton *et al.*[38a,69] by treating CpRuCl(COD) with 1.1 equivalents of NaTp in boiling ethanol. Shortly afterwards, Mann *et al.*[12] published a more general route to Cp and Cp* compounds of the TpRu fragment *via* the reactions of $[CpRu(NCMe)_3]PF_6$ or $[Cp^*Ru(NCMe)_3]PF_6$ with KTp, KTp* or KpzTp in acetonitrile. Oxidation of Tp*CpRu with $AgPF_6$ allowed the isolation of

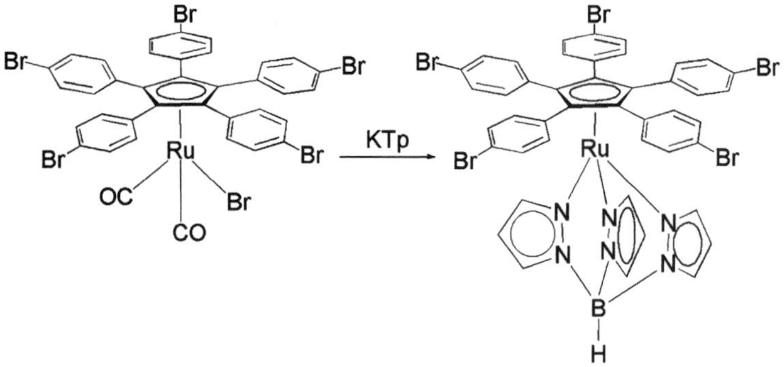

SCHEME 17. Synthesis of an organometallic 'turnstile'.

the corresponding Ru(III) species. Furthermore, strongly π-accepting ligands such as CO and $P(OMe)_3$ were found to react with Tp^*CpRu and (pzTp)CpRu yielding complexes of the type (κ^2-Tp^*)CpRu(CO) and (κ^2-pzTp)CpRu(CO), featuring κ^2-*N*,*N'*-bonded Tp^* and pzTp ligands. It should be mentioned that the synthesis of Tp^*Cp^*Ru failed. The most recent development in this area has been reported by Launay *et al.*[70] These authors describe a short route to a ruthenium Tp complex with a pentaphenyl substituted cyclopentadienyl which can be viewed as an organometallic molecular turnstile (Scheme 17).

O'Sullivan and Lalor[71] and Ferguson *et al.*[72,73] reported the synthesis and the X-ray crystal structures of $[TpRu(\eta\text{-}C_6H_6)]PF_6$ and $[(pzTp)Ru(\eta\text{-}C_6H_6)]PF_6$ *via* reaction of the dimeric species $Ru_2(\mu\text{-}Cl)_2Cl_2(\eta\text{-}C_6H_6)_2$ with KTp or KTp^4 in boiling methanol. Although O'Sullivan and Lalor simply noted that the benzene ligands were susceptible towards nucleophilic attack, Bhambri and Tocher systematically investigated the reaction of these compounds with nucleophiles Nuc (Nuc = H^-, D^-, OH^-, CN^-).[74] All the conversions are smooth giving neutral η^5-cyclohexadienyl compounds of the formula $TpRu(\eta^5\text{-}C_6H_6Nuc)$. The crystal structure of TpRu ($\eta^5\text{-}C_6H_6CN$) confirms that the incoming nucleophile has added in an *exo* fashion. The same group extended this investigation to the related arene ligands *p*-xylene, mesitylene, hexamethylbenzene, *p*-cymene (X-ray structure included) and 1,2,4,5-tetramethylbenzene.[75] Also $[Tp^*Ru(\eta^6\text{-}C_6H_6)]PF_6$ was synthesized.[76] A chiral alkyltris(pyrazolyl)borate *p*-cymene complex viz. $[(RTp)Ru(\eta^6\text{-}{}^iPrC_6H_4Me\text{-}4)]PF_6$ (R = isopinocampheyl) has been prepared by Bailey *et al.*.[77] Tocher *et al.* found that $RuCl_2(NCMe)(\eta^6$-arene) reacts with KTp in CH_2Cl_2 without cleavage of the second Ru–Cl bond to give complexes of the type (κ^2-Tp)RuCl(η^6-arene).[78] The same group prepared and structurally characterized the first TpRu complex containing a π-tetramethylthiophene ligand viz. $[TpRu(\eta^5\text{-}C_4Me_4S)]PF_6$ by treating $Ru_2(\mu\text{-}Cl)_2Cl_2((\eta^5\text{-}C_4Me_4S)_2$ with NaTp in the presence of NH_4PF_6.[78] Landgrafe and Sheldrick[79] prepared the complex $[TpRu([9]aneS_3)]^+$, a formal analogue of $[TpRu(arene)]^+$, by treatment of $[Ru([9]aneS_3)(NCMe)_3]^{2+}$ with KTp.

In an attempt to prepare the neutral compound TpRu(quin)(thf) (quin = quinolin-8-olate), $Ru(\eta^6\text{-}{}^iPrC_6H_4Me\text{-}4)$(quin)Cl has been treated with one equivalent of KTp in thf. This approach failed and instead the unusual complex (κ^1-*N*-Tp)Ru

(η^6-^{i}PrC$_6$H$_4$Me-4)(quin) was obtained as the major product, which features a κ^1-coordinated Tp ligand.[80]

C. *Carbonyl and Nitrosyl Compounds*

$Ru_3(CO)_{12}$ reacts with KTp and halogens to give moderate yields of the compounds TpRuX(CO)$_2$ (X = Cl, Br, I), whereas with Ru$_2$(μ-Cl)$_2$Cl$_2$(CO)$_6$ and TlTp, TpRuCl(CO)$_2$ is obtained directly.[81] Similarly, Ru$_2$(κ^2*C,N*-azp)$_2$(μ-Cl)$_2$CO)$_4$ (Hazp = diphenyldiazene) reacts with KTp to give TpRu(κ^1-*C*-azb)(CO)$_2$, which loses one CO ligand upon irradiation with UV light producing TpRu(κ^2-*C,N*-azb)(CO).[82] New TpRu carbonyl complexes were obtained by Shiu *et al.* upon treatment of TpRuX(CO)$_2$ (X = Br, I) with Me$_3$NO in MeCN to give TpRuX(CO)(NCMe).[83] These compounds reacted with ^{i}PrSH in the presence of base to give the first dimeric thiolate-bridged TpRu complexes of the types Tp$_2$Ru$_2$(μ-X)(μ-S^{i}Pr)(CO)$_2$ and Tp$_2$Ru$_2$(μ-S^{i}Pr)$_2$(CO)$_2$ albeit in rather low yields. The reaction of the polymer *catena*-[Ru(O$_2$CMe)(CO$_2$)]$_n$ with either KTp or KpzTp yields the neutral metal–metal bonded dimers Tp$_2$Ru$_2$(CO)$_4$ and (pzTp)$_2$Ru$_2$(CO)$_4$, respectively, where the metal center is in the formal oxidation +I. The X-ray structure of this dinuclear complex shows that the two Tp ligands, coordinate as κ^3-*N,N′,N″*-tridentate ligands in a *cis*-staggered configuration around the dimetal core. The Ru–Ru distance of 2.882(1) Å corresponds to a metal–metal single bond.[67] Sørlie and Tilset[68] investigated the oxidation of Tp$_2$Ru$_2$(CO)$_4$ leading to cleavage of the Ru–Ru bond and the formation of the solvento complexes [TpRu(CO$_2$)L]$^+$ (L = CH$_3$CN, H$_2$O, THF, acetone). For this process, an associative mechanism is implicated by the large and negative entropies of activation. The dimer is oxidized at $E^\circ = 0.15$ V vs. Cp$_2$Fe/Cp$_2$Fe$^+$ in CH$_3$CN as the solvent. Angelici *et al.*[84] reported the reaction of Tp$_2$Ru$_2$(CO)$_4$ with CF$_3$SO$_3$H yielding the dimeric complex [Tp$_2$Ru$_2$(μ-H)(CO)$_4$}$^+$. A promising route to TpRu complexes containing hydrido, alkyl, vinyl and CO co-ligands is the reaction of complexes of the general type RuR(Cl)(CO)(PPh$_3$)$_2$ (R = H, aryl, alkyl, vinyl, Cl) with KTp. The first example of this conversion was the preparation of TpRu{CH(CN)Me}(CO)(PPh$_3$) by treating [Ru{CH(CN)Me}Cl(CO)(PPh$_3$)$_2$]$_2$ with KTp.[85] Similarly, Ru{κ^2-*C,O*-CH(CO$_2$Me)CH$_2$CO$_2$Me}Cl(CO)(PPh$_3$)$_2$ reacts with NaTp to yield TpRu(CH(CO$_2$Me)CH$_2$CO$_2$Me}(CO)(PPh$_3$).[86] Hill *et al.* extended this reaction to other derivatives with R = vinyl, 1,2-dimethylvinyl, 1,2-diphenylvinyl, 1-phenyl-2-phenylethynylvinyl (X-ray structure determined) and 4-methylphenyl substituents.[87,88] The reaction sequence starts from RuHCl(CO)(PPh$_3$)$_3$, which is first reacted with an alkyne and subsequently treated *in situ* with KTp. Sun and Simpson[65] prepared the complexes of the type [TpRu(L)(CO)(PPh$_3$)]$^+$ (L = CO, ^{t}BuNC, P(OMe)$_3$, PMe$_3$) *via* the reaction of TpRuCl(CO)(PPh$_3$) with NaPF$_6$ in the presence of L. The key starting material TpRuCl(CO)(PPh$_3$) results from the conversion of TpRuH(CO)(PPh$_3$) with chlorinated solvents. The crystal structure of [TpRu(PMe$_3$)(PPh$_3$)(CO)]PF$_6$ has been determined.

Recently, this methodology has been applied by Liu and co-workers to obtain bimetallic ruthenium Tp complexes with σ,σ'-bridging azobenzene chains.[89]

Furthermore, the $TpRu(PPh_3)(CO)$ functionalized ruthenium complexes $[TpRu(CH{=}CHCH_2PPh_3)(CO)(PPh_3)]BPh_4$ and $TpRu(CH{=}CHCHO)(CO)(PPh_3)$ have been prepared.[90] These complexes were used as starting materials for Wittig reactions. The bimetallic complex $\{TpRu(CO)(PPh_3)\}_2(\mu\text{-}CH{=}CH\text{–}CH{=}CH\text{–}C_6H_4\text{–}CH{=}CH\text{–}CH{=}CH)$ was obtained from the reaction of $[TpRu(CH{=}CHCH_2PPh_3)(CO)(PPh_3)]BPh_4$ with $NaN(SiMe_3)_2$ and terephthaldicarboxaldehyde.

$RuHCl(CO)(P^iPr_3)_2$ is reported to react with NaTp to give $(\kappa^2\text{-Tp})RuH(CO)(P^iPr_3)_2$, which upon heating liberates one P^iPr_3 ligand to afford $TpRuH(CO)(P^iPr_3)$.[91] The latter can easily be protonated with HBF_4 giving $[TpRu(H_2)(CO)(P^iPr_3)]^+$. The dihydrogen ligand in turn can easily be displaced by acetone. From the reaction of $RuHCl(CO)(PPh_3)_3$ or $[RuH(NCMe)_2(CO)(PPh_3)_2]BF_4$ with KTp at room temperature in CH_2Cl_2, $(\kappa^2\text{-Tp})RuH(CO)(PPh_3)_2$ could be isolated and crystallographically characterized.[92] Likewise, Hill *et al.* isolated the related thiocarbonyl complexes. At elevated temperatures, the corresponding $TpRuH(CA)(PPh_3)$ (A = O, S) complexes are formed. The complexes *cis,trans*-$Ru(Me)I(CO)_2(PMe_3)_2$ and $Ru(Me)I(CO)_3(PMe_3)$ react with KTp (or NaTp) to give the acyl complexes $(\kappa^2\text{-Tp})Ru\{C({=}O)Me\}(CO)(PMe_3)_2$ and $TpRu\{C({=}O)Me\}(CO)(PMe_3)$, respectively.[22] The κ^2-*N,N′* complex $(\kappa^2\text{-Tp})Ru\{C({=}O)Me\}(CO)(PMe_3)_2$ could not be converted to $TpRu\{C({=}O)Me\}(CO)(PMe_3)$. Recently, Jia *et al.*[57] reported on a new approach for obtaining complexes of the type $TpRu(R)(CO)(PPh_3)$ *via* reactions of $TpRuCl(NCMe)(PPh_3)$ with $NaBH_4$ and primary alcohols RCH_2OH (R = H, Et, ^{n}Pr, Ph, 4-MePh, 4-ClPh). The decarbonylation of the alcohols were studied extensively and a mechanistic rationale involving metal η^2-aldehyde and η^2-dihydrogen intermediates was proposed.

In a series of articles, Gunnoe *et al.* described the synthesis and reactivity of TpRu amine, amido and imido complexes based on $TpRu(OTf)(L)(L')$ (L = PMe_3, $P(OMe)_3$, CO, NH_3; L′ = PMe_3, $P(OMe)_3$, PPh_3).[93–98] The reaction of $TpRu(OTf)(CO)(PPh_3)$ with LiNHPh afforded the anilido complex $TpRu(NHPh)(CO)(PPh_3)$. If $TpRu(OTf)(CO)(PPh_3)$ is treated with an excess of aniline, the salt $[TpRu(CO)(PPh_3)(NH_2Ph)]OTf$ is obtained.[93,94] Reactions of the Ru(II) amido complexes $TpRu(NHPh)L_2$ (L = CO, PMe_3 or $P(OMe)_3$) with AgOTf yield the binuclear salts $[\{TpRuL_2\}_2(\mu\text{-}NHC_6H_4C_6H_4NH)](OTf)_2$ along with the Ru(II) amine complexes $[TpRuL_2(NH_2Ph)]OTf$ in an approximate 1:1 molar ratio.[95] In these reactions, the two ruthenium fragments are coupled *via* C–H bond cleavage and C–C bond formation at the *para* position of the anilido ligands.

Deprotonation of $[TpRu(CO)(PPh_3)(NHPh)](OTf)_2$ yields the thermally unstable Ru(IV) d^4 imido salt $[TpRu(NPh)(CO)(PPh_3)]OTf$ (Scheme 18).[96] This complex is

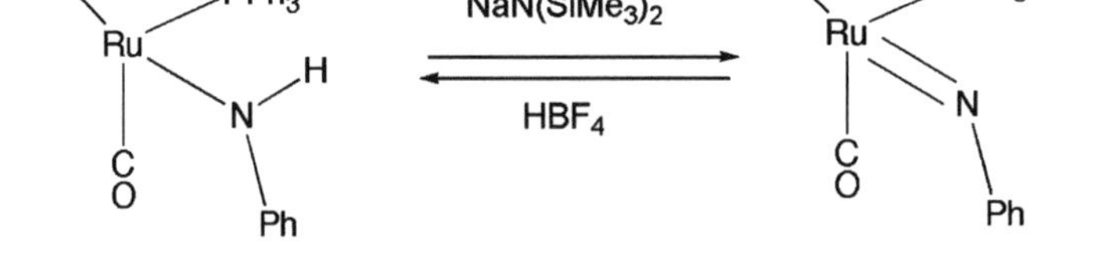

Scheme 18.

only stable below $-50\,^{\circ}C$, and its instability and paramagnetic nature have been attributed to the π-conflict between a filled dπ orbital and electrons on the imido ligand.

Onishi described the preparation of $TpRuCl_2(NO)$ and $Tp^*RuCl_2(NO)$ from $Ru(NO)Cl_3$.[99] Both substances were considered to be Ru(II) species.

D. *Hydride, Dihydrogen and HSiR₃ Complexes*

Chaudret *et al.*[34,35] prepared the complexes TpRuH(COD), Tp^*RuH(COD) and $Tp^{iPr,Br}$RuH(COD) *via* the reactions of RuHCl(bpzm)(COD) with KTp, KTp^* and $KTp^{iPr,Br}$, respectively. TpRuH(COD) can also be synthesized by heating TpRuCl (COD) in methanol in the presence of base. Hydrogenation of Tp^*RuH(COD) and $Tp^{iPr,Br}$RuH(COD) under H_2 (3 bar) yielded $Tp^*Ru(H_2)_2H$ and $Tp^{iPr,Br}Ru(H_2)_2H$. With TpRuH(COD), however, no reaction took place. From $Tp^*Ru(H_2)_2H$, $Tp^{iPr,Br}Ru(H_2)_2H$ and D_2, the corresponding isotopomers $Tp^xRuH_{5-x}D_x$ ($x=1$–4; $Tp^x=Tp^*$, $Tp^{iPr,Br}$) were obtained. The nature of the hydride ligands has been established by T_1 measurements and the J_{HD} coupling is interesting in that the presence of two η^2-bound dihydrogen ligands and one hydride ligand is indicated. Hydrogenation of Tp^*RuH(COD) and $Tp^{iPr,Br}$RuH(COD) in the presence of one equivalent of L (L = PCy_3, tht, pyridine, $HNEt_2$) leads to the formation of the hydrido dihydrogen complexes $Tp^*RuH(L)(H_2)$ and $Tp^{iPr,Br}RuH(L)(H_2)$ (Scheme 19). In the presence of a large excess of L (L = pyridine, tht, CO) over the ruthenium complex, the compounds $Tp^*RuH(L)_2$ were obtained. $Tp^*RuH(H_2)_2$ did not react with MeI or CF_3CO_2H but reactions with $HBF_4 \cdot Et_2O$ and HOTf occur to afford, in the presence of CH_3CN, the complex $[Tp^*Ru(NCMe)_3]^+$. Chaudret *et al.*[100] also reported the syntheses of $TpRuH(PCy_3)(H_2)$ and $Tp^*RuH(PCy_3)(H_2)$ by treating $RuHI(PCy_3)_2(H_2)$ with KTp or KTp^* under a H_2 atmosphere in 40 and 4% yield, respectively.

The decarbonylation reaction mentioned above could proceed *via* a $TpRuH(PPh_3)(H_2)$ intermediate as proposed by Jia *et al.*[56] These authors independently prepared this compound by reacting $TpRuH(PPh_3)$(NCMe) with H_2 (40 atm). We have recently shown that upon treatment of the Ru(III) complex $TpRuCl_2(PR_3)$ ($PR_3 = PPh_3$, $PPh_2{}^iPr$, P^iPr_3, PCy_3) with $NaBH_4$ in EtOH as solvent, dihydrogen compounds of the type $TpRu(PR_3)(H)(H_2)$ also become readily accessible.[51] An analogous dihydrogen complex has been prepared by utilizing the precursor $TpRuCl(P^iPr_2Me)$(NCMe).[61,101]

Akita *et al.* described the synthesis of a series of ruthenium Tp^{iPr_2} complexes.[30,31] The precursor $Tp^{iPr_2}RuH(H_2)_2$ used in this study is protonated by HOTf resulting in

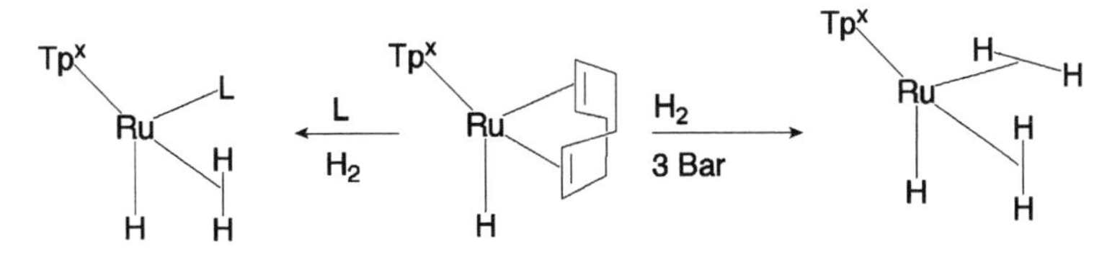

SCHEME 19. $Tp^x = Tp^{iPr2,4R}$; R = H, Br; L = PCy_3, py, $HNEt_2$, THT.

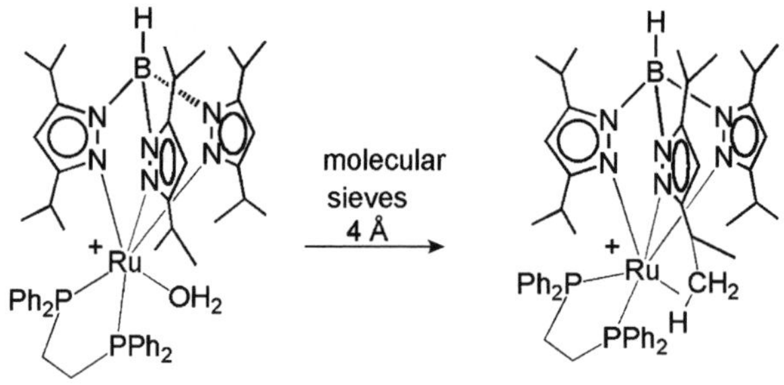

SCHEME 20.

a cationic diaquo adduct, which is shown to be a precursor for several new $Tp^{iPr_2}Ru$ complexes.[102] Worthy of mentioning is the substitution reaction of the cationic aquo-ruthenium complex $[Tp^{iPr_2}Ru(dppe)(OH_2)]OTf$ giving rise to $[Tp^{iPr_2}Ru(dppe)(L)]OTf$ ($L = N_2$, acetone), by way of the dehydrated intermediate $[(\kappa^4\text{-}N,N',N'',H\text{-}Tp^{iPr})Ru(dppe)]OTf$, which has an agostic interaction with a methyl C–H bond in the isopropyl substituent of the Tp^{iPr} ligand as revealed by X-ray crystallography and NMR analysis (Scheme 20).[102b] Furthermore, exposure of the cationic aquo-ruthenium(II) complex, $[Tp^{iPr_2}Ru(dppene)(OH_2)]^+$ (dppene = 1,2-bis(diphenylphosphino)ethene), to O_2 at ambient temperature results in oxidative cleavage of the $C(sp^3)$–$C(sp^3)$ bond in an isopropyl group in the Tp^{iPr_2} ligand to give an acetyl functional group.[102a]

The diastereomerically pure dihydrogen complex $[TpRu(R,R\text{-dippach}(H_2)]BAr_4$ (R,R-dippach = (R,R)-1,2-bis((diisopropylphosphino)amino)cyclohexane) and the non-chiral derivative $[TpRu(dippae)(H_2)]BAr_4$ (dippae = 1,2-bis((diisopropylphosphino)amino)ethane) have been prepared and characterized by Valerga *et al.*[101] Dicationic dihydrogen complexes $[TpRu(R,R\text{-dippachH})(H_2)]^{2+}$ and $[TpRu(dippaeH)(H_2)]^{2+}$ result from the protonation at one of the NH groups of the respective phosphinoamine ligands by an excess of HBF_4. These species undergo slow tautomerization to their monohydrido isomers $[TpRuH(R,R\text{-dippachH}_2)]^{2+}$ and $[TpRuH(dippaeH_2)]^{2+}$.

Cationic dihydrogen complexes of the type $[TpRu(L)(L')(H_2)]^+$ ($L = P(OEt)_3$, $PPh(OEt)_2$, $L' = PPh_3$, $L = L' = P(OEt)_3$) have been utilized to prepare various RuTp diazenido compounds.[103] Treatment of $[TpRu(L)(L')_2(H_2)]^+$ with aryldiazonium salts give aryldiazenido complexes $[TpRu(L)(L')(NNAr)]^{2+}$. Analogous reactions were performed with hydrazines. With monohydrido complexes TpRuH(L)(L′) and aryldiazonium salts, aryldiazene complexes $[TpRu(L)(L')_2(NH{=}NAr)]^+$ were obtained.

Man *et al.* utilized dihydrogen complexes to prepare the first TpRu(III) superoxo compounds.[104] $[TpRu(L_2)(H_2)]^+$ (L_2 = dppm, dppp, $(PPh_3)_2$), prepared *in situ* by protonation of the respective monohydride precursor $TpRuH(L_2)$, reacts with O_2 to yield the paramagnetic Ru(III) superoxo complex $[TpRu(L_2)(O_2)]^+$. In THF, the superoxo moiety of the complex readily abstracts a hydrogen atom from the solvent to generate the hydroperoxo ($-OOH^-$) group, which then changes into the hydroxoligand by transferring an oxygen atom to a phosphine ligand. DFT calculations at the B3LYP level on the model complex $[TpRu(PH_3)_2(O_2)]^+$ show that the Ru(III) superoxo (O_2^-) structure with a triplet state is more stable than the Ru(IV)-peroxo (O_2^{2-}) structure with a singlet state.

Lin and co-workers have found that not only H_2 but also the silanes $HSiR_3$ ($R_3 = Et_3$, $(OEt)_3$, Ph_3, HEt_2, HPh_2, H_2Ph) can displace acetonitrile from TpRuH(NCMe)(PPh_3) to afford complexes of the type TpRuH(PPh_3)(η^2-$HSiR_3$).[105] This formulation is based on NMR spectroscopic data and DFT calculations. These complexes react reversibly with pressurized H_2, MeCN and PPh_3 to give TpRuH(PPh_3)(H_2), TpRuH(NCMe)(PPh_3) and TpRuH$(PPh_3)_2$. The same authors discovered that TpRuH(NCMe)(PPh_3) is able to catalyze H/D exchange reactions between CH_4 and deuterated solvents.[106] The exchange process has also been investigated by means of DFT/B3LYP calculations suggesting that σ-complexes TpRuH(PPh_3)(η^2-H-CH_3) are active species in the H/D exchange reaction.[107]

E. *Ruthenium–Carbon Single Bonds*

Tp^xRu complexes containing Ru–C single bonds are readily available and stable, even in the presence of β-H atoms, provided there is one phosphine and one CO ligand present. Otherwise, it is more difficult to introduce an alkyl ligand into the TpRu scaffold. As shown by Jalon *et al.*,[108] the reaction of $RuCl_2$(bpzm)(COD) (bpzm = bis(pyrazolyl)methane) with MeMgCl yields Ru(Me)Cl(bpzm)(COD), which on treatment with AgOTf affords the complex Ru(Me)(OTf)(bpzm)(COD). The latter complex reacts with KTp to give TpRu(Me)(COD), while with KTp* the compound (κ^3-*N,N′,H*-Tp*)Ru(Me)(COD) is obtained featuring a three-center B–H–Ru bond as proven by X-ray crystallography. According to Ozawa *et al.*,[109] direct alkylation of TpRuCl(COD) with Me_3Al or Et_3Al results in the formation of TpRu(Me)(COD) and TpRu(Et)(COD), respectively, both in high yields. Other alkylating reagents such as Et_2Mg, EtMgBr and EtLi were found to be less efficient.

A series of alkynyl complexes TpRu(C≡CR)Cl(NO) were prepared by Onishi *et al.* from $TpRuCl_2$(NO) and an excess of terminal alkynes in the presence of NEt_3 (Scheme 21).[110] If the same reaction is performed with ethynylbenzene in the presence of CuI as catalyst, the novel bis(alkynyl) complex TpRu$(C{\equiv}CPh)_2$(NO) was obtained.[111] Upon treatment of these σ-acetylide complexes with one equivalent of H_2O in the presence of $HBF_4 \cdot Et_2O$ in distilled MeOH several compounds were obtained in good yields including ketonyl, bisketonyl, acyl and metallacyclopentenone complexes.

The same group has shown that TpRu(C≡CR)Cl(NO) (R = Ph, CMe_2OH, CPh_2OH, CH_2OH) undergo regioselective addition of PPh_3 to give cationic phosphonio-alkenyl, -alkynyl and -allenyl complexes.[112] In similar fashion, the bis(alkynyl) complex TpRu$(C{\equiv}CPh)_2$(NO) was found to react with PPh_3 in the presence of HBF_4 to yield the dicationic bis(β-phosphonio-alkenyl complex $[TpRu(CH{=}CPhPPh_3)_2(NO)](BF_4)_2$.

Lin and co-workers have shown that the alkynyl complex TpRu(C≡CPh)$(PPh_3)_2$ (prepared from TpRuCl$(PPh_3)_2$ and HC≡CPh and NaOMe) reacts with ICH_2CN to afford the cationic vinylidene salt $[TpRu({=}C{=}CPhCH_2CN)(PPh_3)_2]$I, which

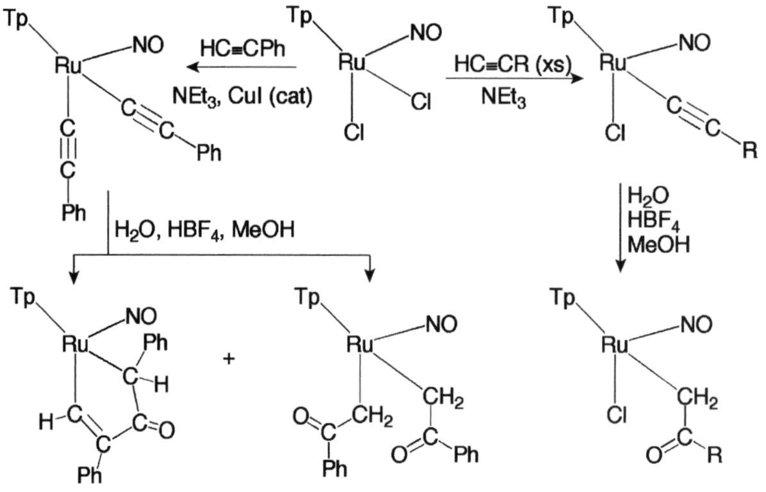

SCHEME 21. Synthesis and reactions of nitrosyl-alkynyl complexes. R = Ph, C_6H_4Me-4, ^tBu, CH_2CH_2OH, CH_2OH, CMe_2OH, CPh_2OH.

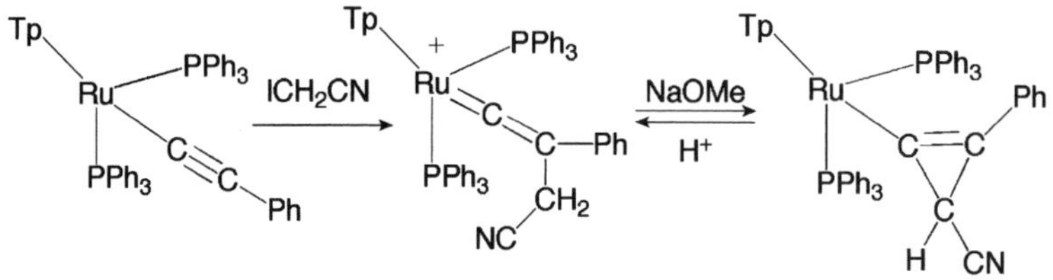

SCHEME 22.

upon treatment with base yields the neutral ruthenium cyclopropenyl complex $TpRu(C{=}CPhCHCN)(PPh_3)_2$ (Scheme 22).[113]

The alkynyl complex $TpRu(C{\equiv}CPh)(\kappa^2\text{-}P,O\text{-}PPh_2CH_2CH_2OMe)$ has been prepared from the reaction of $TpRuCl({=}C{=}CHPh)(\kappa^1\text{-}P\text{-}PPh_2CH_2CH_2OMe)$ with lithium diisopropylamide in THF.[43]

The hydrido complex $Tp^{Ms}RuH(COD)$ reacts with CO to afford the acyl complex $Tp^{Ms}Ru\{C({=}O)C_8H_{13}\}(CO)_2$ *via* intramolecular olefin insertion followed by CO addition and insertion.[32] In contrast, reaction of $Tp^{Ms}RuH(COD)$ with acetonitrile was accompanied by C–H activation of a ligand methyl group and H_2 elimination, ultimately producing $Tp^{Ms@}Ru(NCMe)_2$ as the final product (Scheme 23).

In an attempt to generate neutral azavinylidene derivatives, Puerta *et al.* carried out the deprotonation of $[TpRu(NH{=}CPh_2)(dippe)]BPh_4$ and $[TpRu(NH{=}CPh_2)(PEt_3)_2]BPh_4$ using KO^tBu as base.[114] Unexpectedly, no azavinylidene complexes were obtained. Rather, in both cases *ortho*-metalation of one of the phenyl rings of the benzophenone imine took place, giving rise to species $TpRu(\kappa^2\text{-}C,N\text{-}C_6H_4CPh{=}NH)(dippe)$ and the phenyl complex $TpRu(\sigma\text{-}C_6H_4CPh{=}NH)$

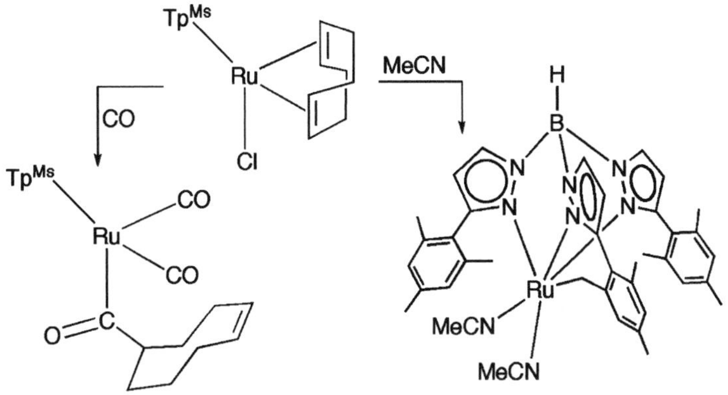

SCHEME 23.

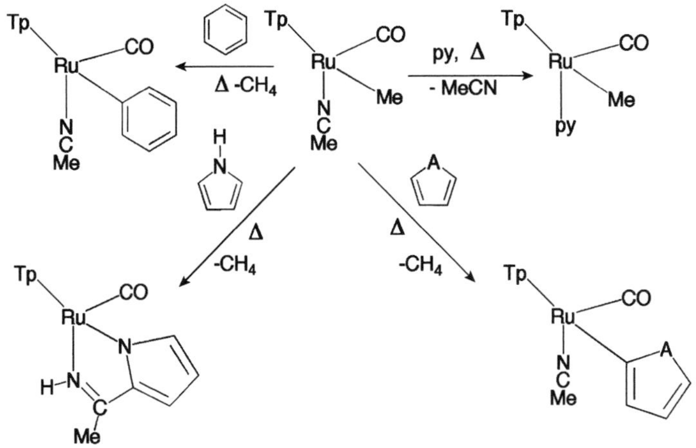

SCHEME 24. C–H activation processes with TpRu(Me)(CO)(NCMe). All reactions at 90 °C. A = O, S.

$(PEt_3)_2$ containing a direct Ru–C bond with the NH group of the imine not being coordinated to ruthenium.

Recently, Gunnoe and co-workers showed in several detailed studies that TpRu (Me)(CO)(NCMe)[66] is a very reactive compound with respect to C–H activation processes giving rise to several unusual transformations of organic substrates. For instance, TpRu(Me)(CO)(NCMe) was found to polymerize styrene and methyl methacrylate in benzene at 90 °C.[66] Furthermore, TpRu(Me)(CO)(NCMe)[115,116] initiates carbon–hydrogen bond activation of benzene and at the 2-position of furan and thiophene to produce methane and TpRu(aryl)(CO)(NCMe) (aryl = phenyl, 2-furyl, 2-thienyl) (Scheme 24). The solid-state structure has been determined for TpRu(2-thienyl)(CO)(NCMe). The complexes TpRu(aryl)(CO)(NCMe) also serve as catalysts for the formation of ethylbenzene, 2-ethylfuran and 2-ethylthiophene from ethylene and benzene, furan, and thiophene, respectively (*vide infra*). DFT/B3LYP calculations are consistent with a non-oxidative addition pathway in which the electrophilic character of Ru(II) dictates the regioselectivity, i.e., the reaction does not proceed through a Ru(IV) intermediate. The reaction of TpRu(Me)(CO)(NCMe) with pyridine under the same reaction conditions affords the product from MeCN/pyridine exchange TpRu(Me)(CO)(py).[116]

In contrast to the reactions with benzene, furan and thiophene, TpRu(Me)(CO)(NCMe) reacts with pyrrole to give the metallacycle TpRu{κ^2-*N*,*N'*-NH=CMe(NC_4H_3-2)}(CO) (Scheme 24).[117] The formation of this complex involves the cleavage of the N–H bond and 2-position C–H bonds of pyrrole as well as a C–C bond forming step between pyrrole and the acetonitrile ligand. Computational studies again support the suggested selectivity for initial N–H bond cleavage in preference to C–H bond activation.

Lee *et al.* investigated reactions of the Ru(II) phenyl complex TpRu(C_6H_5)(CO)(NCMe) with substrates that contain C–N and C–O multiple bonds including carbodiimides and *N*-methylacetamide. Carbodiimides react with TpRu(C_6H_5)(CO)(NCMe) to yield amidinate complexes that result from C–C bond formation between the phenyl ligand and the carbodiimide carbon. With *N*-methylacetamide, benzene and the amidate complex TpRu(κ^2-*N*,*O*-OCMeNMe)(CO) is formed.[118] Very recently, the Gunnoe group also investigated the reactions of TpRu(C_6H_5)(CO)(NCMe) with electron-rich olefins such as ethyl vinyl sulfide and 2,3-dihydrofuran. A key result of this study is shown in Scheme 25.[119]

Single-electron oxidation of the Ru(II) complexes TpRu(R)(L)(L′) (L=CO, L′=NCMe, R=Me, CH_2CH_2Ph; L=L′=PMe_3, R=Me) with AgOTf leads to alkyl elimination reactions that produce TpRu(OTf)(L)(L′) and organic products that likely result from Ru–C_{alkyl} bond homolysis.[120] This reaction apparently proceeds *via* a Ru(III) species. Calculations have shown that homolytic Ru–C bond cleavage is far more facile if the metal center is in the +III oxidation state.

Finally, Feng *et al.* showed that at elevated temperatures, complexes of the type TpRuX(PMe_3)$_2$ (X=OH, OPh, Me, Ph or NHPh) undergo regioselective hydrogen–deuterium (H/D) exchange with deuterated arenes.[121] Mechanistic studies indicate that the likely pathway for the H/D exchange involves ligand dissociation (PMe_3 or MeCN), Ru-mediated activation of an aromatic C–D bond and deuteration of the basic non-dative ligand (hydroxide or anilido) or Tp positions *via* net D^+ transfer.

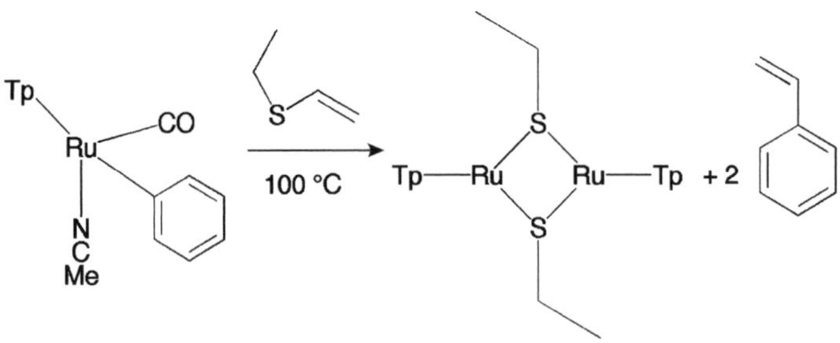

Tp, Ru, CO, N, C, Me, S, 100 °C, Tp—Ru, Ru—Tp + 2, S, S

SCHEME 25.

F. *Ruthenium–Carbon Double and Triple Bonds*

A series of neutral and cationic ruthenium(II) alkylidenes containing Tp have been described by Grubbs *et al.*[122] The reaction of $RuCl_2(=CHPh)(PCy_3)_2$ with KTp afforded $TpRuCl(=CHPh)(PCy_3)$ in 84% yield (Scheme 26). Treatment of this complex with Ag^+ salts in the presence of a variety of coordinating solvents yielded cationic alkylidenes $[TpRu(=CHPh)(PCy_3)(L)]^+$ (L = H_2O, MeCN, py). These compounds were not active in olefin metathesis reactions. However, TpRuCl $(=CHPh)(PCy_3)$ could be activated for ring-closing metathesis by addition of HCl, CuCl and $AlCl_3$.

The Grubbs group prepared the first TpRu alkylidene complex featuring a N-heterocyclic carbene ('NHC') ligand (Scheme 27).[123] The reaction of $RuCl_2$ $(=CHPh)(IMesH_2)(py)_2$ ($IMesH_2$ = 1,3-bis(2,4,6-trimethylphenyl)-4,5-dihydroimidazol-2-ylidene) with KTp afforded quantitatively $TpRuCl(=CHPh)(IMesH_2)$.

A similar reaction had been investigated recently by Slugovc *et al.*[124] The reaction of $RuCl_2(\kappa^2$-*C,O*-2-formylbenzylidene)$(ImesH_2)(Cl)_2$ with KTp in dichloromethane yielded an unusual ruthenium complex (κ^3-*N,N′,N″*-ClTp)RuCl{κ^2-*C,C′*-1-(mesityl)-3-(4,6-dimethylphenyl-2-methylidene)-4,5-dihydroimidazol-2-ylidene), where a chlorotris(pyrazolyl)borate ligand, which had been created during this reaction, binds in the κ^3-*N,N′,N″* mode to the central ruthenium atom (Scheme 28). Additionally, a double C–H activation of a methyl group of the H_2IMes ligand resulted in the formation of a chelating N-heterocyclic biscarbene ligand and liberation of the former 2-formylbenzylidene as 2-methylbenzaldehyde.

As demonstrated by Grubbs and coworkers, $TpRu(\kappa^2\text{-}O_2CCHPh_2)(PPh_3)$ has proven to be a versatile precursor for the preparation of TpRu alkylidenes.[58] While with 3,3-diphenylcyclopropene and phenylacetylene the metallacyclic species TpRu{κ^2-*C,O*-C(=$CHCHPh_2$)OC($CHPh_2$)=O}(PPh_3) (Scheme 29) and TpRu

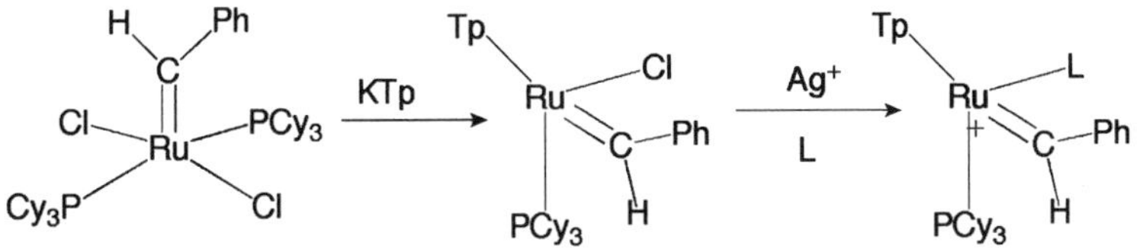

SCHEME 26. L = H_2O, MeCN, py.

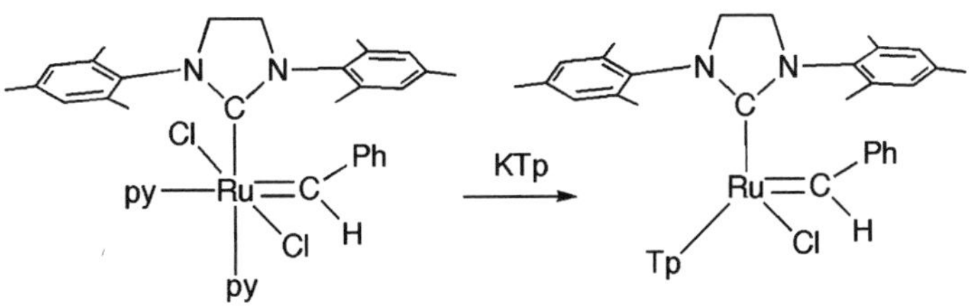

SCHEME 27.

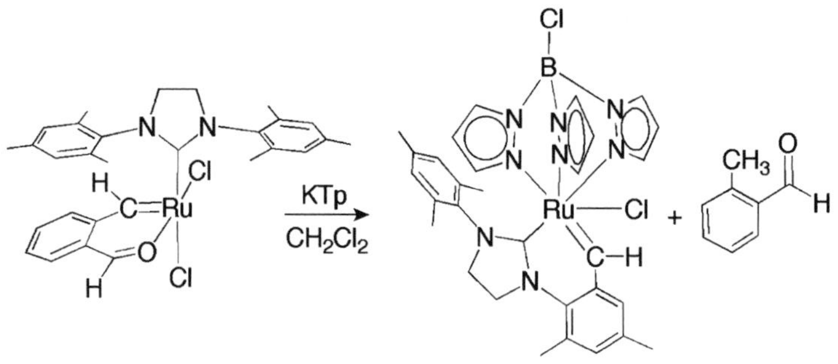

SCHEME 28.

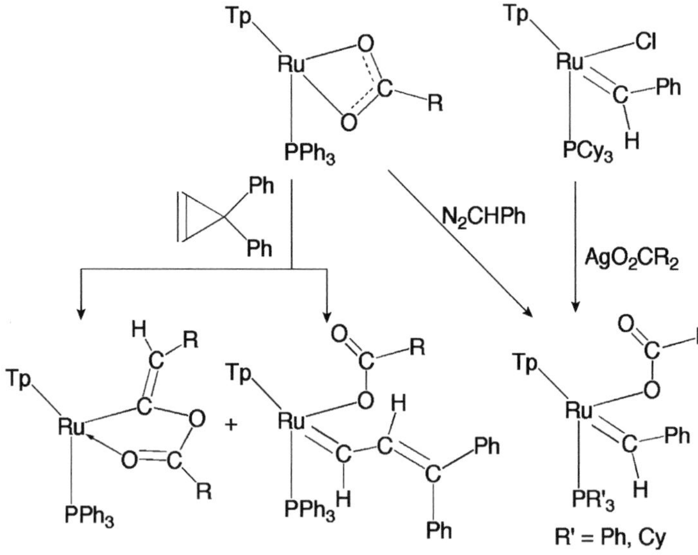

SCHEME 29. R = CHPh$_2$.

{κ^2-*C*,*O*-C(=CHPh)OC(CHPh$_2$)=O}(PPh$_3$) were generated, with phenyldiazomethane the new transition metal benzylidene complex TpRu(κ^1-O$_2$CCHPh$_2$)(=CHPh)(PPh$_3$) were cleanly formed (Scheme 29). A similar species was also available by the reaction between AgO$_2$CCHPh$_2$ and TpRuCl(=CHCH=CMe$_2$)(PCy$_3$), which affords TpRu(κ^1-O$_2$CCHPh$_2$)(=CHCH=CMe$_2$)(PCy$_3$). This compound is active as a single-component catalyst for the ring-opening metathesis polymerization of norbornene. The product polymer is obtained in low yield and is of low molecular weight, suggesting that the active catalyst has a relatively short lifetime under the polymerization conditions.

Cationic vinylidene complexes [TpRu(=C=CHR)(L$_2$)]$^+$ can be obtained from [TpRu(L$_2$)(L$'$)]$^+$ and terminal alkynes, where L$'$ is a labile ligand such as acetone, DMF or H$_2$O. Chelate ligands L$_2$ used are tmeda,[45] pn,[46] Phschiff

(Phschiff = *N*,*N*′-bis-(phenylmethylene)-1,2-ethane-diamine)[46] and tertiary phosphines[30] or are prepared by halide abstraction from $TpRuCl(L_2)$ (L = tertiary phosphines and phosphites) in the presence terminal alkynes.[62,125,126] These complexes are not air-sensitive but are readily deprotonated giving neutral alkynyl complexes. A similar behavior was observed for $[TpRu(=C=CHR)(dippe)]^+$ ($R = CO_2Me$, Ph, H), which was prepared by treating TpRuCl(dippe) with $HC{\equiv}CR$ in the presence of $NaBPh_4$. $[TpRu(=C=CHR)(dippe)]^+$ reacts with KO^tBu to give $TpRu(C{\equiv}CR)(dippe)$, while with CH_3OH, cationic carbene complexes of the type $[TpRu\{=C(OMe)CH_2R)]^+$ are obtained.[125] The reaction of $[TpRu(=C=CHCO_2Me)(PEt_3)_2]BPh_4$ with a further equivalent of $HC{\equiv}CCO_2Me$ yields the *E*-stereoisomer of the vinylvinylidene derivative $[TpRu(=C=C(CO_2Me)–CH=CHCO_2Me)(PEt_3)_2]BPh_4$, which has been structurally characterized.[126] This C–C coupling reaction has been interpreted in terms of a [2+2] cycloaddition of the alkyne to the C_α–C_β bond of the vinylidene ligand to yield a cyclobutenylidene intermediate, followed by a concerted ring opening.

With propargylic alcohols, in general, allenylidenes are formed. However, in the case of $HC{\equiv}CCMe_2(OH)$, an intermolecular dimerization of the allenylidene ligands takes place to yield a vinylidene–carbene complex. In the presence of base, this dimeric complex transforms into the corresponding alkynyl–carbene ruthenium complex.

In sharp contrast to the stability of cationic vinylidene complexes towards air, neutral vinylidene complexes of the type $TpRuCl(=C=CHR)(PR_3)$ convert readily into the corresponding carbonyl compounds $TpRuCl(CO)(PR_3)$ and the aldehyde $RCH=O$ in solution. This process was intensively studied by Bianchini *et al.* using a similar precursor.[127] In addition, neutral vinylidene complexes are substitutionally labile. Thus, the vinylidene moiety is easily replaced by L (L = py, PMe_3, CO). In fact, even metathesis with other terminal alkynes (added in excess) is feasible.[50]

TpRu complexes containing hemilabile ligands were found to facilitate the formation of vinylidene complexes. An example is TpRuCl(κ^2-*P*,*O*-po) ($po = Ph_2PCH_2CH_2OMe$), which reacts with $HC{\equiv}CPh$ to give $TpRuCl(=C=CHPh)$(κ^1-*P*-po).[43] In cases where the hemilabile ligand is (even weakly) nucleophilic, attack at the α-carbon of the vinylidene moiety occurs. Thus, TpRuCl(acpy) (acpy = *N*-acetyl-2-aminopyridine) reacted with $HC{\equiv}CR$ (R = Ph, ^{n}Bu, CH_2Ph, *c*-hexenyl, CO_2Me) to yield neutral amidocarbene complexes.[42] The same reactivity pattern is found with allyl alcohols as ligands. $TpRuCl(HOCH_2–CH=CH_2)$, formed *in situ* by the reaction of TpRuCl(κ^1-*S*-DMSO)$_2$ with allyl alcohol, reacts with terminal alkynes to afford allyloxycarbenes of the type $TpRuCl\{=C(CH_2R)OCH_2CH=CH_2\}Cl$.[44]

The reactions of TpRuCl(COD) with the bidentate ligands 2-aminopyridine (apy), 2-amino-4-picoline (apic) and 2-(methylamino)pyridine (mapy) in the presence of terminal alkynes $HC{\equiv}CR$ (R = Ph, ^{n}Bu, C_6H_9) afford the cyclic aminocarbene complexes $TpRuCl\{=C(CH_2R)apy\}$, $TpRuCl\{=C(CH_2R)apic\}$ and $TpRuCl\{=C(CH_2R)mapy\}$.[39] This reaction proceeds most likely *via* the intermediacy of both a reactive TpRu complex containing a strained, and thus labile, κ^2-*N*,*N*′-coordinated aminopyridine ligand and a vinylidene intermediate.

Pyridine-based aldimines and aminals were shown to act as chelating ligands towards the RuTpCl fragment giving κ^2-*N*,*N'*-coordinated cyclic imine complexes which rearrange into cyclic aminocarbene complexes.[47] TpRu(COD)Cl readily reacts with the imines py-N=CHR (R = Ph, *p*-Ph-OMe, Np) at elevated temperatures to yield the aminocarbene complexes TpRu(=C(R)NH-py)Cl, which is a new synthetic route for obtaining carbene complexes. The mechanism of this novel imine–aminocarbene conversion was analyzed by DFT/B3LYP calculations.

Neutral vinylidene complexes are key intermediates in alkyne insertion reactions into aliphatic C–H bonds. Complexes TpRuCl(κ^2-*P*,*N*-$PPh_2CH_2CH_2NR_2$) (R = Me, Et) and TpRuCl(DMF)(κ^1-*P*-$PPh_2CH_2CH_2N^iPr_2$) react with terminal alkynes $HC{\equiv}CR^2$ (R^2 = Ph, CH_2Ph, CO_2Et) to yield coupling products of the alkyne and the pn ligand. Depending on the substituents of the amino group, two different types of ligand, namely, butadienyldiphenylphosphine (Scheme 30)[128] and 4-diphenylphosphino-1-buten-3-amine were formed.[129] These are coordinated *via* the phosphorus atom and the terminal double bond. A mechanistic rationale for this transformation is as follows: At the first step, cleavage of the Ru–N bond and formation of a vinylidene complex TpRuCl(=C=CHR^2)(κ^1-*P*-$PPh_2CH_2CH_2NR_2$) is suggested. This species is readily deprotonated by the pendant NR_2 group affording the 16e alkynyl complex TpRu($C{\equiv}CR^2$)(κ^1-*P*-$PPh_2CH_2CH_2NHR_2$)]Cl. Concomitant γ-C–H activation and insertion of the alkynyl ligand gives the final products in a highly diastereoselective fashion. Unsaturated alkynyl complexes, generated *in situ* from neutral vinylidene complexes on treatment with strong bases, such as ^{n}BuLi, could be trapped as the CO complex TpRu($C{\equiv}CPh$)(CO)(PPh_3).[48] Furthermore, selective couplings of olefins and terminal acetylenes were reported[130,131] to take place in the coordination sphere of the TpRu scaffold *via* the successive intermediacy of vinylidene and ruthenacyclobutane complexes (Scheme 31). Subsequent deprotonation of one of the β-hydrogen atoms of the latter with NaOEt yields η^3-butadienyl complexes, while in the presence of chloride, rearrangement takes place to give neutral η^2-butadiene complexes *via* a β-hydrogen elimination/reductive elimination sequence. Starting materials were both TpRuCl(COD) and TpRuCl {η^3-*P*,*C*,*C'*-$Ph_2PCH{=}CHCPh{=}CH_2$}, which reacted with $HC{\equiv}CR$ (R = Ph, *c*-hex, ferrocenyl, CH_2Ph, ^{n}Bu).

TpRuCl(DMF)(PR_3) (crystal structure determined for PPh_3), typically prepared *in situ* by the reaction of TpRuCl(COD) with one equivalent of phosphine in refluxing DMF, reacts readily with terminal alkynes, propargylic alcohols and $HC{\equiv}C(CH_2)_nOH$ (*n* = 2, 3, 4) to afford the neutral vinylidene, allenylidene and

Tp Cl Ru PPh$_2$ Me$_2$N —H–C≡C–R / - HNMe$_2$→ Tp Cl Ru PPh$_2$ R

SCHEME 30. R = CO_2Et, Ph, CH_2Ph.

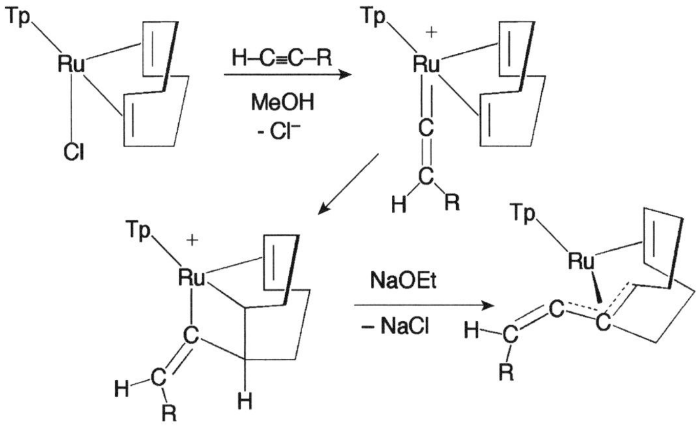

SCHEME 31.

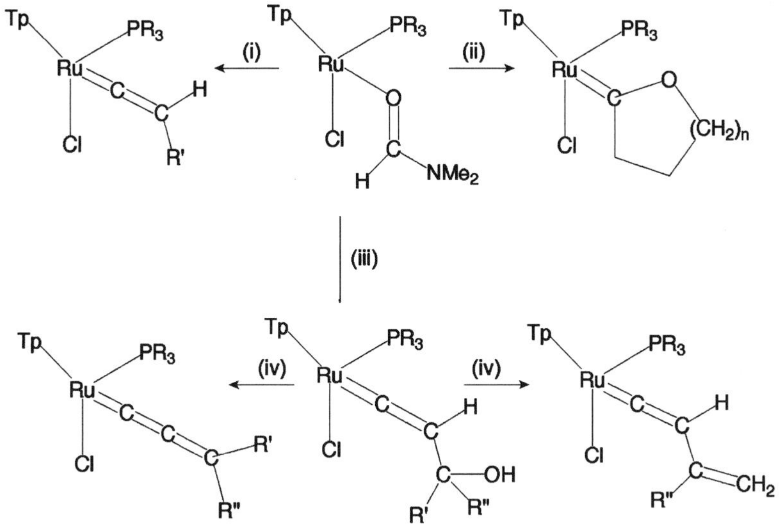

SCHEME 32. (i) $HC{\equiv}CR$; (ii) $HC{\equiv}CCH_2(CH_2)_nOH$ ($n = 1, 2, 3$); (iii) $HC{\equiv}CC(OH)R'R''$; (iv) $-H_2O$; (v) $-H_2O$, $R' = CH_3$.

cyclic oxycarbene complexes $TpRuCl({=}C{=}CHR')(PR_3)$, $TpRuCl({=}C{=}C{=}CR'R'')(PR_3)$, and five-, six- and seven-membered cyclic oxycarbene complexes $TpRuCl({=}C_4H_6O)(PR_3)$, $TpRuCl({=}C_5H_8O)(PR_3)$ and $TpRuCl({=}C_6H_{10}O)(PR_3)$ (Scheme 32).[48–50,52,132,133] Depending on the phosphine as well as the alkyne substituents, in some cases 3-hydroxyvinylidene complexes and vinylvinylidene complexes were also obtained.

While many allenylidene complexes, especially if they are cationic, add nucleophiles either at the C_α or C_γ carbon atom, electron-rich allenylidene

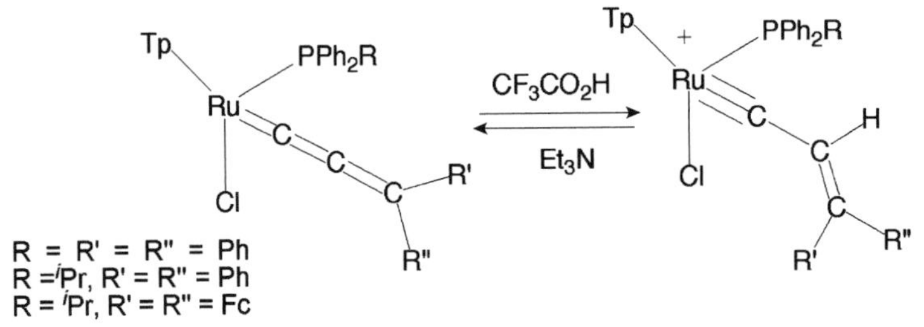

R = R' = R'' = Ph
R = ^{i}Pr, R' = R'' = Ph
R = ^{i}Pr, R' = R'' = Fc

SCHEME 33.

complexes, particularly neutral ones, are capable of adding electrophiles at the C_β carbon atom, thereby forming vinylcarbyne complexes.[134] Accordingly, we have investigated the reaction of TpRu allenylidene complexes with CF_3CO_2H. Addition of CF_3CO_2H to a solution of TpRuCl(=C=C=CR_2)(PR_3) (PR_3 = PPh_3, $PPh_2{}^iPr$; R = Ph, Fc) resulted in quantitative formation of the novel vinylcarbyne complexes [TpRu(≡C–CH=CPh_2)Cl($PPh_2{}^iPr$)]$^+$, [TpRu(≡C–CH=CFc_2)Cl($PPh_2{}^iPr$)]$^+$ and [TpRu(≡C–CH=CPh_2)Cl(PPh_3)]$^+$ (Scheme 33). Similar results have been obtained by Spivak *et al.* who obtained TpRu carbyne complexes upon protonation of vinylidene complexes TpRuCl(=C=CHR) (PPh_3) (R = Ph, ^{t}Bu, ^{n}Bu) with $HBF_4 \cdot Et_2O$ at low temperature.[135]

Finally, Hill *et al.* reported the synthesis of vinylidene and allenylidene complexes of the types [TpRu(=C=CHR)$(L)_2$]$^+$ (L = PPh_3 or L_2 = 1,1′-bis(diphenylphosphino)-ferrocene; R = Ph, C_6H_4Me), [TpRu(=C=C=CR_2)$(L)_2$]$^+$ and TpRuCl(=C=C=CPh_2)(PPh_3).[136–138] The cationic complexes underwent an unprecedented coupling with dithiocarbamates, which led to metallacyclic vinyl and allenyl complexes (Scheme 34). Similar allenylidene complexes but featuring redox-active substituents and ligands based on ferrocene were reported by Hartmann *et al.*[139]

G. *Catalytic Reactions*

1. Hydrogenations

Onishi *et al.*[140] performed the hydrogenation of methylacrylate and 3-phenyl-propene with TpRuCl$(NCPh)_2$ and (pzTp)RuCl$(NCPh)_2$ (substrate to catalyst ratio is 200:1 (mol:mol)) in MeOH in the presence of NEt_3 under a H_2 pressure of 50 kg cm^{-2} at 50 °C. Under these conditions, methylacrylate were quantitatively converted to methylpropionate (turnover number per catalysis is 200) with both catalyst precursors. In the case of 3-phenylpropene, 100% conversion was observed with (pzTp)RuCl$(NCPh)_2$, but only 57% with TpRuCl$(NCPh)_2$. In either case, *E*- and *Z*-1-phenylpropenes were formed as by-products *via* an olefin double bond migration. The complexes Tp*RuH$(H_2)_2$, Tp*RuH(COD) and TpRuH(COD) H were catalytically active in the reduction of unactivated ketones to alcohols either by dihydrogen or by hydrogen transfer from alcohols in basic media, as shown by Chaudret *et al.*[36] Both Tp*RuH$(H_2)_2$ and Tp*RuH(COD) exhibit similar

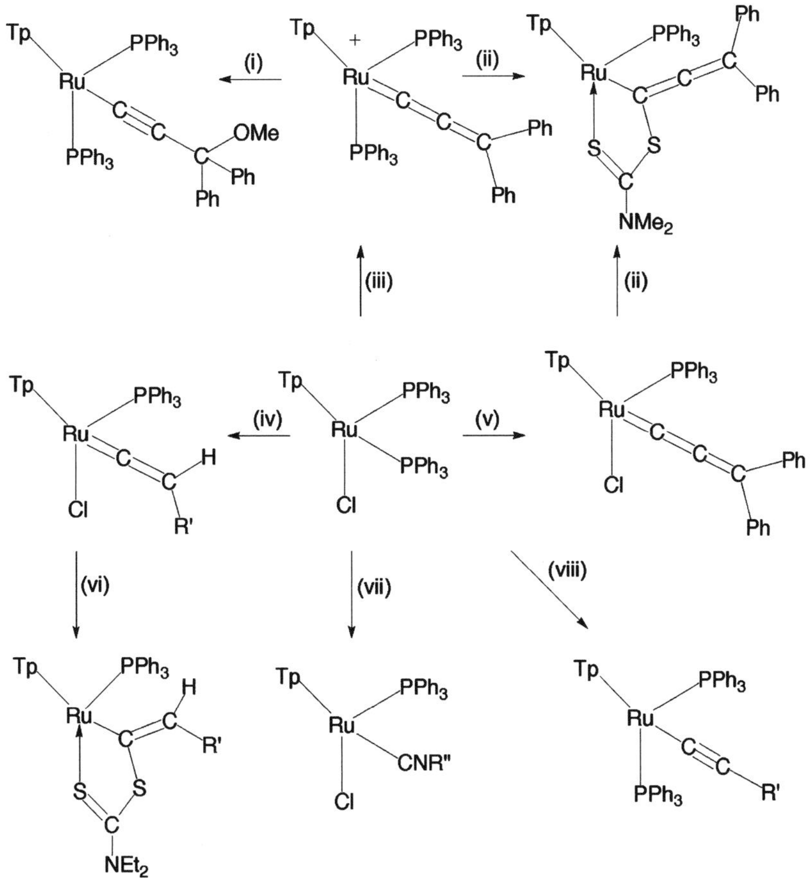

SCHEME 34. (i) NaOMe; (ii) $Na[S_2CNMe_2]$; (iii) $HC{\equiv}CCPh_2OH$, $AgPF_6$; (iv) $HC{\equiv}CR'$ ($R'=C_6H_4Me$-4), THF; (v) $HC{\equiv}CCPh_2OH$, THF; (vi) $[Et_2NH_2][S_2CNEt_2]$; (vii) CNR'' ($R''={}^tBu$, $C_6H_3Me_2$-2,6); (viii) $HC{\equiv}CR'$, MeOH, THF.

activities in the reduction of cyclohexanone suggesting that, under the reaction conditions employed, Tp*RuH(COD) is rapidly hydrogenated. High pressures of H_2 inhibit the catalytic reaction, whereas the presence of olefins has a promoting effect. Also, briefly discussed in the article is transfer hydrogenation of ketones and olefins. Jia *et al.*[56] reported on catalytic hydrogenation of olefins with $[TpRu(NCMe)(PPh_3)_2]^+$ and $[TpRu(NCMe)_2(PPh_3)]^+$ as catalyst precursors. Conversions could be increased in the presence of H_2O or NH_3. The authors demonstrated that dihydrogen complexes are involved in the catalytic cycle.

Lau *et al.* investigated the promoting effect of water in the catalytic hydrogenation of CO_2 to formic acid with $TpRuH(NCMe)(PPh_3)$.[141] A strong promoting effect of water in the catalytic hydrogenation of CO_2 to formic acid with the metal hydride species $TpRuH(NCMe)(PPh_3)$ was observed. High-pressure NMR monitoring of the catalytic reaction showed that CO_2 readily inserts into the

Ru–H bond to form the metal formate TpRu(κ^1-OCHO)(NCMe)(PPh$_3$)$\cdot$H$_2$O, in which the formate ligand is intermolecularly hydrogen-bonded to a water molecule. Theoretical calculations carried out at the B3LYP level show that reaction barrier of the CO_2 insertion is significantly reduced in the presence of water. The same authors also studied the same reaction in various alcohols.[142]

2. Coupling Reactions

TpRuCl(COD), TpRuCl(py)$_2$ and TpRuCl(tmeda) were found to catalyze the coupling of HC≡CPh with benzoic acid. *E*- and *Z*-vinylesters were formed in a ratio varying from 1:1 to 2:1. Similar effects were observed with TpRuCl(py)$_2$ in the coupling of HC≡CPh with allyl alcohol leading to the corresponding *Z*-vinylether and 1-phenyl-3-butenal in a 1:1 ratio. The latter is formed *via* a Claisen rearrangement of the *E*-vinylether.[29] TpRuCl(PPh$_3$)$_2$ and especially TpRuH(PPh$_3$)$_2$ are effective catalysts for the dimerization of terminal alkynes to give enynes. These reactions were performed in refluxing toluene for 20 h. With TpRuCl(COD) as the catalyst precursor, oligo- and polymerization products were obtained. The same reaction is also catalyzed by TpRu(C≡CPh)(κ^2-*P*,*O*-po)[43] and TpRu(CPh=CPhC≡CPh)(PMe^{i}Pr$_2$) featuring an *E*-1,4-enynyl ligand.[126] Ozawa *et al.* showed[143] that the neutral ruthenium vinylidene complex TpRuCl(=C=CHPh)(PPh$_3$) is active in the ring-opening metathesis polymerization (ROMP) of norbornene. The efficiency of the catalyst is enhanced by adding Lewis acids such as BF$_3\cdot$Et$_2$O to the system.

Gunnoe *et al.* have recently demonstrated that TpRu(Me)(CO)(NCMe) and TpRu(Ph)(CO)(NCMe) are active catalysts for the addition of arenes and heteroaromatics to olefins.[115,116,144] For instance, under a dinitrogen atmosphere, the combination of 0.1 mol% TpRu(Ph)(CO)(NCMe) in dry benzene under 25 psi of ethylene at 90 °C results in the formation of ethylbenzene [Eq. (2)]. After 4 h of reaction, approximately 51 catalytic turnovers are observed per mole of catalyst, with TOF = 3.5×10^{-3} mol^{-1} s^{-1}. Placing the benzene/catalyst solution under 25 psi of propene results in the addition of benzene to propene within 30 min at 90 °C, giving rise to the formation of a 1.6:1.0 ratio of *n*-propyl to isopropylbenzene, with 14 catalytic turnovers after 19 h.

0.1mol% [Ru], 90°C, 25 psi (propene): benzene → isopropylbenzene (1.0) + n-propylbenzene (1.6); 0.1mol% [Ru], 90°C, 25 psi (ethylene): benzene → ethylbenzene (2)

Similar results were obtained with TpRu(Me)(CO)(NCMe) and furan yielding, catalytically, 2-ethyl furan [Eq. (3)] and thiophene yielding 2-ethyl thiophene.[116]

The mechanism of these hydroarylations was extensively studied by means of DFT/B3LYP calculations.[145]

0.1mol% [Ru]
90°C
25 psi C_2H_4 (3)

3. Hydration of Nitriles

Fung *et al.* studied the hydration of nitriles to amides by using the indenylruthenium hydride complex (η^5-C_9H_7)RuH(dppm) as a catalyst.[146] In the course of this study, TpRuH(NCMe)(PPh_3) was also tested as a catalyst for the hydration of MeCN to give acetamide. It was found that the Tp complex is less active than the indenyl complex exhibiting a smaller TON by a factor of about 4. The chloro analogue TpRuCl(NCMe)(PPh_3) showed no catalytic activity at all.

4. Isomerization and Cyclization Reactions

In recent years, Liu and co-workers have studied a variety of catalytic organic reactions using, primarily, [TpRu(NCMe)$_2$(PPh_3)]PF_6 and also TpRuCl(py)$_2$ and TpRuCl(NCMe)(PPh_3) as catalysts. Some examples are described in the following paragraphs, others may be found in Refs. 147–150. TpRuCl(py)$_2$ was found to be a new catalyst for *cis–trans* isomerization of various functionalized epoxides.[151] The isomerization of chiral epoxides was enantiospecific without loss of enantiopurity, and epimerization occurred only at the epoxide carbon of the activating group. The salt [TpRu(NCMe)$_2$(PPh_3)]PF_6 turned out to be a very efficient catalyst for a variety of unusual complex organic transformations involving primarily alkyne and olefin functionalities. Examples include the cycloisomerization of 1-ethynyl-3-ols and *cis*-3-en-1-ynes into cyclopentadiene and related derivatives [Eqs. (4) and (5)];[152] the cyclization of iodoalkyne-epoxide functionalities [Eq. (6)],[153] alkyne-epoxide motifs [Eq. (7)],[154] and 3-en-1-ynyl imines with nucleophiles *via* tandem 5-*exo-dig* cyclization and nucleophilic addition;[155] the cycloisomerization of 2-(ethynyl)phenylalkenes to afford diene derivatives *via* skeletal rearrangements;[156] the aromatization of enediynes *via* highly regioselective nucleophilic additions on a π-alkyne functionality [Eq. (8)];[157] and a 1,3-regioselective methylene transfer in the cycloisomerization of 3,5-dien-1-ynes, which involves cleavage of two σ-carbon–carbon bonds of the aliphatic type [Eq. (9)].[158]

Ph OH Ar
[Ru]
10mol%
80 °C
Ph Ar (4)

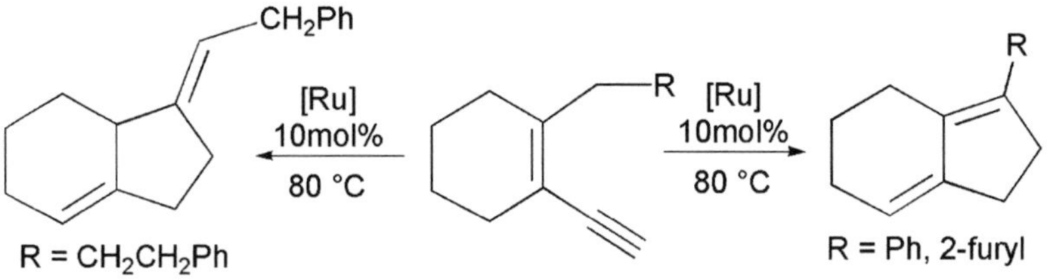
CH2Ph
[Ru]
10mol%
80 °C
R
[Ru]
10mol%
80 °C
R
R = CH2CH2Ph
R = Ph, 2-furyl

(5)

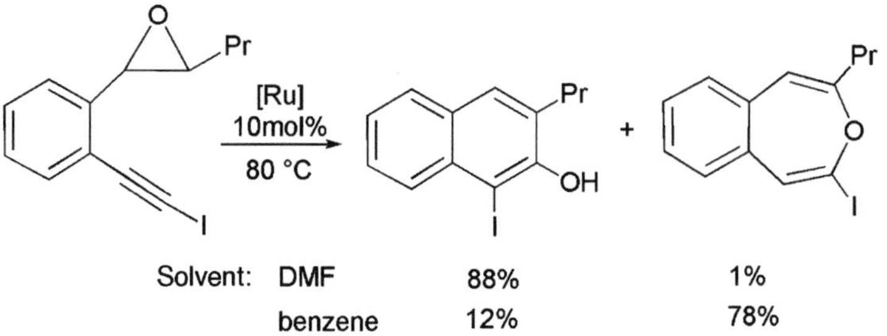
O
Pr
I
[Ru]
10mol%
80 °C
Pr
OH
I
+
Pr
O
I
Solvent: DMF 88% 1%
benzene 12% 78%

(6)

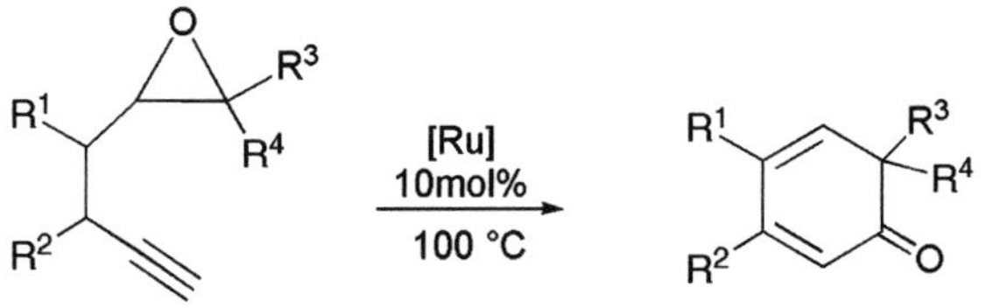
O
R3
R1
R4
R2
[Ru]
10mol%
100 °C
R1
R3
R4
R2
O

(7)

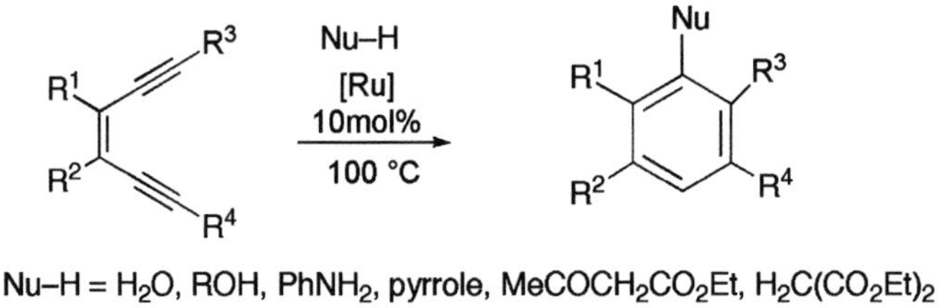
R3
R1
R2
R4
Nu–H
[Ru]
10mol%
100 °C
Nu
R1
R3
R2
R4
Nu–H = H2O, ROH, PhNH2, pyrrole, MeCOCH2CO2Et, H2C(CO2Et)2

(8)

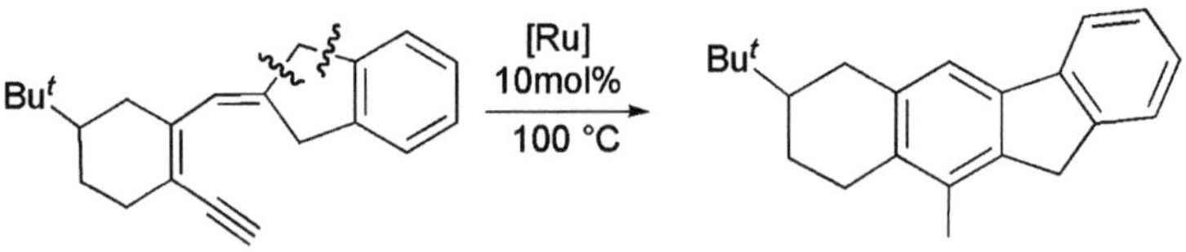
But
[Ru]
10mol%
100 °C
But

(9)

The salt $[TpRu(NCMe)_2(PPh_3)]PF_6$ also catalyzes the efficient rearrangement of α, β-epoxyketones to 1,2-diketones [Eq. (10)].[159] The substrates include monosubstituted, 1,2-disubstituted and even trisubstituted α,β-epoxy ketones. An interesting feature of this reaction is that the trend in the catalytic activity of these epoxides is opposite to that seen with common acid catalysts.

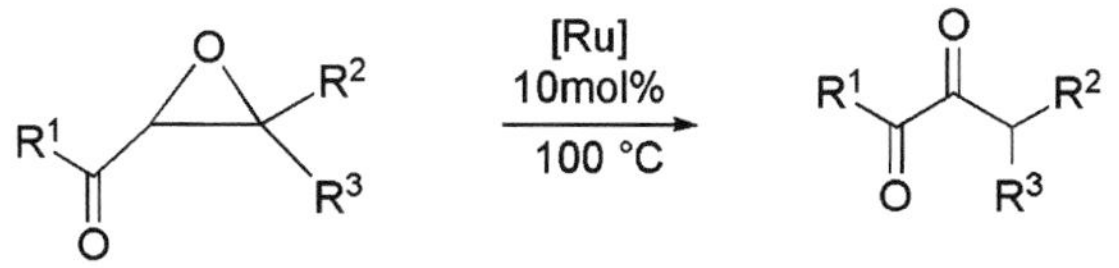

(10)

Finally, Tenorio *et al.* showed[160] that the σ-enynyl complex TpRu(CPh=CPhC≡CPh)(PMe^{i}Pr$_2$) efficiently catalyses the regioselective cyclization of α,ω-alkynoic acids to yield predominantly endocyclic enol lactones having ring size up to 12 atoms [Eq. (11)].

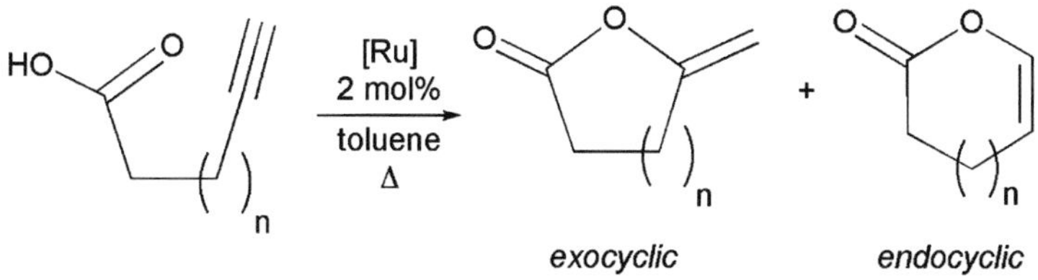

(11)

IV
OSMIUM TRIS(PYRAZOLYL)BORATE CHEMISTRY

The first osmium Tp complex was prepared by Singleton *et al.* in 1990 *via* the reaction of the μ-carboxylato dimer $Os_2(\mu\text{-}O_2CMe)_2(CO)_6$ with KTp in refluxing methanol to give $Tp_2Os_2(CO)_4$ (*Os–Os*).[67] Treatment of this dinuclear compound with Br_2 affords the mononuclear Os(II) complex $TpOsBr(CO)_2$ *via* oxidative cleavage of the Os–Os bond (Scheme 35).

By analogy with the well-known general route to ruthenium-mixed sandwich complexes of the type CpTpRu,[12] the corresponding osmium salt $[CpOs(NCMe)_3]PF_6$ reacts readily with KTp in refluxing acetonitrile to afford CpTpOs in high yield.[161] LaPointe and Schrock used NaTp to prepare the osmium carbyne complex $TpOs(\equiv C^tBu)(CH_2{}^tBu)_2$ from $Os(\equiv C^tBu)(CH_2{}^tBu)_2(py)_2(OTf)$.[162] The structure of this air and water stable compound was confirmed crystallographically. The

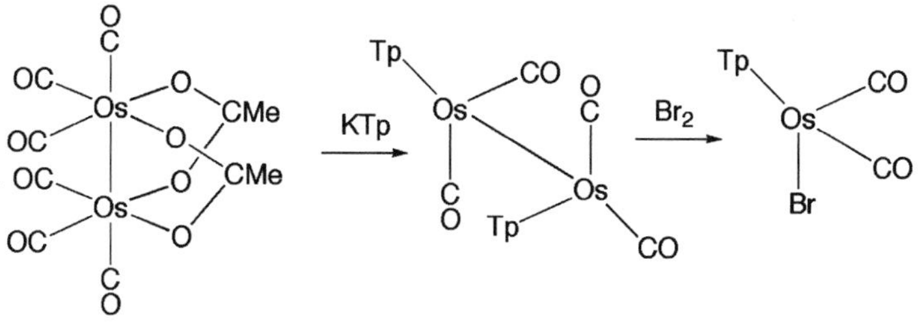

SCHEME 35.

first Tp^*Os complex $Tp^*Os(\equiv N)Ph_2$ was synthesized by Hunt and Shapley *via* the reaction of the nitrido salt $[Os(\equiv N)Ph_2(py)_2]BF_4$ with KTp^* in THF.[163] Reactions with KTp also afford osmium Tp complexes bearing phosphine ligands starting from various phosphine-containing precursors. Treatment of $OsCl_2(PPh_3)_3$ with KTp gives the neutral Os(II) bisphosphine complex $TpOsCl(PPh_3)_2$,[136,164] whereas the reaction of $OsHCl(CO)(P^iPr_3)_2$ and $OsPhCl(CO)(PPh_3)_2$ with NaTp or KTp afforded the complexes $(\kappa^2\text{-}Tp)OsH(CO)(P^iPr_3)_2$[91] and $(\kappa^2\text{-}Tp)Os(Ph)(CO)(PPh_3)_2$, respectively,[92] where, in each case, the Tp ligand is bound to the metal center in a bidentate fashion. The latter two complexes can be converted into the κ^3-Tp variants $TpOsH(CO)(P^iPr_3)$ and $TpOs(Ph)(CO)(PPh_3)$ by heating solutions of the isolated products in toluene, thereby releasing one phosphine ligand. Protonation of $TpOsH(CO)(P^iPr_3)$ with HBF_4 yielded the dihydrogen complex containing salt $[TpOs(\eta^2\text{-}H_2)(CO)(P^iPr_3)]BF_4$ (Scheme 36). The phosphine precursor $TpOsCl(PPh_3)_2$ can be converted into the monohydride $TpOsH(PPh_3)_2$ by reaction with NaOMe or KOH.[136,164] Further protonation with HBF_4 leads to the cationic dihydrogen species $[TpOs(\eta^2\text{-}H_2)(PPh_3)_2]BF_4$ (Scheme 37).[164]

One of the most used osmium Tp precursors is the osmium(VI) nitrido complex $TpOs(\equiv N)Cl_2$,[165] which was prepared by the reaction of $K[Os(\equiv N)(=O)_3]$ with KTp in ethanolic HCl. The complex serves as the starting material for numerous osmium Tp species as demonstrated especially by the studies of Mayer and co-workers.[165] The chloride ligands of this latter complex can be metathesised to give complexes of the type $TpOs(\equiv N)X_2$ (X = acetate, trihaloacetate, bromide, nitrate and oxalate).[166] Treatment of the nitrido complex with Grignard reagents and phosphines results in direct addition to the *electrophilic* nitride. The reaction of $TpOs(\equiv N)Cl_2$ with PhMgCl leads to formation of $[TpOs(=NPh)Cl_2]^-$, which on protonation gives the fully characterized amide complex $TpOs(=NHPh)Cl_2$.[166] The amido ligand in this complex does not undergo protonation or electrophilic addition reactions,[167] whereas its reduced form $[TpOs(=NHPh)Cl_2]^-$ can be easily protonated to afford the aniline complex $[TpOs(NH_2Ph)Cl_2]$.[168] Reaction of $TpOs(\equiv N)Cl_2$ with triarylphosphines PAr_3 ($Ar = C_6H_4R\text{-}4$; R = H, CH_3, CF_3) affords the phosphiniminato complexes $TpOs(N{=}PAr_3)Cl_2$ in almost quantitative yields.

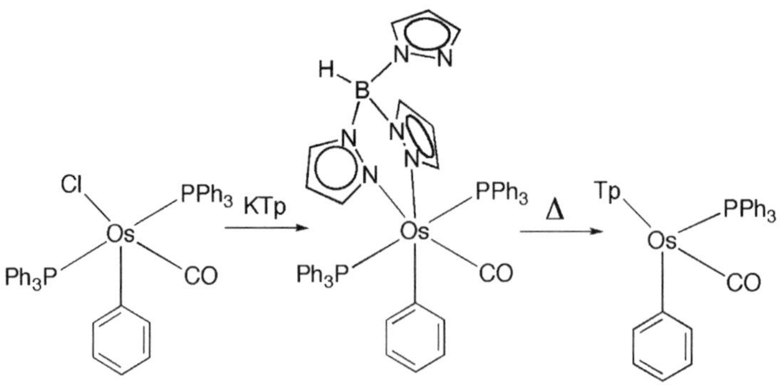

SCHEME 36.

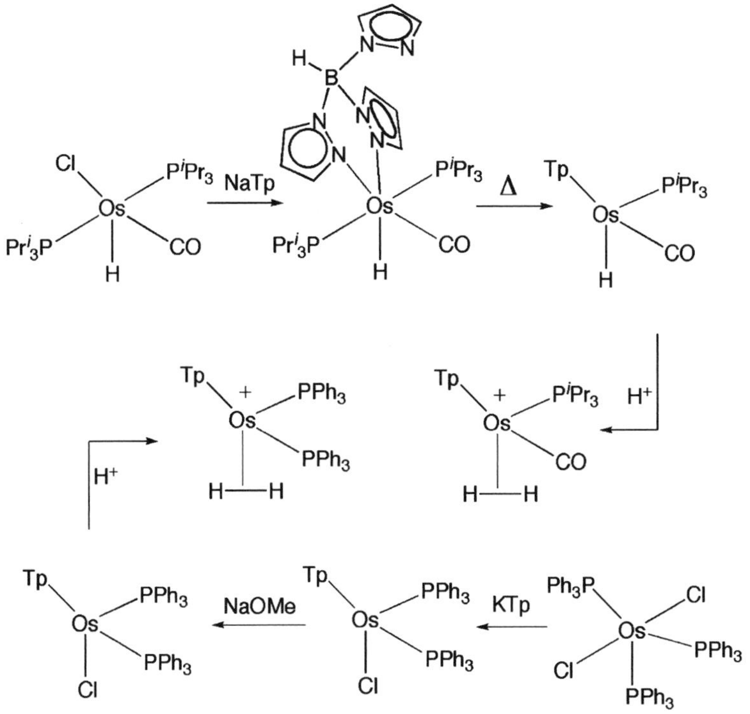

SCHEME 37.

Protonation of these compounds with trifluoromathan sulfonic acid in acetonitrile results in the formation of the salt [TpOs(NH$=$PAr$_3$)Cl$_2$]OTf.[169] On addition of excess of HOTf to TpOs($=$NPPh$_3$)Cl$_2$, the substituted nitride ligand is lost and the Os(IV) complexes TpOs(OTf)Cl$_2$ and TpOsCl$_3$ are formed.[170]

The triflato complex TpOs(OTf)Cl$_2$ did not undergo substitution reactions, but on addition of nitriles and ammonia, the precursor nitride complex TpOs($\equiv$N)Cl$_2$ is recovered.[171] Another intriguing reaction is the formation of the osmium(III) salt [Cp$_2^*$Fe][OsTp(OTf)Cl$_2$], which was obtained by the reduction of TpOs(OTf)Cl$_2$ with Cp$_2^*$Fe. This compound reacted with various Lewis bases to give Os(III) complexes of the type TpOs(L)Cl$_2$ (L$=$NCMe, NCPh, PPh$_3$, py, imidazole and NH$_3$), which can be oxidized by means of [NO]BF$_4$ to afford the corresponding osmium(IV) salts of the general formula [TpOs(L)Cl$_2$]BF$_4$.[170] On addition of BPh$_2$R (R$=$Ph, OBPh$_2$) to the nitrido nitrogen of TpOs($\equiv$N)Cl$_2$, the boron-containing complexes TpOs{NPh(BPhR)}Cl$_2$ are formed. The adduct derived from BPh$_3$ features one Cl coordinated to boron, forming a four-membered ring.[166] In Scheme 38 an overview of the most important reactions is presented.

Starting from the nitrido complex TpOs($\equiv$N)Cl$_2$, several unusual polynuclear compounds have been obtained. For example, TpOs($\equiv$N)Cl$_2$ reacts readily with cobaltocene (Cp$_2$Co) to give the mixed-valence dinitrogen-bridged anionic complex [Tp$_2$Os$_2$(μ:σ(N),(N')-N$_2$)Cl$_4$][Cp$_2$Co].[172] Upon reaction of TpOs($\equiv$N)Cl$_2$ with the

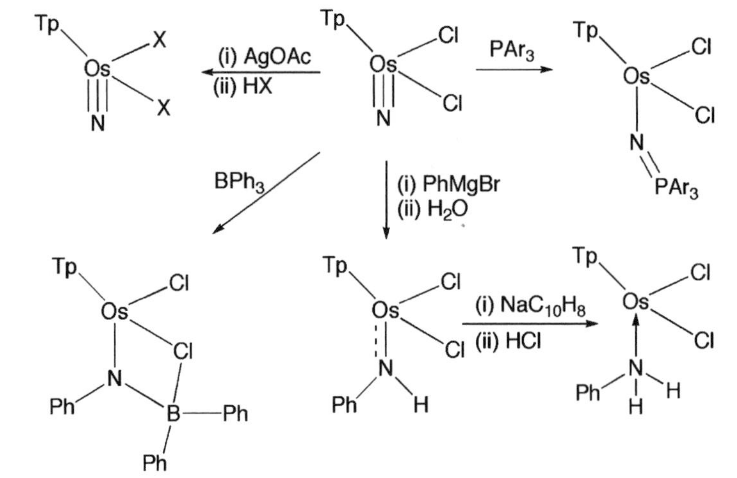

SCHEME 38. Ar = C_6H_4R-4; R = H, CH_3, CF_3.

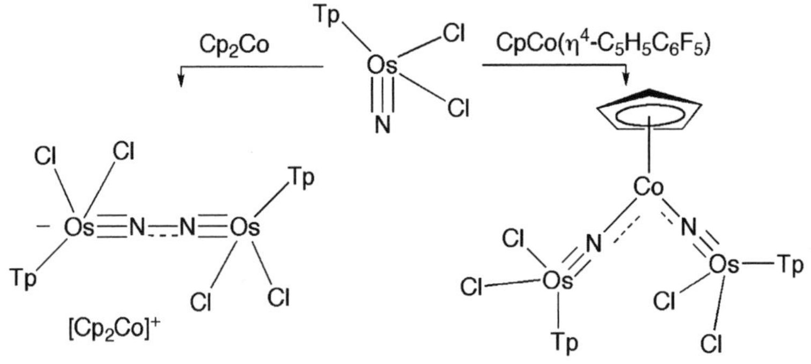

SCHEME 39.

cobalt diene complex $CpCo(\eta^4\text{-}C_5H_5\text{–}C_6F_5)$, the interesting trimetallic species $\{TpOs(\equiv N)Cl_2\}_2CoCp$ were isolated (Scheme 39).[173] The structure of the latter complex was determined by X-ray crystallography and showed that the Os–N–Co bonds are almost linear.

Osmium complexes have also been shown to be active in carbon-to-metal hydrogen atom transfer reactions.[174] Photolysis of the osmium dimer $Tp_2Os_2(CO)_4$ in THF results in carbon-to-osmium hydrogen atom transfer, and this reaction can be used synthetically as a route to $TpOsH(CO)_2$. Very recently, Dickinson and Girolami[175] reported the synthesis of TpOsH(COD), analogous to the well-known ruthenium complex TpRuCl(COD) by treating the polymer $[OsBr_2(COD)]_x$ with KTp. Starting from this precursor several complexes of the type TpOsX(COD) (X = Br, OTf, Me) were obtained *via* reactions with $CHBr_3$, MeOTf or $MgMe_2$ (Scheme 40).

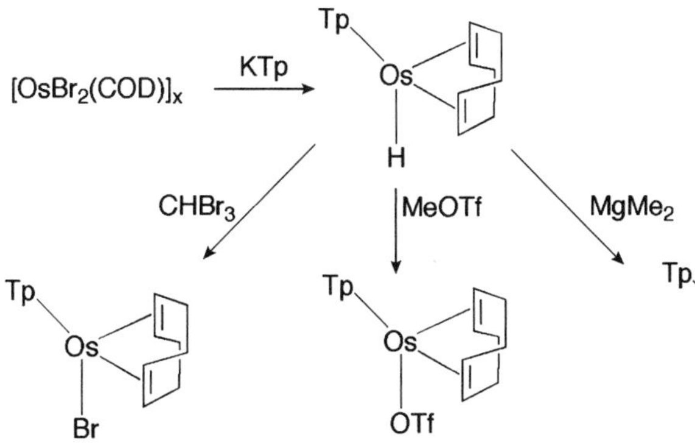

SCHEME 40.

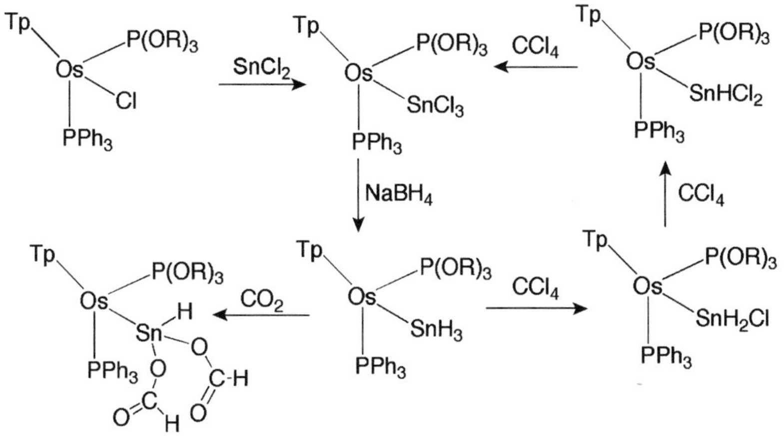

SCHEME 41. R = Me, Et.

Treatment of the bromide complex TpOsBr(COD) with NaOMe in methanol afforded the methoxide compound TpOs(OMe)(COD). In contrast to ruthenium, attempts to replace the COD ligand with phosphines failed. However, by using the previously described OsTp precursor $TpOsCl(PPh_3)_2$,[136,164] it was possible to exchange PPh_3 against dppm to afford $TpOsCl(\kappa^1\text{-}P\text{-dppm})(PPh_3)$ and TpOsCl (κ^2-dppm), the former being the kinetically favored product.

Recently, the Tp osmium stannyl complexes $TpOs(SnH_3)(PPh_3)(POR_3)$ (R = Me, Et) were prepared containing SnH_3 as co-ligand and their reactivity has been explored.[176] The reaction of $TpOsTpCl(PPh_3)\{P(OR)_3\}$ with $SnCl_2$ gives the trichlorostannyl derivatives $TpOs(SnCl_3)(PPh_3)\{P(OR)_3\}$, which upon reaction with $NaBH_4$ afford $TpOs(SnH_3)(PPh_3)\{P(OR)_3\}$ (R = Me, Et), the Os–Sn bond remaining intact during the transformation. Investigations of the reactivity included the reaction with CCl_4 and insertion of CO_2 into the Sn–H bond (Scheme 41).

V
CONCLUDING REMARKS

In contrast to, for example, groups 6 and 9, the organometallic chemistry of group 8 poly(pyrazole)borates has been slow to emerge and in the case of iron and osmium still remains comparatively unexplored. The established stability of the d^6-ML_6 octahedral geometry, so favored by Tp^x ligands, appears to dominate group 8 chemistry with the divalent metal state being most commonly encountered. Although higher oxidation state chemistry is also beginning to emerge, at present this is more advanced for osmium than the lighter elements. In the case of ruthenium, there has been rapid recent growth, with the availability of compounds of the forms $TpRuCl(PR_3)_2$, TpRuCl(COD) and TpRu(R)(CO)(NCMe) each showing a rich chemistry. Perhaps more than any other metal to date, ruthenium has been shown to mediate a diversity of catalytic transformations supported by Tp and related ligands. The field of C–H activation chemistry has also in recent times included important results based on the TpRu scaffold, hinting at parallels with the more mature chemistry of group 9 metals ligated by Tp^x ligands, which is included in the following chapter.

ACKNOWLEDGEMENTS

Financial support by the 'Fonds zur Förderung der wissenschaftlichen Forschung' is gratefully acknowledged (Project No. P16600-N11). The dedication and expertise of the co-workers cited in the references is also gratefully acknowledged.

REFERENCES

(1) Trofimenko, S. *J. Am. Chem. Soc.* **1966**, *88*, 1842.
(2) (a) Trofimenko, S. *The Coordination Chemistry of Polypyrazolyborate Ligands*, Imperial College Press, **1999**. (b) Trofimenko, S. *Chem. Rev.* **1993**, *93*, 943. (c) Trofimenko, S. *Chem. Rev.* **1972**, *72*, 497.
(3) Slugovc, C.; Schmid, R.; Kirchner, K. *Coord. Chem. Rev.* **1999**, *185–186*, 109.
(4) Rüba, E.; Simanko, W.; Mereiter, K.; Schmid, R.; Kirchner, K. *Inorg. Chem.* **2000**, *39*, 382.
(5) Trofimenko, S. *J. Am. Chem. Soc.* **1967**, *89*, 3170.
(6) Trofimenko, S. *J. Am. Chem. Soc.* **1967**, *89*, 6288.
(7) (a) Hannay, C.; Thissen, R.; Briois, V.; Hubin-Franskin, M.-J.; Grandjean, F.; Long, G. J.; Trofimenko, S. *Inorg. Chem.* **1994**, *33*, 5983. (b) LeCloux, D. D.; Keyes, M. C.; Osawa, M.; Reynolds, V.; Tolman, W. B. *Inorg. Chem.* **1994**, *33*, 6316. (c) Rheingold, A. L.; Ostrander, R. L.; Haggerty, B. S.; Trofimenko, S. *Inorg. Chem.* **1994**, *33*, 3666. (d) Sohrin, Y.; Kokusen, H.; Matsui, M. *Inorg. Chem.* **1995**, *34*, 3928. (e) Jaekle, F.; Polborn, K.; Wagner, M. *Chem. Ber.* **1996**, *129*, 603. (f) Ito, M.; Amagai, H.; Fukui, H.; Kitajima, N.; Moro-oka, Y. *Bull. Chem. Soc. Jpn.* **1996**, *69*, 1937. (g) Fabrizi de Biani, F.; Jaekle, F.; Spiegler, M.; Wagner, M.; Zanello, P. *Inorg. Chem.* **1997**, *36*, 2103. (h) Rheingold, A. L.; Haggerty, B. S.; Yap, P. A.; Trofimenko, S. *Inorg. Chem.* **1997**, *36*, 5097. (i) Lobbia, G. G.; Bovio, B.; Santini, C.; Cecchi, P.; Pettinari, C.; Marchetti, F. *Polyhedron* **1998**, *17*, 17. (j) Rheingold, A. L.; Yap, P. A.; Liable-Sands, L. M.; Guzei, I. A.; Trofimenko, S. *Inorg. Chem.* **1997**, *36*, 6261. (k) Arulsamy, N.; Bohle, D. S.; Hansert, B.; Powell, A. K.; Thomson, A. J.; Wocaldo, S. *Inorg. Chem.* **1998**, *37*, 746. (l) Janiak, C.; Temizdemir, S.; Dechert, S.; Deck, W.; Girgsdies, F.; Heinze, J.; Kolm, M. J.; Scharmann, T.; Zipffel, O. M. *Eur. J. Inorg. Chem.* **2000**, 1229. (m) Guo, S.; Peters, F.; Fabrizi de Biani, F.; Bats, J. W.; Herdtweck, E.;

Zanello, M.; Wagner, M. *Inorg. Chem.* **2001**, *40*, 4928. (n) Cecchi, P.; Berrettoni, M.; Giorgetti, M.; Lobbia, G. G.; Calogero, S.; Stievano, L. *Inorg. Chim. Acta* **2001**, *318*, 67.
(8) Loehr, J. S.; Wheeler, W. D.; Shiemke, A. K.; Averill, B. A.; Loehr, T. M. *J. Am. Chem. Soc.* **1989**, *111*, 8084.
(9) Kitajima, N.; Tamura, N.; Amagai, H.; Fufui, H.; Moro-oka, Y.; Mizutani, Y.; Kitagawa, T.; Mathur, R.; Heerwegh, K.; Reed, C. A.; Randall, C. R.; Que, L. J.; Tatsumi, K. *J. Am. Chem. Soc.* **1994**, *116*, 9071.
(10) See for example: Mehn, M. P.; Fujisawa, K.; Hegg, E. L.; Que, L. *J. Am. Chem. Soc.* **2003**, *125*, 7828 (and references therein).
(11) Li, D.; Parkin, S.; Wang, G.; Yee, G. T.; Holmes, S. M. *Inorg. Chem.* **2006**, *45*, 1951 (and references therein).
(12) McNair, A. M.; Boyd, D. C.; Mann, K. R. *Organometallics* **1986**, *5*, 303.
(13) Trofimenko, S. *Acc. Chem. Res.* **1971**, *4*, 17.
(14) Brunker, T. J.; Cowley, A. R.; O'Hare, D. *Organometallics* **2002**, *21*, 3123.
(15) King, R. B.; Bond, A. *J. Am. Chem. Soc.* **1974**, *96*, 1343.
(16) King, R. B.; Bond, A. *J. Am. Chem. Soc.* **1974**, *96*, 1334.
(17) Anderson, S.; Hill, A. F.; White, A. J. P.; Williams, D. J. *Organometallics* **1998**, *17*, 2665.
(18) Cotton, F. A.; Frenz, B. A.; Shaver, A. *Inorg. Chim. Acta* **1973**, *7*, 161.
(19) Anderson, S.; Hill, A. F.; Slawin, A. M. Z.; White, A. J. P.; Williams, D. J. *Inorg. Chem.* **1998**, *37*, 594.
(20) Anderson, S.; Hill, A. F.; Ng, Y. T. *Organometallics* **2000**, *19*, 15.
(21) Graziani, O.; Toupet, L.; Hamon, J.-R.; Tilset, M. *J. Organomet. Chem.* **2003**, *669*, 200.
(22) Bellachioma, G.; Cardaci, G.; Gramlich, V.; Macchioni, A.; Pieroni, F.; Venanzi, L. M. *J. Chem. Soc., Dalton Trans.* **1998**, 947.
(23) Gorrell, I. B.; Parkin, G. *Inorg. Chem.* **1990**, *29*, 2452.
(24) Kisko, J.; Hascall, T.; Parkin, G. *J. Am. Chem. Soc.* **1998**, *120*, 10561.
(25) Shirasawa, N.; Akita, M.; Hikichi, S.; Moro-oka, Y. *J. Chem. Soc., Chem. Commun.* **1999**, 417.
(26) Shirasawa, N.; Nguyet, T. T.; Hikichi, S.; Moro-oka, Y.; Akita, M. *Organometallics* **2001**, *20*, 3582.
(27) Akita, M.; Shirasawa, N.; Hikichi, S.; Moro-oka, Y. *J. Chem. Soc., Chem. Commun.* **1998**, 974.
(28) Albers, M. O.; Crosby, S. F. A.; Liles, D. C.; Robinson, D. J.; Shaver, A.; Singleton, E. *Organometallics* **1987**, *6*, 2014.
(29) Gemel, C.; Trimmel, G.; Slugovc, C.; Kremel, S.; Mereiter, K.; Schmid, R.; Kirchner, K. *Organometallics* **1996**, *15*, 3998.
(30) Takahashi, Y.; Akita, M.; Hikichi, S.; Moro-oka, Y. *Inorg. Chem.* **1998**, *37*, 3186.
(31) Takahashi, Y.; Akita, M.; Hikichi, S.; Moro-oka, Y. *Organometallics* **1998**, *17*, 4888.
(32) Pariya, C.; Incarvito, C. D.; Rheingold, A. L.; Theopold, K. H. *Polyhedron* **2004**, *23*, 439.
(33) Jalon, F. A.; Otero, A.; Rodriguez, A. *J. Chem. Soc., Dalton Trans.* **1995**, 1629.
(34) Moreno, B.; Sabo-Etienne, S.; Chaudret, B.; Rodriguez, A.; Jalon, F.; Trofimenko, S. *J. Am. Chem. Soc.* **1994**, *116*, 2635.
(35) Moreno, B.; Sabo-Etienne, S.; Chaudret, B.; Rodriguez, A.; Jalon, F.; Trofimenko, S. *J. Am. Chem. Soc.* **1995**, *117*, 7441.
(36) Vicente, C.; Shul'pin, G. B.; Moreno, B.; Sabo-Etienne, S.; Chaudret, B. *J. Mol. Catal. A Chem.* **1995**, *98*, 15.
(37) Wilson, D. C.; Nelson, J. H. *J. Organomet. Chem.* **2003**, *682*, 272.
(38) (a) Albers, M. O.; Robinson, D. J.; Shaver, A.; Singleton, E. *Organometallics* **1986**, *5*, 2199. (b) Fagan, P. J.; Mahoney, W. S.; Calabrese, J. C.; Williams, I. D. *Organometallics* **1990**, *9*, 1843.
(39) Rüba, E.; Hummel, A.; Mereiter, K.; Schmid, R.; Kirchner, K. *Organometallics* **2002**, *21*, 4955.
(40) Standfest-Hauser, C. M.; Mereiter, K.; Schmid, R.; Kirchner, K. *Eur. J. Inorg. Chem.* **2003**, 1883.
(41) Slugovc, C.; Gemel, C.; Shen, J.-Y.; Doberer, D.; Schmid, R.; Kirchner, K.; Mereiter, K. *Monatsh. Chem.* **1999**, *130*, 363.
(42) Slugovc, C.; Mereiter, K.; Schmid, R.; Kirchner, K. *Organometallics* **1998**, *17*, 827.
(43) Pavlik, S.; Gemel, C.; Slugovc, C.; Mereiter, K.; Schmid, S.; Kirchner, K. *J. Organomet. Chem.* **2001**, *617–618*, 301.
(44) Rüba, E.; Slugovc, C.; Gemel, C.; Mereiter, K.; Schmid, R.; Kirchner, K. *Organometallics* **1999**, *18*, 2275.

(45) Gemel, C.; Wiede, P.; Mereiter, K.; Sapunov, V. N.; Schmid, R.; Kirchner, K. *J. Chem. Soc., Dalton Trans.* **1996**, 4071.
(46) Trimmel, G.; Slugovc, C.; Wiede, P.; Mereiter, K.; Sapunov, V. N.; Schmid, R.; Kirchner, K. *Inorg. Chem.* **1997**, *36*, 1076.
(47) Standfest-Hauser, C. M.; Mereiter, K.; Schmid, R.; Kirchner, K. *Organometallics* **2004**, *23*, 2194.
(48) Slugovc, C.; Doberer, D.; Gemel, C.; Schmid, R.; Kirchner, K.; Winkler, B.; Stelzer, F. *Monatsh. Chem.* **1998**, *129*, 221.
(49) Slugovc, C.; Mereiter, K.; Zobetz, E.; Schmid, R.; Kirchner, K. *Organometallics* **1996**, *15*, 5275.
(50) Slugovc, C.; Sapunov, V. N.; Wiede, P.; Mereiter, K.; Schmid, R.; Kirchner, K. *J. Chem. Soc., Dalton Trans.* **1997**, 4209.
(51) Pavlik, S.; Puchberger, M.; Mereiter, K.; Kirchner, K. *Eur. J. Inorg. Chem.* **2006**, 4137.
(52) Gemel, C.; Kickelbick, G.; Schmid, R.; Kirchner, K. *J. Chem. Soc., Dalton Trans.* **1997**, 2117.
(53) Kremel, S.; Mereiter, K.; Slugovc, C.; Gemel, C.; Pfeiffer, J.; Schmid, R.; Kirchner, K. *Monatsh. Chem.* **2001**, *132*, 551.
(54) Gemel, C.; Mereiter, K.; Schmid, R.; Kirchner, K. *Organometallics* **1997**, *16*, 2623.
(55) (a) Alcock, N. W.; Burns, I. D.; Claire, K. S.; Hill, A. F. *Inorg. Chem.* **1992**, *31*, 2906. (b) Hill, A. F.; Wilton-Ely, J. D. E. T. *Inorg. Synth.* **2002**, *33*, 206.
(56) Chan, W. C.; Lau, C. P.; Chen, Y. Z.; Fang, Y.-Q.; Ng, S. M.; Jia, G. *Organometallics* **1997**, *16*, 34.
(57) Chen, Y.-Z.; Chan, W. C.; Lau, C.-P.; Chu, H. S.; Jia, G. *Organometallics* **1997**, *16*, 1241.
(58) Sanford, M. S.; Valdez, M. R.; Grubbs, R. H. *Organometallics* **2001**, *20*, 5455.
(59) Pavlik, S.; Mereiter, K.; Puchberger, M.; Kirchner, K. *Organometallics* **2005**, *24*, 3561.
(60) Tenorio, M. J.; Tenorio, M. A. J.; Puerta, M. C.; Valerga, P. *Inorg. Chim. Acta* **1997**, *259*, 77.
(61) Tenorio, M. A. J.; Tenorio, M. J.; Puerta, M. C.; Valerga, P. *J. Chem. Soc., Dalton Trans.* **1998**, 3601.
(62) Albertin, G.; Antoniutti, S.; Bortoluzzi, M.; Zanardo, M. *J. Organomet. Chem.* **2005**, *690*, 1726.
(63) Burns, I. D.; Hill, A. F.; Williams, D. J. *Inorg. Chem.* **1996**, *35*, 2685.
(64) Ng, S. M.; Fang, Y. Q.; Lau, C. P.; Wong, W. T.; Jia, G. *Organometallics* **1998**, *17*, 2052.
(65) Sun, N.-Y.; Simpson, S. J. *J. Organomet. Chem.* **1992**, *434*, 341.
(66) (a) Arrowood, B. N.; Lail, M.; Gunnoe, T. B.; Boyle, P. D. *Organometallics* **2003**, *22*, 4692. (b) Foley, N. C.; Lail, M.; Lee, J. P.; Gunnoe, T. B.; Petersen, J. L. *J. Am. Chem Soc.* **2007**, *129*, 6765. (c) Lee, J. P.; Jimenez-Halla, J. O. C.; Cundari, T. R.; Gunnoe, T. B. *J. Organomet. Chem.* **2007**, *692*, 2175.
(67) Steyn, M. M. de V.; Singleton, E.; Hietkamp, S.; Liles, D. C. *J. Chem. Soc., Dalton Trans.* **1990**, 2991.
(68) Sørlie, M.; Tilset, M. *Inorg. Chem.* **1995**, *34*, 5199.
(69) Albers, M. O.; Oosterhuizen, H. E.; Robinson, D. J.; Shaver, A.; Singleton, E. *J. Organomet. Chem.* **1985**, *282*, 49.
(70) Carella, A.; Jaud, J.; Rapenne, G.; Launay, J.-P. *Chem. Commun.* **2003**, 2434.
(71) O'Sullivan, D. J.; Lalor, F. J. *J. Organomet. Chem.* **1973**, *57*, 58.
(72) Restivo, R. J.; Ferguson, G. *J. Chem. Soc., Chem. Commun.* **1973**, 847.
(73) Restivo, R. J.; Ferguson, G.; O'Sullivan, D. J.; Lalor, F. *Inorg. Chem.* **1975**, *14*, 3046.
(74) Bhambri, S.; Tocher, D. A. *J. Organomet. Chem.* **1996**, *507*, 291.
(75) Bhambri, S.; Tocher, D. A. *J. Chem. Soc., Dalton Trans.* **1997**, 3367.
(76) Bhambri, S.; Tocher, D. A. *Polyhedron* **1995**, *16*, 2763.
(77) Bailey, P. J.; Pinho, P.; Pardsons, S. *Inorg. Chem.* **2003**, *42*, 8872.
(78) Birri, A.; Steed, J. W.; Tocher, D. A. *Polyhedron* **1999**, *18*, 1825.
(79) Landgrafe, C.; Sheldrick, W. S. *J. Chem. Soc., Dalton Trans.* **1994**, 1885.
(80) Gemel, C.; John, R.; Slugovc, C.; Mereiter, K.; Schmid, R.; Kirchner, K. *J. Chem. Soc., Dalton Trans.* **2000**, 2607.
(81) Bruce, M. I.; Sharrocks, D. N.; Stone, F. G. A. *J. Organomet. Chem.* **1971**, *31*, 269.
(82) Bruce, M. I.; Iqbal, M. Z.; Stone, F. G. A. *J. Chem. Soc. A* **1971**, 2820.
(83) Shiu, K.-B.; Chen, J.-Y.; Yu, S.-J.; Wang, S.-L.; Liao, F.-L.; Wang, Y.; Lee, G.-H. *J. Organomet. Chem.* **2002**, *648*, 193.
(84) Nataro, C.; Thomas, L. M.; Angelici, R. J. *Inorg. Chem.* **1997**, *36*, 6000.
(85) Hiraki, K.; Ochi, N.; Kitamura, T.; Sasada, Y.; Shinoda, S. *Bull. Chem. Soc. Jpn.* **1982**, *55*, 2356.

(86) Hiraki, K.; Ochi, N.; Takaya, H.; Fuchita, Y.; Shimokawa, Y.; Hayashida, H. *J. Chem. Soc., Dalton Trans.* **1990**, 1679.
(87) Hill, A. F. *J. Organomet. Chem.* **1990**, *395*, 35.
(88) Alcock, N. W.; Hill, A. F.; Melling, R. P. *Organometallics* **1991**, *10*, 3898.
(89) Yin, J.; Yu, G.-Y.; Guan, J.; Mei, F.; Liu, S. H. *J. Organomet. Chem.* **2005**, *690*, 4265.
(90) Liu, S. H.; Xia, H.; Wan, K. L.; Yeung, R. C. Y.; Hu, Q. Y.; Jia, G. *J. Organomet. Chem.* **2003**, *683*, 331.
(91) Bohanna, C.; Esteruelas, M. A.; Gomez, A. V.; Lopez, A. M.; Martinez, M.-P. *Organometallics* **1997**, *16*, 4464.
(92) Burns, I. D.; Hill, A. F.; White, A. J. P.; Williams, D. J.; Wilton-Ely, J. D. E. T. *Organometallics* **1998**, *17*, 1552.
(93) Jayaprakash, K. N.; Gunnoe, T. B.; Boyle, P. D. *Inorg. Chem.* **2001**, *40*, 6481.
(94) Jayaprakash, K. N.; Conner, D.; Gunnoe, T. B.; Boyle, P. D. *Organometallics* **2001**, *20*, 5254.
(95) Conner, D.; Jayaprakash, K. N.; Gunnoe, T. B.; Boyle, P. D. *Organometallics* **2002**, *21*, 5265.
(96) Jayaprakash, K. N.; Gillepsie, A. M.; Gunnoe, T. B.; White, D. P. *Chem. Commun.* **2002**, 372.
(97) Conner, D.; Jayaprakash, K. N.; Wells, M. B.; Manzer, S.; Gunnoe, T. B.; Boyle, P. D. *Inorg. Chem.* **2003**, *42*, 4759.
(98) Conner, D.; Jayaprakash, K. N.; Gunnoe, T. B.; Boyle, P. D. *Inorg. Chem.* **2002**, *41*, 3042.
(99) Onishi, M. *Bull. Chem. Soc. Jpn.* **1991**, *64*, 3039.
(100) Halcrow, M. A.; Chaudret, B.; Trofimenko, S. *J. Chem. Soc., Chem. Commun.* **1993**, 465.
(101) Tenorio, M. J.; Palacios, M. D.; Puerta, M. C.; Valerga, P. *Organometallics* **2005**, *24*, 3088.
(102) (a) Takahashi, Y.; Hikichi, S.; Akita, M.; Moro-oka, Y. *Chem. Commun.* **1999**, 1491. (b) Takahashi, Y.; Hikichi, S.; Akita, M.; Moro-oka, Y. *Organometallics* **1999**, *18*, 2571. (c) Akita, M.; Takahashi, Y.; Hikichi, S.; Moro-oka, Y. *Inorg. Chem.* **2001**, *40*, 169. (d) Takahashi, Y.; Hikichi, S.; Moro-oka, Y.; Akita, M. *Polyhedron* **2004**, *23*, 225.
(103) Albertin, G.; Antoniutti, S.; Bortoluzzi, M.; Castro-Fojo, J.; Garcia-Fontan, S. *Inorg. Chem.* **2004**, *43*, 4511.
(104) Man, M. L.; Zhu, J.; Ng, S. M.; Zhou, Z.; Yin, C.; Lin, Z.; Lau, C. P. *Organometallics* **2004**, *23*, 6214.
(105) Ng, S. M.; Lau, C. P.; Fan, M.-F.; Lin, Z. *Organometallics* **1999**, *18*, 2484.
(106) Ng, S. M.; Lam, W. H.; Mak, C. C.; Tsang, C. W.; Jia, G.; Lin, Z.; Lau, C. P. *Organometallics* **2003**, *22*, 641.
(107) Lam, W. H.; Jia, G.; Lin, Z.; Lau, C. P.; Eisenstein, O. *Chem. Eur. J.* **2003**, *9*, 2775.
(108) Corrochano, A. E.; Jalon, F. A.; Otero, A.; Kubicki, M. M.; Richard, P. *Organometallics* **1997**, *16*, 145.
(109) Maruyama, Y.; Ikeda, S.; Ozawa, F. *Bull. Chem. Soc. Jpn.* **1997**, *70*, 689.
(110) Arikawa, Y.; Nishimura, Y.; Kawano, H.; Onishi, M. *Organometallics* **2003**, *22*, 3354.
(111) Arikawa, Y.; Nishimura, Y.; Ikeda, K.; Onishi, M. *J. Am. Chem. Soc.* **2004**, *126*, 3706.
(112) (a) Nishimura, Y.; Arikawa, Y.; Inoue, T.; Onishi, M. *Dalton Trans.* **2005**, 930. (b) Arikawa, Y.; Asayama, T.; Onishi, M. *J. Organomet. Chem.* **2007**, *692*, 194.
(113) (a) Lo, Y.-H.; Lin, Y.-C.; Lee, G.-H.; Wang, Y. *Organometallics* **1999**, *18*, 982. (b) Lo, Y.-H.; Lin, Y.-C.; Lee, G.-H.; Wang, Y. *Eur. J. Inorg. Chem.* **2004**, 4616.
(114) Tenorio, M. A. J.; Tenorio, M. J.; Puerta, M. C.; Valerga, P. *Inorg. Chim. Acta* **2000**, *300–302*, 869.
(115) Lail, M.; Arrowood, B. N.; Gunnoe, T. B. *J. Am. Chem. Soc.* **2003**, *125*, 7506.
(116) Pittard, K. A.; Lee, J. P.; Cundari, T. R.; Gunnoe, T. B.; Petersen, J. L. *Organometallics* **2004**, *23*, 5514.
(117) Pittard, K. A.; Cundari, T. R.; Gunnoe, T. B.; Day, C. S.; Petersen, J. L. *Organometallics* **2005**, *24*, 5015.
(118) Lee, J. P.; Pittard, K. A.; DeYonker, N. J.; Cundari, T. R.; Gunnoe, T. B.; Petersen, J. L. *Organometallics* **2006**, *25*, 1500.
(119) Goj, L. A.; Lail, M.; Pittard, K. A.; Riley, K. C.; Gunnoe, T. B.; Petersen, J. L. *Chem. Commun.* **2006**, 982.
(120) Lail, M.; Gunnoe, T. B.; Barakat, K. A.; Cundari, T. R. *Organometallics* **2005**, *24*, 1301.
(121) (a) Feng, Y.; Lail, M.; Barakat, K. A.; Cundari, T. R.; Gunnoe, T. B.; Petersen, J. L. *J. Am. Chem. Soc.* **2005**, *127*, 14174. (b) Feng, Y.; Lail, M.; Foley, N. A.; Gunnoe, T. B.; Barakat, K. A.; Cundari, T. R.; Peteresen, J. L. *J. Am. Chem. Soc.* **2006**, *128*, 7982.

(122) Sanford, M. S.; Henling, L. M.; Grubbs, R. H. *Organometallics* **1998**, *17*, 5384.
(123) Sanford, M. S.; Love, J. A.; Grubbs, R. H. *Organometallics* **2001**, *20*, 5314.
(124) Burtscher, D.; Perner, B.; Mereiter, K.; Slugovc, C. *J. Organomet. Chem.* **2006**, *691*, 5423.
(125) Tenorio, M. A. J.; Tenorio, M. J.; Puerta, M. C.; Valerga, P. *Organometallics* **1997**, *16*, 5528.
(126) Tenorio, M. A. J.; Tenorio, M. J.; Puerta, M. C.; Valerga, P. *Organometallics* **2000**, *19*, 1333.
(127) Bianchini, C.; Casares, J. A.; Peruzzini, M.; Romerosa, A.; Zanobini, F. *J. Am. Chem. Soc.* **1996**, *118*, 4585.
(128) Slugovc, C.; Wiede, P.; Mereiter, K.; Schmid, R.; Kirchner, K. *Organometallics* **1997**, *16*, 2768.
(129) Slugovc, C.; Mauthner, K.; Kacetl, M.; Mereiter, K.; Schmid, R.; Kirchner, K. *Chem. Eur. J.* **1998**, *4*, 2043.
(130) Slugovc, C.; Mereiter, K.; Schmid, R.; Kirchner, K. *J. Am. Chem. Soc.* **1998**, *120*, 6175.
(131) Slugovc, C.; Mereiter, K.; Schmid, R.; Kirchner, K. *Eur. J. Inorg. Chem.* **1999**, 1141.
(132) Pavlik, S.; Mereiter, K.; Schmid, R.; Kirchner, K. *Monatsh. Chem.* **2004**, *135*, 1349.
(133) Pavlik, S.; Puchberger, M.; Mereiter, K.; Kirchner, K. *J. Organomet. Chem.* **2005**, *690*, 5497.
(134) (a) Rigaut, S.; Touchard, D.; Dixneuf, P. H. *Organometallics* **2003**, *22*, 3980. (b) Castarlenas, R.; Dixneuf, P. H. *Angew. Chem. Int. Ed. Engl.* **2003**, *42*, 4524. (c) Bustelo, E.; Jimenez-Tenorio, M.; Mereiter, K.; Puerta, M. C.; Valerga, P. *Organometallics* **2002**, *21*, 1903.
(135) Beach, N. J.; Williamson, A. E.; Spivak, G. J. *J. Organomet. Chem.* **2005**, *690*, 4640.
(136) Buriez, B.; Burns, I. D.; Hill, A. F.; White, A. J. P.; Williams, D. J.; Wilton-Ely, J. D. E. T. *Organometallics* **1999**, *18*, 1504.
(137) Buriez, B.; Cook, D. J.; Harlow, K. J.; Hill, A. F.; Welton, T.; White, A. J. P.; Williams, D. J.; Wilton-Ely, J. D. E. T. *J. Organomet. Chem.* **1999**, *578*, 264.
(138) Harlow, K. J.; Hill, A. F.; Ely, J. D. E. T. *Dalton Trans.* **1999**, 285.
(139) Hartmann, S.; Winter, R. F.; Brunner, B. M.; Sarkar, B.; Knödler, A.; Hartenbach, I. *Eur. J. Inorg. Chem.* **2003**, 876.
(140) Onishi, M.; Ikemoto, K.; Hiraki, K. *Inorg. Chim. Acta* **1991**, *190*, 157.
(141) Yin, C.; Xu, Z.; Yang, S.-Y.; Ng, S. M.; Wong, K. Y.; Lin, Z.; Lau, C. P. *Organometallics* **2001**, *20*, 1216.
(142) Ng, S. M.; Yin, C.; Yeung, C. H.; Chan, T. C.; Lau, C. P. *Eur. J. Inorg. Chem.* **2004**, 1788.
(143) Katayama, H.; Yoshida, T.; Ozawa, F. *J. Organomet. Chem.* **1998**, *562*, 203.
(144) Lail, M.; Bell, C. M.; Conner, D.; Cundari, T. R.; Gunnoe, T. B.; Petersen, J. L. *Oganometallics* **2004**, *23*, 5007.
(145) (a) Oxgaard, J.; Goddard, W. A. III, *J. Am. Chem. Soc.* **2004**, *126*, 442. (b) Oxgaard, J.; Periana, R. A.; Goddard, W. A. III, *J. Am. Chem. Soc.* **2004**, *126*, 11658.
(146) Fung, W. K.; Man, M. L.; Ng, S. M.; Hung, M. Y.; Lin, Z.; Lau, C. P. *J. Am. Chem. Soc.* **2003**, *125*, 11544.
(147) Shen, H.-C.; Tang, J.-M.; Chang, H.-K.; Yang, C.-W.; Liu, R.-S. *J. Org. Chem.* **2005**, *70*, 10113.
(148) Yeh, K.-L.; Liu, B.; Lai, Y.-T.; Li, C.-W.; Liu, R.-S. *J. Org. Chem.* **2004**, *69*, 4692.
(149) Datta, S.; Chang, C.-L.; Yeh, K.-L.; Liu, R.-S. *J. Am. Chem. Soc.* **2003**, *125*, 9294.
(150) Shen, H.-C.; Su, H.-L.; Hsueh, Y.-C.; Liu, R.-S. *Organometallics* **2004**, *23*, 4332.
(151) Lo, C.-Y.; Pal, S.; Odedra, A.; Liu, R.-S. *Tetrahedron Lett.* **2003**, *44*, 3143.
(152) Datta, S.; Odedra, A.; Liu, R.-S. *J. Am. Chem. Soc.* **2005**, *127*, 11606.
(153) Lin, M.-Y.; Maddirala, J.; Liu, R.-S. *Org. Lett.* **2005**, *9*, 1745.
(154) Ming-Yuan, L.; Madhushaw, R. J.; Liu, R.-S. *J. Org. Chem.* **2004**, *69*, 7700.
(155) Shen, H.-C.; Li, C.-W.; Liu, R.-S. *Tetrahedron Lett.* **2004**, *45*, 9245.
(156) Madhushaw, R. J.; Lo, C.-Y.; Hwang, C.-W.; Su, M.-D.; Shen, H.-C.; Pal, S.; Shaikh, I. R.; Liu, R.-S. *J. Am. Chem. Soc.* **2004**, *126*, 15560.
(157) Odedra, A.; Wu, C.-J.; Pratap, T. B.; Huang, C.-W.; Ran, Y.-F.; Liu, R.-S. *J. Am. Chem. Soc.* **2005**, *127*, 3406.
(158) Lian, J.-J.; Odedra, A.; Wu, C.-J.; Liu, R.-S. *J. Am. Chem. Soc.* **2005**, *127*, 4186.
(159) Chang, C.-L.; Kumar, M. P.; Liu, R.-S. *J. Org. Chem.* **2004**, *69*, 2793.
(160) Tenorio, M. J.; Puerta, M. C.; Valerga, P.; Moreno-Dorado, F. J.; Guerra, F. M.; Massanet, G. M. *Chem. Commun.* **2001**, 2324.
(161) Freedman, D. A.; Gill, T. P.; Blough, A. M.; Koefod, R. S.; Mann, K. R. *Inorg. Chem.* **1997**, *36*, 95.

(162) LaPointe, A. M.; Schrock, R. R. *Organometallics* **1993**, *12*, 3379.
(163) Hunt, J. L.; Shapley, P. A. *Organometallics* **1997**, *16*, 4071.
(164) Ng, W. S.; Jia, G.; Hung, M. Y.; Lau, C. P.; Wong, K. Y.; Wen, L. *Organomteallics* **1998**, *17*, 4556.
(165) (a) Crevier, T. J.; Mayer, J. M. *J. Am. Chem. Soc.* **1998**, *120*, 5595. (b) Crevier, T. J.; Myer, J. M. *Angew. Chem. Int. Ed. Engl.* **1998**, *37*, 1891.
(166) (a) Crevier, T. J.; Bennett, B. K.; Soper, J. D.; Bowman, J. A.; Dehestani, A.; Hrovat, D. A.; Lovell, S.; Kaminsky, W.; Mayer, J. M. *J. Am. Chem. Soc.* **2001**, *123*, 1059. (b) Dehestani, A.; Kaminsky, W.; Mayer, J. M. *Inorg. Chem.* **2003**, *42*, 605.
(167) Soper, J. D.; Bennett, B. K.; Lovell, S.; Mayer, J. M. *Inorg. Chem.* **2001**, *40*, 1888.
(168) Soper, J. D.; Rhile, I. J.; DiPasquale, A. G.; Mayer, J. M. *Polyhedron* **2004**, *23*, 323.
(169) Bennett, B. K.; Saganic, E.; Lovell, S.; Kaminsky, W.; Samuel, A.; Mayer, J. M. *Inorg. Chem.* **2003**, *42*, 4127.
(170) Bennett, B. K.; Pitteri, S. J.; Pilobello, L.; Lovell, S.; Kaminsky, W.; Mayer, J. M. *J. Chem. Soc., Dalton Trans.* **2001**, 3489.
(171) Bennett, B. K.; Lovell, S.; Mayer, J. M. *J. Am. Chem. Soc.* **2001**, *123*, 4336.
(172) Demadis, K. D.; El-Samanody, E.-S.; Coia, G. M.; Meyer, T. J. *J. Am. Chem. Soc.* **1999**, *121*, 535.
(173) Crevier, T. J.; Lovell, S.; Mayer, J. M. *J. Chem. Soc., Chem. Commun.* **1998**, 2371.
(174) Zhang, J.; Grills, D. C.; Huang, K.-W.; Fujita, E.; Bullock, R. M. *J. Am. Chem. Soc.* **2005**, *127*, 15684.
(175) Dickinson, P. W.; Girolami, G. S. *Inorg. Chem.* **2006**, *45*, 5215.
(176) Albertin, G.; Antoniutti, S.; Bacchi, A.; Bortoluzzi, M.; Pelizzi, G.; Zanardo, G. *Organometallics* **2006**, *25*, 4236.

The Organometallic Chemistry of Group 9 Poly(pyrazolyl)borate Complexes ☆

IAN R. CROSSLEY*

Research School of Chemistry, Institute of Advanced Studies, Australian National University, Canberra, ACT 0200, Australia

I
INTRODUCTION

During the 40 years of poly(pyrazolyl)borate chemistry, complexes with group 9 metals have provided a particularly fertile area of research. Over 400 publications describe aspects ranging from the synthesis of homoleptic and heteroleptic coordination compounds to the application of complex organometallic molecules in catalysis, and the development of advanced specialist materials.

This review is concerned specifically with those materials that conform to the generally accepted definition of 'organometallic', but also considers some materials

☆To the best of the author's knowledge this review is comprehensive to the end of 2005 and includes selected material from the earlier 2006 literature.

*Corresponding author. Department of Chemistry, University of Sussex, Brighton, BN1 9QJ, United Kingdom. Tel.: +44 (0) 1273 678650.
E-mail: i.crossley@sussex.ac.uk (Ian R. Crossley).

ADVANCES IN ORGANOMETALLIC CHEMISTRY
VOLUME 56 ISSN 0065-3055/DOI 10.1016/S0065-3055(07)56004-1

that, while not in the strictest sense 'organometallic', are or have potential to be of significant interest to the organometallic chemist. Thus, by convention, metal hydrides, carbonyls, isonitriles and phosphine complexes are unequivocally included. So too are those simple heteroleptic poly(pyrazolyl)borate coordination complexes that have been exploited in the development of organometallic systems, in contrast to those pursued more arbitrarily, such as the plethora of $Tp^{x}CoX$ (Tp^{x} = generic poly(pyrazolyl)borate, X = halide) complexes that have become standard synthetic targets for each successive generation of 'scorpionate' ligand. Many such 'standard' derivatives become apparent upon perusal of the literature, which in the case of rhodium and iridium often bear carbonyl, alkene or diene ligands. Where these compounds are merely derivative examples with little or no subsequent reported chemistry, they will be reported in tabular form. Individual attention will be given only to systems that show significant individuality in either structure or chemistry.

In organising the review it is logical to segregate cobalt from its heavier counterparts, given the limited extent to which the chemistry of 3d metals overlaps that of the 4d and 5d elements, which are most conveniently considered in unison. Beyond this, material is sub-divided according to the ligands of interest, prioritising 'true' organometallic π (η^n) and σ (η^1) complexes over metal hydrides and complexes of inorganic ligands (e.g. phosphines) and selected 'non-organometallic' complexes (e.g. those with dinitrogen, imido or oxo ligands) that are deemed of interest and not considered elsewhere. In general, where a complex embodies two or more *types* of ligand, it is classified on the basis of the greater chemical significance, and cross-referenced as appropriate.

Throughout, the standard conventions of poly(pyrazolyl)borate abbreviated nomenclature will be applied; i.e. Tp and Bp for the parent hydrotris- and dihydrobis(pyrazolyl)borate ligands, with pyrazole substituents denoted in the form $Tp^{R,4R',R''}$ where superscripts R, 4R′ and R″ indicate substituents in the 3, 4 and 5 positions respectively (omitted if none). Similarly, Tp^{R_2} denotes equivalent substituents in the 3/5 positions and Tp^{R_3} all three equivalents. The non-systematic but commonly employed abbreviations Bp* and Tp* will be used to indicate ligands derived from 3,5-dimethylpyrazole, i.e. Bp^{Me_2} and Tp^{Me_2}. More complex systems will be defined where appropriate.

II
COBALT

A. *Homoleptic L_2Co and Heteroleptic LCoX and LL′Co Complexes*

The pursuit of simple homoleptic cobalt complexes has become a ubiquitous means of structurally characterising each 'novel' derivative of the basic tetrakis-, hydrotris- or dihydrobis-(pyrazolyl)borate ligand motifs. Complexes of this type are inaccessible only for the most sterically demanding Tp ligands, in which case reports of the isolation of LCoX (X = halide, $^{-}$OR, $^{-}$SR…) complexes have instead proliferated. Examples of these are now known for essentially every poly(pyrazolyl)borate ligand yet developed, often being sought in order to assess the scorpionates' capacity to act as 'tetrahedral enforcers'.

Among this expansive body of work are several compounds that have fundamental importance in organometallic chemistry, either as synthetic precursors or in representing novel structural motifs. Most significant in the former respect are the complexes $Tp^{x}CoX$ ($Tp^{x} = Tp^{^tBu,Me}$, X = Cl (**1**), I (**2**);[1] $Tp^{x} = Tp^{^iPr,Me}$, X = Cl (**3**)[2]), $Tp^{^iPr_2}CoCl$[3] (**4**) and $Tp^{Me_3}CoCl$ (**5**) (Chart 1), prepared from the respective cobaltous halide and KTp^{x}, which have been extensively utilised as precursors to a range of remarkably stable alkyl, allyl, alkenyl and alkynyl complexes (see later). It is interesting to note that the ubiquitous 3,5-dimethylpyrazole derived ligand Tp^{Me_2} (historically Tp*) seems rarely used in the pursuit of organocobalt compounds, which presumably reflects a perceived necessity for more bulky substituents in the 3-position; i.e. that a strong 'tetrahedral enforcer' is required for attaining such potentially reactive species. Indeed, even the use of the Tp^{Me_3} ligand is a relatively recent development.

From a structural standpoint, a handful of complexes of the type LL′Co are noteworthy for the fact that both $H_2B(pz)_2$ and $HB(pz)_3$ type ligands have been observed to engage in 3-centre 2-electron (3c-2e) B–H–Co bridging interactions. More commonly studied for the metals of groups 6, 8 and 10, such 'agostic' interactions have been suggested to exist for the ligand 'L' within the complexes $Tp^{^iPr,4Br}CoL$ (L = Tp (**6**), Tp^{Me_2} (**7**), Bp (**8**)), on the basis of complex patterns observed in the B–H stretching region of the IR spectra (2500–2200 cm^{-1}) and by analogy to related nickel complexes.[4] Moreover, for $Tp^{^iPr,4Br}CoL$ (L = Bp^{Ph} (**9**), Tp^{Ph} (**10**)) these interactions were observed crystallographically, both complexes exhibiting pseudo-octahedral geometries with one pendant pyrazolyl group and Co–H separations of 2.36 and 2.26 Å respectively. These lie well within the sum of contact radii (3.23 Å).[5] A similar situation has been claimed for $Tp^{CHPh_2}CoL$ (L = Bp^{Ph} (**11**), Tp^{Ph} (**12**)),[6] though while the Co–H separations in the two crystallographically unique molecules of **11** (2.273, 2.268 Å) are conclusive, structural data for **12** have yet to appear.

Bridging B–H–M interactions have also been observed for the homoleptic complexes $(Bp^{^tBu})_2Co$ (**13**)[7] and $(Bp^{^tBu,^iPr})_2Co$ (**14**),[8] which adopt octahedral geometries ($r_{Co–H}$ = 1.95 Å), and the adventitiously isolated mixed-ligand complexes

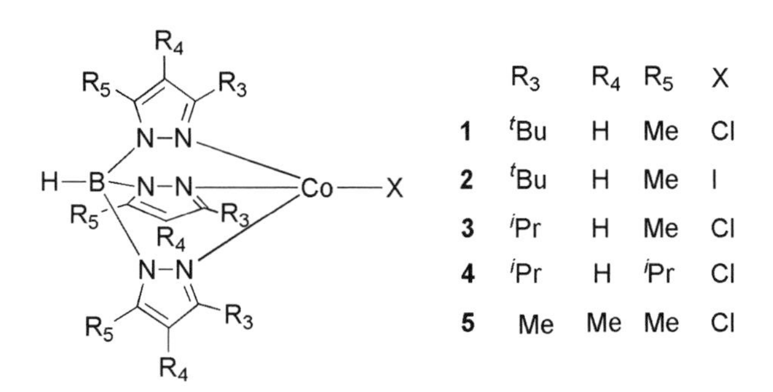

	R3	R4	R5	X
1	^{t}Bu	H	Me	Cl
2	^{t}Bu	H	Me	I
3	^{i}Pr	H	Me	Cl
4	^{i}Pr	H	^{i}Pr	Cl
5	Me	Me	Me	Cl

CHART 1. [$Tp^{x}CoX$] complexes.

$(Bp^{Ph_2})Co\{HB(pz^{Me_2})_2(pz^{Ph_2})\}$ (**15**)[9] and $(Bp^{Ph_2})Co\{HB(pz^{Ph_2})_2(pz^{Me_2})\}$ (**16**)[10] ($r_{Co–H(Bp)}$ = 2.03 and 2.035 Å respectively). The inferred strength of these interactions has led authors to conclude that they significantly influence the adopted structure, rather than being a sterically enforced consequence thereof.[8] It should be noted that at the time of writing there appear to have been no attempts made to establish any manifestation of hemi-lability in these cobalt-centred linkages.

L

11 Tp^{Ph2}

12 Bp^{Ph2}

N-N = third pz ring

Classical C–H–Co agostic interactions have also been reported, for the ligand 9,9-$(pz)_2$BBN (BBN = 9-borabicyclo-[3.3.1]-nonyl), the potassium salt of which was obtained from $(9\text{-HBBN})_2$ and Kpz.[11] The complex $Co\{(pz)_2BBN\}_2$ (**17**), best prepared from $Tl[(pz)_2BBN]$ and $Co(NO_3)_2$, adopts a pseudo-octahedral geometry (Fig. 1) in which the pyrazolyl nitrogen donors occupy a square plane, with two apical C–H–Co interactions ($r_{Co–H}$ = 2.16 Å) to the bridgehead carbons. These interactions are also apparent from a characteristic infrared absorption (ν_{CHCo} = 2690 cm^{-1}). Thermal displacement of the agostic linkages has been inferred on the basis of a colour change that was deemed consistent with the

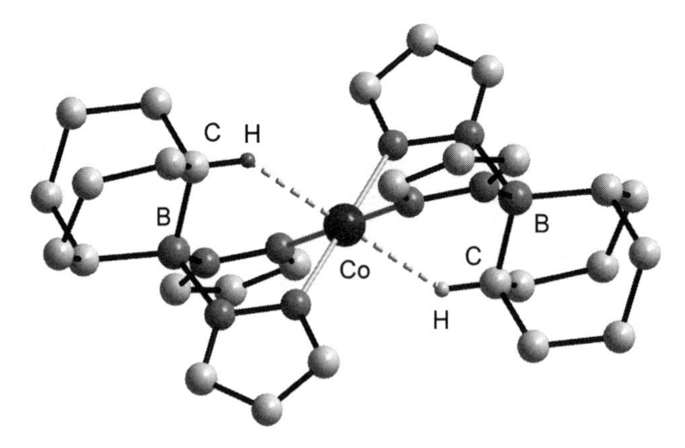

FIG. 1. Molecular structure of $Co\{(pz)_2BBN\}_2$ (**17**). Non-agostic hydrogen atoms omitted for clarity.

adoption of a tetrahedral structure. However, no unequivocal evidence exists, since the interactions are apparently restored upon cooling. The steric requirements for engaging in this agostic bonding were assessed by preparing $Tp^{iPr,4Br}Co\{(pz)_2BBN\}$ (**18**), in which the steric shielding of the three *iso*propyl substituents prohibits coordination of the bridgehead C–H function.[12]

Complex **17** has also been the subject of magnetic and spectroscopic studies intended to identify unusual and/or special effects inherent from this unusual geometry for cobalt(II).[13] Magnetic measurements and UV/Visible spectra were consistent with low-spin Co(II) in a square-planar environment (i.e. the CoN_4 unit). Unusually for Co(II), the EPR spectrum was observable at ambient temperature, though broad line-shapes necessitated recourse to doping within lattices of the Zn and Ni analogues. The data obtained were deemed unusual and attributed to the presence of the C–H–Co interactions.

B. *Complexes with σ-, π-Donor/π-Acceptor Ligands*

Relatively few compounds of this type exist. Only two simple alkene π-complexes are known, both prepared from the cobalt(I) complex $Tp^{tBu,Me}Co(N_2)$ (**19**, obtained by magnesium reduction of $Tp^{tBu,Me}CoX$ (X = Cl, I) under dinitrogen)[14] with excess $H_2C{=}CHR$ (R = H, CH_3 Scheme 1).[15] The ethylene complex **20** was readily isolated from solution at −30 °C, while the propene analogue **21** was only observed spectroscopically, since it proved unstable toward loss of propene upon removal of excess alkene and exposure to N_2, leading to the regeneration of **19**. This lability of the propene ligand was attributed to the significant steric hindrance imposed by the $Tp^{tBu,Me}$ ligand.

The study of anionic ligands has been more extensive, though once again for the simplest case (i.e. η^3-allyl) only one example exists: the 17-electron complex $Tp^{iPr_2}Co(\eta^3\text{-}C_3H_5)$ (**22**),[16] which is prepared from $Tp^{iPr_2}CoCl$ and allylmagnesium chloride. This complex adopts a square pyramidal structure in the solid state and was found to undergo instantaneous oxygenation by O_2 in toluene solution, yielding exclusively $CH_2{=}CHCHO$. The migratory insertion of CO into the $Co\text{–}C_{allyl}$ linkage has also been observed, carbonylation in hexane affording the acyl complex $Tp^{iPr_2}Co\{C({=}O)CH_2CH{=}CH_2\}(CO)$ (**23**) (Scheme 2).[3]

Somewhat more prevalent has been the study of mixed-sandwich complexes, comprising both a hydrotris(pyrazolyl)borate and cyclopentadienyl ligand, despite many early synthetic attempts being unsuccessful.[17] The first successful synthesis was effected by the reaction of $(\eta^5\text{-}C_5R_5)Co(CO)I_2$ and Tp or pzTp salts to yield the

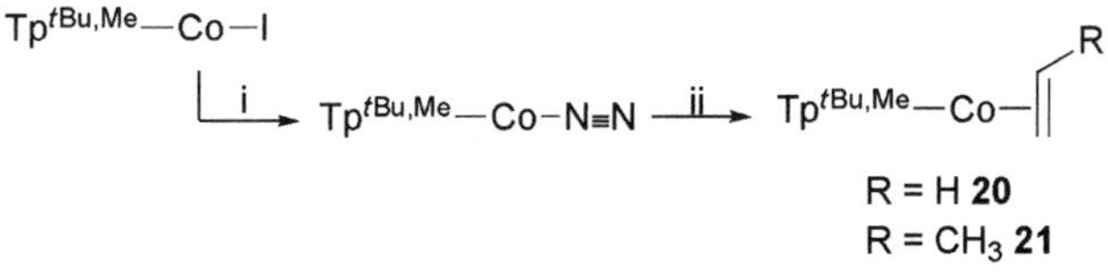

$Tp^{tBu,Me}$—Co—I → (i) $Tp^{tBu,Me}$—Co—N≡N → (ii) $Tp^{tBu,Me}$—Co—(η²-$H_2C{=}CHR$)

R = H **20**
R = CH_3 **21**

SCHEME 1. (i) Mg, THF, N_2; (ii) excess $H_2C{=}CHR$.

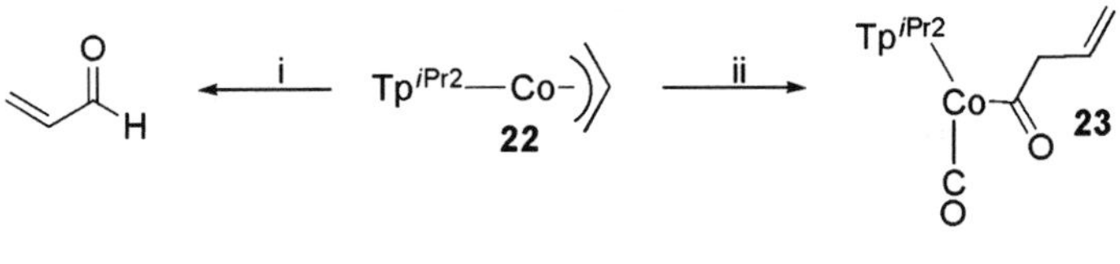

SCHEME 2. (i) O_2, toluene; (ii) CO 1 atmosphere, hexane.

complexes $[Tp^x(\eta^5\text{-}C_5R_5)Co]I$ ($Tp^x = Tp$, R = H ($\mathbf{24}^+$), Me ($\mathbf{25}^+$); Tp^x = pzTp; R = H ($\mathbf{26}^+$)) in near quantitative yield.[18] This approach was also used to prepare $TpCo(\eta^4\text{-}C_4Ph_4)$ (**27**) and several rhodium complexes (Section III-B.3), though characterising data was limited. Other workers later described the analogous synthesis of $\mathbf{24}^+$, reported initially as the I_3^- salt[19] but later as the simple monoiodide,[20] which was then reduced to the neutral complex TpCpCo (**24**). The synthesis of TpCp*Co (**25**) was reported to proceed directly from the reaction of $Cp^*_2Co_2(\mu\text{-}Cl)_2$ and KTp, its subsequent oxidation by ferrocinium ion affording the salt $\mathbf{25}\cdot\mathbf{PF_6}$. Remarkable features of **24** include a well-defined ^{1}H-NMR spectrum, despite its paramagnetism, and considerably greater stability (as a solid and in solution) than its Cp analogue.

Extensive structural, spectroscopic and magnetic investigations of $\mathbf{24}/\mathbf{24}^+$ and **25** have revealed several noteworthy facets. Foremost is the difference in spin-state of the two neutral mixed-sandwiches **24** and **25**, the latter adopting the anticipated low-spin configuration, while the former is unexpectedly high-spin.[19–21] In this, **24** was the first example of an organometallic cobalt complex to preferentially adopt a maximum spin configuration over a potential low-spin alternative.[21] Structurally, **24** and **25** differ in the coordination mode of the Tp ligand, the former exhibiting κ^3 binding (formally a 19-electron complex), while the latter is 17-electron (κ^2) in the solid state. This disparity was attributed to the steric demand of the Cp* ligand, relative to Cp. Magnetic and EPR measurements in solution are, however, consistent with the presence of a complex with one unpaired electron, interpreted as a result of conformational exchange *via* a formally κ^3 intermediate.[21]

C. Complexes with σ-Donor (Alkyl, Aryl) Ligands

1. σ-Alkyls

Simple σ-alkyl complexes are again rare, however, their chemistry has been quite extensively explored. Thus, while the only examples are $Tp^{^tBu,Me}CoR$ (R = Me (**28**), Et (**29**), ^{n}Bu (**30**)) and $Tp^{^tBu}CoMe$ (**31**)[15] (each obtained from the respective Tp^xCoCl and RLi), Tp^xCoEt ($Tp^x = Tp^{^iPr_2}$ (**32**),[22,3] Tp^{Me_3} (**33**)[3]) and $Tp^{^iPr_2}CoR$ (R = $CH_2CH{=}CHCH_3$ (**34**); $CH_2CH{=}C(CH_3)_2$ (**35**))[23] (synthesised from Grignard reagents), a relatively rich and diverse chemistry has been established. The 15-electron species **28**–**30** are remarkably stable under anaerobic conditions in the absence of protic solvents, remaining unchanged even upon prolonged heating at 90 °C. They are also inert toward hydrogenolysis under comparable conditions. Significantly, the β-hydride elimination pathway appears inaccessible for these species; indeed, complex **29** has alternatively been prepared *via* insertion of C_2H_4 into the respective metal hydride (Section II-E). In contrast, complex **28** did not react with C_2H_4 (90 °C, 2 weeks), but was found to catalyse the polymerisation of methyl methacrylate (70 °C, benzene); a process believed to be initiated by methyl radicals derived from Co–C bond homolysis. The treatment of **28** with CO (50 °C, 1 day) resulted in quantitative conversion to $Tp^{^tBu,Me}Co(CO)$ (**36**), while the analogous reaction of **31** is complete within minutes at ambient temperature. This differential reactivity was attributed to the greater steric encumbrance imparted to the fourth coordination site by $Tp^{^tBu,Me}$, relative to $Tp^{^tBu}$.[15]

Complex **32** is also resistant to β-elimination, but hydrogenolysis (2 atm, 16 h, ambient temperature) and protonolysis (HCl) afford ethane, in 91 and 96% yields respectively. The remarkable reactivity traits of **32** (a 15-electron complex) were explored theoretically using extended Huckel molecular orbital (EHMO) calculations, which suggested a high-spin electronic configuration for cobalt, enforced by the tripodal $Tp^{^iPr_2}$ ligand. In this configuration all non-bonding d-orbitals have either full or half occupancy, thus no completely 'vacant' orbitals (i.e. coordination sites) are available. Nonetheless, the electronic structure is deemed flexible, such that appropriate substrates would lead to a change in electronic configuration, affording a suitable coordination site. This has been exemplified by the reaction of **32** with CO (1 atm, 1 h), which affords the acyl complex $Tp^{^iPr_2}Co\{C(=O)Et\}(CO)$ (**37**) *via* migratory insertion into the Co–C linkage, comparable to that observed with the allyl complex **22** (Section II-B).[22,3] Complex **32** also undergoes oxygenation by O_2, though the presence of β-hydrogen atoms results in a mixture of products that include ethanol, acetaldehyde and acetic acid.

The interaction of **32** with unsaturated organic substrates has also been explored, once again revealing reluctance for insertion of alkenes or alkynes into the Co–C linkage. Only for phenylacetylene was any reaction observed, affording after hydrolysis an isomeric mixture of *cis*- and *trans*-1-phenyl-1-butene in 68% yield (Scheme 3). These products were presumed to arise from initial alkyne insertion into the Co–Et bond, with subsequent protonolysis of the Co–C alkenyl linkage.[3]

Also noteworthy are the *p*-methylbenzyl and α-naphthylmethyl complexes $Tp^xCo(CH_2R)$ ($Tp^x = Tp^{^iPr_2}$, $R = C_6H_4Me$-4 (**38**), 1-$C_{10}H_7$ (**39**); Tp^{Me_3}, $R = C_6H_4Me$-4 (**40**), 1-$C_{10}H_7$ (**41**)), prepared in the same study to assess both the influence of β-hydrogen substituents upon reactivity, and the relative steric influence of the bulky $Tp^{^iPr_2}$ ligand. No significant structural distinction between the $Tp^{^iPr_2}$ and Tp^{Me_3} analogues was observed, implying negligible dependence on steric shielding for attaining a pseudo-tetrahedral geometry. However, small deviations from perfect C_{3v} symmetry were observed for **40**, and attributed to a resonance-hybridised structure with contributions from η^1 (major) and η^2 (minor) binding of

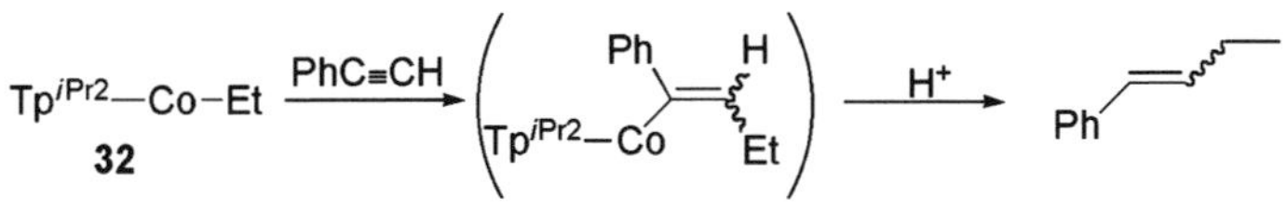

Tp^{iPr2}–Co–Et (**32**) $\xrightarrow{PhC\equiv CH}$ (Tp^{iPr2}–Co(Ph)C=C(H)Et) $\xrightarrow{H^+}$ PhCH=CHEt

SCHEME 3. Conditions, 70 °C, 1 h.

the benzyl group. The reactivity of compounds **38** and **40** was found to be comparable to that of the allyl complex discussed previously.

2. σ-Perfluoroalkyls

A single report documents the study of perfluoroalkyl poly(pyrazolyl)borate cobalt complexes; however, their breadth and significance is appreciable.[24] The compounds $\{H_xB(pz)_{4-x}\}CpCo(R_f)$ ($x = 0, 1, 2$) were prepared by the interaction of the respective $K[H_xB(pz)_{4-x}]$ salt with $CpCo(R_f)(CO)I$, and in each case exhibit bidentate coordination of the poly(pyrazolyl)borate ligand, regardless of the number of pyrazole donors potentially available.

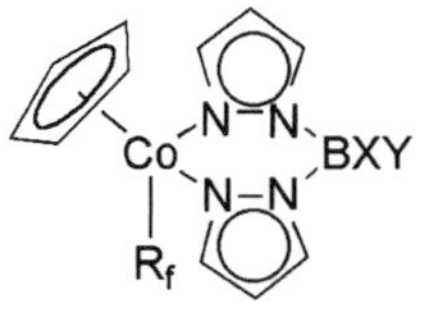

R_f	X=Y=H	X=pz Y=H	X=Y=pz
CF_3	**42**	**46**	
C_2F_5	**43**	**47**	**49**
C_3F_7	**44**	**48**	**50**
$CF(CF_3)_2$	**45**		

The inherent non-planarity of the Co–N_4–B chelate ring allows for stereoisomers of these compounds, which in the simplest case (i.e. Bp or pzTp derivatives) are distinguished by the relative orientation (axial or equatorial) of the Cp and fluoroalkyl groups (Fig. 2a). Both isomers were obtained for compounds **44** and **49** and chromatographically separated. The isomers were clearly distinguished by the differing ^{1}H-NMR chemical shifts of the Cp rings, and for **49** of the two un-coordinated pyrazole rings. The situation for the tris(pyrazolyl)borate derivatives is more complex, due to the additional scope for axial/equatorial orientation of the un-coordinated pyrazolyl and hydride substituents (Fig. 2b). For **47** an isomeric mixture was again obtained, though this defied chromatographic separation. Each isomer did, however, exhibit a distinctly different ^{1}H-NMR chemical shift for the Cp resonance, which by comparison to those of **42–44**, **49** and **50**, led to the conclusion that the minor isomer has axial Cp and pyrazole rings (isomer ab), while in the major isomer the uncoordinated pyrazole is equatorial (isomer aa). No conclusions were drawn as to why most of the complexes were obtained isomerically pure, though one can speculate as to likely steric and electronic influences.

A further anomaly of the reactions between $CpCo(R_f)(CO)I$ ($R_f = C_2F_5$, $^nC_3F_7$, $^iC_3F_7$) and $KB(pz)_4$, is that the major products were not the respective tetrakis(pyrazolyl)borate complexes. Rather, these materials were formulated as $CpCo(R_f)(pz_2H)$ (Chart 2), on the basis of their monomeric and diamagnetic nature, and NMR spectroscopic data.

3. σ-Aryls

At the time of writing, there is a single relevant complex, viz. $Tp^{^iPr_2}Co(C_6H_4Me\text{-}4)$ (**51**), obtained *via* metathesis of $BrMgC_6H_4Me\text{-}4$ and $Tp^{^iPr_2}CoCl$.[25] This compound is moisture sensitive, readily hydrolysing to toluene and the hydroxo complex

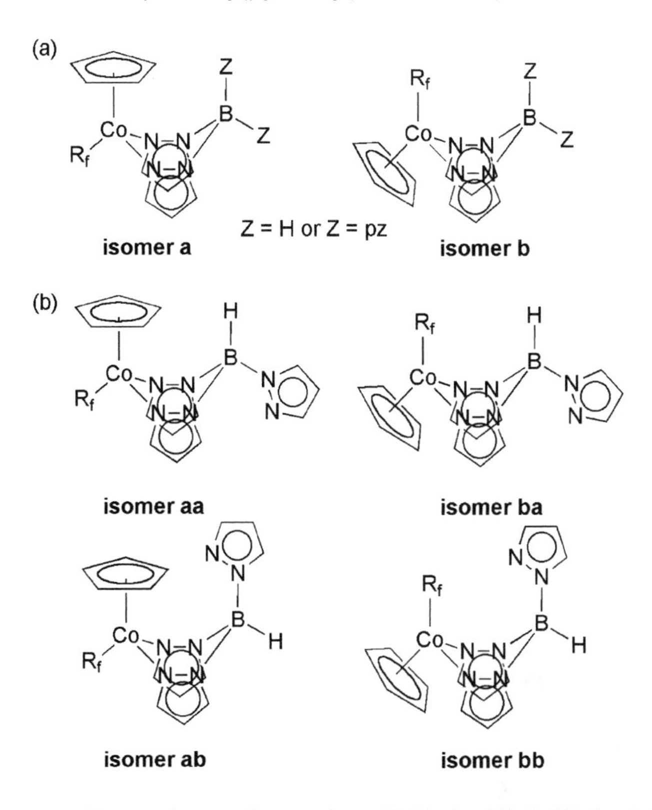

FIG. 2. Possible stereoisomers for complexes $\{H_xB(pz)_{4-x}\}CpCo(R_f)$ ($x = 0, 1, 2$).

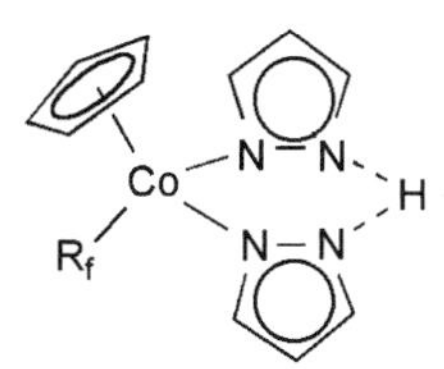

CHART 2. Proposed structure of $CpCo(R_f)(pz_2H)$ complexes.

$(Tp^{^iPr2})_2Co_2(\mu\text{-}OH)_2$. It is also decomposed by O_2, presumably in a similar manner to the analogous alkyl compounds (*vide supra*) though no tractable products were isolated. The carbonylation of **51** is effected readily, though in contrast to the alkyl systems the initial product is believed to be $Tp^{^iPr2}Co(C_6H_4Me\text{-}4)(CO)$ (**52**), rather than a metal-acyl resulting from migratory insertion, since no infrared band consistent with a metal-acyl function was observed. This has been attributed to the reduced nucleophilicity of C(sp^2) over C(sp^3). The isolation of **52** proved impracticable, and all efforts to acquire X-ray quality crystals yielded the κ^2-carboxylato complex $Tp^{^iPr2}Co(\kappa^2\text{-}O_2CC_6H_4Me\text{-}4)$, the formation of which implicates migratory insertion of CO, and subsequently O_2, into the Co–tolyl linkage (Scheme 4).

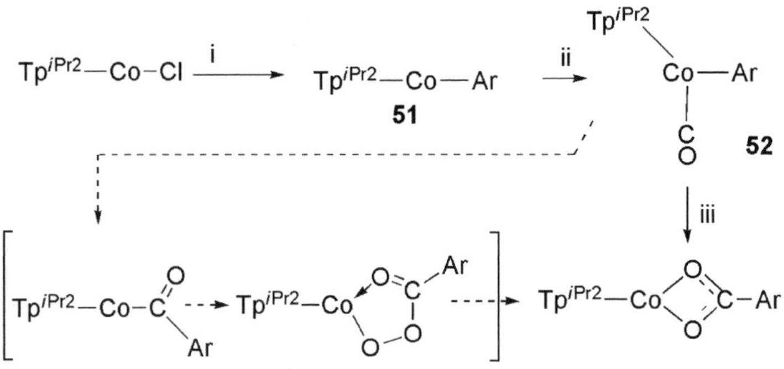

SCHEME 4. $Ar = C_6H_4Me$-4; Reagents (i) BrMgAr; (ii) CO; (iii) O_2.

Attempts to react **51** with a range of unsaturated hydrocarbons were unsuccessful, though it has been passively implied that terminal alkynes undergo metathesis, affording $Tp^{iPr_2}Co(C{\equiv}CR)$ (Section II-D.1) and presumably toluene.

D. *Complexes with σ-Donor/π-Acceptor Ligands*

1. Metal–Alkenyl and Alkynyl Compounds

It is an intriguing omission that no relevant alkenyl complexes have been isolated. Even the suggested initial product of phenylacetylene insertion into the Co–Et linkage of $Tp^{iPr_2}CoEt$ (**32**, *vide supra*), $Tp^{iPr_2}Co(CPh{=}CHEt)$,[3] has not been directly observed, even *in situ*. It is equally remarkable that other than further (unsuccessful) alkyne insertions, apparently no attempts have been made to synthesise such materials.

A similar dearth of alkynyl species is apparent, though this can be directly attributed to a lack of synthetic methodology, the simple approach, from $Tp^{x}CoX$ and organolithium or Grignard reagents, having proven ineffective. The handful of current examples were prepared only recently, *via* the dehydrative condensation of terminal alkynes with $(Tp^{iPr_2})_2Co_2(\mu\text{-}OH)_2$ (Scheme 5).[25,26] The efficiency of this route directly depends on the acidity of the terminal alkyne; thus, while alkynes bearing acyl substituents react rapidly and quantitatively, the less acidic alkyl- and aryl-1-alkynyls have not been obtained in analytically pure form. An additional complication arose with $Ph_3SiC{\equiv}CH$, since the silicon atom is more electrophilic than that of $Me_3SiC{\equiv}CH$, and is thus preferentially attacked by the hydroxo ligand, so affording the triphenylsiloxo complex $Tp^{iPr_2}Co(OSiPh_3)$.

The electronic structure of compounds **53**–**59** has been studied in some detail, using vibrational spectroscopy, crystallographic data and EHMO calculations. It was thus concluded that for these electron-deficient (15-electron) species, the extent of retrodonation from the metal to the alkynyl π-system is not significant, and is in fact negligible when R = silyl alkyl, aryl. A small contribution is noted where R comprises the electron-withdrawing carbonyl moiety.

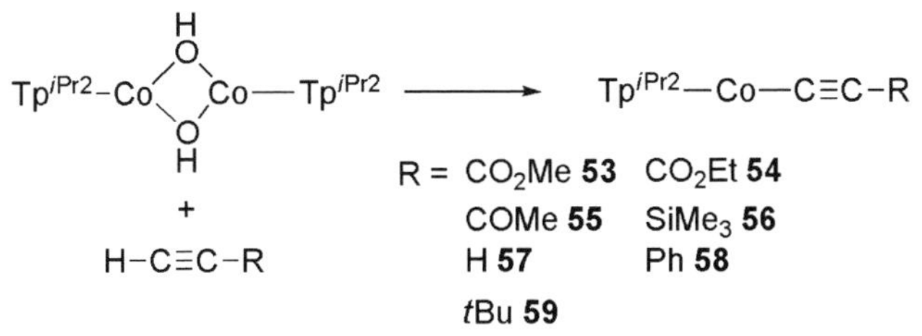

Tp^{iPr2}–Co(μ-OH)$_2$Co–Tp^{iPr2} + H–C≡C–R ⟶ Tp^{iPr2}–Co–C≡C–R

R = CO_2Me **53** CO_2Et **54**
COMe **55** $SiMe_3$ **56**
H **57** Ph **58**
*t*Bu **59**

SCHEME 5. Conditions: molecular sieves (4 Å), hexane.

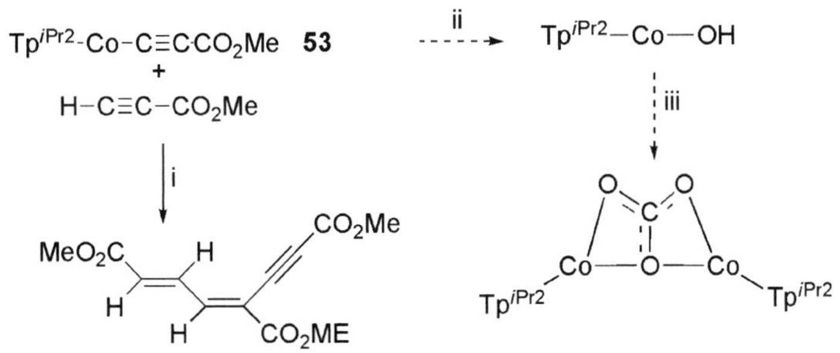

SCHEME 6. Solvents and reagents, (i) hexane; (ii) H^+; (iii) adventitious CO_2.

These metal–alkynyl complexes can be protonated to afford the free alkynes and parent cobalt hydroxo complex (comparable reactivity to their alkyl and aryl congeners), but have proven inert toward oxygenation and carbonylation. They are also thermally stable up to 100 °C. Attempts to explore the reactions of these compounds with unsaturated hydrocarbons were typically fruitless. The one exception is the reaction between **53** and its parent alkyne (HC≡CO_2Me, Scheme 6), which under benzene reflux effects catalytic, stereospecific, linear trimerisation of the alkyne to afford (*E,E*)-buta-1,3-dien-5-yne. The reaction was, however, slow (4.5 turnovers in 20 h) and suffered from catalytic deactivation due to hydrolysis of **53**, which subsequently reacted with adventitious CO_2 to irreversibly form an inert μ-carbonato complex. The catalytic cycle was concluded to involve initially a double coordination-insertion of the C≡C bond of methylpropiolate into the Co–C_{alkyne} linkage. Subsequent hydrolysis of the Co–C bond by a third equivalent of HC≡CCO_2Me would then afford the observed product and regenerate **53**. However, a definitive explanation for the stereospecificity of the process was not established.

2. Metal Carbonyls

The synthesis of several such complexes has been achieved during investigations of the reactivity of allyl, alkyl and aryl species. Thus, the compounds $Tp^xCo\{C(=O)R\}(CO)$ ($Tp^x = Tp^{iPr_2}$, R = $CH_2CH{=}CH_2$ (**23**), $CH_2C_6H_4Me$-4 (**60**), Et (**37**);[22] Tp^{Me_3}, R = $CH_2C_6H_4Me$-4 (**61**), Et (**62**)),[3] $Tp^{iPr_2}Co(C_6H_4Me\text{-}4)$(CO) (**52**)[25] and $Tp^{iPr_2}Co(CH_2CH{=}CHMe)(CO)_2$ (**63**)[23] were each obtained by carbonylation of the respective Tp^xCoR compounds. Similarly, the Co(I) compounds $Tp^xCo(CO)$ ($Tp^x = Tp^{^tBu}$ (**64**), $Tp^{^tBu,Me}$ (**36**))[15] were obtained by carbonylative cleavage of Tp^xCoR (R = Me, H).

However, a range of cobalt(I) carbonyls had previously been prepared by magnesium reduction of Tp^xCoI under CO, affording the compounds $Tp^xCo(CO)$ ($Tp^x = Tp^{^iPr,Me}$ (**65**),[27] $Tp^{^tBu,Me}$ (**36**), $Tp^{^tBu}$ (**64**)),[28] which can alternatively be obtained by treating $(Tp^x)_2Co_2(\mu\text{-}N_2)$ with CO. The $Tp^{^iPr,Me}Co$ fragment holds particular interest, since it can bind either one or two carbonyls, depending upon the reaction pressure. Thus, while the monocarbonyl **65** is most stable under ambient conditions, an over-pressure of CO affords $Tp^{^iPr,Me}Co(CO)_2$ (**66**), which is sufficiently stable at 0 °C to allow for crystallographic characterisation,[29] but loses CO spontaneously in ambient temperature solutions. Detailed thermochemical experiments revealed a remarkably low dissociation energy for the second carbonyl (13 kcal mol^{-1}), which was attributed to the steric encumbrance of the $Tp^{^iPr,Me}$ ligand. This was supported by the fact that bulkier systems (e.g. **36**) showed no evidence for binding a second molecule of CO. Variable temperature NMR studies of the equilibrium binding/dissociation process also revealed a remarkably low activation barrier, and negligible kinetic barrier, for this process, which converts the triplet-state monocarbonyl **65** into the singlet state dicarbonyl **66**. Detailed theoretical investigations (DFT) established that this results from **66** possessing a triplet-state that corresponds to a minor geometric perturbation (i.e. elongated Co–C bonds) with no inherent coordinative change. This state lies close to the minimum energy crossing point for the triplet (**65**) and singlet (**66**) potential surfaces, thus rendering CO binding a barrier-less process.[30]

One impetus for the study of the carbonyl compounds has been to establish the factors that cause formally 16-electron complexes of the type TpCoL (L = 2-electron neutral donor, e.g. N_2, O_2), to adopt a low-symmetry C_s structure in which the ligand L is ‘bent’ away from the C_3 B–Co axis. This is exemplified by the d^8 complex $Tp^{Np}Co(CO)$ (**67**, Np = neopentyl = $CH_2{}^tBu$), wherein the ‘bend’ is 26°. That the d^9 complex $\{Tp^{Np}Co(\mu\text{-}CO)\}_2Mg(THF)_4$ (**68**, obtained by Mg reduction of **67**) instead exhibits a classical ‘linear’ geometry implied that the ‘bend’ is an integral facet of d^8 systems. Exhaustive DFT and EHMO calculations of the complexes TpCoL (L = CO (Co^I d^8), COLi (Co^0, d^9), I (Co^{II}, d^7)) support this notion and attribute the thermodynamically preferable ‘bent’ geometry to the relative contributions of orbital mixing derived from different combinations of metal-valence electron density and ligand donor/acceptor character.[28,31] It is interesting to note that TpCo(CO) has, itself, never been prepared. Its use as a ‘model’ complex has been justified in terms of ‘a negligible electronic influence of substituents in the 3-position,’ in spite of the contribution that bulky substituents in this position are claimed to make to the overall (presumably kinetic) stability of the complex.

3. Metal Acyls

Aside from the handful of examples described in the preceding section, each obtained by carbonylation of an alkyl, allyl or aryl compound, acyl complexes have been neglected. Even among these few examples no reactivity investigations have been documented.

4. Metal Nitrosyls

A single such complex is known; $Tp^{^tBu,Me}Co(NO)$ (**69**), obtained by the treatment of $Tp^{^tBu,Me}Co(N_2)$ with an excess of gaseous NO.[32] This compound was crystallographically characterised, revealing that the NO ligand resides upon a molecular threefold axis; a stark contrast to the carbonyl analogues, in which the CO ligand is 'bent' away from the C_3 axis (*vide supra*). Though hailed as only the second example of a $\{Co(NO)\}^9$ complex, the chemistry of **69** has not been further explored.

E. *Metal Hydrides*

Given the classical utility of cobalt hydrides in synthetic and catalytic applications, one might anticipate a wealth of chemistry for poly(pyrazolyl)borate derivatives. However, there is, surprisingly, a single documented example, viz. $Tp^{^tBu,Me}CoH$ (**70**).[15] This is particularly remarkable given the usual prevalence of metal hydrides as organometallic decomposition products, a context in which such species have seemingly never been encountered within the poly(pyrazolyl)borate chemistry of cobalt. The synthesis of **70** has been effected *via* several routes, including (i) reaction of $Tp^{^tBu,Me}CoI$ with hydride-donor reagents, (ii) treatment of $Tp^{^tBu,Me}Co(BH_4)$ with 4-(dimethylamino)pyridine (DMAP), (iii) hydrogenation of $Tp^{^tBu,Me}Co(\eta^2\text{-}C_2H_4)$ (**20**) and most efficiently (iv) thermal hydrogenation of $Tp^{^tBu,Me}Co(N_2)$.[14] The formation of **70** is *not* observed from decomposition of $Tp^{^tBu,Me}CoR$ (R = Et (**29**), ^{n}Bu (**30**)), which are both resistant to β-Co–H elimination. Rather, **29** can be generated by the migratory insertion of ethylene into the Co–H bond of **70**. However, the insertion of other unsaturated hydrocarbons has not been explored. The carbonylation of **70** has been investigated, and found to yield $Tp^{^tBu,Me}Co(CO)$ (**36**).

The interaction of **70** with dioxygen has been investigated in some detail,[33] as the outcome differs between reaction in solution (C_6D_6) and the solid state. In solution, the only isolated products are $Tp^{^tBu,Me}Co(O_2)$ and $Tp^{^tBu,Me}Co(OH)$, while in the solid state $Tp^{^tBu,Me}Co(OH)$ forms in admixture with the novel alkoxide chelate **71**.

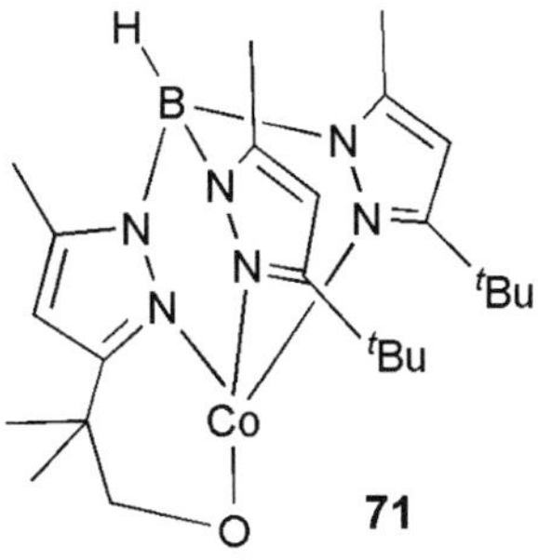

Low-temperature *in situ* NMR and infrared spectroscopic studies led the workers to conclude that the reaction commences with migratory insertion of O_2 into the

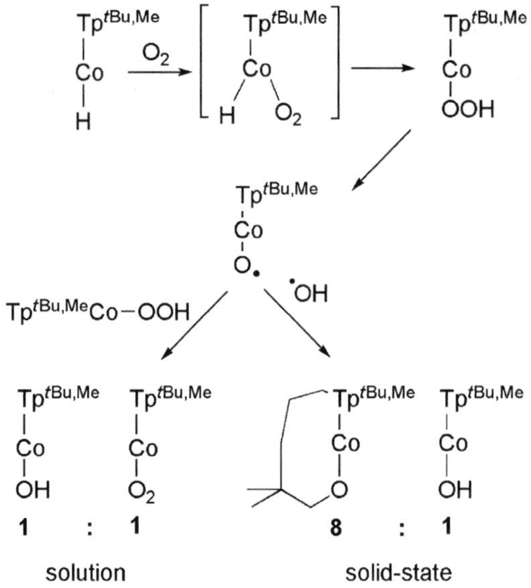

SCHEME 7. Interaction of $Tp^{^tBu,Me}CoH$ (**70**) with dioxygen in the solid state and in solution.

Co–H linkage, affording $Tp^{^tBu,Me}CoOOH$, which is susceptible to homolytic cleavage of the O–O bond. In solution, free motion of cobalt-oxo ($Tp^{^tBu,Me}Co{=}O \leftrightarrow Tp^{^tBu,Me}CoO^{-}$) and hydroxyl radicals allows for bimolecular abstraction of hydrogen from $Tp^{^tBu,Me}CoOOH$ by $Tp^{^tBu,Me}CoO$, thus affording the hydroxo- and dioxo-species. In contrast, restricted motion in the solid state results in the hydroxyl radical abstracting hydrogen from a proximal methyl group, the resulting primary radical then recombining with the cobalt-oxo to yield **71** (Scheme 7).

F. *Non-Organometallic Coordination Compounds*

These constitute the bulk of the literature surrounding cobalt poly(pyrazolyl) borates, and most are of negligible interest for the organometallic chemist. However, several notable materials are included.

One intriguing class of compounds are the xenophilic bimetallic complexes $Tp^{^iPr_2}M–Co(CO)_3L$ (M = Mn, Fe, Co, Ni; L = CO, PPh_3)[34,35] in which '$Tp^{^iPr_2}M$' behaves as the 'hard' component. These are also rare examples of complexes bearing unsupported, polar metal–metal bonds. Synthetically, $Tp^{^iPr_2}Co–Co(CO)_4$, for instance, is accessed by the interaction of PPN[$Co(CO)_4$] with [$Tp^{^iPr_2}Co(NCMe)_3$]PF_6, which is one of several such species, prepared by conventional means,[36,37] that one might anticipate to be versatile precursors for organometallic chemistry.

One area of cobalt poly(pyrazolyl)borate chemistry that has yet to fully develop, but that will undoubtedly furnish a wealth of organometallic opportunities, is that of terminal imido complexes. The first 'stable' examples of such species have

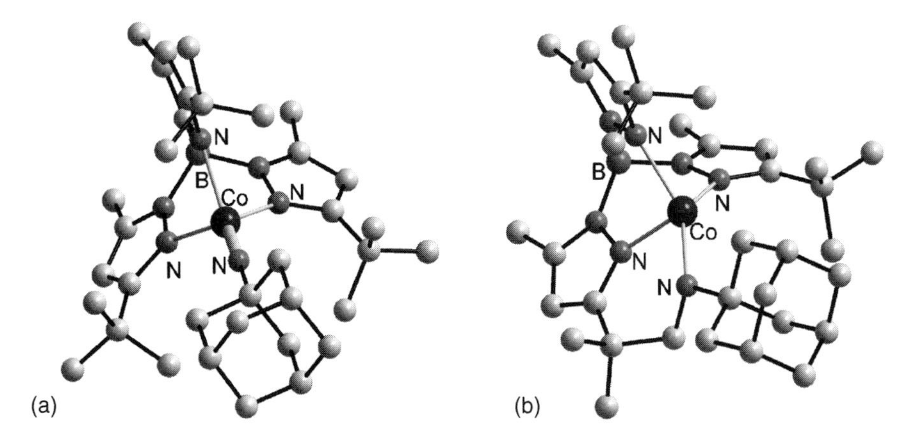

FIG. 3. Molecular structure of (a) $Tp^{^tBu,Me}Co{=}NAd$ and (b) a rearrangement product involving C–H activation of one pyrazole ^{t}Bu substituent.

recently been obtained from the reaction of $Tp^{^tBu,Me}Co(N_2)$ with organic azides RN_3 (R = adamantyl (Ad), ^{t}Bu), affording $Tp^{^tBu,Me}Co{=}NAd$ (Fig. 3) as the exclusive product in 78% yield, while $Tp^{^tBu,Me}Co{=}N^tBu$ was obtained in admixture with the azido complex $Tp^{^tBu,Me}Co(N_3)$, a fact attributed to the greater stability of the ^{t}Bu radical.[38]

III

RHODIUM AND IRIDIUM

A. *Homoleptic $[L_2M]X$ and Heteroleptic $LMX_{3-n}(sol)_n$ ($n = 0,1$) Complexes*

In contrast to cobalt, homoleptic poly(pyrazolyl)borate complexes of the heavier group 9 elements are rare $[Tp_2Rh]X$ (**72 · X**, X = PF_6,[39] Cl[40]) alone having been reported, without characterisational data. The pursuit of simple heteroleptic halide complexes has also been limited, presumably due to the greater range of organometallic precursors, and synthetic routes, available for rhodium and iridium. Nonetheless, the complexes $[Tp^xMCl_2]_2$ (M = Rh, Tp^x = Tp (**73**), Tp^{Me_2} (**74**); M = Ir, $Tp^x = Tp^{Me_2}$ (**75**))[41] have been reported to arise from the reaction of $RhCl_3 \cdot 3H_2O$ and $NaTp^x$ in methanol, though in the case of **74** prolonged reaction times instead afford the methanol-stabilised monomer $Tp^{Me_2}RhCl_2(MeOH)$ (**76**). This final step does *not* occur for **73**; however, $TpRhCl_2(NCMe)$ (**77**) is readily obtained from refluxing acetonitrile. Both **73** and **74** readily react with neutral donors (L), including phosphines (Section III-H) to afford the respective $Tp^xRhCl_2(L)$ complexes.

The structure of **73**, originally assigned from ^{1}H-NMR spectroscopic and microanalytical data, was subsequently reformulated as the salt $Na[TpRhCl_3]$ (**Na · 78**),[42] on the basis of its conversion by ion exchange to the unequivocally characterised **$[PPh_4]$ · 78**. Several other analogues have been prepared by the addition of $[PPh_4]Cl$ to $Tp^xRhCl_2(NCMe)$ ($Tp^x = Tp^{Me}$, Tp^{Me_2}, Tp^{Me_3}, $Tp^{Me_2,4Cl}$, $Tp^{^iPr}$), which are in turn prepared in the same manner as **77**. It is, however, interesting to note that $Tp^{CF_3,Me}RhCl_2(NCMe)$ (**79**) can only be prepared from isolated $RhCl_3(MeCN)_3$ and $NaTp^{CF_3,Me}$, since ligand fragmentation prevails when commencing from $RhCl_3 \cdot 3H_2O$.

Also noteworthy are the mixed halide salts $[Me_4N][Tp^xRhCl_2X]$ ($Tp^x = Tp$, X = Br (**80**); Tp^{Me_2}, X = Br (**81**), F (**82**)), obtained by treatment of **74** and **76** with $[Me_4N]X$ (X = Br, F).[41] However, few of these compounds have received extended study in an organometallic context.

B. *Complexes with σ-, π-Donor/π-Acceptor Ligands*

1. Alkene and Alkyne Complexes

In contrast to cobalt, simple alkene complexes of rhodium and iridium have been the subjects of prolific research, much of it ultimately directed toward the development of catalysts for hydrogenation and C–H bond activation processes. In view of the expansive literature on these materials, it would seem appropriate to consider their synthesis and structural facets separately from their reactivity.

a. Synthesis and Structures

Rhodium

The first example of this type of compound was $TpRh(\eta^2\text{-}C_2H_4)_2$ (**83**),[43] reported by Trofimenko in 1969 as the product of the reaction between $Rh_2(\mu\text{-}Cl)_2(\eta^2\text{-}C_2H_4)_4$ and two equivalents of KTp; a now commonly exploited synthetic route to such materials. Complex **83** was proposed to adopt a square-planar 16-electron geometry with a κ^2 bound Tp ligand, the equivalence of the three pyrazole groups on the ^{1}H-NMR time-scale being attributed to dynamic exchange *via* a 5-coordinate (18-electron) intermediate. In order to verify this suggestion several related compounds were prepared, viz. $(pzTp)Rh(\eta^2\text{-}C_2H_4)_2$ (**84**), TpRh(1,5-COD) (**85**), (pzTp)Rh(1,5-COD) (**86**), BpRh(1,5-COD) (**87**),[40] (pzTp)Rh(NBD) (**88**), (pzTp)Rh(DQ) (**89**),[44] Tp*Rh(1,5-COD) (**90**), Bp*Rh(1,5-COD) (**91**)[45] (COD = cyclooctadiene, NBD = norbornadiene, DQ = duroquinone).

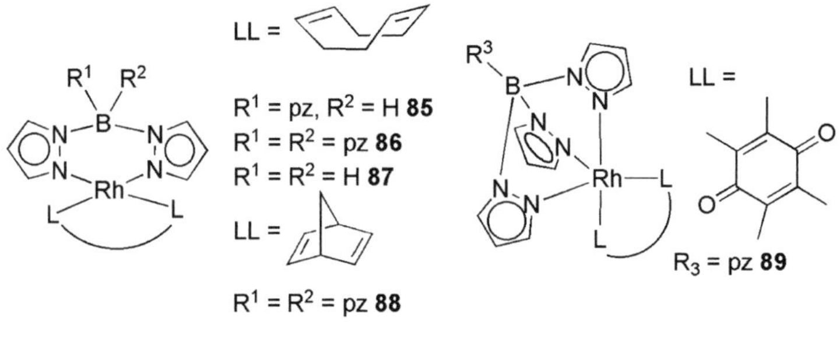

LL =

R^1 = pz, R^2 = H **85**
R^1 = R^2 = pz **86**
R^1 = R^2 = H **87**

LL =

R^1 = R^2 = pz **88**

LL =

R_3 = pz **89**

Simple ^{1}H-NMR spectroscopic studies confirmed that for most cases rapid exchange of coordinated and uncoordinated pyrazole groups occurs at ambient temperature, thereby ruling out 'static' 5-coordinate, 18-electron structures. The exception was complex **89**, for which a 5-coordinate geometry (κ^3-pzTp) was suggested, with the inequivalent axial and equatorial pyrazoles undergoing rapid dynamic exchange; hence the observation of only two resonances, in 1:3 ratio. A second (much slower) exchange of coordinated and pendant pyrazoles was deemed responsible for broadening of these resonances.[44] These conclusions were supported by crystallographic studies, which confirmed that **86** and **88** adopt square-planar geometries in the solid state, while **89** is trigonal-bipyramidal.[46] For **86** the solution-phase dynamic exchange process can be 'frozen out' at $-90\,^{\circ}$C, under which conditions the complex appears to adopt a 5-coordinate geometry, despite being 4-coordinate in the solid state.[47] The comparable exchange within the norbornadiene complex **88**, however, could not be frozen out, from which a direct correlation was inferred between the rate of exchange and electron donor capacity of the diene ligand; i.e. stronger donor = faster exchange, thus the relative rates of exchange: **89** (DQ) < **86** (COD) < **88** (NBD).

Determining the favoured geometry for complexes Tp^{x}Rh(LL) (LL = (C_2H_4), diene, $(CO)_2$ (Section II-D.3), $(PR_3)_2$), or more formally the position of the **A**/**B**/**C** equilibrium (Chart 3), is a recurrent problem to which increasingly sophisticated spectroscopic methods have been applied. For a range of complexes Tp^{x}RhLL (**92**–**100**) (Table I), exhaustive 1- and 2-D ^{1}H-, ^{13}C- and ^{103}Rh-NMR spectroscopic data have been collected, from which it has been concluded that: (i) geometries of type **A** are apparent from symmetry considerations in ^{1}H- and ^{13}C-NMR spectra of the pyrazolyl units; (ii) to distinguish between geometries **B** and **C** the ^{103}Rh-NMR chemical shifts (from ^{1}H–^{103}Rh correlation) are diagnostic.[48] Specifically, for 1,5-COD complexes adopting a κ^2-Tp^{x} chelation mode (i.e. 16-electron, **A** or **B**) the ^{103}Rh shifts lie in the range 947–1205 ppm (or 1034–1374 ppm for NBD). The precise shift is dependent upon the Tp^{x} ligand; for instance, moving from Tp to Tp^{Me} decreases δ_{Rh} by ca. 200 ppm. However, for complexes that possess appreciable contributions from the κ^3-Tp^{x} mode (18-electron **C**) these values are appreciably shifted to higher frequency. On this basis compounds **90**, **92**–**97** and **99** were assigned geometries **A** and/or **B** exclusively, while **87** and **98**–**100** were deemed to adopt, in part, form **C**.

Additional diagnostic data were obtained from the ^{13}C-NMR chemical shifts for the olefin fragments, on the basis of variations associated with the extent of

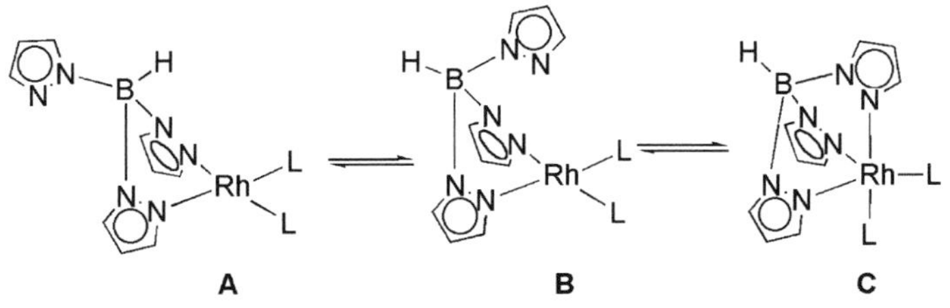

A B C

CHART 3. Possible geometries for [Tp^{x}RhLL] complexes.

metal→alkene retrodonation, which would necessarily be most pronounced for the κ^3-Tp^x geometry (18-electron, **C**). It was thus concluded that for 1,5-COD complexes, shifts of ca. 80 ppm are typical of 4-coordinate systems, while for 5-coordinate compounds (e.g. **87**) these appear at lower frequencies (ca. 73 ppm). A similar correlation was noted for NBD complexes; i.e. 57 ppm for **100** (4-coordinate), cf. 41.7 ppm for **98** (5-coordinate).

Though no diagnostic trends were determined from ^{15}N-NMR data, recent work has established the ^{11}B-NMR chemical shift as a general method for determining poly(pyrazolyl)borate hapticity, viz.: $\delta_B(\kappa^2) >> \delta_B(\kappa^3)$ (e.g. for Rh: $\delta_B(\kappa^2) \sim -6$, $\delta_B(\kappa^3) \sim -10$).[49] Notably, this correlation was established, in part, by reference to Akita's suggestion[50] that the frequency of the B–H stretching mode could be used as a guide to hapticity, κ^3 chelation leading to higher-energy absorptions (ca. 2550 cm^{-1}) than for κ^2 (ca. 2470 cm^{-1}).

On the basis of these early investigations it was concluded that complexes with bulky co-ligands and large substituents at the 3-position of the pyrazole groups preferentially adopt 16-electron square-planar geometries, while smaller substituents and co-ligands favour 18-electron trigonal-bipyramidal arrangements.[48,51] Investigations have, however, continued in the hope of elucidating the precise geometric influence of steric and electronic variation of substituents at both boron and the pyrazolyl groups. Numerous 1,5-COD, NBD and bis(ethylene) complexes (**101**–**122**, Table I, Chart 4) have thus been prepared and exhaustively characterised, though at the time of writing there remains no definitive answer to this question.

In some cases the ligands employed in these studies, and resulting adopted structures, have proven more complex. Several examples comprising Tp ligands

TABLE I

SUMMARY OF GENERIC COMPLEXES OF THE TYPE Tp^{x}Rh(LL) AND BpRh(LL) WHERE LL = 1,5-COD, NBD, 2 C_2H_4

Tp^x	Number	Reference	Tp^x	Number	Reference	Tp^x	Number	Reference
Tp^{x}Rh(1,5-COD)								
Tp	**85**	40	PzTp	**86**	40	Tp^{Me2}	**90**	40
Tp^{Me}	**92**	48	Tp^{Me2,4Cl}	**93**	48	Tp^{Me3}	**94**	48
Tp$^{^iPr,4Br}$	**95**	48	Tp^{CF3,Me}	**96**	48	Tp^{Ph,Me}	**97**	48,52
Tp$^{(CF3)2}$	**101**	51	Tp$^{^iPr2}$	**102**	50	Tp^{Ph}	**103**	53
Tp^{Ph2}	**104**	54	Tp^{a}	**105**	54	Tp^{a,Me}	**106**	54
Tp^{b}	**107**	54	Tp^{4Bo,3Me}	**108**	55	Tp^{a*}	**109**	55
Tp^{a*,3Me}	**110**	55	Tp^{Me,mt4}	**111**	56	Tp^{p-An}	**112**	57
Tp^{x}Rh(NBD)								
Tp^{Me}	**98**	48	Tp^{Me2}	**99**	48	Tp$^{^iPr,4Br}$	**100**	48
Tp$^{(CF3)2}$	**113**	51	Tp^{CF3,Me}	**114**	51	Tp^{Ph}	**115**	54
Tp$^{^iPr2}$	**116**	50	Tp^{p-An}	**117**	57			
Tp^{x}Rh(η^2-C_2H_4)$_2$								
Tp	**83**	43	PzTp	**84**	40	Tp^{Me2}	**118**	58
Bp^{x}Rh(η^2-C_2H_4)$_2$								
Bp	**119**	59	Bp^{Me2}	**120**	59			
Bp^{x}Rh(1,5-COD)								
Bp	**87**	40	Bp^{Me2}	**91**	54	Bp$^{(CF3)2}$	**121**	54
Bp^{4CN}	**122**	60						

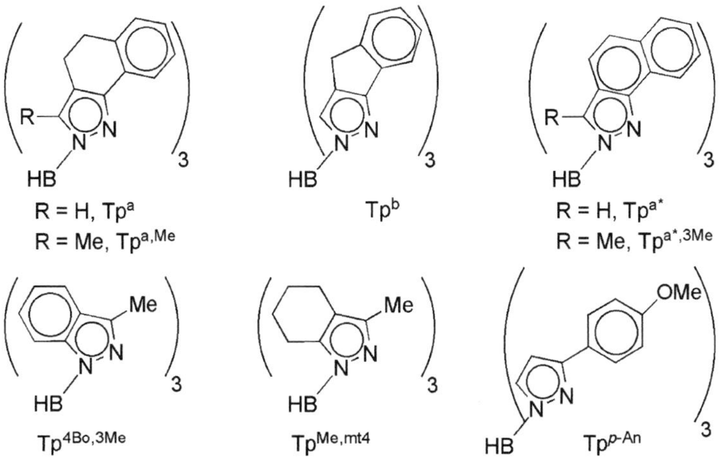

R = H, Tp^{a}
R = Me, $Tp^{a,Me}$
Tp^{b}
R = H, Tp^{a*}
R = Me, $Tp^{a*,3Me}$
$Tp^{4Bo,3Me}$
$Tp^{Me,mt4}$
$Tp^{p\text{-}An}$

CHART 4. Polycyclic scorpionate ligands.

with mixed pyrazolyl groups, i.e. HB(pz′)$_2$(pz″), have been prepared and their solution and solid-state structures determined. The compounds $\{HB(pz^{Ph,Me})_2(pz^{Me,Ph})\}$Rh(1,5-COD) (**123**),[61] $\{HB(pz^{Me,Ph})_2(pz^{Et2})\}$Rh(1,5-COD) (**124**)[61] and $\{HB(pz^{Ph,{}^iPr})_2(pz^{Me_2})\}$Rh(1,5-COD) (**125**)[62] are all square-planar (κ^2-Tp^x), as determined from detailed VT-NMR studies, and in the cases of **123** and **124** verified (in the solid state, at least) crystallographically.[62] Once again (for **123**) two dynamic exchange processes were elucidated, viz.: (i) exchange of the 2-coordinated pyrazoles and (ii) the more rapid exchange of coordinated and uncoordinated $pz^{Ph,Me}$ groups.[61] In contrast, $\{HB(pz^{Ph,{}^iPr})_2(pz^{Me_2})\}$Rh(NBD) (**126**),[62] exhibits κ^3 binding in the solid state, and this was believed to be maintained in solution, on the basis of NMR spectroscopic data. It was, however, noted that the observation of two B–H stretching frequencies (ν 2534, 2472 cm^{-1}) both in solution and the solid state might imply the presence of both κ^3 and κ^2 modes. The preparation of $\{HB(pz^{Mes})_2(pz^{5Mes})\}$Rh(1,5-COD) (**127**)[63] has also been reported, as was the serendipitous formation of $\{HB(pz^{{}^iPr,4Br})_2(pz^{4Br,5{}^iPr})\}$Rh(1,5-COD) (**128**)[48] from the reaction of $[Tp^{{}^iPr_2,4Br}]^-$ with $Rh_2(\mu\text{-}Cl)_2(1,5\text{-COD})_2$; the apparent result of a 1,2-boron shift between nitrogen atoms. For neither compound was a geometry assigned, though the complexity of the ^{1}H-NMR spectrum for **127** implied that it lacks a plane of symmetry[63] (in contrast to Tp^{Mes}Rh(1,5-COD) (**129**)). This *perhaps* implies κ^2-coordination through two differently substituted pyrazole groups, a situation believed to be reflected in **128**.

A more limited study has been undertaken of complexes in which the poly(pyrazolyl)borate ligand bears organic substituents at boron. The simple complexes $MePhBp^{Me}$Rh(LL) (LL = 1,5-COD (**130**); NBD (**131**)) exist in solution as stereochemically rigid single isomers (with pseudo-axial B–Ph), a geometry believed to be maintained in the solid state, though no data have been presented.[64] In contrast $MeTp^{Me}$Rh(LL) (LL = 1,5-COD (**132**); NBD (**133**)) are, predictably, fluxional in solution, and exist as mixtures of four- (κ^2-RTp) and 5-coordinate

(κ^3-RTp) species, the former predominating. In the solid state, however, **133** assumes exclusively the 5-coordinate κ^3 geometry.

LL = (1,5-COD)

R = H, **134**; Me **135**

LL = (NBD)

R = H, **136**; Me **137**

More significant structural facets are observed for $\{(pz^R)_2BBN\}Rh(LL)$ (**134**–**137**),[65] diamagnetic relatives of $\{(pz)_2BBN\}_2Co$ (**17**, Section II-A),[11,12] which exhibited agostic C–H···M interactions in the solid state. A similar situation is observed for **134** (Fig. 4), though the Rh···H interaction (r_{Rh-H} = 2.42(4) Å) is somewhat long, and thus described as pre-agostic.[65] Comprehensive NMR spectroscopic studies (^1H, ^{13}C, COSY, HMQC, NOESY, ^1H–^{103}Rh correlation) for **134**–**137** enabled the full spatial assignment of all spin active nuclei. This revealed that all four adopt square-planar geometries in which the chelate assumes a static boat-like conformation that is resistant to inversion at temperatures to 100 °C; a feature attributed to restricted flexibility of the bicyclic BBN unit. A comparison study of $Ph_2BpRh(LL)$ (LL = 1,5-COD (**138**), NBD (**139**)) illustrated a different scenario, these compounds exhibiting rapid boat-inversion at ambient temperature.

From a structural viewpoint one final compound of note is the tris(cycloheptatrienyl)phosphine chelate $Tp^{Me_2}Rh\{P(C_7H_7)_3\}$ (**140**), prepared from either Tp^{Me_2} Rh(1,5-COD) and $P(C_7H_7)_3$ or $RhCl\{P(C_7H_7)_3\}$ and KTp^{Me_2}.[66] This compound adopts the rare, though previously established,[67] κ^2-*N,H*-Tp^{Me_2} coordination mode (**140a**) ($r_{H\cdots Rh}$ = 1.789(7) Å (molecule A), 1.899(7) Å (molecule B)), with the remaining coordination sites occupied by phosphorus, and two η^2-olefinic functions. This structure, determined crystallographically in the solid

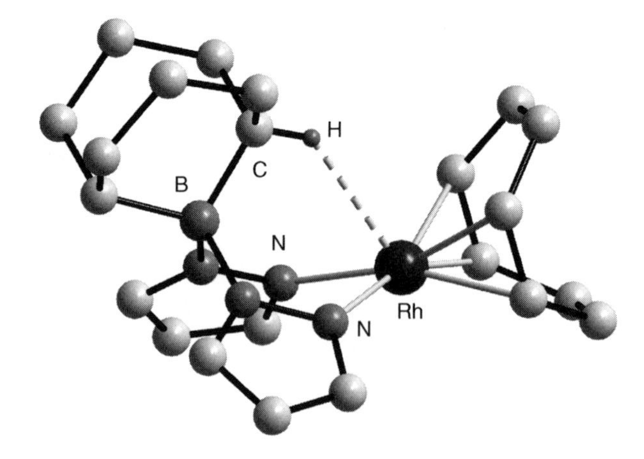

FIG. 4. Molecular structure of $\{(pz)_2BBN\}Rh(1,5\text{-}COD)$ (**134**); (non-agostic) hydrogen atoms omitted.

state, is apparently retained in solution, as demonstrated by comprehensive NMR spectroscopic data (i.e. ^{1}H, ^{11}B, $^{11}B\{^{1}H\}$, ^{13}C, ^{31}P, ^{103}Rh). However, trace amounts of two coordination isomers (**140b** and **140c**) were also observed, and both of these species have been implicated in the slow (several months) decomposition of **140** that ultimately yields the dioxygen adduct **141** (Scheme 8).

A broader range of materials have been prepared purely for their chemical interest, *or* through chemical investigation of other alkene complexes. Thus, the compounds $Tp^{x}Rh(\eta^{2}\text{-}C_2H_4)(L)$ (Tp^{x} = Tp, L = PMe_2Ph (**142**), PEt_3 (**143**),[58] PPh_3 (**144**);[68,69] Tp^{Me_2}, L = PMe_3 (**145**),[58,70,71] PMe_2Ph (**146**),[58] PEt_3 (**147**)[58]) have been prepared from the respective bis(ethylene) complexes *via* reactions with donors, as have $Tp^{Me_2}Rh(\eta^{2}\text{-}C_2H_4)(L)$ (L = CO (**148**),[72] CNCy (**149**),[58] CN^{t}Bu (**150**),[58,70] $CNC_6H_3Me_2$-2,6 (**151**),[72] $CNCH_2{}^{t}Bu$ (**152**)[73]). The η^{2}-propene complex $Tp^{Me_2}Rh(\eta^{2}\text{-}H_2C{=}CHMe)(CNCH_2{}^{t}Bu)$ (**153**[72]) has also been prepared, but by thermolysis of the rhodacyclobutane complex **154**, which is derived from C–C bond activation of a σ-cyclopropyl moiety (Section II-C.1).

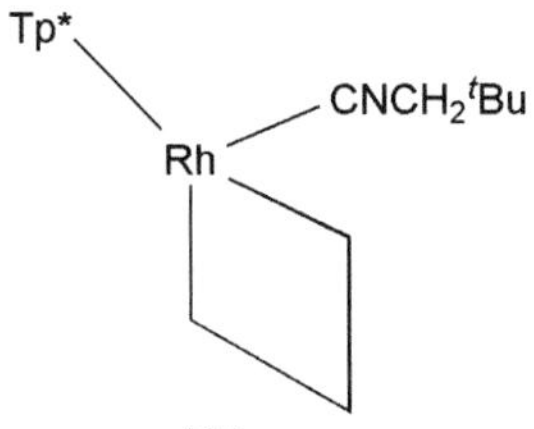

Tp*
CNCH$_2{}^{t}$Bu
Rh

154

The thermal rearrangement of an organorhodium(III) complex has also been implicated in the synthesis of $Tp^{*}Rh(\eta^{2}\text{-}C_2H_4)(NCMe)$ (**155**) from $Tp^{*}Rh(Et)(CH{=}CH_2)(NCMe)$ (**156**), itself a kinetic product of the reaction between

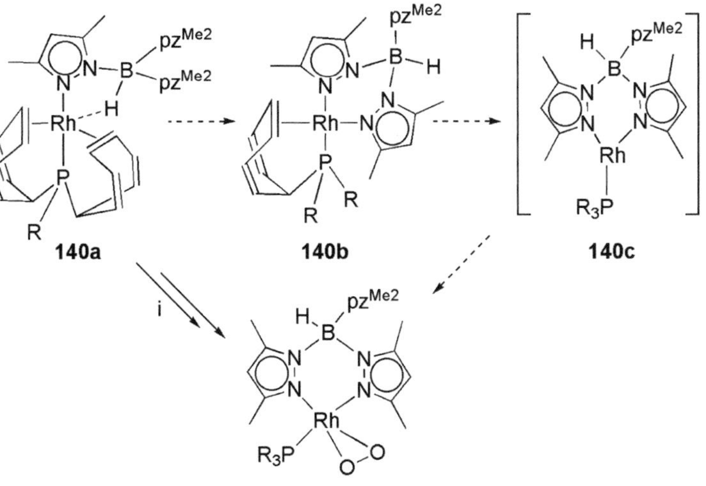

pz^{Me2} N–B pz^{Me2} H Rh P R — **140a**
pz^{Me2} B–H N Rh–N P R R — **140b**
H pz^{Me2} B N N Rh R$_3$P — **140c**
i
H B pz^{Me2} N N Rh R$_3$P O O — **141**

SCHEME 8. R = C_7H_7. Conditions: (i) solid or solution, several months, O_2.

$Tp^*Rh(\eta^2\text{-}C_2H_4)_2$ (**118**) and acetonitrile.[70] A similar kinetic product arises when **118** is treated with pyridine. However, the analogous reaction of $Bp^*Rh(\eta^2\text{-}C_2H_4)_2$ (**120**) instead affords $Bp^*Rh(\eta^2\text{-}C_2H_4)(py)$ (**157**).[59] The compounds $Bp^*Rh(C_2H_4)(L)$ ($L = PMe_3$ (**158**), NH_2Et (**159**)) are similarly prepared, while the same reaction with NH_3 preferentially yields $Bp^*Rh(C_2H_4)(NH_3)_2$ (**160**). Analogous bis(ethylamine) (**161**) and bis(pyridine) (**162**) complexes can be prepared from **157** and **159**, respectively. The acetonitrile complex $Tp^{^iPr_2}Rh(\eta^2\text{-COE})(NCMe)$ (**163**, COE = cyclooctene) is obtained from the one-pot reaction of $Rh_2(\mu\text{-Cl})_2(COE)_4$ and $KTp^{^iPr_2}$ in $CH_2Cl_2/MeCN$,[74] a variation of which is presumably employed to prepare the Tp* analogue **164**, though this synthesis does not appear to have been reported, despite the prolific use of **164** as a synthetic precursor.[50,74–76]

1,3-Diene Complexes. A small range of these complexes has been prepared from the respective 1,3-dienes and $Tp^*Rh(\eta^2\text{–}C_2H_2)_2$ (**118**), or in a single pot from $Rh_2(\mu\text{-Cl})_2(\eta^2\text{-}C_2H_4)_4$, KBp* or KTp* and the diene. Thus, the compounds Tp*Rh(LL) ($LL = \eta^4\text{-}C_4H_6$ (**165**), $\eta^4\text{-2,3-}Me_2C_4H_4$ (**166**), $\eta^4\text{-1-}EtC_4H_5$ (**167**)) and $Bp^*Rh(\eta^4\text{-2,3-}Me_2C_4H_4)$ (**168**) were prepared in high yield and purity.[58] Alternatively, **165** can be obtained from the ambient temperature reaction of **118** with excess ethylene, a reaction that at −40 °C also affords the *cis–cis* isomer of **167** and ethyl–allyl complex **169**. Interestingly, the allyl complex $Tp^*RhH(\eta^3\text{-}syn\text{-}C_3H_4Me)$ (**170**, Section III-B.2), obtained by thermal decomposition of **118**, also yields **167** under prolonged ambient-temperature photolysis (Scheme 9).

An allylic intermediate has also been proposed for the photolytic isomerisation of the COD ligand in Tp*Rh(1,5-COD) (**90**), affording Tp*Rh(1,3-COD) (**171**, Scheme 10).[77,78] This reaction is believed to occur by dechelation of the diene ligand to afford an η^2-complex that undergoes C–H activation to give, transiently, $Tp^*RhH(\eta^2{:}\eta^3\text{-}C_8H_{12})$, intramolecular reductive elimination then yields **171**. The photochemistry of **171** has also been explored, and is discussed later (Sections II-C.2 and IV-A.2).

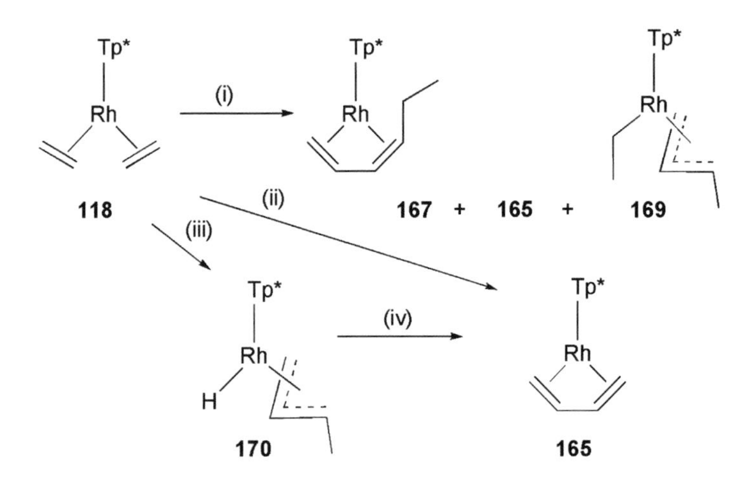

SCHEME 9. Conditions (i) C_2H_4, −40 °C; (ii) C_2H_4, 25 °C; (iii) 60 °C; (iv) *hv* (400 nm), 7 h, 25 °C.

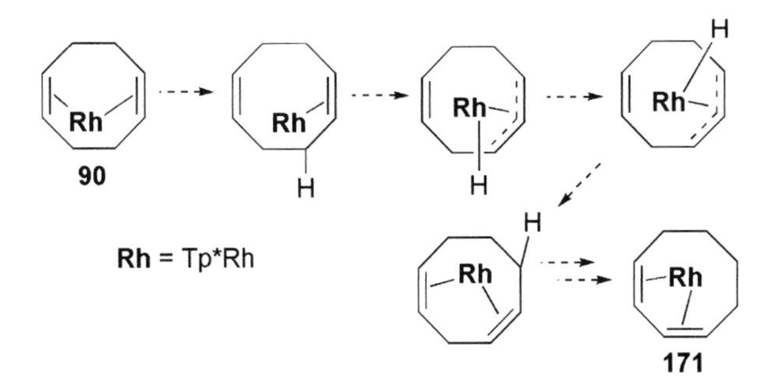

SCHEME 10. Conditions: $h\nu$ (400 nm).

Alkyne Complexes. A single such complex exists, specifically TpRh $\{\eta^2\text{-}C_2(CO_2Me)_2\}(PPh_3)$ (**172**),[79] generated by treatment of $TpRh(PPh_3)_2$ (**173**, obtained from Wilkinson's catalyst, Section III-H) with $C_2(CO_2Me)_2$ under ambient conditions. The chemistry of **172** is largely unexplored, though it has been shown to add HCl, affording the σ-vinyl $TpRhCl\{C(CO_2Me){=}CHCO_2Me\}$ (**174**), in which the ester groups are presumed to lie *cis*. Though perhaps not strictly relevant to this section, a further material derived from **173** is the $\eta^2\text{-}CS_2$ complex $TpRh(\eta^2\text{-SCS})(PPh_3)$ (**175**), which is the only example of its kind in group 9 poly(pyrazolyl)borate chemistry. The alkylation of **175** with $[Et_3O]BF_4$ to provide $[TpRh(PPh_3)(\eta^2\text{-SCSEt})]BF_4$ (**176 · BF₄**) has also been reported.

Iridium

In contrast to rhodium, the majority of work with iridium has focussed upon the chemistry of these materials, with little effort expended on attempts to elucidate structural systematics, though a small number of cyclooctadiene complexes (**177–188**, Table II) have been prepared from $Ir_2(\mu\text{-}Cl)_2(1,5\text{-COD})_2$ and MTp^x (M = Na, K, or Tl) largely to this end.

For ligands bearing only a 3-substituent on pyrazole 3/5-scrambling was, once again, encountered. Thus, the synthesis of **178** also yielded analogues containing $HB(pz^{Me})_2(pz^{5Me})$ (**189**) and $HB(pz^{Me})(pz^{5Me})_2$ (**190**), while with $Tp^{^iPr}$ a small amount of $\{HB(pz^{^iPr})_2(pz^{5^iPr})\}Ir(1,5\text{-COD})$ (**191**) was obtained in addition to **183**.[80] A further unsymmetrical ligand was obtained by protonating **179** with triflic acid, affording $[\{\kappa^2\text{-}HB(pz^{Me_2})_2(pz^{Me_2}H)\}Ir(1,5\text{-COD})]OTf$ (**192 · OTf**),[81,82] in which the Tp^xH ligand is neutral.

Given the impetus to explore reactivity, it is perhaps unsurprising that the bulk of work has focussed on bis(ethylene) and 1,3-butadiene compounds. It is, however, surprising that relatively few such materials have been prepared, given the chemical diversity available from varying the electronic and steric properties of the pyrazole substituents. Nonetheless, only three bis(ethylene) complexes are known, viz.: $TpIr(\eta^2\text{-}C_2H_4)_2$ (**193**),[84,85] $Tp^{Me_2}Ir(\eta^2\text{-}C_2H_4)_2$ (**194**)[81,86] and $Tp^{Br_3}Ir(\eta^2\text{-}C_2H_4)_2$ (**195**).[87] Attempts to prepare an analogous complex with the Tp^{Ph} ligand have so far been fruitless, the reaction of $Ir_2(\mu\text{-}Cl)_2(\eta^2\text{-}C_2H_4)_4$ (formed *in situ* from

TABLE II
SUMMARY OF 1,5-COD COMPLEXES OF IRIDIUM

Tp^x	Number	Reference	Tp^x	Number	Reference	Tp^x	Number	Reference
$Tp^xIr(1,5\text{-COD})$								
Tp	**177**	80	Tp^{Me}	**178**	80	Tp^{Me_2}	**179**	81
Tp^{Me_3}	**180**	82	$Tp^{Me_2,4Cl}$	**181**	80	$Tp^{Me_2,4Br}$	**182**	80
$Tp^{^iPr}$	**183**	80	$Tp^{^iPr_2}$	**184**	80	$Tp^{CF_3,Me}$	**185**	80
$Tp^{Ph,Me}$	**186**	80	PzTp	**187**	83	[BpIr(1,5-cod)]	**188**	83

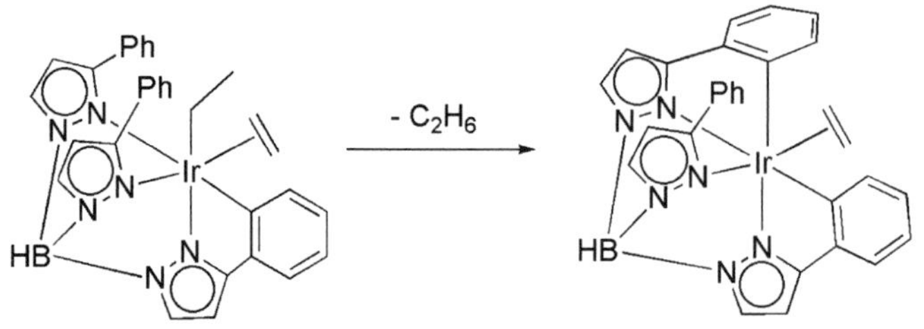

SCHEME 11. Conditions 80 °C, benzene solution.

$Ir_2(\mu\text{-}Cl)_2(\eta^2\text{-}COE)_4$) with $TlTp^{Ph}$ yielding instead the novel complex **196**, in which one phenyl substituent has undergone cyclometallation, with transfer of the hydride to an ethylene ligand.[88] Interestingly, thermolysis (80 °C, Scheme 11) of **196** in benzene solution leads to a second cyclometallation, with elimination of ethane, to afford **197**. Further examples of Tp^{Aryl} cyclometallation are described later.

The bis(cyclooctene) complex $TpIr(COE)_2$ has also been sought, but has surprisingly eluded isolation. Moreover, the nature of the products obtained would seem to be dependant on the conditions employed. Thus, the treatment of $Ir_2(\mu\text{-}Cl)_2(COE)_4$ with NaTp in tetrahydrofuran yielded exclusively TpIrH (η^2-C_8H_{14})(σ-C_8H_{13}) (**198**), the protonation of which (with HBF_4) afforded $[TpIrH(\eta^2\text{-}C_8H_{14})_2]BF_4$ (**199 · BF$_4$**).[85] In contrast, an analogous reaction with KTp in dichloromethane was independently reported to yield exclusively the allyl complex $TpIrH(\eta^3\text{-}C_8H_{13})$ (**200**), as determined from NMR spectroscopic and mass spectrometric data.[84] Though at variance, these results both implicate intramolecular C–H activation; a process that is now widely researched and exploited (Section IV-A). Indeed, photochemical or thermal rearrangement of **193**, **194** and **195** to their iridium(III) vinyl isomers $Tp^xIrH(CH{=}CH_2)(\eta^2\text{-}C_2H_4)$ ($Tp^x = Tp$ (**201**),[84] Tp^{Me_2} (**202**),[86] Tp^{Br_3} (**203**)[87]), *via* the C–H activation of ethylene, has been implicated in numerous synthetic processes, and the specific case of **193**→**201** studied at length. Moreover, the one-pot reaction between $Ir_2(\mu\text{-}Cl)_2(COE)_4$, KTp* and $CH_2{=}CHR$ (R = Me, Et) yields directly the hydrido-alkenyl complexes $Tp^*IrH(CH{=}CHR)(\eta^2\text{-}CH_2{=}CHR)$ (R = Me (**204**), Et (**205**)), *via* the (speculated) intermediacy of the bis(alkene) complexes, though these have not been detected.[89]

Complex **202** undergoes high-pressure (500 atmospheres) hydrogenation to afford $Tp^*IrH_2(\eta^2\text{-}C_2H_4)$ (**206**).[81] Under similar conditions, $Tp^xIrH_2(\eta^2\text{-COE})$ ($Tp^x = Tp^*$ (**207**), Tp^{Me_3} (**208**)) can also be prepared, commencing from the respective 1,5-cod complexes $Tp^xIr(\eta^4\text{-1,5-COD})$ in which cases the diene becomes partially hydrogenated. In contrast, at lower pressures (200 atmospheres), diene hydrogenation is avoided, and $Tp^*IrH_2(\eta^2\text{-1,5-COD})$ (**209**) is obtained.

The reaction chemistry of the hydridovinyl complexes has, however, often proven unexpectedly complicated, owing to their propensity to effect *intra* and *inter*molecular C–H activations. For instance, under an atmosphere of ethylene, benzene solutions of **202** undergo a thermal (60 °C) reaction to afford exclusively **210**. This reaction is presumed to involve the intermediacy of $Tp^*Ir(C_6H_5)_2(\eta^2\text{-}C_2H_4)$, an unobserved product of the C–H activation of benzene.[90] This proposal is supported by the fact that $Tp^*Ir(C_6H_5)_2(N_2)$ (**211**) can be isolated (Section II-C.2), and converts cleanly to **210** under comparable conditions.

Complex **203** exhibits similar reactivity, effecting photolytic C–H activation of pentane to afford $Tp^{Br_3}IrH(CH_2{}^nBu)(\eta^2\text{-}C_2H_4)$ (**212**) in admixture with the double C–H activation product $Tp^{Br_3}IrH_2(\eta^2\text{-}H_2C{=}CH^nPr)$ (**213**),[87] which can also be generated thermally from **212**. Analogous reactions proceed with diethylether to afford $Tp^{Br_3}IrH(CH_2CH_2OEt)(\eta^2\text{-}C_2H_4)$ (**214**) and $Tp^{Br_3}IrH_2(\eta^2\text{-}CH_2{=}CH_2OEt)$ (**215**). It was also noted that under thermolysis, **203** inserts ethylene into the Ir–vinyl linkage to afford the hydrido-butenyl complex **216**. This forms in admixture with the hydrido-crotyl complex **217**, which is believed to result from rearrangement of the undetected product of ethylene insertion into the Ir–H bond, i.e. $Tp^{Br_3}Ir(Et)(CH{=}CH_2)$.

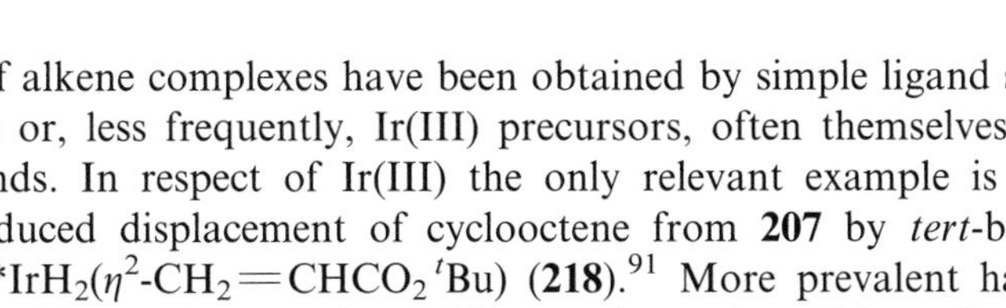

210 **216** **217**

A variety of alkene complexes have been obtained by simple ligand substitution of either Ir(I) or, less frequently, Ir(III) precursors, often themselves containing π-alkene ligands. In respect of Ir(III) the only relevant example is the photochemically induced displacement of cyclooctene from **207** by *tert*-butylacrylate to afford $Tp^*IrH_2(\eta^2\text{-}CH_2{=}CHCO_2{}^tBu)$ (**218**).[91] More prevalent has been the generation of Ir(I) complexes under ambient or thermal conditions, first illustrated in the reaction of $TpIr(\eta^2\text{-}C_2H_4)_2$ (**193**) with methyl acrylate to afford $TpIr(\eta^2\text{-}C_2H_4)(\eta^2\text{-}CH_2{=}CHCO_2Me)$ (**219**).[84] Thereafter, the reactions of $Tp^xIr(\eta^2\text{-}C_2H_4)_2$ with single equivalences of tertiary phosphines were widely employed to prepare $Tp^xIr(\eta^2\text{-}C_2H_4)(PR_3)$ ($Tp^x = Tp$, $PR_3 = PPh_3$ (**220**),[68] PCy_3 (**221**),[68] PMe_2Ph (**222**),[92] PEt_3 (**223**);[92] $Tp^x = Tp^*$, $PR_3 = PMe_3$ (**224**),[92] PEt_3 (**225**),[92] PMe_2Ph (**226**)[92]), while the reaction of $Tp^*Ir(\eta^2\text{-}C_2H_4)_2$ (**194**) with half an equivalent of dmpe yields the phosphine-bridged complex **227**.[92] It should be noted that **220** reacts further in the presence of excess PPh_3 to afford a cyclometallated complex that reverts to **220** under an atmosphere of ethylene.[93]

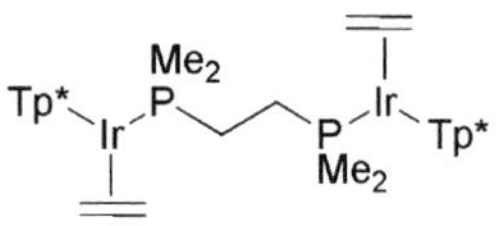

Tp*–Ir–P(Me_2)–CH$_2$CH$_2$–P(Me_2)–Ir–Tp*

227

Carbonylation of the bis(ethylene) complexes tends to afford the respective dicarbonyls, which have thus served as precursors to TpIr(CO)(η^2-C_2H_4) (**228**)[94] and presumably $Tp^{CF_3,Me}$Ir(CO)(η^2-C_2H_4) (**229**),[95] upon treatment with ethylene, though for the latter no experimental details have appeared. Complex **228** has been subjected to extensive NMR spectroscopic studies to determine its structural facets, revealing a 5-coordinate (κ^3-Tp, trigonal-bipyramidal) geometry with the anticipated thermal equilibration of the pyrazole groups. Somewhat unusually, however, this dynamic process was easily slowed, coalescence being observed around 0 °C. It was thus established that rotation of the ethylene ligand is slow on the NMR timescale (AA′XX′ spin system). Hence, the magnetic equivalence observed for the ethylenic carbon nuclei strongly implied an equatorial disposition of this ligand; a notion supported by NOE experiments.

1,3-Diene Complexes. Several complexes of η^4-conjugated 1,3-dienes (**230**–**241**, Chart 5) have been prepared, and studied for their capacity to activate C–H bonds (see later). Each was prepared from the one-pot reaction of $Ir_2(\mu$-$Cl)_2(COE)_4$ with the respective diene and Tp^x ligands, as either thallium ($Tp^x = Tp^{Ph}$, Tp^{Th}) or potassium (Tp^x = Tp, Tp^{Me_2}) salts.

For complexes of Tp and Tp* some effort was expended in determining structural facets for both the diene and Tp^x ligand. It was thus concluded that the dienes adopt the anticipated η^4-*S-cis* conformations, with an appreciable contribution from the metallacyclopent-3-ene resonance form, apparent from particularly low-frequency ^{13}C-NMR chemical shifts for the terminal alkenic nuclei (ca. 3.9 ppm). However, since the J_{CH} values were more typical of π-coordination, this representation was adopted. In respect of the Tp^x ligands, considerable fluxionality was observed, the rate of exchange being found to be heavily dependant upon the

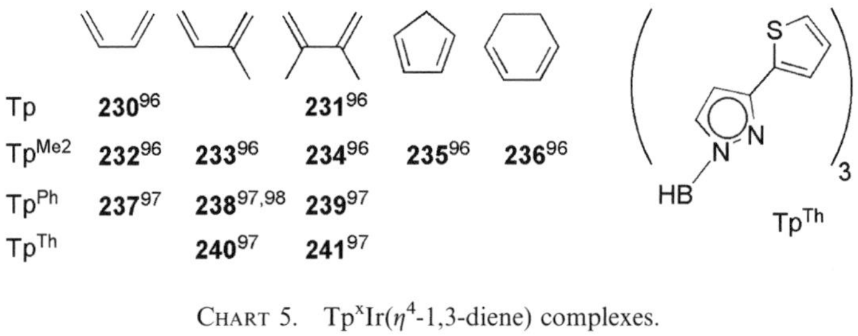

Tp	**230**[96]		**231**[96]		
Tp^{Me2}	**232**[96]	**233**[96]	**234**[96]	**235**[96]	**236**[96]
Tp^{Ph}	**237**[97]	**238**[97,98]	**239**[97]		
Tp^{Th}		**240**[97]	**241**[97]		

HB(pz-thienyl)$_3$ = Tp^{Th}

CHART 5. Tp^xIr(η^4-1,3-diene) complexes.

extent of steric congestion. Thus, at ambient temperature the least congested complex (**230**) exhibited two unique pyrazolyl environments (2:1 ratio), implying slow exchange, while the most congested (**241**), exhibits only one environment; consistent with more rapid exchange. On this basis, an 18-electron, 5-coordinate ground state was suggested, with the 16-electron, square-planar form being readily accessible, the ease of interconversion being enhanced by increasing steric pressure on the ligands.

Other η^2-Alkene Complexes. Before concluding this section a small range of more complex molecules comprising an η^2-alkene ligand must be considered, having arisen during the investigation of the chemistry of the Tp*Ir fragment, in particular with respect to C–H activation processes. For instance, while studying the formation and chemistry of iridapyrroles of type **D**, obtained thermally from Tp*Ir(η^2-C_2H_4)$_2$ (**194**) and nitriles (Scheme 12), it was found that the nucleophilicity of C_β in the canonical form $\mathbf{D_c}$ could be exploited by protonation.[99,100] In this way, ***242-cis***[99] was obtained, kinetically, under mild conditions (CH_2Cl_2, 20 °C) in admixture with small amounts of the *trans* isomer. Complete conversion to the thermodynamically favoured ***242-trans*** was effected on prolonged heating (80 °C, 12 h). In each case, the structures were assigned on the basis of 1D and 2D-NMR spectroscopic studies, and in the case of ***242-trans*** confirmed crystallographically.[100]

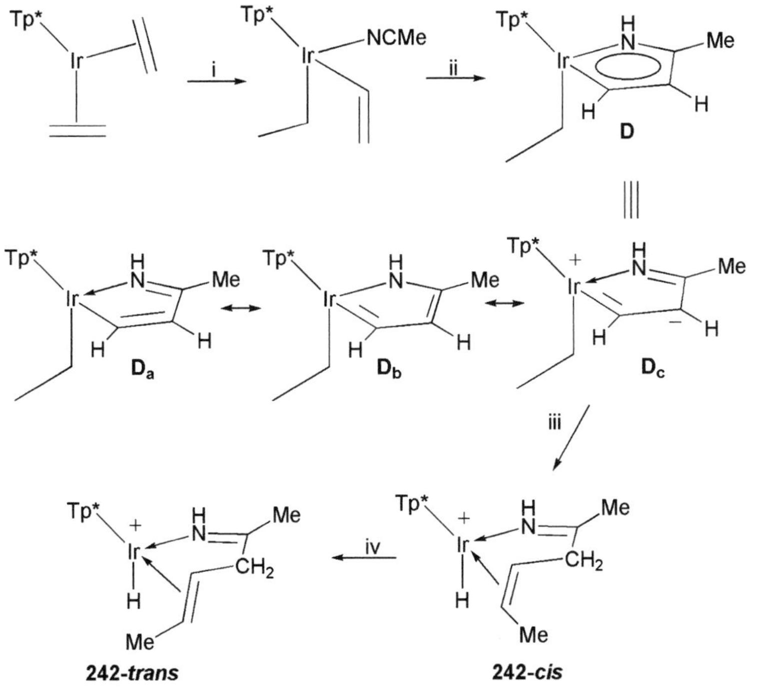

Tp* Ir NCMe i ii Tp* H N Me H H D
Tp* Ir H N Me H H D_a ↔ D_b ↔ D_c
iii iv
Tp* + Ir H N Me CH_2 H Me **242-trans** **242-cis**

SCHEME 12. Conditions: (i),(ii) NCMe, Δ; (iii) $[H(OEt_2)_2][BAr^f_4]$; (iv) 80 °C.

In comparable fashion **243-*cis*** and ***trans*** were also obtained, from the propylene hydrido-alkenyl complex Tp*IrH(CH=CHMe)(η^2-CH_2=CHMe) (**204**) (Scheme 13).

In an extension of this chemistry the reactions of **194** with a series of aldehydes were investigated and found to directly yield the alkoxide-alkene chelates (**244–247**, Scheme 14), with no evidence of iridafurans being obtained.[100] Protonation of the alkoxide oxygen was also effected, thus affording the respective alcohol complexes (**248–251**, Scheme 15), the Ir–O(H)R functionalities of which are remarkably stable toward substitution by extraneous donors, a feature attributed to the chelating nature of the ligand. It is interesting to note that in this instance the *trans* isomers alone are obtained, with no evidence for the *cis*.

The remaining examples were obtained in a more serendipitous fashion, while investigating the 'Tp*Ir' mediated C–H activation of substituted thiophenes. Thus, the reaction of Tp*Ir(η^2-C_2H_4)$_2$ (**194**) with 2,5-dimethylthiophene (3,5-$Me_2C_4H_2S$) afforded the anticipated C(sp^2)–H activation product **252**, in admixture with the chelate complex **253** (Scheme 16).[101] Though no definitive evidence was found, **253**

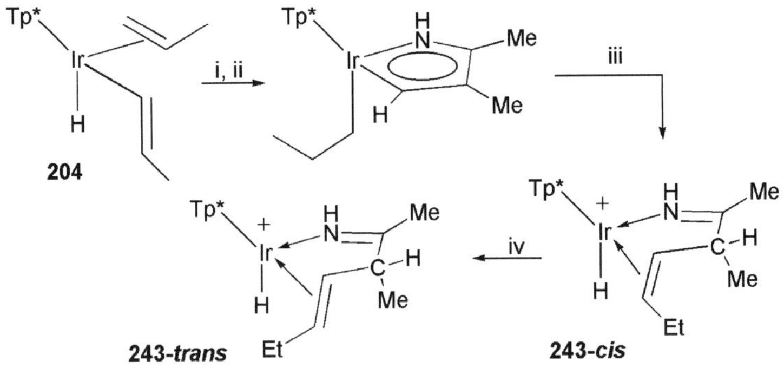

SCHEME 13. Conditions:(i),(ii) NCMe,Δ; (iii) $[H(OEt_2)_2][BAr^f_4]$; (iv) 80 °C.

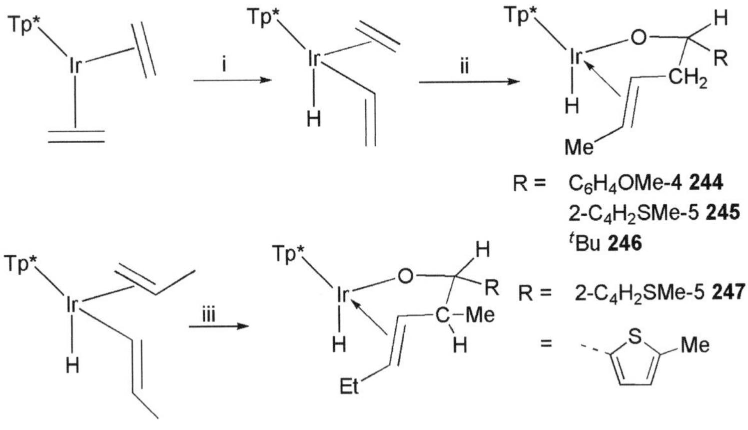

SCHEME 14. Conditions and reagents (i) 60 °C, (ii) O=CHR, 60 °C, (iii) O=CHR, 60 °C.

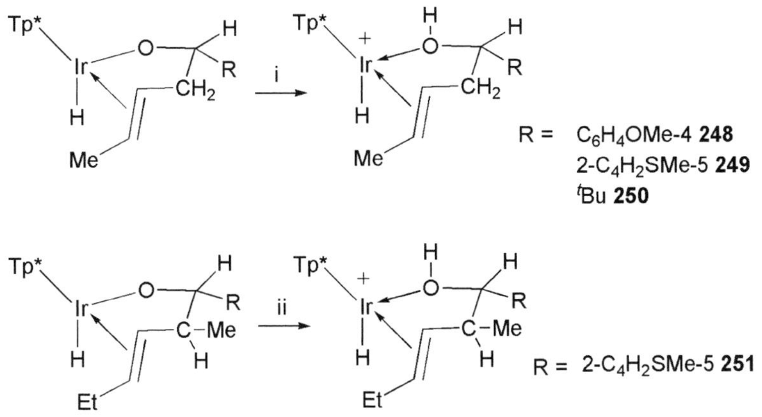

R = C_6H_4OMe-4 **248**
2-C_4H_2SMe-5 **249**
^tBu **250**

R = 2-C_4H_2SMe-5 **251**

SCHEME 15. Conditions: (i), (ii) $[H(OEt_2)_2][BAr^f_4]$.

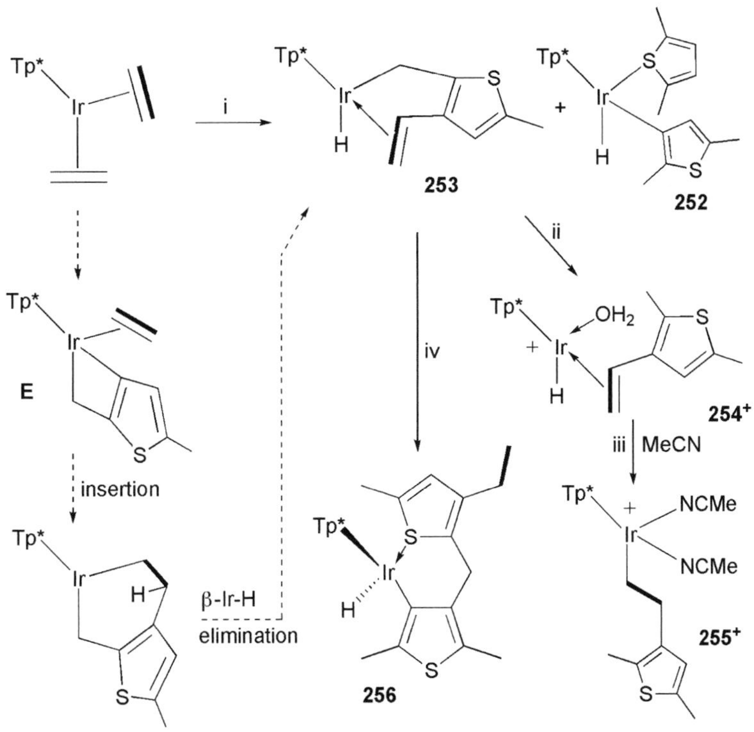

SCHEME 16. Conditions: (i) 2,5-dimethylthiophene, 60 °C; (ii) $[H(OEt_2)_2][BAr^f_4]$, CH_2Cl_2, −50 °C; (iii) 90 °C; (iv) 85 °C, 24 h.

was proposed to form *via* intermediate **E**, which derives from double C–H activation of one 3,5-$Me_2C_4H_2S$ molecule and concomitant loss of ethane (observed by ^{1}H-NMR spectroscopy).[102] Migratory insertion of ethylene, followed by *β*-H elimination then affords **253**.

The Ir–C_{alkyl} linkage of **253** has been found to be susceptible to protonolysis, which affords the cationic complex $\mathbf{254}^+$, the vacant coordination site being occupied by adventitious water. Displacement of the alkene from $\mathbf{254}^+$ by Lewis bases is surprisingly non-facile, induced migration of the hydride to the alkene instead predominating to afford the alkyl complex $\mathbf{255}^+$. It is also noteworthy that **253** can activate a second molecule of the thiophene to afford the complex chelate **256**, which will be discussed later (Section III-D.1).

b. Reactivity

Ligand Substitutions

The interaction of the complexes $Tp^xM(LL)$ (M = Rh, Ir; LL = diene, $(\eta^2\text{-}C_2H_4)_2$) with extraneous donors (e.g. PR_3, CO, CNR) is, unsurprisingly, an important and widely exploited reactivity. In this way, compounds **142–152**; **157–162**; **165–168**; **219–227** (Chart 6) have each been prepared.[58,59,68–72,84,92] The disubstitution of bis-ethylene complexes, as in the formation of **165–168**, is also observed for Bp*Rh $(\eta^2\text{-}C_2H_4)_2$ (**120**) upon treatment with depe, dppe or excess PR_3, affording Bp*Rh (P–P) (P–P = depe (**257**), dppe (**258**), 2 PPh_3 (**259**), 2 PMe_3 (**260**)). The analogue $BpRh(PPh_3)_2$ (**261**) has been similarly obtained from **119**.[59]

With stronger donors complete displacement of ethylene is often encountered exclusively (i.e. regardless of stoichiometry) as in the formation of $Bp^*Rh(CO)_2$ (**262**)

$TpRh(\eta^2\text{-}C_2H_4)(L)$	L = PMe_2Ph **142**; PEt_3 **143**; PPh_3 **144**
$Tp^*Rh(\eta^2\text{-}C_2H_4)(L)$	L = PMe_3 **145**; PMe_2Ph **146**; PEt_3 **147**; CO **148**; CNCy **149**; CN^tBu **150**; $CNC_6H_3Me_2$-2,6 **151**; $CNCH_2{}^tBu$ **152**
$Bp^*Rh(\eta^2\text{-}C_2H_4)(L)$	L = py **157**; PMe_3 **158**; NH_2Et **159**
$Bp^*Rh(\eta^2\text{-}C_2H_4)(L)_2$	L = NH_3 **160**; py **161**; NH_2Et **162**
Tp*Rh(LL)	LL = **165** **166** **167**
Bp*Rh(LL)	LL = **168**
$TpIr(\eta^2\text{-}C_2H_4)(L)$	L = $CH_2{=}CHCO_2Me$ **219**; PPh_3 **220**; PCy_3 **221**; PMe_2Ph **222**; PEt_3 **223**
$Tp^*Ir(\eta^2\text{-}C_2H_4)(L)$	L = PMe_3 **224**; PEt_3 **225**; PMe_2Ph **226**

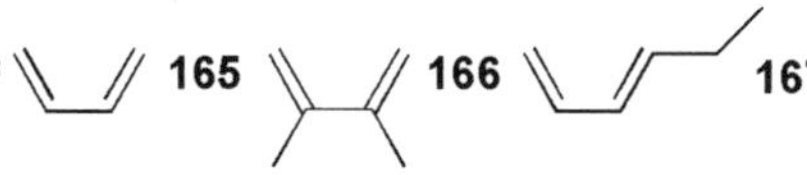

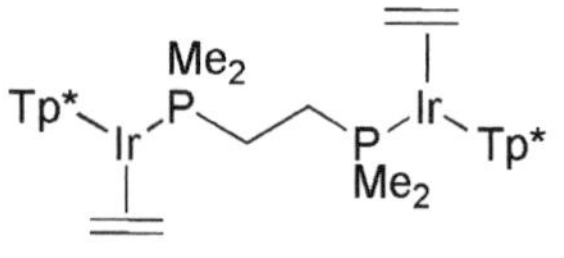

Tp* Ir P Me_2 P Me_2 Ir Tp* **227**

CHART 6. Summary of mono-alkene complexes prepared by ligand substitution.

and $Bp^xRh(CN^tBu)_2$ (Bp^x = Bp (**263**), Bp* (**264**)).[59] A similar scenario is encountered upon carbonylation of $TpM(\eta^2\text{-}C_2H_4)_2$ (M = Rh (**83**), Ir (**193**)), affording for iridium $TpIr(CO)_2$ (**265**),[94] while for rhodium the sole product is formulated as $Tp_2Rh_2(CO)_3$ (**266**) (Section II-D.2).[40] This latter outcome contrasts the comparable reaction of $Tp^*Rh(\eta^2\text{-}C_2H_4)_2$ (**118**), which yields exclusively the monocarbonyl complex **148**;[72] a tangible illustration of the influence of pyrazole substitution upon reactivity. Carbonylation of the iridium complex $Tp^*Ir(\eta^2\text{-}C_2H_4)_2$ (**194**) has *not* been reported, but given its otherwise comparable behaviour to **118** one might anticipate a similar result.

A further example of the tendency toward multiple substitution was observed upon treatment of $Tp^*Rh(\eta^2\text{-}C_2H_4)(L)$ (L = C_2H_4 (**118**), PMe_3 (**145**)) with excess (five to six equivalents) PMe_3, which affords the tris(phosphine) complex $(\kappa^1\text{-}N\text{-}Tp^*)Rh(PMe_3)_3$ (**267**),[103] in which one pyrazole donor has been displaced in addition to the ancillary ligands. The identity of **267** was established crystallographically (Fig. 5), while variable temperature $^{31}P\{^1H\}$-NMR spectroscopic studies revealed thermal equilibration of the pyrazolyl groups, a process that could be 'frozen out' at −80 °C.[104]

The intermediacy of $(\kappa^2\text{-}Tp^*)Rh(PMe_3)_2$ (**268**) in the generation of **267** was established by its controlled synthesis from **145**, and subsequent treatment with excess PMe_3, to afford exclusively **268**. On this basis, the first stepwise dechelation of a 'Tp^x' ligand was claimed (Scheme 17), the final step, i.e. extrusion the $[Tp^*]^-$ anion, being effected at 120 °C in the presence of excess PMe_3 to afford, presumably, the salt $[Rh(PMe_3)_4][Tp^*]$ (**269 · Tp^x**), though the instability of this material has precluded its unequivocal characterisation. A formally related reaction

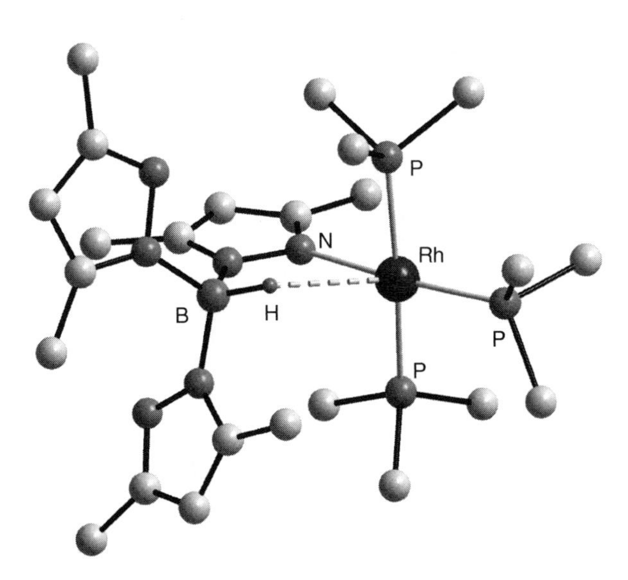

FIG. 5. Single crystal X-ray structure of $(\kappa^1\text{-}N\text{-}T^*)Rh(PMe_3)_3$ (**267**). Pyrazolyl and phosphine. Hydrogen atoms omitted; $r_{H...Rh} = 2.59$ Å.

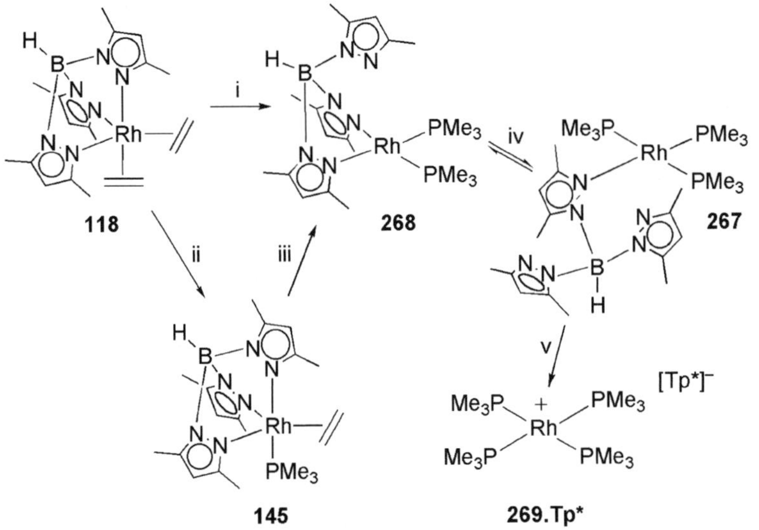

SCHEME 17. PMe_3 induced stepwise dechelation of the Tp* ligand from rhodium. Reagents and conditions: (i) 5 PMe_3; (ii) PMe_3, (iii) 5 PMe_3; (iv) PMe_3; (v) 120 °C, PMe_3.

that also leads to extrusion of the Tp^- anion, is observed upon treating $TpIr(C_2H_4)_2$ (**193**) with dppe, which affords $[Ir(dppe)_2]$ [Tp] (**270 · Tp**),[105] in admixture with **193**.

Oxidative Substitution

This approach has been widely exploited to convert M(I) π-alkene precursors into M(III) complexes, both with and without retention of the alkene. Thus, $TpRh(C_2H_4)_2$ (**83**) adds R_fI to afford $TpRh(R_f)I(\eta^2\text{-}C_2H_4)$ ($R_f = CF_2C_6F_5$ (**271**), $CF_2CF_2CF_3$ (**272**)), from which the remaining ethylene is displaced by CO or PMe_3 to afford $TpRh(R_f)I(L)$ ($R_f = CF_2C_6F_5$, L = CO (**273**), PMe_3 (**274**); $R_f = {}^nC_3F_7$, L = CO (**275**), PMe_3 (**276**)).[106] The iridium complex $TpIr({}^nC_3F_7)I(\eta^2\text{-}C_2H_4)$ (**277**) was similarly prepared, but in this case subsequent treatment with PMe_3 affords the salt $[TpIr({}^nC_3F_7)(\eta^2\text{-}C_2H_4)(PMe_3)]I$ (**278 · I**), rather than displacing the ethylene ligand. Analogues of **273–276** were instead obtained from alkene-free precursors (Section II-C.1).

The acetonitrile complex $Tp^*Rh(\eta^2\text{-COE})(NCMe)$ (**164**) reacts with disulfides under ambient conditions to afford $Tp^*Rh(SR)_2(NCMe)$ (R = Ph (**279**), Tol (**280**)), a reaction that is believed to involve initial dissociation of cyclooctene and subsequent oxidative addition of RSSR.[76] The analogous Tp precursor also reacts with $S_2(C_6H_4Me\text{-}4)_2$ to afford $TpRh(SC_6H_4Me\text{-}4)_2(NCMe)$ (**281**), which is the exclusive product obtained at −20 °C. However, under milder conditions **281** forms in admixture with the thiolato-bridged complex $Tp_2Rh_2(\mu\text{-}SC_6H_4Me\text{-}4)_2(SC_6H_4Me\text{-}4)_2$ (**282**, Scheme 18), the accessibility of which is attributed to the reduced steric encumbrance of the Tp ligand, relative to Tp*. A more complex reaction ensues between **164** and 2-pyridyldisulfide (Py_2S_2), in which both the

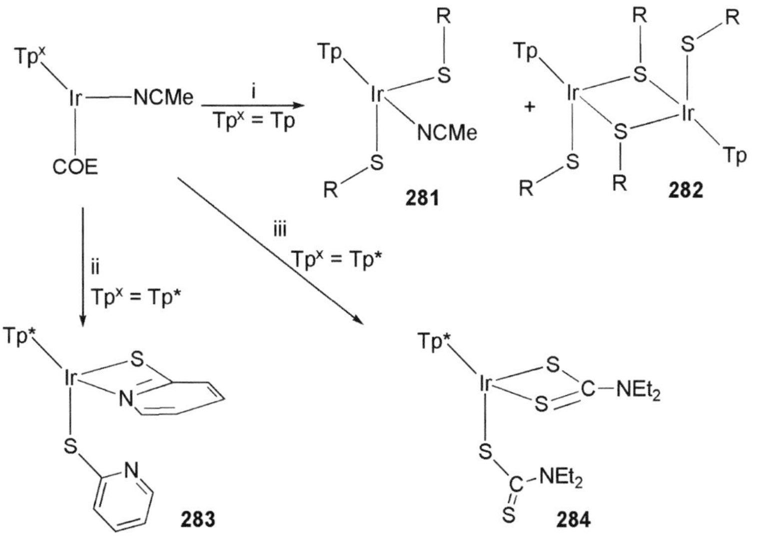

SCHEME 18. Reagents and conditions: (i) S_2R_2 ($R = C_6H_4Me$-4), THF, r.t.; (ii) Py_2S_2, THF, r.t.; (iii) $(Et_2NCS_2)_2$, THF, r.t.

cyclooctene and NCMe ligands are displaced to afford Tp*Rh(κ^2-*S,N*-SPy)(κ^1-*S*-SPy) (**283**), identified from both spectroscopic and crystallographic data. Similarly, **164** reacts with tetraethylthiuram to afford the bis(dithiocarbamato) complex Tp*Rh(κ^2-S_2CNEt_2)(κ^1-S_2NEt_2) (**284**, Scheme 18). Curiously, the comparatively simple reaction between **164** and benzyldisulfide proceeds only at elevated temperature, and affords only intractable materials.

Relatively little attention has been paid to the hydrogenation of alkene complexes, though several exceptions are notable. As previously described, high-pressure (500 atm) hydrogenation of Tp^xIr(1,5-COD) (Tp^x = Tp* (**179**), Tp^{Me_3} (**180**)) has been employed to prepare the respective cyclooctene complexes $Tp^xIrH_2(\eta^2$-coe) (Tp^x = Tp* (**207**), Tp^{Me_3} (**208**)), while at lower pressure (200 atm) Tp*$IrH_2(\eta^2$-1,5-COD) (**209**) was similarly obtained.[81] Interestingly, Tp*IrH(CH$=$CH$_2$)(η^2-C_2H_4) (**202**) was also hydrogenated at 500 atm pressure to afford Tp*$IrH_2(\eta^2$-C_2H_4) (**206**). This is notable in that Tp*Ir(η^2-$C_2H_4)_2$ (**194**), the photochemical precursor to **202** (*vide infra*), was subsequently hydrogenated under ambient conditions (1–2 atm, 20 °C) to give **206** in admixture with small amounts of Tp*IrH(CH_2CH_3)(η^2-C_2H_4) (**285**).[92] The latter product is kinetic in origin, and is favoured at low temperature; i.e. the extrusion of C_2H_4 is more favourable at higher temperature.

The analogous hydrogenation of TpIr(η^2-$C_2H_4)_2$ (**193**) is similarly facile, though affords only the ethyl complex TpIrH(CH_2CH_3)(η^2-C_2H_4) (**286**). On the basis of these data (and those for a series of $Tp^xIr(\eta^2$-C_2H_4)(PR_3) complexes (Section III-H) that yield exclusively phosphine-dihydrides) it was suggested that hydrogenation proceeds associatively, i.e. oxidative trapping of a coordinately unsaturated 16-electron complex by H_2. Where the ancillary ligand is C_2H_4, its insertion into an Ir–H bond is kinetically competitive, thus accounting for the product mixture so obtained. These suggestions were supported by the demonstration that in isolation **285** reversibly eliminates ethylene at 60 °C to yield **206**. Under an atmosphere of ethylene, however, **285** rapidly undergoes thermally induced σ-bond metathesis to afford the vinyl complex **202**. Moreover, both **285** and **206** undergo migratory insertion in the presence of PMe_3, yielding $Tp^*Ir(Et)_2(PMe_3)$ (**287**) and $Tp^*IrH(Et)(PMe_3)$ (**288**) respectively.

Reactions Involving C–H Bond Activation

The capacity of the 'Tp^xM' (M = Rh, Ir) fragment to effect C–H bond activation is a generic phenomenon. Whilst this might be entirely expected by analogy to Cp^xM congeners, it might be argued that the 'Tp^xM' scaffolds have dramatically surpassed Cp^xM systems, both in terms of the versatility and the understanding such studies have brought to the field. The study of these processes in compounds of the type $[Tp^xM(\eta^2\text{-alkene})(L)]$ have, in particular, fundamentally contributed to our current understanding. Prominent among this work was the seminal report of the thermally induced isomerisation of $Tp^{CF_3,Me}Ir(CO)(\eta^2\text{-}C_2H_4)$ (**229**) to the hydridovinyl $Tp^{CF_3,Me}IrH(CH{=}CH_2)(CO)$ (**289**),[95] which proceeds cleanly at 60 °C. This proved surprising, given that the generally perceived order of stability is more typically $L_nM(\eta^2\text{-}C_2H_4) > L_nMH(CH{=}CH_2)$;[107–110] e.g. $Cp^*Ir(PMe_3)(\eta^2\text{-}C_2H_4)$,[111,112] was found to be too stable to be implicated as an intermediate in the formation of $Cp^*IrH(CH{=}CH_2)(PMe_3)$. The anomalous nature of the **229** → **389** isomerisation is more apparent when one considers that for the related rhodium systems $Tp^*Rh(\eta^2\text{-}C_2H_4)(CO)$ (**148**) and $Tp^*RhH(CH{=}CH_2)(CO)$ (**290**) the order of stability is reversed. Indeed, **290** (generated photolytically from the dicarbonyl) isomerises completely to **148** at 25 °C in the dark ($T_{1/2}$ 3.2 min).[95]

Several more hydridovinyl isomers of 'Tp^xIr(alkene)' complexes have been generated thermally, photolytically or chemically, in seeking to understand the processes involved. Thus, $TpIrH(CH{=}CH_2)(\eta^2\text{-}C_2H_4)$ (**201**) was generated by UV irradiation of $TpIr(\eta^2\text{-}C_2H_4)_2$ (**193**),[84] and remained the sole product even in the presence of other alkenes, thus proving direct C–H activation of a coordinated ethylene molecule. The complex $Tp^*IrH(CH{=}CH_2)(\eta^2\text{-}C_2H_4)$ (**202**), though not initially isolated, was observed spectroscopically as an intermediate in the thermally induced (60 °C) conversion of $Tp^*Ir(\eta^2C_2H_4)_2$ (**194**) into $Tp^*IrH(\eta^3\text{-}C_3H_4Me)$ (**291**).[86] It was subsequently, and independently, established that in the presence of acetonitrile **202** undergoes insertion of the remaining π-alkene into the M–H linkage, and coordinates NCMe to afford $Tp^*Ir(Et)(CH{=}CH_2)(NCMe)$ (**292**).[113] Rhodium analogues of **292**, $Tp^*Rh(Et)(CH{=}CH_2)(L)$ (L = NCMe (**156**), py (**293**)), have also been obtained, directly from the interaction of $Tp^*Rh(\eta^2\text{-}C_2H_4)_2$ (**118**) with these 'hard' bases (cf. the reaction of **118** with softer bases, which effects substitution (*vide supra*)).[70] On the basis of these observations it was concluded that, though undetectable, $Tp^*RhH(CH{=}CH_2)(\eta^2\text{-}C_2H_4)$ (**294**) is kinetically accessible even at ambient temperature, under which conditions the interaction of hard donors with the Rh(I) species is kinetically disfavoured, their interaction with **294** thus predominating. Moreover, this step was attributed to the kinetic accessibility (trace amounts) of the coordinately unsaturated complex Tp*Rh(Et)($CH=CH_2$), which has, however, not been directly observed.

During the earlier studies two suggestions were made regarding the influence of the Tp^x ligand in such processes, viz.: (i) the presence of methyl substituents (e.g. Tp*, $Tp^{CF_3,Me}$) is essential to the thermal stability of both π-alkene and hydridovinyl isomers,[86] and (ii) hapticity change (κ^2-$Tp^x \rightarrow \kappa^3$-$Tp^x$) stabilises the d^6-octahedral M(III) product of C–H activation and thus serves as a driving-force.[114] This latter notion has been disputed,[95] and it is noted that many Tp^xM (π-alkene)(L) complexes preferentially engage in κ^3-Tp^x coordination in the ground state.

Greater insight into both influences was provided by *ab initio* studies of intramolecular C–H activation in a series of model compounds of varying complexity, including the 'real' complexes $Tp^xM(C_2H_4)_2$ (M = Rh, Ir; $Tp^x = Tp^{Me}$, Tp^{CF_3}).[115] It was thus concluded that 3-substituents on the pyrazole rings exert a steric influence on the stability of the π-complex (i.e. greater steric bulk in either κ^3- or κ^2-isomers of $Tp^xM(C_2H_4)_2$ weakens the M–C_{alkene} bond). Moreover, it was found that for rhodium κ^2-chelation was favoured over κ^3, and that (κ^2-Tp^x)Rh(η^2-C_2H_4)$_2$ are 20–28 kcal mol^{-1} more stable than the respective hydridovinyl isomers. In contrast, for iridium all κ^2 isomers are unstable with respect to κ^3, while the relative stabilities of (κ^3-Tp^x)Ir(η^2-C_2H_4)$_2$ and (κ^3-Tp^x)IrH($CH=CH_2$)(η^2-C_2H_4) depend upon the sterics of the Tp ligand; i.e. greater bulk favours the hydridovinyl. For the conversion **229** → **289** ($Tp^{CF_3,Me}$) these effects render C–H activation exothermic. That these isomerisations are more endothermic for rhodium than iridium was explained in terms of the preference for late second-row elements to adopt a d^{n+1} ground state with high-energy d^ns^1 excited state, cf. d^ns^1 ground state and low-lying d^ns^1 excited state for the third row; i.e. oxidative addition is easier for third row elements.

Hydridovinyl complexes of the type discussed have been widely implicated as the active species in a variety of C–H activation reactions observed for simple bis(π-alkene) complexes. However, since these reactions typically commence from the bis(π-alkene) complexes, a summary of these processes is warranted here, though many are more appropriately discussed elsewhere.

Thermally induced (60 °C) activation of benzene proceeds with Tp*Rh(η^2-C_2H_4)(CO) (**158**) to afford Tp*RhH(C_6H_5)(CO) (**295**),[114] while the related Tp*Rh(Et)(C_6H_5)(NCMe) (**296**) is similarly obtained from the ethyl–vinyl complex **156**.[116] In contrast, **194** thermally activates two molecules of benzene to yield the unstable, 16-electron bis-phenyl complex 'Tp*Ir(C_6H_5)$_2$' (**297**), which is isolable as the dinitrogen adduct Tp*Ir(C_6H_5)$_2$(N_2) (**211**).[90,117] It is, however, noted that Tp*IrH(C_6H_5)(η^2-C_2H_4), an analogue of **295**, does arise during ambient-temperature photolysis of **194** in benzene, though the bulk product is the hydridovinyl complex **202**.[89]

Both **211** and **297** (see also Section II-C.2) have been found to effect double C–H activation of cyclic ethers to afford Fischer-type carbene complexes (Section II-D.2), leading in some instances to C–C coupling reactions.[118] Comparable double C–H activations have also been effected using **194**,[119] and the η^4-isoprene complex Tp^{Ph}Ir(η^4-2-MeC_4H_5) (**238**) with several substrates (MeOPh, Me_2NPh,

THF, EtOPh, Et_2O).[98] In the latter case an additional, ligand-centred, C–H activation is observed, viz.: orthometallation of one 3-phenyl substituent, as was previously observed during the attempted synthesis of $Tp^{Ph}Ir(C_2H_4)_2$, which instead afforded mono- (**196**) and di- (**197**) orthometallated complexes (Scheme 11).[88]

Photochemical C–H activation has also been effected with a variety of Tp^xMLL' complexes, often with surprising results. For instance, activation of diethyl ether by $Tp^{Br_3}IrH(CH{=}CH_2)(\eta^2\text{-}C_2H_4)$ (**203**) (UV irradiation), affords not a Fischer carbene, but rather a mixture of $Tp^{Br_3}IrH(CH_2CH_2OEt)(\eta^2\text{-}C_2H_4)$ (**214**) and $Tp^{Br_3}IrH_2(\eta^2\text{-}H_2C{=}CH_2OEt)$ (**215**), products analogous to those obtained from the thermal activation of *n*-pentane.[87]

More extensive studies have focussed on thermal and photochemical activation of thiophenes. It has thus been established that at 60 °C, **194** reacts, *via* its hydridovinyl isomer **202**, with excess thiophene or methylthiophene to afford complexes **298**–**300** respectively (Scheme 19). Under ambient conditions the thiophene ligand is in each case labile, leading to slow decomposition, though this is impeded at low temperature, or by exchanging it with (stronger) extraneous donors.[120] Complexes **298**–**300** have been similarly generated from $Tp^*Ir(\eta^4\text{-}2,3\text{-}Me_2C_4H_4)$ (**234**). In contrast, the iridium(III) complex $Tp^*IrH_2(\eta^2\text{-}C_2H_4)$ (**206**) activates a single molecule of thiophene to afford **301**, in which the thiophene ligand is again labile.[121]

More intriguing are the analogous reactions of the bulkier 2,5-dimethylthiophene, which undergoes a complicated series of C–C bond-forming processes to afford a range of chelating 4-vinyl-2-thienyl complexes (**253**–**254**, *vide supra*, Scheme 16), which react further to afford novel cage structures (e.g. **256**, *vide infra*).[102]

When the thiophene activations were performed with **234**, a transient intermediate of type **F** (Chart 7) was observed,[121] presumed to arise through interaction of the thiophene donor with **234**. This was confirmed by the synthesis and isolation of a range of such materials (**302**–**316**, Chart 7), generated thermally

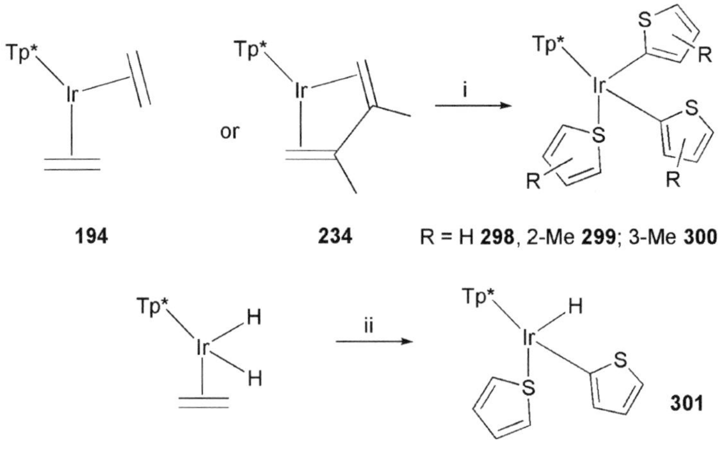

SCHEME 19. Conditions: (i) 60 °C, SC_4H_3R (R = H, 2-Me, 3-Me); (ii) 60 °C, SC_4H_4.

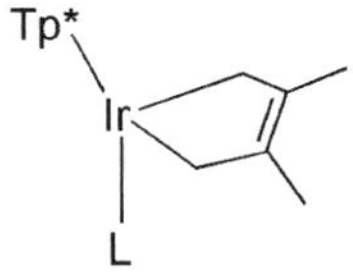

L = PMe_3 **302**; CO **303**; SC_4H_8 **304** (F); C_2H_4 **305**
NCMe **306**; py **307**; $C_6H_5C(H)$=NMe **308**

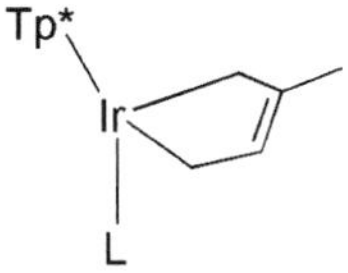

L = PMe_3 **309**; CO **310**; SC_4H_8 **311** (F); NCMe **312**;
py **313**

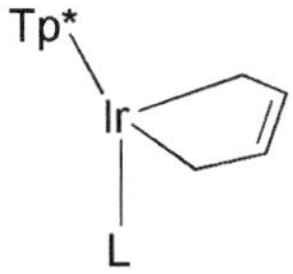

L = PMe_3 **314**; CO **315**; py **316**

CHART 7. 1-4-σ^2-but-2-ene complexes.

(20–80 °C) from Tp*Ir(η^4-2-R,3-R′C_4H_4) (R = R′ = H (**232**), R = H, R′ = Me (**233**), R = R′ = Me (**234**)) and the respective donors.[122] These materials were identified spectroscopically and, in the case of **311**, crystallographically characterised metallocycle. The mechanism by which these materials form (i.e. associative or dissociative) was concluded to be donor dependant.

In contrast to the exclusive selectivity for thiophene C–H activation exhibited by Tp^RIr systems, the ambient temperature photolysis of Tp*Rh(η^2-C_2H_4)(PMe_3) (**145**) and thiophene results in competitive metal insertion into the C–S bond, affording a 3:1 mixture of **317** and **318**.[71] This concurs with other observations of preferential C–S over C–H cleavage at rhodium,[123,124] exemplified by the irreversible conversion of several rhodium-2-thienyls to the respective metallacycles.[125–128] However, the 'Tp*Rh(PMe_3)' scaffold seems to reverse this thermodynamic preference, the 2-thienyl **318** being favoured in the thermal (90 °C) reaction[71] and upon heating of the photolytic products. This effect is enhanced in the PEt_3 analogue **147**, which affords a 1:1 mixture (**319**:**320**) photolytically, with 95% of the 2-thienyl (**337**) obtained thermally.

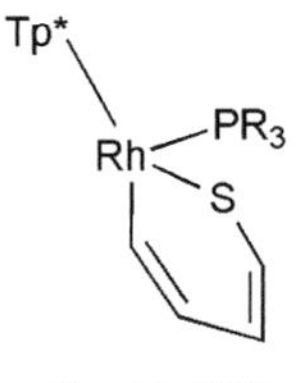

R = Me **317**
Et **319**

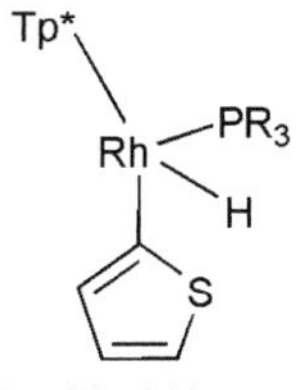

R = Me **318**
Et **320**

A significant series of studies have focussed on the iridium(III) complex $Tp^*IrH_2(COE)$ (**207**), which in the presence of $P(OMe)_3$ photolytically

($\lambda = 335$ nm) activates benzene to afford $Tp^*IrH(C_6H_5)\{P(OMe)_3\}$ (**321**).[91] Mechanistic studies led the authors to propose initial photolytic dissociation of cyclooctene, with subsequent oxidative addition of benzene to afford the 7-coordinate Ir(V) intermediate $Tp^*IrH_3(C_6H_5)$. Photolytically induced reductive elimination of H_2 then generates a coordinately unsaturated species that is trapped by $P(OMe)_3$ to afford **321**.

Interestingly, when photolysis is conducted in methanol a 1:1 mixture of Tp^*IrH_4 (**322**) and $Tp^*IrH_2(CO)$ (**323**) is obtained.[129] Both of these are believed to result from the degradation of methanol by the initial photoproduct, 'Tp^*IrH_2', through a progressive oxidative-addition, reductive-elimination sequence, the final step of which is either loss of CO (**322**) or H_2 (**323**). The same study reports largely comparable photochemistry for $Tp^*Ir(\eta^4$-1,5-COD) (**179**). In contrast, photolysis (λ 400 nm) of $Tp^*Rh(\eta^4$-1,5-COD) (**90**) in benzene results in isomerisation of the cyclooctadiene ligand, *via* a formal 1,3-H shift, to afford $Tp^*Rh(\eta^4$-1,3-COD) (**171**).[77,78,130] The photochemistry of **171** is generally comparable to that of **207**, in that its irradiation in benzene in the presence of $P(OMe)_3$ affords $Tp^*RhH(C_6H_5)\{P(OMe)_3\}$ (**324**), while in methanol alone the major product is $Tp^*RhH_2(CO)$ (**325**).

The activation of benzene by **171** has also been reported in the presence of *tert*-butylacrylate, in this case affording $Tp^*Rh(C_6H_5)(CH_2CH_2CO_2{}^tBu)$ (**326**), believed to arise from alkene insertion into the intermediate rhodium hydride $Tp^*RhH(C_6H_5)(\eta^2\text{-}CH_2{=}CHCO_2{}^tBu)$. It is also reported that irradiation of **173** in ether/$P(OMe)_3$ yields $Tp^*RhH(OEt)\{P(OMe)_3\}$ (**327**), presumed to result from $P(OMe)_3$ intercepting the initial photoproduct 'Tp*Rh', with subsequent oxidative addition of an ethereal C–O bond and loss of ethylene.[78]

Intramolecular photolytic activation of the 1,3-diene complexes **232**–**234** has also been explored (Scheme 20), resulting in a series of η^3-1,2,3-butadienyl, complexes (**328**–**331**).[96] The study revealed that activation of internal alkenic C–H bonds is exclusive over terminal linkages, but that the activation of a 2-methyl substituent is preferred. Thus, while prolonged photolysis of **232** yielded only small amounts of **328**, the activation of **233** and **234** was more facile, yielding predominantly species with a vinyl substituent at C2. Besides the main study, the thermolysis (120 °C) of **232** in benzene under dinitrogen was found to activate benzene and afford the N_2-bridged complex **332** (Scheme 20), as formulated on the basis of spectroscopic (NMR, IR, Raman) data.

Two final, serendipitous, C–H activation reactions are also noteworthy. The treatment of $TpIr(\eta^2\text{-}C_2H_4)(PPh_3)$ (**220**) (in dichloromethane) with excess (six equivalents) PPh_3 effects displacement of ethylene and affords the novel bis-phosphine complex **333** (Scheme 21), in which one pyrazole arm has been C–H activated at the 5-position.[93,131] This process, which is reversible in the presence of excess ethylene, was the first documented example of intramolecular C–H activation of a 'Tp^x' ligand.

In a related reaction, treating $TpIr(\eta^2\text{-}C_2H_4)_2$ (**193**) with dppm results in displacement of both ethylene ligands *and* proton transfer from the bridging

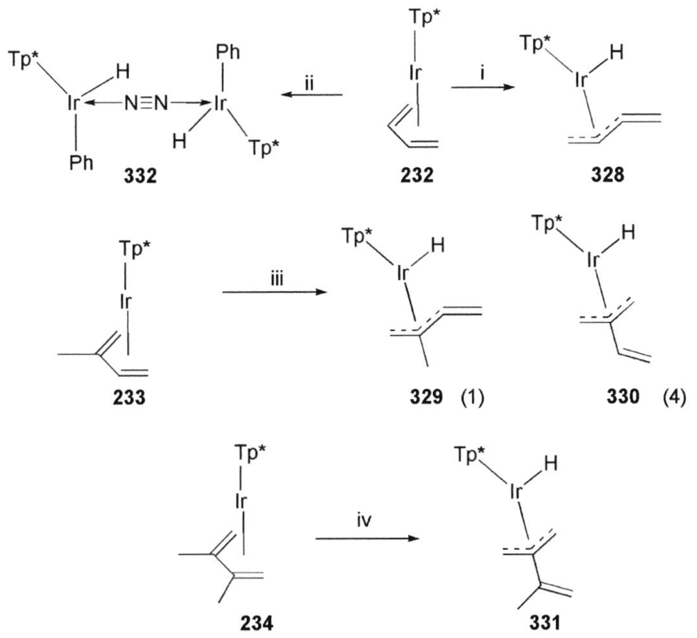

SCHEME 20. Conditions (i) *hv*, 50 h; (ii) 120 °C, benzene, 1 h; (iii) *hv*, 20 h; (iv) *hv*, 10 h.

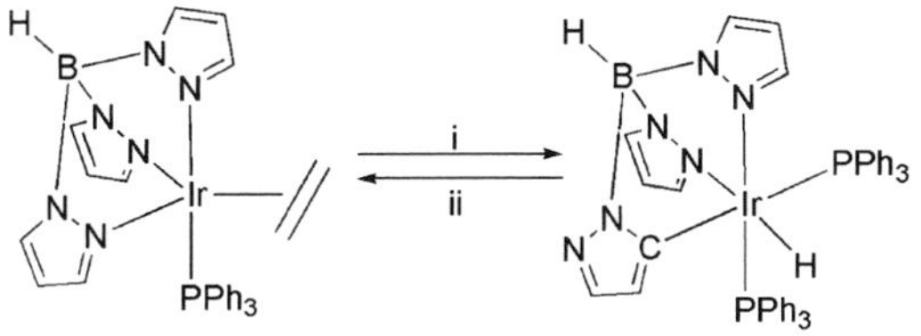

SCHEME 21. Reagents: (i) xs PPh_3; (ii) C_2H_4.

methylene group to the, presumably strongly basic, iridium centre. This affords the hydrido-bis(phosphino)methanido complex **334**,[105] the identity of which was established from spectroscopic, crystallographic and reactivity data.

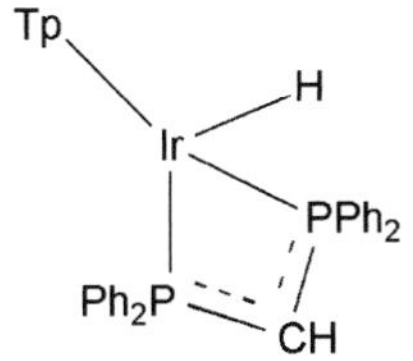

334

Reactions Involving C–C Coupling Processes

Relatively few reactions involving direct coupling of π-alkenes in the coordination sphere of a group 9 pyrazolylborate complex have been reported, possibly a result of such complexes often being merely transient intermediates in the catalytic scheme, or precursors to the active species. Nonetheless, several examples exist, such as the previously described (Section III-B.2) thermal (60 °C), solid-state decomposition of Tp*Rh($\eta^2C_2H_4)_2$ (**118**) leading to Tp*RhH(η^3-C_3H_4Me) (**170**) (Scheme 9).[58] This process is believed to occur *via* the hydrido-vinyl isomer Tp*RhH(CH$=$CH$_2$)(η^2-C_2H_4) (**294**), as for the analogous iridium case (**194**→**291**),[86] since both reactions afford similar isomeric mixtures.[89] It is also noted that the Tp^{Br_3} analogue **203** affords Tp^{Br_3}IrH(σ,η^2-$CH_2CH_2CH{=}CH_2$) (**216**) in addition to an allyl species analogous to **170**.

Complex **118** will also incorporate ethylene (1 atm) to afford either Tp*Rh (η^4-C_4H_6) (**165**) (25 °C), or at low temperature a mixture of the hexadiene complex Tp*Rh(η^4-1-EtC_4H_5) (**167**) and the Rh(III) allyl Tp*Rh(η^3-C_3H_3Me)(Et) (**169**).[58]

Though comparable reactions of iridium have *not* been reported, TpIr(η^2-$C_2H_4)_2$ (**193**) does incorporate DMAD in the presence of MeCN to afford the iridacyclopent-2-ene complex **335** (Scheme 22).[132] This reaction was proposed to proceed by displacement of one ethylene ligand by DMAD and subsequent cyclodimerisation of intermediate **G**, the existence of which was supported by spectroscopic data. The THF analogues of **335** (**336** and **336-d_8**) were also isolated and found to afford **335** upon heating in acetonitrile. In contrast, heating **336** with DMAD led to incorporation of a second equivalent of the alkyne (presumably *via* cyclo-addition/cyclo-reversion), with liberation of C_2H_4, which remains in the coordination sphere of **337**.

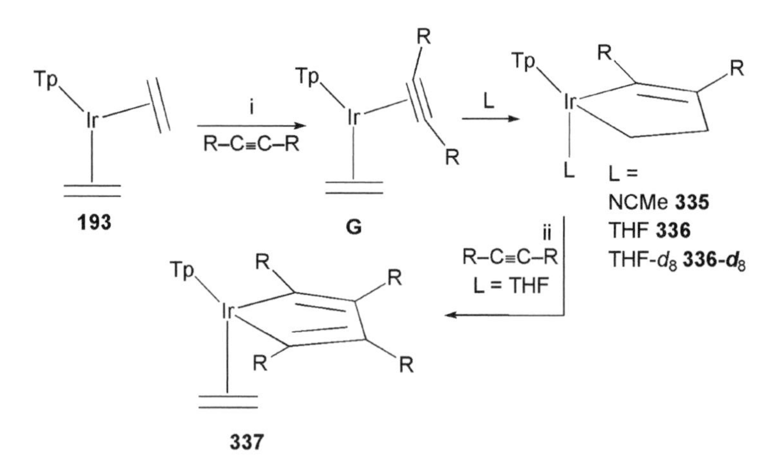

SCHEME 22. R = CO_2Me. Conditions: (i) MeCN, 25 °C; (ii) –L, 60 °C.

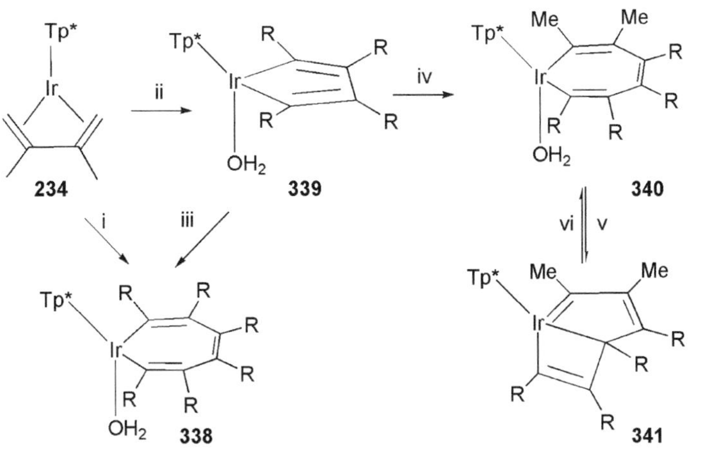

SCHEME 23. R = CO_2Me. Reagents and conditions: (i) 3 DMAD, 60 °C, C_6H_{12}; (ii) 2 DMAD, 60 °C, C_6H_{12}, xs H_2O; (iii) DMAD, 90 °C; (iv) MeC≡CMe, 90 °C; (v) $-H_2O$; (vi) H_2O.

Attempts to mimic this chemistry with the butadiene complex **234** (featuring the bulkier Tp* ligand) instead afforded the iridacycloheptatriene **338** (Scheme 23),[133] derived from three molecules of the alkyne and one of adventitious water. With excess water, the intermediate iridacyclopentadiene complex **339**, arguably an analogue of **337**, was instead obtained (quantitatively) then converted to **338** by reaction with excess DMAD; a process clearly retarded by excess water.[134] Alternatively, treatment with but-2-yne in place of DMAD affords **340** as an equilibrium mixture with **341**, the latter being favoured in the absence of water. The water molecule in each of **338**–**340** can be displaced by extraneous donors.

Curiously, attempts to synthesise **338** from the ethylene complex **194** instead resulted in coupling of two molecules of DMAD and one of ethylene. This ultimately afforded the 'tethered' allyl complex **342**, proposed to arise *via* isomerisation (stereospecific hydrogen shift) of the intermediate iridacylcoheptadiene **H** (Scheme 24).[133]

2. Allyl Complexes

Surprisingly few such compounds are known and of these only Tp*Rh(η^3-C_3H_5)(η^1-C_3H_5) (**343**)[41] and Tp*Rh(η^3-C_3H_5)Br (**344**)[75] were prepared intentionally. Complex **343** was obtained *via* the reaction of $Rh_2(\mu\text{-Cl})_2(\eta^3\text{-}C_3H_5)_4$ with NaTp*,[41] while **344** is generated thermally (60 °C) from Tp*Rh(η^1-C_3H_5)Br(NCMe) (**345**). The latter is prepared by oxidative addition of allyl bromide to Tp*Rh(η^2-COE)(NCMe) (**164**),[75] which proceeds similarly (albeit more slowly) with Tp*Rh(η^2-C_2H_4)(NCMe) (**155**), but fails with Tp*Rh(1,5-COD) (**90**), even at elevated temperature. It is reported that crotyl-, cinnamyl- and 2-methylallyl bromides will also add to **164**, though more slowly, such that products have not been isolated. Prolonged exposure (2 days) of **344** to excess acetonitrile results in

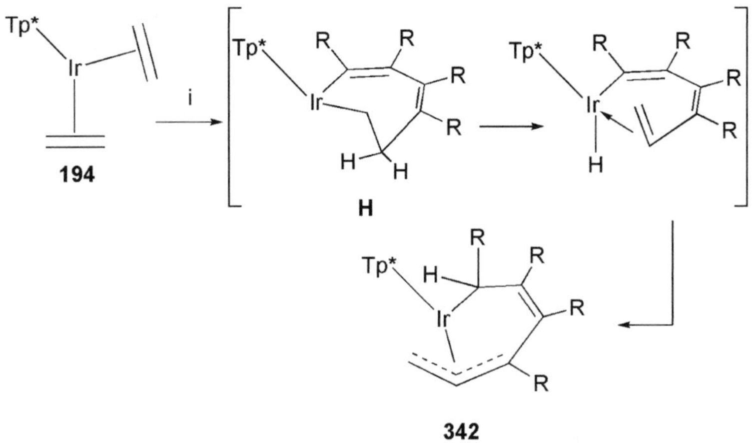

SCHEME 24. R = CO_2Me. Reagents and conditions: (i) 2 DMAD, 60 °C.

regeneration of the η^1-allyl complex **345**, while similar treatment with PMe_2Ph affords Tp*Rh(η^1-C_3H_5)Br(PMe_2Ph) (**346**).

The triflate salt [Tp*Rh(η^3-C_3H_5)]OTf (**347 · OTf**) was obtained by AgOTf metathesis of **344**, like which it was also found to be unreactive toward weak nucleophiles (e.g. dimethyl malonate). However, **344** does react with MeMgBr to afford Tp*Rh(η^3-C_3H_5)(Me) (**348**) as a mixture of *endo* and *exo* isomers (7:1) that converts completely to the *endo* isomer on heating in benzene. Similarly, the reaction of **344** with Li[$BHEt_3$] affords *endo*-Tp*RhH(η^3-C_3H_5) (**349**), while with Li[$BDEt_3$] the deuterium is incorporated into the allyl ligand (**350**, Scheme 25). The hydride of **350** easily exchanges with allyl bromide to afford **351**, which then reacts with Li[$BHEt_3$] to give **352**.

The remaining examples of allyl complexes were encountered serendipitously during the study of simple alkene complexes. For example, the solid-state decomposition of Tp*Rh(C_2H_4)$_2$ (**118**), which affords Tp*RhH(η^3-*syn*-C_3H_4Me) (**170**) as a kinetic mixture (1:1) of *endo*- and *exo* isomers (Section III-B.2), that convert to a thermodynamic ratio of 6:1 after 24 h in benzene (20 °C).[58] This particular decomposition pathway does not prevail in solution, the butadiene complex Tp*Rh(η^4-C_4H_6) (**165**) forming instead, though this latter complex has also been obtained upon photolysis of **170** (Scheme 26). Alternatively, under an ethylene atmosphere incorporation of an extra molecule of C_2H_4 affords hexa-1,3-diene (**167**) and ethyl-η^1:η^2-butadiene **169** complexes (Section III-B.1).

Similar decomposition was observed for Tp*Ir(C_2H_4)$_2$ (**194**) in cyclohexane solution, which thermally (60 °C)[86] degrades to a kinetic mixture of *exo–anti*, *endo–syn* and *exo–syn* isomers of Tp*IrH(η^3-C_3H_4Me) (**291**) (20:2:1).[89] Prolonged heating (3 days, 120 °C) yields a thermodynamic mixture of *endo*- and *exo–syn* isomers (2:1) that interconvert above 80 °C. This mixture undergoes hydride exchange with $CDCl_3$ at 60 °C to afford a mixture of *endo–syn* and *exo–syn*-**353** (8:2, Scheme 27).

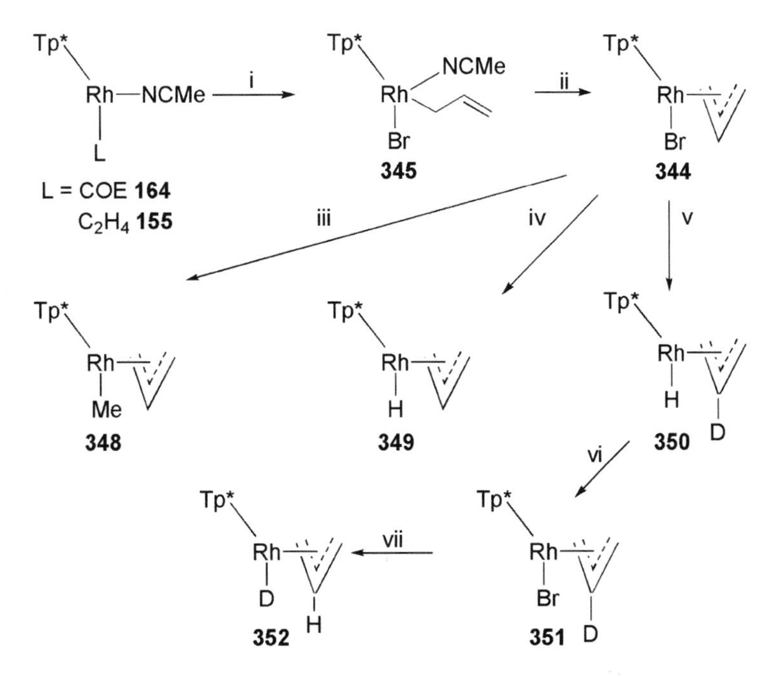

SCHEME 25. Conditions: (i) $CH_2{=}CHCH_2Br$, benzene, 1 h 25 °C; (ii) 60 °C; (iii) MeMgBr, THF, 25 °C; (iv) Li[$BHEt_3$], THF, 25 °C; (v) Li[$BDEt_3$], THF, 25 °C; (vi) $CH_2{=}CHCH_2Br$; (vii) Li[$BHEt_3$].

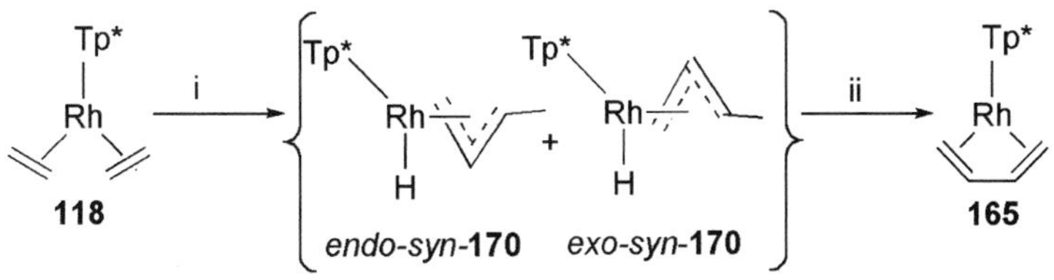

Tp*
Rh
118
i
Tp* Rh H + Tp* Rh H
endo-syn-**170** *exo-syn*-**170**
ii
Tp*
Rh
165

SCHEME 26. Conditions: (i) solid-state, 60 °C, 16 h; (ii) *hν*, C_8H_{12}, 25 °C, 7 h.

Similar mixtures of hydrido-allyl compounds are obtained from complexes of 1-propene and 1-butene (Scheme 28), the difference being that discrete Tp*Ir (alkene)$_2$ complexes have not been isolated, only their hydridovinyl isomers. It is interesting to note that the analogous reaction of $Tp^{Br_3}IrH(CH{=}CH_2)(\eta^2\text{-}C_2H_4)$ (**203**) affords (kinetically) the 'tethered' butene complex $Tp^{Br_3}IrH(\sigma\text{-}\pi\text{-}CH_2CH_2CH{=}CH_2)$ (**216**, 80%, *vide supra*), and a small quantity of the hydrido-crotyl complex $Tp^{Br_3}IrH(\eta^3\text{-}C_3H_4Me)$ (**217**),[87] which becomes the sole product upon prolonged heating (100 °C, 20 h). A series of related hydrido-allyl complexes (**328**–**331**) are obtained from ambient temperature photolysis of the 1,3-butadiene complexes **232**–**234** (Section III-B.2).

The remaining examples of allyl complexes are somewhat anomalous. For instance, $TpIrH(\eta^3\text{-}C_8H_{13})$ (**200**), the first such material to be reported,[84] was obtained serendipitously from the reaction between $Ir_2(\mu\text{-}Cl)_2(\eta^2\text{-}COE)_4$ and KTp.

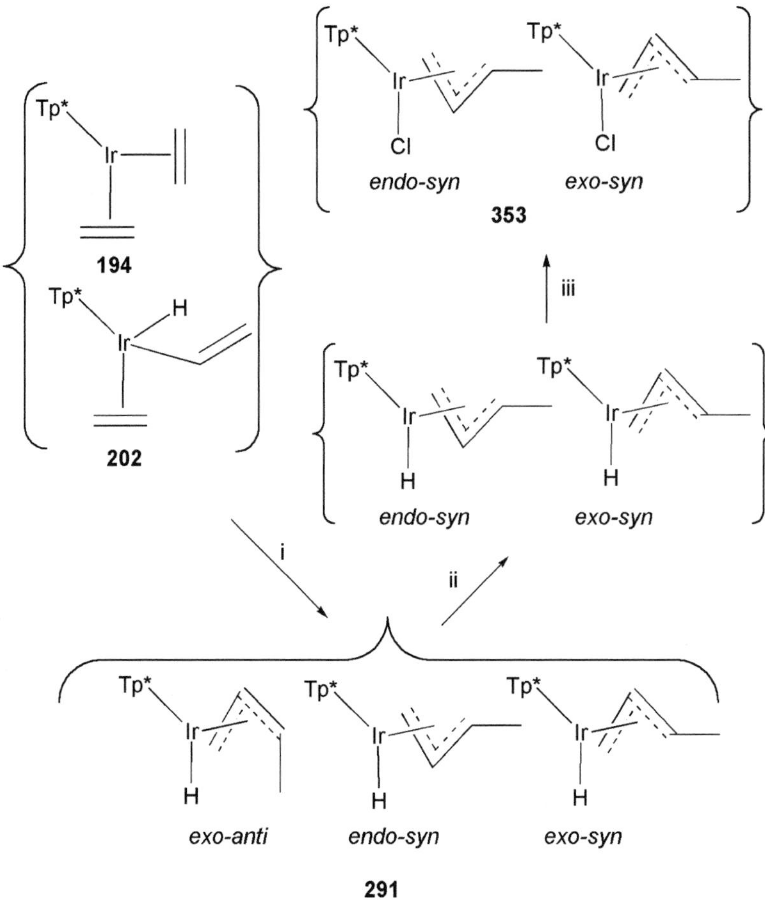

SCHEME 27. Conditions: (i) 60 °C, C_6H_{12}; (ii) 120 °C; (iii) $CDCl_3$, 60 °C.

This remains the only report of **200** being obtained in this manner, though it was subsequently established that $TpIrH(\eta^1\text{-}C_8H_{13})(\eta^2\text{-}C_8H_{14})$ (**198**), which was independently reported to result from same reaction,[85] thermally rearranges to **200** (100 °C, C_6H_{12}).[89] More remarkable is the formation of **342** from $Tp^*Ir(\eta^2\text{-}C_2H_4)_2$ (**194**) and two equivalents of DMAD (Scheme 24, Section III-B.1), believed to arise from an intermediate iridacylcohepta-2,4-diene undergoing a stereospecific hydride shift.

3. Cyclopentadienyl Complexes

Only a handful of 'mixed-sandwich' complexes exist that comprise both Cp^x and Tp^x ligands; indeed, the synthesis of such materials proved to be a significant challenge. The first examples were prepared from $Cp^*{}_2Rh_2(\mu\text{-}Cl)_2Cl_2$ and $[RB(pz)_3]^-$ (R = H, pz), with *in situ* anion exchange affording $[Cp^*TpRh]PF_6$

(**356** · $\mathbf{PF_6}$) and $[Cp^*(pzTp)Rh]PF_6$ (**357** · $\mathbf{PF_6}$).[18] The analogous tris(pyrazolyl)-methane (Tpm) salt $[Cp^*(Tpm)Rh)](PF_6)_2$ (**358** · $\mathbf{PF_6}$) was also reported, but Tp* analogues proved elusive, a fact attributed to the steric bulk of this ligand favouring (highly unstable) coordinately unsaturated products. Indeed, the crystallographic characterisation of **356** · $\mathbf{PF_6}$ (Fig. 6),[135] revealed a structure in which four of the Cp* methyl groups are significantly displaced from the mean ring plane, as a result of close contacts with the pyrazole carbon atoms. This also results in a twisting of the pyrazole rings relative to the metallo [2.2.2] cage.

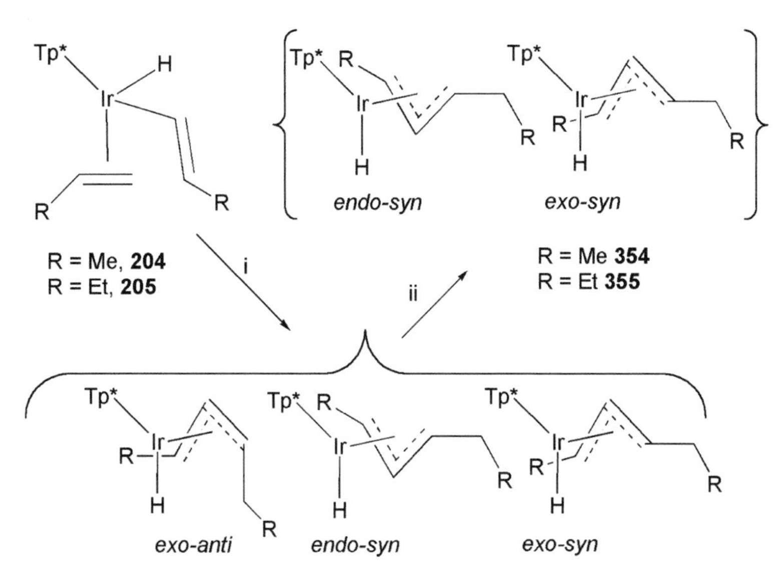

SCHEME 28. Conditions: (i) 60 °C, C_6H_{12}; (ii) 120 °C.

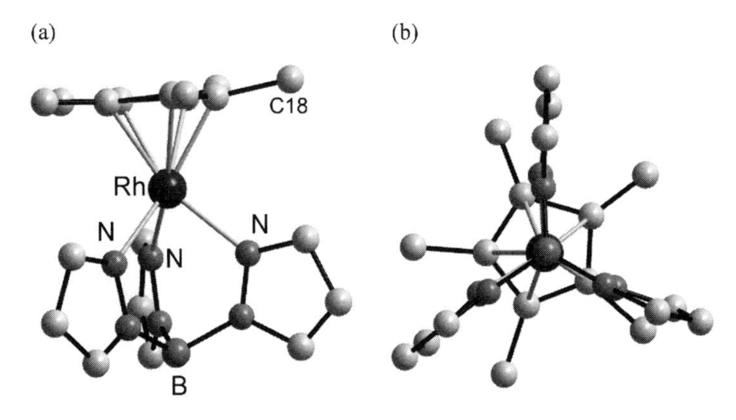

FIG. 6. Molecular structure of the complex $\mathbf{356}^+$ (a) projected to illustrate deformation of methyl C18 from ring plane and (b) normal to the cyclopentadienyl plane.

Comparable syntheses were subsequently employed for the complexes [Cp*Tp^{x}MCl] (M = Rh, Tp^x = Tp (**359**), Tp^{Me_2} (**360**); M = Ir, Tp^x = Tp (**361**), Tp^{Me_2} (**362**), Scheme 29). These were found to be air/moisture stable in the solid state, but sensitive in solution, though they exhibit indefinite stability under inert atmospheres.[136] In each case the structures were formulated as η^5-Cp*, κ^2-Tp^x on the basis of NMR and infrared spectroscopic data: viz. one doublet ($J_{RhC} \approx 8$ Hz) resonance in the $^{13}C\{^1H\}$-NMR spectra for the Cp* moiety and infrared B–H absorptions consistent with κ^2-Tp^x, as defined by Akita[50] (Section III-B.1). These formulations were confirmed crystallographically for **360**–**362**. In the case of **360**, the anomalous observation of two B–H absorptions was established, on the basis of X-ray crystallography, to result from the presence of two conformers, which are distinguished by the orientation of the pendant pyrazole group.

For each of **359**–**362** VT-NMR spectroscopy revealed slow-exchange of coordinated and pendant pyrazolyl groups, the spectra remaining unchanged (i.e. 2:1 pyrazolyl environments) up to 328 K. For conformational exchange of the uncoordinated functions (i.e. pz and H), the slow-exchange regime was located around 218 K. It was also noted that over extended periods (48 h) at 293 K, presumably in air, partial degradation of **359**–**362** occurs, a feature that has been attributed to hydrolytic B–N cleavage of the uncoordinated pyrazolyl groups.[136,80]

For complexes **359** and **361**, anion metathesis with $AgNO_3$ afforded products characterised as the salts **356**·**NO$_3$** and **363**·**NO$_3$** respectively, on the basis of spectroscopic data (^{1}H-NMR, ν_{BH} 2490 cm^{-1}) that imply a tripodal (κ^3) Tp ligand, and the presence of free NO_3^- (infrared, conductivity). In contrast, complexes **360** and **362** are degraded upon treatment with $AgNO_3$, consistent with the earlier assertion that the steric profile of Tp* prohibits κ^3-chelation in the presence of η^5-Cp*, resulting in low stability. Bis(pyrazolyl)borate analogues of **359** and **361** were also sought, and Cp*BpRhCl (**364**) duly obtained and fully characterised, including crystallographically. However, the reaction of $Cp*_2Ir_2(\mu\text{-}Cl)_2Cl_2$ with KBp instead affords $Cp*_2Ir_2(\mu\text{-}Cl)(\mu\text{-}pz)Cl_2$ (**365**) and $Cp*IrCl_2(pzH)$ (**366**), the rhodium analogue of which has been encountered from the reaction of $Cp*_2Rh_2(\mu\text{-}Cl)_2Cl_2$ with KTp* in wet solvent.

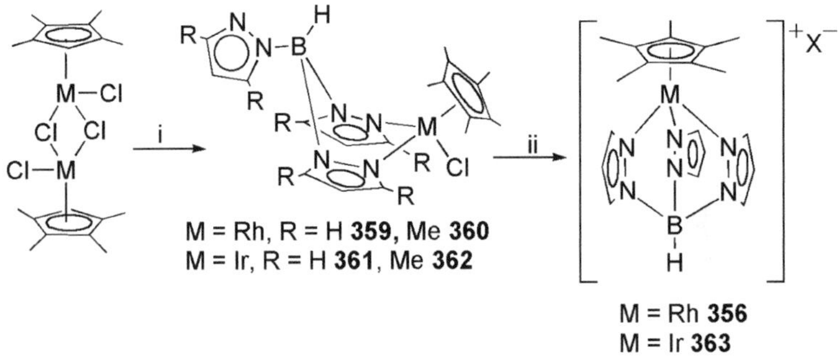

M = Rh, R = H **359,** Me **360**
M = Ir, R = H **361**, Me **362**

M = Rh **356**
M = Ir **363**

SCHEME 29. Conditions: (i) 2 KTp^x, (ii) AgX, CH_2Cl_2.

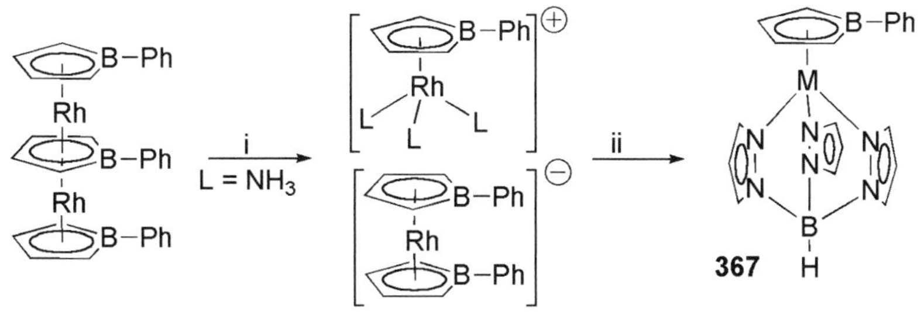

SCHEME 30. Conditions: (i) NH_3, CH_2Cl_2; (ii) KTp.

One final compound relevant to this section is the 1-phenylborole complex TpRh(η^5-C_4H_4BPh) (**367**), the only example of its kind.[137] Obtained during the study of nucleophilic degradation of the triple-decker sandwich $Rh_2\{\mu{:}\eta^5,\eta^5$-$C_4H_4BPh)(\eta^5$-$C_4H_4BPh)_2$ and its fragments, **367** formed upon reaction of $[Rh(NH_3)_3(\eta^5$-$C_4H_4BPh)]^+$ with KTp, and was assigned the η^5-κ^3 geometry on the basis of spectroscopic data (Scheme 30).

C. *Complexes with σ-Donor Ligands*

1. Alkyl Complexes

Numerous complexes comprising a formal σ-alkyl-metal linkage are known, most having been obtained during the detailed investigation of C–H activation and C–C coupling processes mediated by 'Tp^xM' fragments. A handful were, however, prepared purely for academic merit, the earliest examples being $(Et_2Bp)Rh(Me)I(CNR)_2$ (R = C_6H_4Me-4 (**368**), ^{t}Bu (**369**), $CH_2SO_2C_6H_4Me$-4 (**370**)),[138] derived from the oxidative addition of CH_3I to the parent bis(isonitrile) complexes. Surprisingly, few other examples have utilised oxidative addition of RX, the exception being the addition of allyl bromide to Tp*Rh(η^2-COE)(NCMe) (**164**) to yield Tp*Rh(η^1-C_3H_5)Br(NCMe) (**345**), an isolable intermediate *en route* to the η^3-allyl complex Tp*Rh(η^3-C_3H_5)Br (**344**).[75] The synthesis of TpRh(Me)I(PPh_3) (**371**) from the parent diiodide and CH_3I has also been reported,[139,140] though this is not an oxidative process, and full details have yet to appear.

A more prevalent approach has been the nucleophilic alkylation of a metal halide, as in the metathesis of **344** with methyllithium to afford Tp*Rh(Me)(η^3-C_3H_5) (**348**).[75] This approach was also employed to prepare the iridium complex Tp*Ir(Me)Br(PMe_3) (**372**) from the respective dibromide.[141] Similarly, a range of alkyl halides of general formula Tp*Rh(R)X($CNCH_2{}^t$Bu) (**374**–**385**, Scheme 31)[142,143] were prepared from Tp*$RhCl_2$($CNCH_2{}^t$Bu) (**373**,) and the respective Grignard reagents. For RMgX (X = Br, I) mixtures of the chloro- and halo- complexes were initially obtained, and converted to the chlorides *via* a two-step halide exchange.

The majority of relevant σ-alkyls have been obtained *via* intra- and intermolecular C–H activation. Prominent amongst these are the complexes

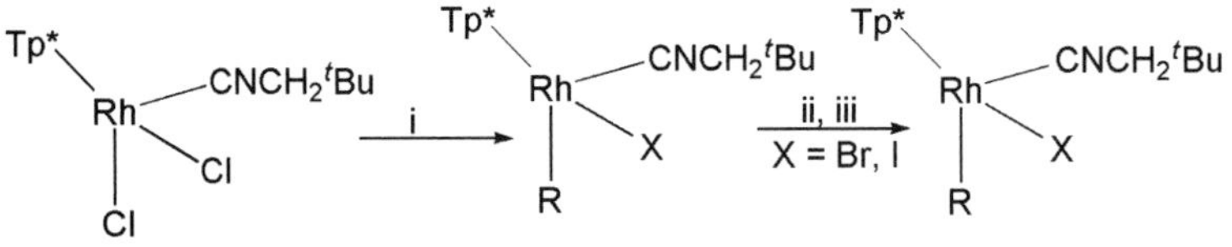

Tp*Rh(CNCH$_2$^{t}Bu)Cl$_2$ **373** $\xrightarrow{\text{i}}$ Tp*Rh(CNCH$_2$^{t}Bu)(R)X $\xrightarrow[\text{X = Br, I}]{\text{ii, iii}}$ Tp*Rh(CNCH$_2$^{t}Bu)(R)X

RMgX': R = CH_3, X' = Cl
R = CD_3, X' = I
R = ^{n}Pr, X' = Cl
R = ^{i}Pr, X' = Cl
R = ^{c}Pr, X' = Br
R = Et, X' = Cl
R = ^{n}Bu, X' = Cl
R = ^{n}Pn, X' = Cl
R = ^{n}Hex, X' = Cl
R = ^{s}Bu, X' = Cl

R = CH_3, X = Cl **374**
R = CD_3, X = Cl **375**, I **376**
R = Et, X = Cl **377**
R = ^{n}Pr, X = Cl **378**
R = ^{i}Pr, X = Cl **379**
R = ^{c}Pr, X = Cl **380**, Br **381**
R = ^{n}Bu, X = Cl **382**
R = ^{s}Bu, X = Cl **383**
R = ^{n}Pn, X = Cl **384**
R = ^{n}Hex, X = Cl **385**

SCHEME 31. Reagents and Conditions: (i) RMgX′, (ii) AgOTf, thf, r.t.; (iii) $[Bu_4N]Cl$, THF, reflux.

$Tp^xM(Et)(CH{=}CH_2)(L)$ (Tp^x = Tp*, M = Rh, L = NCMe (**156**), $NCCD_3$ (**156-*d*$_3$**),[70] py (**293**),[70] PMe_3 (**386**);[70] Ir, L = NCMe (**292**),[113] NC^{t}Bu (**387**),[100] PMe_3 (**388**),[119] dmso (**389**)[89]; Tp^x = Tp, M = Ir, L = NCMe (**390**)[89]). These σ-alkyls were obtained *either* from the hydridovinyls $Tp^*MH(CH{=}CH_2)(\eta^2\text{-}C_2H_4)$ (M = Rh (**294**), Ir (**202**)) or bis(ethylene) complexes $Tp^*M(C_2H_4)_2$ (M = Rh (**118**), Ir (**194**)) or $TpIr(\eta^2C_2H_4)_2$ (**193**) (thermally or photochemically) and have each been widely implicated in solvent C–H activation, particularly that of benzene. For instance, **156** reversibly affords $Tp^*Rh(Et)(C_6H_5)(NCMe)$ (**296**),[116] which can be thermally converted to $Tp^*Rh(\eta^4\text{-}H_2C{=}CH\text{–}CH{=}CH_2)$ (**165**) under an atmosphere of ethylene. Comparable conditions also convert **118** into a mixture of **165** and the alkyl–allyl complex $Tp^*Rh(\eta^3\text{-}C_3H_4Me)(Et)$ (**169**, Scheme 9).[58]

The iridium complex **194** activates benzene to afford $Tp^*Ir(C_6H_5)_2(N_2)$ (**211**) *via* the coordinately unsaturated intermediate '$Tp^*Ir(C_2H_5)(CH{=}CH_2)$', which has been trapped as the PMe_3 adduct $Tp^*Ir(Et)(CH{=}CH_2)(PMe_3)$ (**388**).[117,119] This experiment also afforded the zwitterionic complex $Tp^*Ir(C_6H_5)_2(CH_2CH_2PMe_3)$ (**391**),[119] as a result of PMe_3 addition to the ethylene ligand of a second intermediate, i.e. '$Tp^*Ir(C_6H_5)_2(C_2H_4)$'. The reversible loss of ethylene from **391** has been observed to afford $Tp^*Ir(C_6H_5)_2(PMe_3)$ (**392**).[90] Attempts to trap the intermediate '$Tp^*Ir(C_6H_5)_2(C_2H_4)$' with ethylene, rather than PMe_3, afforded the metallation product $Tp^*Ir\{\sigma,\sigma'\text{-}C_6H_4\text{-}2\text{-}(CH_2)_2\}(\eta^2\text{-}C_2H_4)$ (**210**), which was also obtained by thermolysis of **211** under an ethylene atmosphere.

Solvent activation by complexes such as **194**, **202** and **211** is not restricted to benzene, and has become a ubiquitous route to a variety of compounds, many of which are discussed elsewhere. Relevant to the present discussion is the thermally induced (60 °C) double C–H activation of THF by **202**, to yield the Fischer-carbene complex **393** (Scheme 32)[90] that incorporates a σ-butyl ligand. Comparable

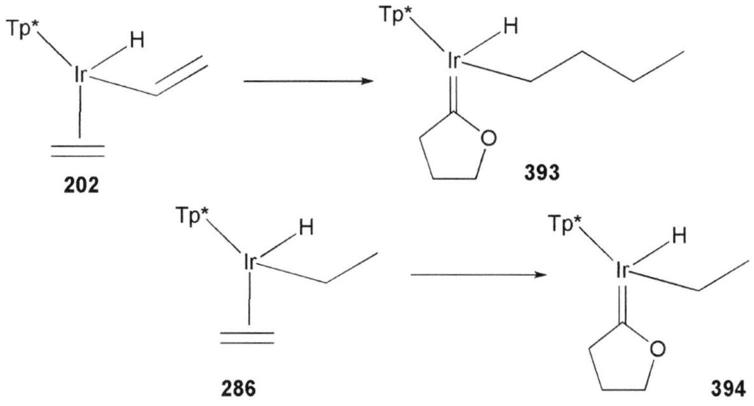

SCHEME 32. Conditions: 60 °C, THF.

reactions are observed for Tp*IrH(Et)(η^2-C_2H_4) (**286**), to yield **394**, and other $Tp^xIrR'R''(L)$ complexes, with a range of cyclic ethers (Section II-D.2).

Significantly, in none of these studies was the activation of linear-chain ethers, observed. However, with the complex $Tp^{Br_3}IrH(CH{=}CH_2)(\eta^2C_2H_4)$ (**203**) diethyl ether undergoes (photochemical) terminal activation to afford $Tp^{Br_3}IrH(CH_2CH_2OEt)(\eta^2\text{-}C_2H_4)$ (**214**), an intermediate *en route* to $Tp^{Br_3}IrH_2$ (η^2-$CH_2{=}CHOEt$) (**215**), which is also obtained.[87] Comparable results were achieved with pentane, affording a mixture of $Tp^{Br_3}IrH(CH_2\,{}^nBu)(\eta^2\text{-}C_2H_4)$ (**212**) and $Tp^{Br_3}IrH_2(\eta^2\text{-}CH_2{=}CH^nPr)$ (**213**). However, the comparable activation of cyclohexane is *not* observed, rather in this solvent the vinyl and ethylene ligands couple to afford the but-3-en-1-yl chelate $Tp^{Br_3}IrH(\sigma,\eta^2\text{-}CH_2CH_2CH{=}CH_2)$ (**216**).

A wide range of rhodium(III) alkyls of general formula Tp*RhH(R) ($CNCH_2\,{}^tBu$) has been prepared by photolysis of the carbodiimide adduct **395** (from which carbodiimide dissociates) in the respective hydrocarbons.[144] In this fashion the hydrides **396–410** (Scheme 33) were prepared and each characterised spectroscopically, their chromatographic instability precluding isolation as pure materials. This difficulty was circumvented by their conversion (with CCl_4) to the respective chlorides, which were isolated and fully characterised. The methyl complexes Tp*RhX(Me)($CNCH_2\,{}^tBu$) (X = H (**422**), Cl (**374**)) have also been obtained, but from the cyclohexyl analogue **400**, which exchanges the alkyl under pressure of methane.[145]

Both toluene and mesitylene afford mixtures of products due to competitive activation of benzylic and aromatic C–H bonds. In the case of toluene approximately equal proportions of the benzyl (**401**), C_6H_4Me-3 (**402**) and C_6H_4Me-4 (**403**) complexes are obtained kinetically, redistributing (Δ, 12 h) to a 2:1 thermodynamic mixture of **402**:**403**. It was noted[145] that this kinetic preference for benzylic activation is at variance with that observed during activation by cyclopentadienyl–iridium complexes.[107,148] With mesitylene, a 3:1 mixture of the 3,5-dimethylbenzyl (**405**) and mesityl (**404**) complexes is obtained, though the latter was inconclusively characterised since it fails to convert to the chloride, treatment with CCl_4 instead affording Tp*$RhCl_2$($CNCH_2\,{}^tBu$) (**373**).

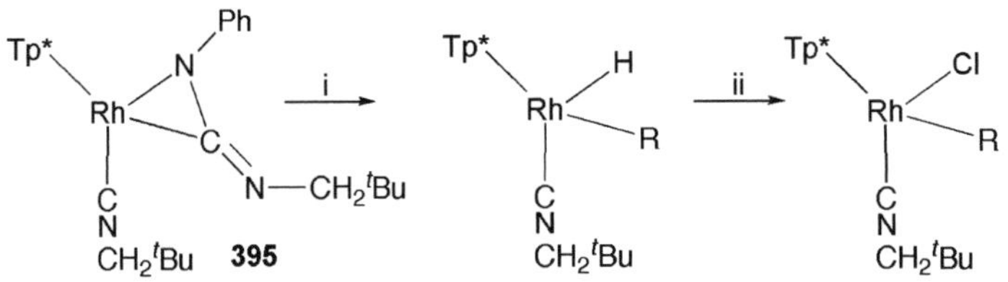

Hydrides: R = ^{n}Pr **396**,[145] ^{c}Pr **397**,[72] ^{n}Pn **398**,[145] ^{c}Pn **399**,[145] Cy **400**,[145] CH_2Ph **401**,[145] C_6H_4Me-3 **402**,[145] C_6H_4Me-4 **403**,[145] $C_6H_2Me_3$-2,4,6 **404**,[145] $CH_2C_6H_3$-Me_2-3,5, **405**,[145] ^{i}Bu **406**,[146] decyl **407**,[146] $(CH_2)_5Cl$ **408**,[147] $(CH_2)_2CH(Cl)Et$ **409**,[147] $(CH_2)_3CH(Cl)CH_3$ **410**,[147]

Chlorides: R = ^{n}Pr **378**,[145] ^{c}Pr **380**,[72] ^{n}Pn **384**,[145] ^{c}Pn **411**,[145] Cy **412**,[145] CH_2Ph **413**,[145] C_6H_4Me-3 **414**,[145] C_6H_4Me-4 **415**,[145] $C_6H_2Me_3$-2,4,6 **416**,[145] ^{i}Bu **417**,[146] decyl **418**,[146] $(CH_2)_5Cl$ **419**,[147] $(CH_2)_2CH(Cl)Et$ **420**,[147] $(CH_2)_3CH(Cl)CH_3$ **421**,[147]

SCHEME 33. Reagents and Conditions: (i) $h\nu$ (345 nm), R–H; (ii) CCl_4.

The activation of chloroalkanes proceeds exclusively at the terminal methyl groups, with no evidence for oxidative addition of the C–Cl bonds. In the case of 2-chloropentane, for which two isomers (2-chloro, and 4-chloro) might be anticipated, only the latter (**410**) is obtained, as a mixture of two diastereoisomers.[147] It was concluded that the 2-chloro-isomer does form, but undergoes spontaneous β-chloride elimination to afford Tp*RhHCl($CNCH_2{}^tBu$) (**422**), obtained as a by-product (28%), and pent-1-ene. This is supported by the observation that UV irradiation of **395** in 2-chloropropane affords only **422** and Tp*$RhCl_2$ ($CNCH_2{}^tBu$) (**373**).

Analytically pure samples of the hydride complexes **396**–**410** are accessible (quantitatively) from the respective chlorides upon their treatment with Cp_2ZrH_2;[143] an approach that has also been used to isotopically label the methyl complexes **374** and **375** by treatment with Cp_2ZrD_2 and Cp_2ZrH_2, respectively. The resulting complexes Tp*RhX(R)($CNCH_2{}^tBu$) (R = CH_3, X = D (**423**); R = CD_3, X = H (**424**)) were both found to undergo isotopic exchange in solution to afford Tp*RhH(CH_2D)($CNCH_2{}^tBu$) and Tp*RhD(CD_2H)($CNCH_2{}^tBu$), respectively.[149]

A curious reactivity of the cyclopropyl complex **397** is its tendency, under ambient conditions, to rearrange intramolecularly to afford the rhodacyclobutane complex **154** ($t_{1/2} = 95$ min).[72] In turn, **154** rearranges thermally to the η^2-propene complex **153**, which upon prolonged heating in C_6D_6 loses propene to afford **425-d_6**, a conversion that is quantitative under photolytic condition (Scheme 34). In the presence of excess neopentyl-isocyanide, thermolysis of **154** instead results in the sequential insertion of two equivalents of $CNCH_2{}^tBu$ to afford **426** (observed spectroscopically) and ultimately the metallacyclohexane **427**. The parent carbodiimide complex **395**, also inserts an additional equivalent of neopentylisonitrile to afford a metallacyclic product, though the identity of this material has not been firmly established (Section II-D.4).[150] Complex **153** has also been observed as the

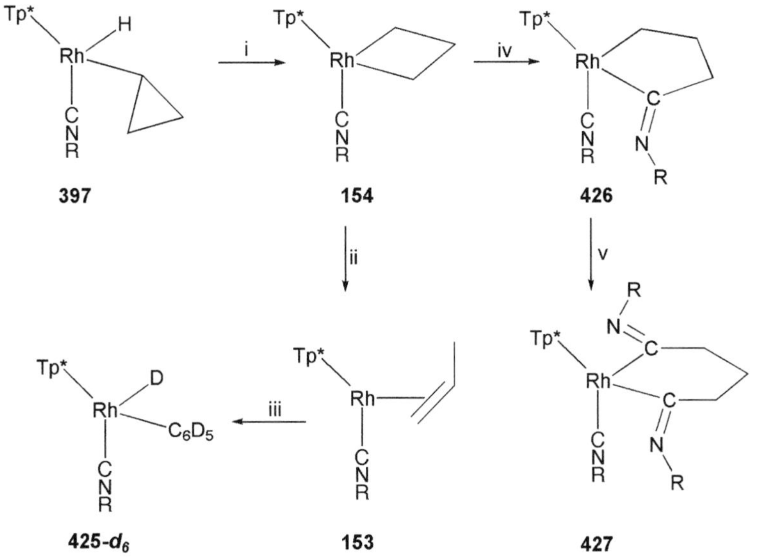

SCHEME 34. R = $CH_2{}^tBu$. Conditions: (i) 22 °C, $t_{1/2}$ = 95 min; (ii) Δ; (iii) C_6D_6, *hν*; (iv) CNR, Δ; (v) CNR, Δ.

thermal decomposition product (in benzene) of Tp*RhH(σ-^{c}Pr)($CNCH_2{}^tBu$) (**428**),† which is obtained by photolytic C–H activation of propene by **395** at −78 °C.[73] The methallyl complex Tp*RhH(σ-$CH_2CMe{=}CH_2$)($CNCH_2{}^tBu$) (**429**) was similarly prepared (−20 °C) and is also unstable, though in this instance reductive elimination is followed by the loss of isobutylene and subsequent activation of benzene to afford **425**.

The photochemical activation of hydrocarbons by Tp*Rh$(CO)_2$ (**430**) has also been explored, leading to the isolation of Tp*Rh(R)Cl(CO) (R = ^{n}Pn (**431**),[151] ^{n}Hex (**432**),[151] Cy (**433**),[152] ^{n}Hep (**434**),[151] ^{i}Oct (**435**)[151]), upon treatment of the respective hydrides, generated *in situ*, with CCl_4. Once again, under an over-pressure of methane the cyclohexyl hydride complex undergoes exchange, subsequent chlorination affording Tp*Rh(Me)Cl(CO) (**436**).[152] Interestingly, when **430** is photolysed in the presence of Et_3SiH, Si–H activation predominates over C–H activation of the solvent, implying enhanced thermodynamic stability for the Si–H addition product. Indeed, Tp*RhH($SiEt_3$)(CO) (**437**) is sufficiently stable to be isolated in its own right, without conversion to the chloride.[151]

The photolysis of Tp*Rh(η^2-C_2H_4)(CO) (**148**) in benzene was reported to afford Tp*RhH(C_6H_5)(CO) (**295**) in admixture with Tp*Rh(Et)(C_6H_5)(CO) (**438**).[153]

†The rearrangement of Tp*RhH($CH_2CH{=}CH_2$)(CO) to Tp*Rh(η^2-$MeCH{=}CH_2$)(CO), and the formation and chemistry of the rhodacyclobutane complex Tp*Rh(σ-1,3-$CH_2CH_2CH_2$)(CO) have also been documented in: Ghosh, C. K. PhD Thesis, University of Alberta, Edmonton, AB, Canada, 1988; though details have not appeared in the primary literature.

Though these two products cannot be interconverted (i.e. no evidence of ethylene insertion or elimination), an over-pressure of CO does induce migratory insertion for **438** yielding Tp*Rh{C(=O)Et}(C_6H_5)(CO) (**439**), which reductively eliminates PhC(=O)Et when treated with $ZnBr_2$.

In contrast to the case of rhodium, C–H activation has not been exploited in the synthesis of 'simple' Tp^x-ligated iridium alkyls, though their potential for C–H activation has received some attention. Thus, Tp*Ir(Me)(OTf)(PMe_3) (**440**), obtained by metathesis of the bromide analogue (**372**) with AgOTf,[141] was investigated as an analogue of Cp*Ir(Me)(OTf)(PMe_3), which readily activates C–H bonds under mild conditions.[154] Surprisingly, **440** is inert to hydrocarbons and triflate substitution,[141] implying that this function is more tightly bound than in the Cp* analogue. This was supported by crystallographic data ($r_{Ir–O}$ 2.128(5) Å **440** cf. 2.216(10) Å for the Cp* analogue[154]) and attributed to Tp* being a poorer donor than Cp*, hence, **440** has a more electrophilic ('harder') iridium(III) centre.

The more labile salts [Tp*Ir(Me)(L)(PMe_3)]BAr^f_4 ($Ar^f = C_6H_2(CF_3)_2$-3,5; L = N_2 (**441** · $\mathbf{BAr^f_4}$), CH_2Cl_2 (**442** · $\mathbf{BAr^f}$)) were prepared from dichloromethane solutions of **440** and $NaBAr^f_4$ in the presence or absence of N_2 respectively. These were found to react with CO, $AsPh_3$ and CH_3CN to afford [Tp*Ir(PMe_3)(Me)(L)]BAr_f (L = CO (**443** · $\mathbf{BAr_f}$),[141] $AsPh_3$ (**444** · $\mathbf{BAr_f}$),[155] NCMe (**445** · $\mathbf{BAr_f}$)[141]), while $\mathbf{441}^+$ also coordinates aldehydes (Scheme 35).[141,155] Activation of the coordinated aldehydes O=CHR, is not observed under ambient conditions but can be thermally induced, resulting in addition of R and CO to the metal with loss of methane. In contrast, the activation of benzene by **441** · $\mathbf{BAr^f_4}$ proceeds readily at 25 °C.

A curious reaction was reported between **441** · $\mathbf{BAr^f_4}$ and PMe_3 in dichloromethane, which instead of a simple bis-phosphine complex yields Tp*Ir(CH_3)Cl(PMe_3) (**449**) and the phosphonium salt [Me_3PCH_2Cl]BAr^f_4.[155] Labelling experiments (PMe_3-d_9 and CD_2Cl_2) confirmed that the phosphonium salt forms from the extraneous PMe_3 and solvent, thus the reaction was rationalised to proceed in accordance with Scheme 36, a consequence of facile N_2 dissociation, pointing to a dissociative mechanism for ligand exchange.

The Tp analogues of **440–442** have also been prepared, but *via* a different route, commencing from [Cp*Ir(Me)$_3$(PMe_3)]OTf, which dissolves in acetonitrile to give [Ir(Me)$_2$(NCMe)$_2$(PMe_3)]OTf. The reaction of this acetonitrile complex with KTp affords TpIr(Me)$_2$(PMe_3) (**450**), subsequent protonolysis liberating methane to yield TpIr(Me)(OTf)(PMe_3) (**451**).[156] When treated with $NaBAr^f_4$ this complex

SCHEME 35. Conditions: (i) O=CHR (R = Me, C_6H_4Me-4); (ii) Δ (R = Me, 75 °C, C_6H_4Me-4 105 °C).

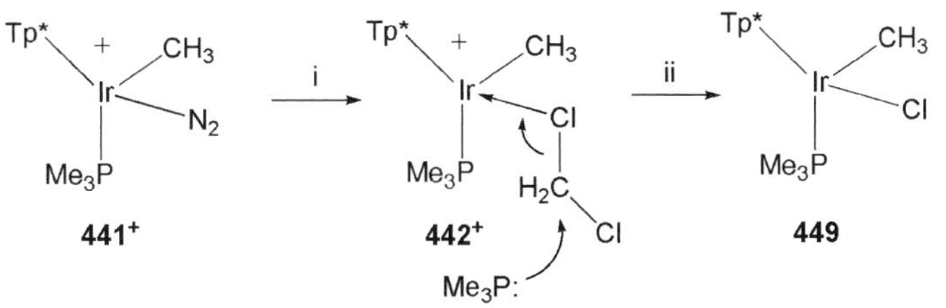

SCHEME 36. Reagents and conditions: (i) PMe_3, CH_2Cl_2.

undergoes anion exchange to afford $[TpIr(PMe_3)(Me)(CH_2Cl_2)]BAr_f$ (**452 · BAr$^f_4$**). It is noted that the CH_2Cl_2 molecule is more strongly bound than in **442 · BAr$^f_4$** (*vide supra*), such that the respective N_2 complex (**453 · BAr$^f_4$**) can only be obtained under 50 atm pressure of N_2. Since the loss of CH_2Cl_2 from **452**$^+$ is irreversible (i.e. **453 · BAr$^f_4$** is significantly more thermodynamically stable than is **452 · BAr$^f_4$**), the difficulty in obtaining the dinitrogen adduct has been attributed to a large kinetic barrier to dichloromethane loss. This barrier has been circumvented by performing the anion exchange in chloroform, which has an appreciably lower binding affinity than CH_2Cl_2, thus enabling **453 · BAr$^f_4$** to be generated at atmospheric pressure.[157] The relative binding affinity for chloromethanes in this context has been determined as: $CH_3Cl > CH_2Cl_2 > CHCl_3$; specifically **452 · BAr$^f_4$** is 2.6 kcal mol^{-1} less stable than $[TpIr(Me)(ClCH_3)(PMe_3)]BAr_f$ (**454 · BAr$^f_4$**), and 6.0 kcal mol^{-1} more stable than the $CHCl_3$ analogue, which is not isolable.

The kinetics of ligand exchange between **453 · BAr$^f_4$** and CH_3CN were studied, and thus confirmed to involve a dissociative mechanism (i.e. initial loss of CH_2Cl_2), which is consistent with the hindered formation of **453 · BAr$^f_4$** in dichloromethane.[156] A comparable mechanism can be assumed for the Tp^{Me_2} analogues, though their greater reactivity, relative to **452 · BAr$^f_4$**, precluded their study. The reduced reactivity of the Tp systems over Tp^{Me_2}, and of both over Cp*, is attributed to the relative donor capacity of the ligands decreasing in the order $Cp^* > Tp^{Me_2} > Tp$.[155] Thus, Tp imposes a greater dissociation energy for the labile ligand (e.g. N_2, CH_2Cl_2) than does Tp^{Me_2} or, in turn, Cp*. The influence upon the transition state of C–H activation is not believed to be influential in this case.

Simpler iridium(I) complexes have also been observed to activate aldehydes. For instance $[Tp^{Me_2}Ir(\eta^4\text{-}C_4H_6)]$ (**234**) activates 4-anisaldehyde to ultimately afford the iridium(III) carbonyl complex **455** (Scheme 37).[158] Intermediates in this process were observed spectroscopically and identified as the $\sigma{:}\sigma'$ butadiene complexes **456** and **458** by performing an analogous reaction with 4-dimethylaminobenzaldehyde, for which rearrangement is less facile thus enabling the isolation of **457** and **459**. It was also determined, by monitoring (^{1}H-NMR) the reaction of **234** with $4\text{-}Me_2NC_6H_4C(=O)H$, that the simple aldehyde adducts are intermediates in formation of the Fischer-type carbene complexes **458** and **459**.

The treatment of **458** with an additional equivalent of the aldehyde was reported to afford the metallabicyclic carbene **460** (Scheme 38). This is believed to proceed

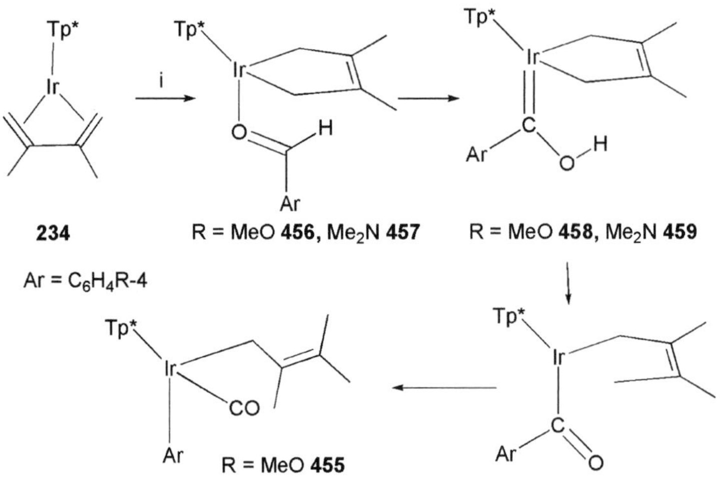

SCHEME 37. Reagents and conditions: (i) $O{=}CHC_6H_4R$-4.

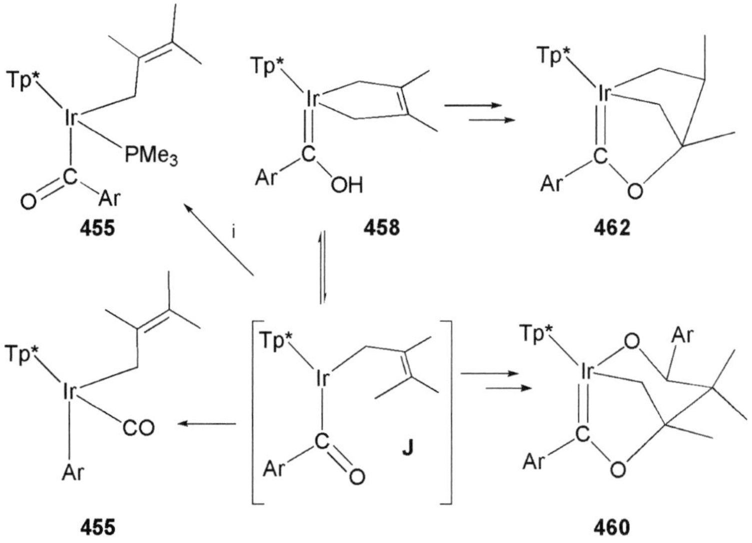

SCHEME 38. $Ar{=}C_6H_4OMe$-4. Reagents: (i) PMe_3.

via a series of postulated intermediates, including the coordinately unsaturated complex **J**, which was deemed kinetically accessible from **458**, and could be trapped as **461** using PMe_3. Both **J** and **458** were also implicated in the thermally induced rearrangement of **460** to the new metallabicyclic carbene **462**, which involves loss of one equivalent of aldehyde.

The σ:σ' (butene-1,4-diyl) binding mode of the butadiene molecule in **456–459** is relatively rare, though has been observed in a range of complexes of the type Tp*Ir(σ:σ'-butadiene)(L) (**302–316**, Section II-C.1, Chart 7), obtained by treating

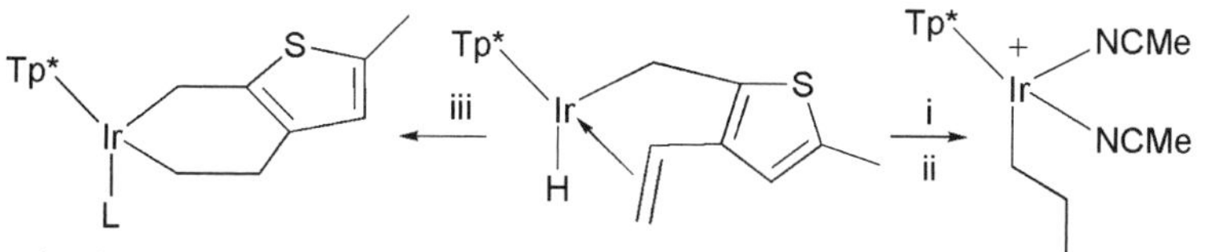

SCHEME 39. Conditions: (i) $[H(OEt_2)_2][BAr^f_4]$, CH_2Cl_2, −50 °C; (ii) 90 °C; (iii) L, C_6H_{12}, 60 °C.

the respective Tp*Ir(η^4-butadiene) (**232**–**234**) complexes with a donor L.[122] The impetus for this study was the observation of a transient intermediate of the type Tp*Ir(σ:σ'-butadiene)(SC_4H_4) during the activation of thiophenes by complex **234**.[121] Thiophene activation is considered fully elsewhere (Sections III-B.1 and III-D.1), however, several relevant metal alkyls have been so obtained, specifically during activation of 2,5-dimethylthiophene by Tp*Ir(C_2H_4)$_2$ (**194**). This reaction affords, *inter alia*, the chelate complex **253** (Scheme 16), derived from the thiophene and one ethylene ligand. The σ-alkyl linkage of **253** can be cleaved by protonolysis, subsequent heating in MeCN effecting insertion of the η^2-alkene into the Ir–H bond. Alternatively, treatment of **253** with extraneous donors (60 °C) causes insertion of the chelating alkene into the Ir–H linkage to afford the substituted iridacyclohexane complexes **463**–**466** (Scheme 39).[102]

The remaining examples of σ-alkyls are largely unremarkable; indeed, in several cases the alkyl ligand is merely coincidental. Thus, for instance, the allyl complex Tp*Rh(σ-C_3H_5)(η^3-C_3H_5) (**343**, Section III-B.2) forms exclusively from the reaction of $Rh_2(\mu\text{-}Cl)_2(\eta^3\text{-}C_3H_5)_4$ and NaTp*,[41] while [TpIr(Et)(L)]BF_4 (L = CO (**467** · **BF$_4$**), η^2-C_2H_4 (**468** · **BF$^f_4$**)) are obtained by protonation of TpIr(L)(η^2-C_2H_4) (L = CO (**228**), η^2-C_2H_4 (**193**)).[94] The photochemical activation of benzene by Tp*Rh (1,3-COD) (**171**) in the presence of *tert*-butylacrylate results in insertion of the acrylate into the Rh–H bond of the undetected intermediate Tp*RhH(C_6H_5) (η^2-CH_2=CHCO$_2$ ^{t}Bu), thereby affording Tp*Rh(C_6H_5)($CH_2CH_2CO_2$ ^{t}Bu) (**326**).[78,130]

Photochemical activation of diisopropylamine has been observed upon irradiation of pentane solutions of $Tp^{Me_2,4Cl}Rh(CO)_2$ (**469**) with excess of the amine, resulting in formation of the novel metallacycle **470**.[159] This reaction is reported to proceed diastereoselectively, but efforts to develop the chemistry of **470** were fruitless, and it has not been further studied.

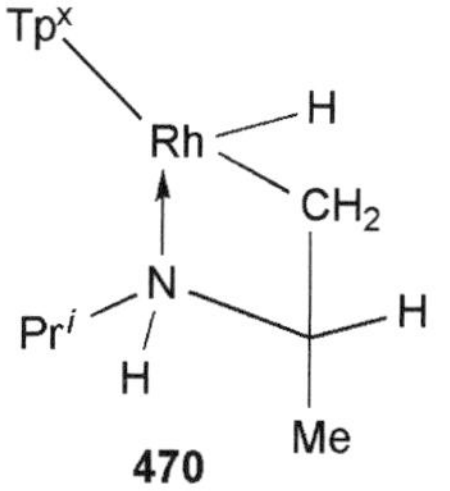

Tp^x = $Tp^{Me2,4Cl}$

470

Intramolecular activation of an isopropyl group has also been observed, during oxygenation of $Tp^{^iPr_2}Rh(dppe)$ (**471**) in the presence of free pyrazole, affording the metallated complex **472**.[160,161] This reaction is not unique; a similar activation of the $Tp^{^iPr_2}$ ligand has been observed for cobalt (Section I), while intramolecular activation of Tp^{Ph} has precluded isolation of $Tp^{Ph}Ir(\eta^2\text{-}C_2H_4)_2$, the octahedral complex (κ^4-*C,N,N′,N″*-$Tp^{Ph}Ir(C_2H_5)(\eta^2\text{-}C_2H_4)$) (**196**, Section II-C.1) being obtained instead.[88,97]

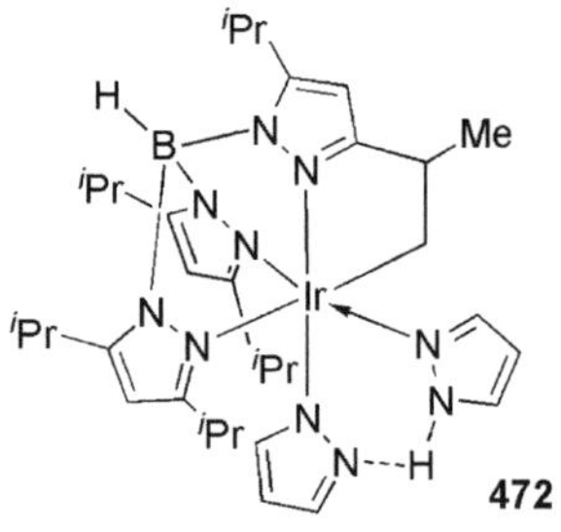

472

The final examples are obtained from the complexes $Tp^*Ir(CH_2CH_2R)(CH{=}CHR)(NCMe)$ (R = H (**292**), Me (**473**)), not *via* C–H activation (*vide supra*), but rather by intramolecular cycloaddition of the acetonitrile and vinyl ligands. Thus, under thermal activation the iridapyrroles **474** and **475** can be prepared, each of which retains the σ-alkyl ancillary ligand (Schemes 12 and 40).[113]

σ-Fluoroalkyls

Tp^x-based fluorocarbon chemistry of group 9 is in general underdeveloped, and only two reports document fluoroalkyl complexes of rhodium or iridium relevant to this discussion. However, as with cobalt, these materials do warrant separate consideration.

The first such complex, $(pzTp)Rh(C_3F_7)I(CO)$ (**476**), was (briefly) reported to arise from the oxidative addition of perfluoro-*n*-propyl iodide to $(pzTp)_2Rh_2(CO)_3$ (**477**, Section II-D.3),[44] and was formulated on the basis of infrared spectroscopic and microanalytical data. More recently, the same approach has been used to

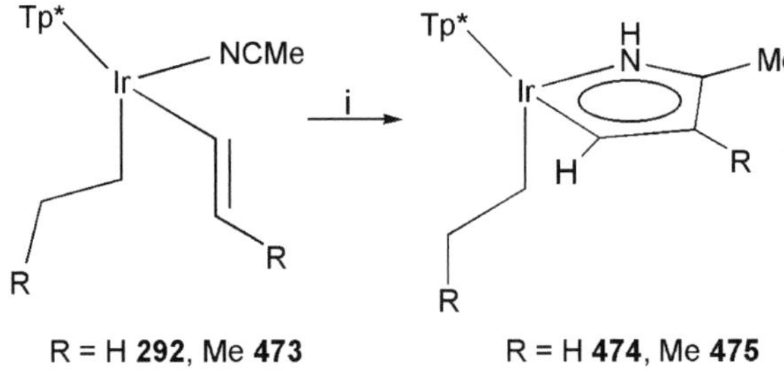

R = H **292**, Me **473** R = H **474**, Me **475**

Scheme 40. Conditions: (i) 100 °C.

prepare the series of complexes of general formula $Tp^xM(R_f)I(CO)$ ($M = Rh$, Ir; $R_f = CF_2C_6F_5$; C_3F_7) from the respective dicarbonyls (Scheme 41).[106] The exception is where $Tp^x = Tp$, since the appropriate dicarbonyl is unknown, and $[Tp_2Rh_2(CO)_3]$ (**266**) is a poor substrate for oxidative addition.[41] These Tp^x complexes (**273** and **275**) were instead prepared by carbonylation of their ethylene analogues (**271** and **272**), derived from $TpRh(C_2H_4)_2$ (**118**). The PMe_3 adducts $TpRh(R_f)I(PMe_3)$ ($R_f = CF_2C_6F_5$ (**274**), C_3F_7 (**276**)) were similarly prepared, though it is noted that the iridium complex $TpIr(^nC_3F_7)I(\eta^2\text{-}C_2H_4)$ (**277**) reacts with PMe_3 differently, affording the salt $[TpIr(CF_2CF_2CF_3)(\eta^2\text{-}C_2H_4)(PMe_3)]I$ (**278 · I**); a reaction for which precedent exists.[69]

Significantly, neither **271** nor **272** can be prepared under ambient conditions, only intractable mixtures being isolated. However, commencing from −70 °C and allowing to warm slowly *does* allow for their isolation, albeit in low yield, as secondary-products, their precursors being observed by low-temperature NMR spectroscopy. Though these intermediates remain inconclusively characterised, they

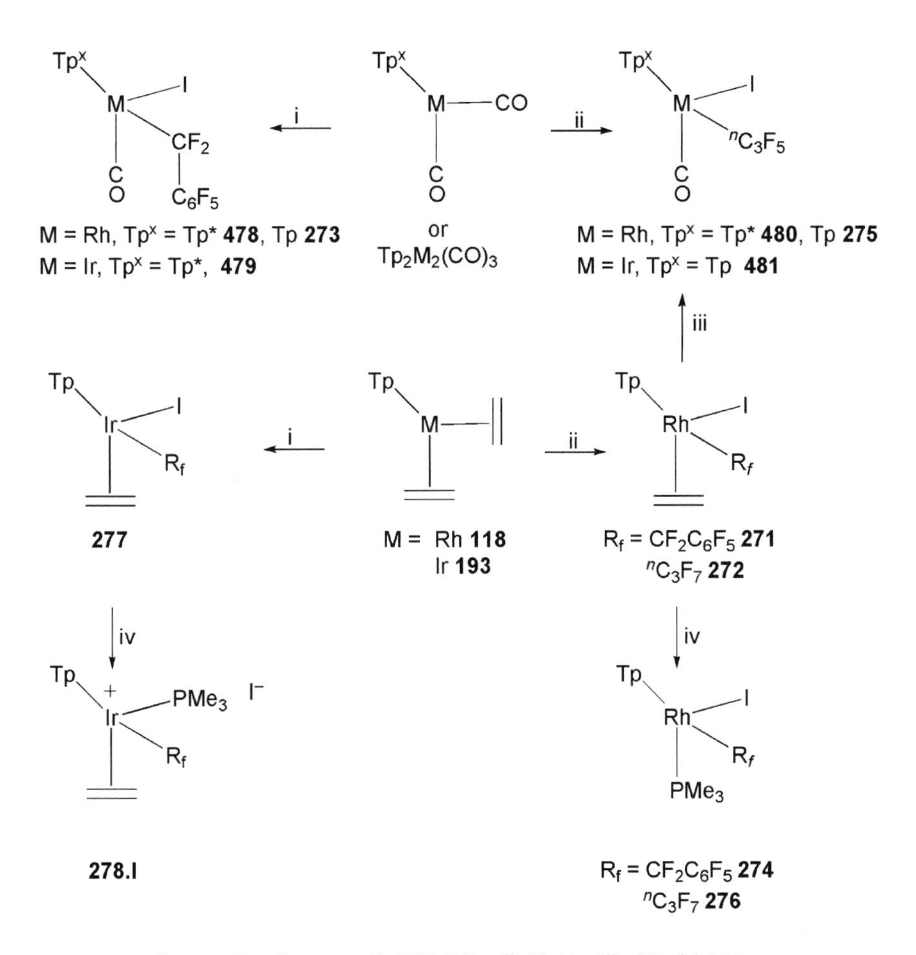

SCHEME 41. Reagents: (i) $ICF_2C_6F_5$, (ii) IC_3F_7; (iii) CO; (iv) PMe_3.

are believed to be the salts $[TpRh(R_f)(THF)(\eta^2\text{-}C_2H_4)]I$, the result of R_fI addition to **118** proceeding, non-concertedly, *via* initial electron-transfer and subsequent radical recombination.

A further key observation was that the *ortho*-aromatic ^{19}F-NMR resonances of the iridium perfluorobenzyl complex **479** are broad and decoalesce at low temperature, implying hindered rotation about the C_6F_5–CF_2 bond (in contrast to the Cp* analogue[162,163]). For this process $\Delta G^{\ddagger}$ was calculated as $10.1 \pm 1.9\,\text{kcal mol}^{-1}$ by simulated line-shape analysis. Curiously, the same process in rhodium complex **478** is rapid at room temperature (sharp ^{19}F-NMR resonances) despite the greater bulk of the Tp* ligand. A similar situation was seen for **271**.

2. Aryl Complexes

These materials can be broadly classified as three types, viz. phenyl, heteroaromatic and chelating biaryl.

a. Phenyl Derivatives

Typically resulting from the C–H activation of benzene, phenyl complexes are the most prolifically studied of the relevant σ-aryls. The first example, Tp*RhH(C_6H_5)(CO) (**295**), was originally obtained photolytically from Tp*Rh$(CO)_2$ (**430**) in benzene,[152] but has also been prepared by purging benzene solutions of **430** with N_2O,[164] which serves to remove CO yielding the reactive intermediate 'Tp*Rh(CO)'. Photolysis of Tp*Rh$(\eta^2\text{-}C_2H_4)$(CO) (**148**) in benzene is similarly effective,[114] though also generates Tp*Rh(Et)(C_6H_5)(CO) (**438**).[153] Under pressure of CO, **438** undergoes migratory insertion to afford Tp*Rh{C(=O)Et}(C_6H_5)(CO) (**439**) from which PhC(=O)Et can be extruded by treating with $ZnBr_2$.

Complex **438** has also been obtained by carbonylation of Tp*Rh(Et)(C_6H_5)(NCMe) (**296**) at 60 °C;[116] an unexpected outcome given that Tp*Rh(Et)$(CH{=}CH_2)$(NCMe) (**156**), a precursor to **296**,[70] carbonylates to afford exclusively the dicarbonyl **430**. Interestingly, the formation of **296** from **156** does itself contrast the case of the iridium analogue Tp*Ir(Et)$(CH{=}CH_2)$(NCMe) (**292**), which sequentially activates two molecules of benzene to afford the unstable 'Tp*Ir$(C_6H_5)_2$' (**297**), which is isolable as the monomeric, Tp*Ir$(C_6H_5)_2(N_2)$ (**211**), or dimeric, $Tp^*_2Ir_2(\mu\text{-}N_2)(C_6H_5)_4$ (**482**), dinitrogen adducts.[90,117] It was thus concluded that different C–H activation mechanisms must operate for rhodium vs. iridium.[116] Specifically, for iridium the active species in C–H activation is the Ir(III) complex **292**, while for rhodium it is the postulated Rh(I), η^2-benzene complex Tp*Rh$(\eta^2\text{-}C_2H_4)(\eta^2\text{-}C_6H_6)$ (**K**), which necessarily precludes activation of a second molecule of benzene.

The thermal activation of benzene by Tp*Rh$(\eta^2\text{-}C_2H_4)(PR_3)$ has also been reported, affording Tp*RhH$(C_6H_5)(PR_3)$ (R = Me (**483**), Et (**484**))[116] of which no direct iridium analogues exist at present. However, the related compounds Tp*Ir$(C_6H_5)_2$(L) (L = PMe_3 (**392**), CO (**485**), CN^{t}Bu (**486**))[117,90] have been obtained from either **211** or **482** and the respective donors. Additionally, **392** is obtained by thermal extrusion of C_2H_4 from Tp*Ir$(C_6H_5)_2(CH_2CH_2PMe_3)$ (**391**, Section II-C.1),[119] which is reversible under pressure (2 atm) of ethylene.[90] The incorporation of ethylene has also been observed for **194**, **292** and **211** (60 °C, benzene) in all cases affording the iridacycle Tp*Ir{σ,σ'-C_6H_4-2-$(CH_2)_2$}(η^2-C_2H_4) (**210**) *via* ligand coupling. Complex **211** also mediates in the double C–H activation of cyclic and acyclic ethers to afford Fischer-type carbene complexes, as observed, for instance, upon prolonged heating (60 °C) of THF solutions to yield **487** (Section II-D.2).[117]

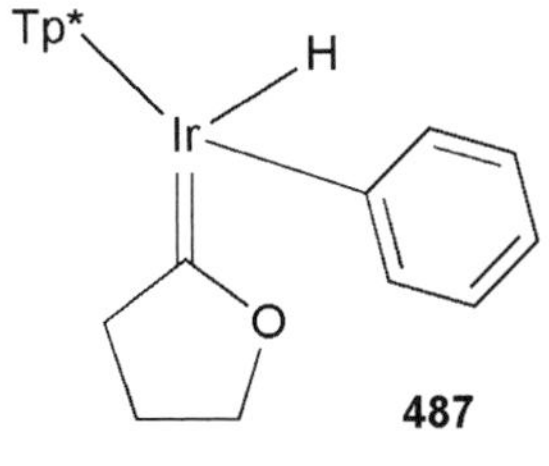

487

Numerous Ir(I) and Ir(III) Tp^x complexes effect comparable conversions, the most notable to the present section being Tp^{Ph}Ir(η^4–isoprene) (**238**), which activates THF to afford **488** (Scheme 42), in which the aryl ligand results from intramolecular orthometallation of Tp^{Ph}.[98] This orthometallation is comparable to that observed during the attempted synthesis of Tp^{Ph}Ir(η^2-C_2H_4) that instead afforded (κ^4-*N,N′,N″,C*-Tp^{Ph})Ir(Et)(η^2-C_2H_4) (**196**, Section III-B.1, Scheme 11).[88] In contrast, the activation of MeOPh and Me_2NPh by **238** preferentially affords the cyclic carbenes **489** and **490**, though trace amounts of the κ^4-Tp^{Ph} complex **491** are obtained in the latter case (Scheme 42).[98]

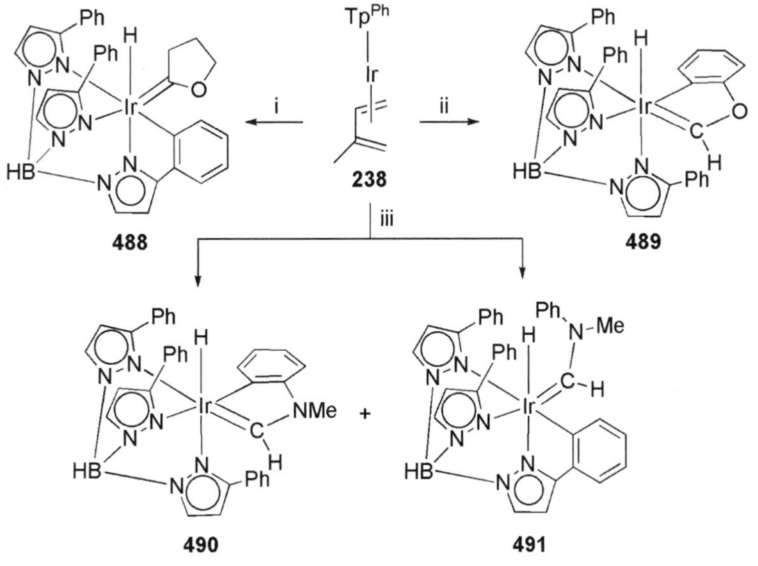

488 **238** **489** **490** **491**

SCHEME 42. Conditions: (i) THF, 60 °C, (ii) MeOPh, 60 °C, (iii) Me_2NPh, 60 °C.

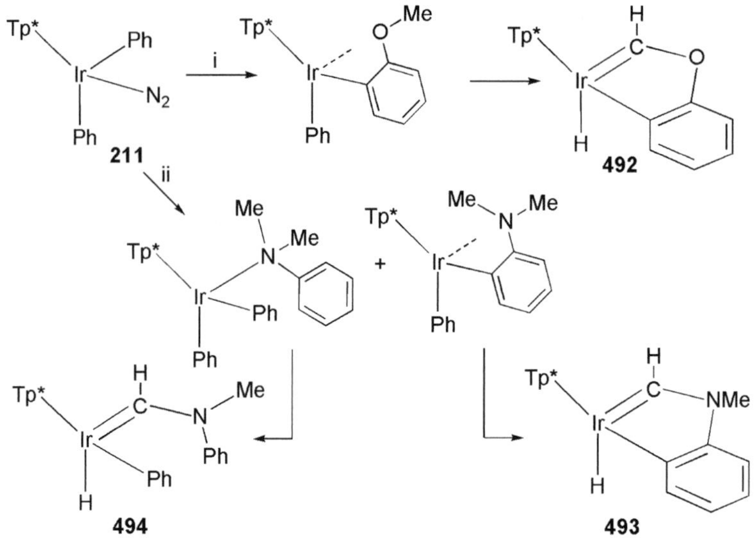

SCHEME 43. Reagents: (i) MeOPh; (ii) Me_2NPh.

The preference for metallation of the carbene-based phenyl ring, over that of the Tp^{Ph} ligand, was attributed to the former resulting in less ring-strain.[98] This assumes that the initial step of the reaction is coordination of the heteroatom, followed by double C–H activation at the α-carbon (i.e. carbene formation precedes aryl activation). However, an independent study of the activation of anisole by $Tp^*Ir(C_6H_5)_2(N_2)$ (**211**) to afford **492** (Scheme 43),[165] suggested that aryl C–H activation may in fact be the initial step, given that this is typically more facile than alkyl C–H activation. This suggestion has credence, given the observation that Me_2NPh is activated to afford a mixture of **493** and **494**, which must arise from competitive pathways (Scheme 43). A definitive mechanism remains to be determined.

Both studies, however, illustrated an unusual propensity for activation of α over β-hydrogens, reporting that both Et_2O[98] and EtOPh[98,165] afford exclusively the carbenes **495**–**498** (Scheme 44), rather than π-alkene complexes (from β-elimination).

The activation of 1,2-dimethoxyethane (dme) has also been effected with **292**, yielding the carbene complex **499**, which undergoes a thermally induced 1,2-hydride shift to afford **500**,[118] in admixture with **501** ($<10\%$, Scheme 45). The latter product is proposed to derive from a σ,σ'-benzyne intermediate, and is formed exclusively upon prolonged thermolysis. All three species have been implicated as intermediates *en route* to the dihydride **502**, which forms, alongside the C–C coupling product C_6H_5–$CH_2O(CH_2)_2OMe$, after prolonged heating in dme/benzene. A mechanism for this coupling process has not been elaborated.

Somewhat more unusual is the reaction of **292** with three equivalents of DMAD, which affords the iridacycloheptatriene complexes **503** and **504**, from which the adventitious water can be displaced by PMe_3 (Scheme 46).[133] These complexes are

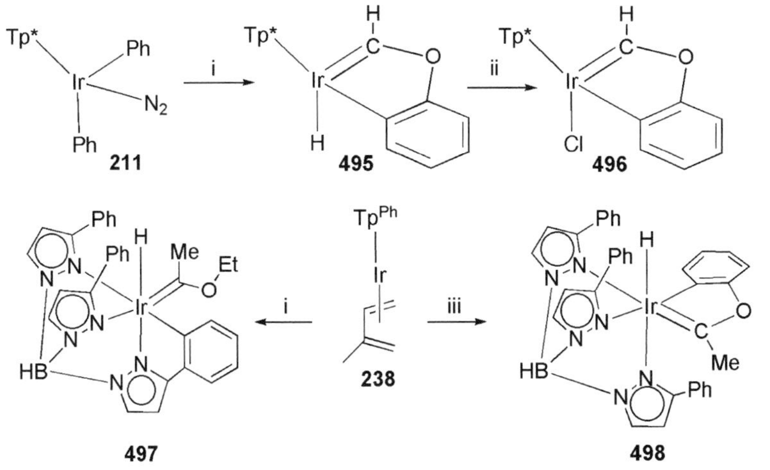

SCHEME 44. Conditions: (i) Et_2O, 60 °C; (ii) CCl_4; (iii) EtOPh, 60 °C.

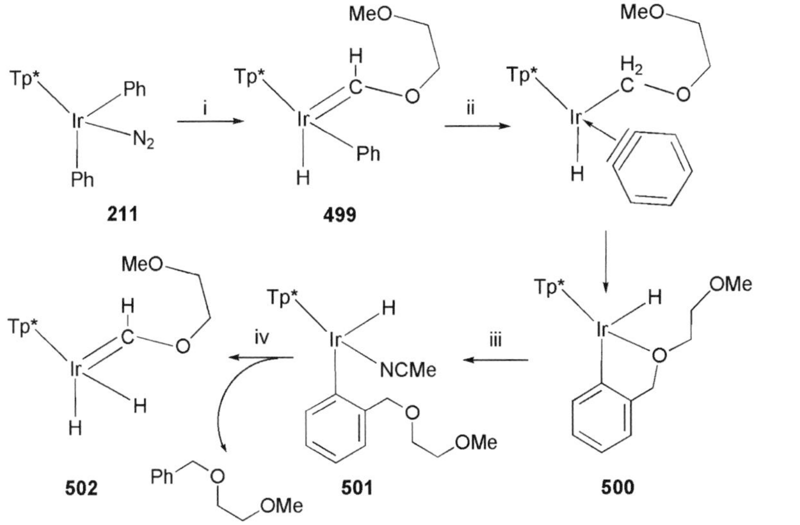

SCHEME 45. Conditions and reagents: (i) C_6H_6/dme, 60 °C, 6 h; (ii) 80 °C; (iii) NCMe; (iv) Δ, C_6H_6/dme.

readily oxidised to the chelating aldehydes **505** and **506** respectively, the latter undergoing further oxidation to afford the iridanaphthalene complex **507**; the first example of its kind (Scheme 46).[166]

The remaining examples of phenyl complexes are simpler, though again typically arise *via* C–H activation. The iridium(III) salt $[Tp^*Ir(C_6H_5)(N_2)(PMe_3)]BAr^f_4$ (**508** · $\mathbf{BAr^f_4}$) results from the ambient-temperature activation of benzene by the analogous methyl complex (**441** · $\mathbf{BAr^f_4}$), which will also thermally activate aldehydes (Scheme 35), the means by which $[Tp^*Ir(C_6H_4Me\text{-}4)(CO)(PMe_3)]BAr^f_4$ (**449** · $\mathbf{BAr^f_4}$)

SCHEME 46. R = CO_2Me. Conditions: (i) 3 DMAD, CH_2Cl_2, 60 °C; (ii) $^tBuO_2H$; (iii) excess $^tBuO_2H$.

TABLE III

SUMMARY OF COMPLEXES Tp*MH(Ar)(L) (Ar = ARYL), OBTAINED BY C–H ACTIVATION OF SOLVENT, AND THE RESPECTIVE CHLORIDES, GENERATED BY CHLORINATION WITH CCl_4

M	X	Ar	L	Number	M	X	Ar	L	Number
Ir	H	Ph	$P(OMe)_3$	**321**	Rh	H	*m*-Tol	$CNCH_2{}^tBu$	**402**
Rh	H	Ph	$P(OMe)_3$	**324**	Rh	Cl	*m*-Tol	$CNCH_2{}^tBu$	**414**
Rh	H	Ph	$CNCH_2{}^tBu$	**424**	Rh	H	*p*-Tol	$CNCH_2{}^tBu$	**403**
Rh	Cl	Ph	$CNCH_2{}^tBu$	**510**	Rh	Cl	*p*-Tol	$CNCH_2{}^tBu$	**415**
Rh	H	Ph	CNMe	**513**	Rh	H	Mes	$CNCH_2{}^tBu$	**404**
Rh	H	Ph	$CNC_6H_3Me_2$	**514**	Rh	Cl	Mes	$CNCH_2{}^tBu$	**416**

Note: Mes = $C_6H_2Me_3$-2,4,6; Tol = C_6H_4Me.

was obtained from 4-tolualdehyde.[141,155] Activation of 4-anisaldehyde by Tp*Ir(η^4-2,3-$Me_2C_4H_4$) (**234**) has also been reported, and affords Tp*Ir(C_6H_4OMe-4){σ-$CH_2C(Me){=}C(Me)_2$}(CO) (**455**, Scheme 37, Section II-C.1).[158]

The photochemical activation of solvents has afforded several complexes of the type $Tp^xM(Ar)X(L)$ (Table III). Thus, in the presence of $P(OMe)_3$, Tp*IrH_2(η^2-COE) (**207**) photolytically activates benzene to afford **321**, the mechanism of which was explored at some length (Section III-B.1 and III-B.2).[91] The rhodium analogue of **321** (**324**) was obtained under comparable conditions, but from the rhodium(I) complex Tp*Rh(η^4-1,3-COD) (**171**),[77,78,149] which also activates benzene in the presence of *tert*-butylacrylate to afford **326**. The bis(isonitrile) complex Tp*Rh($CNCH_2{}^tBu$)$_2$ (**509**) photolytically activates benzene to afford

425,[145,167] which has also been obtained from the carbodiimide complex $Tp^*Rh(CNR)(\eta^2\text{-}PhN{=}CNR)$ ($R{=}CH_2{}^tBu$ (**395**)) at a significantly enhanced rate.[150] Comparable activation is observed for the methyl and 2,6-xylyl carbodiimide complexes (R = Me (**511**), 2,6-Xyl (**512**)) to afford **513** and **514**, though at a reduced rate. Toluene and mesitylene are similarly activated by **395**, but in competition with benzylic activation (Section II-C.1).[145]

Finally, in the solid-state **395** undergoes photolytically induced, intramolecular aryl activation of the carbodiimide phenyl group to afford **515**,[150] *via* an undetermined mechanism.

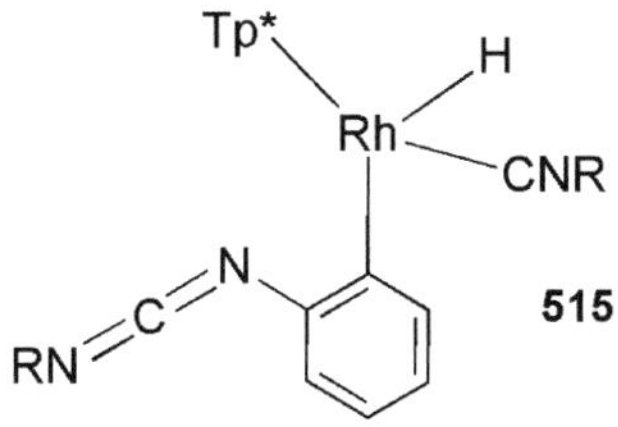

515 $R = CH_2C(CH_3)_3$

b. σ-Heteroaromatics

The simplest examples are the complexes $Tp^*RhH(2\text{-}C_6H_4N)(PEt_3)$ (**516**)[116] and $Tp^*RhH(2\text{-}C_4H_3S)(PR_3)$ (R = Me (**318**), Et (**320**)),[71,116] formed from the reactions of $Tp^*Rh(\eta^2\text{-}C_2H_4)(PR_3)$ (R = Me (**145**), Et (**147**)) with pyridine and thiophene respectively. As previously noted (Section III-B.1), formation of **318** and **320** is in competition with Rh insertion into the C–S bond, which predominates under ambient conditions, though **318** and **320** are the thermodynamic products.

The remaining examples result from thermal activation of (neat) substituted thiophenes by $Tp^*Ir(C_2H_4)_2$ (**194**) these being reactions of some complexity. Thus, C_4H_4S,[120,121] 2-MeC_4H_3S and 3-MeC_4H_3S[121] are each activated to afford (**298–300**, Scheme 47), in the latter two cases as rotameric mixtures. In each case, the thiophene ligand is labile and readily displaced by other donors to give **517–520**; alternatively, hydrogenation (60 °C) effects loss of both 2-thienyl ligands to afford $Tp^*IrH_2(SC_4H_3R)$ (**521–523**).[120,121] Continued heating of **521**, in the absence of H_2, results in dimerisation to afford the bridged complexes **524** and **525**.

The hydrido complex $Tp^*IrH(SC_4H_4)(2\text{-}C_4H_3S)$ (**301**) has been obtained through thiophene activation by $Tp^*IrH_2(\eta^2\text{-}C_2H_4)$ (**206**). The thiophene ligand is again readily displaced by donors to afford $Tp^*IrH(2\text{-}C_4H_3S)(L)$ ($L = PMe_3$ (**526**), NCMe (**527**)),[121] while **526** rearranges on protonation to afford $[Tp^*IrH(SC_4H_4)(PMe_3)]^+$ (**528**$^+$).[102] Similarly, protonation of the dimethylthiophene complex **252** (formed in admixture with **253**) induces its rearrangement to **529**$^+$ (Schemes 16 and 48).[101]

c. Chelating Biaryls

A handful of such materials have been prepared during investigations of the photophysical properties of cyclometallated Ir(III) complexes of the type

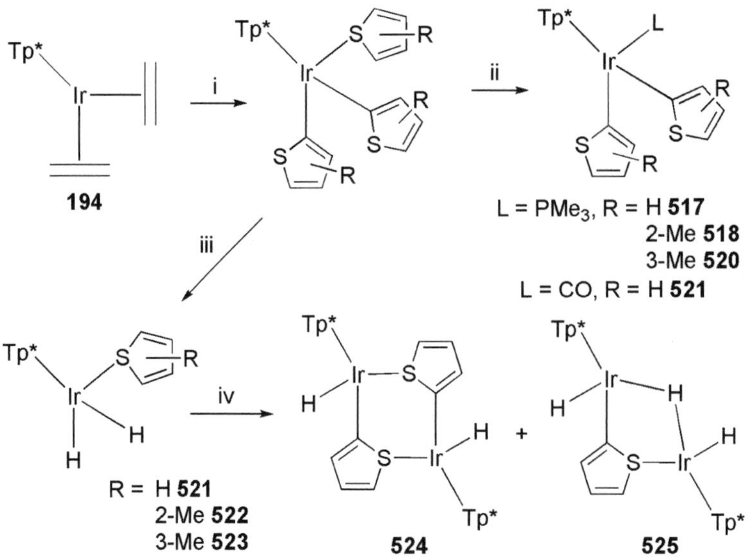

SCHEME 47. Conditions: (i) Δ, SC_4H_3R (R = H, Me); (ii) L; (iii) H_2, 60 °C; (iv) 60 °C.

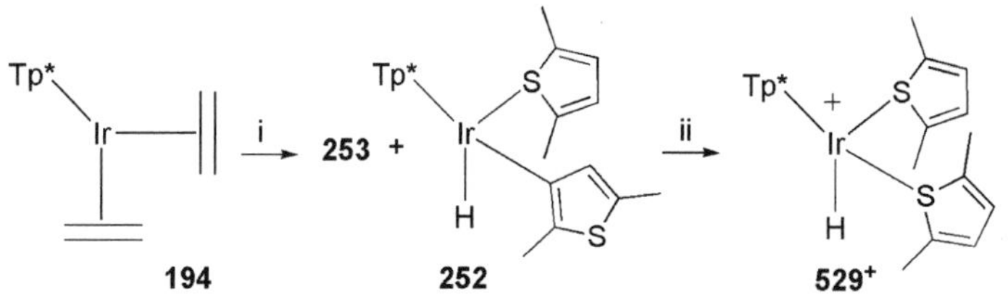

SCHEME 48. Conditions: (i) $SC_4H_2Me_2$-2,5, 60 °C; (ii) H^+.

$(C–N)_2Ir(L–X)$ (C–N = cyclometallating ligand, L–X = anionic co-ligand). Surprisingly, complexes **530–537** (Chart 8) can only be prepared from the triflate salts $[(C–N)_2Ir(OH_2)_2]OTf$, the dimers $[(C–N)_2Ir(Cl)]_2$ affording only pyrazole complexes upon reaction with K[RTp] or $K[R_2Bp]$.[168,169] The related hydrotris (pyrazolyl)methane and tetrakis(triazolyl)borate complexes $[(tpy)_2Ir(\kappa^2\text{-}pz_3CH)]$ OTf (**538 · OTf**) and $[(tpy)_2Ir(\kappa^2\text{-}tz_4B)]$ (**539**) were similarly obtained.[170]

Photophysical studies revealed that the nature of the pyrazolylborate ligand has significant influence in complexes **530–535**, by virtue of it 'tuning' the metal HOMO energies,[168–171] making these materials highly emissive at room temperature. For instance, **535** is utilised in blue and white phosphorescent OLEDs.[168,169,172,173] Complexes **536** and **537** have particular utility in blue-to-green phosphorescent materials, since their emission spectra can be 'tuned' by varying the R and X substituents.[170] These materials feature in rapidly growing patent literature.[174–178]

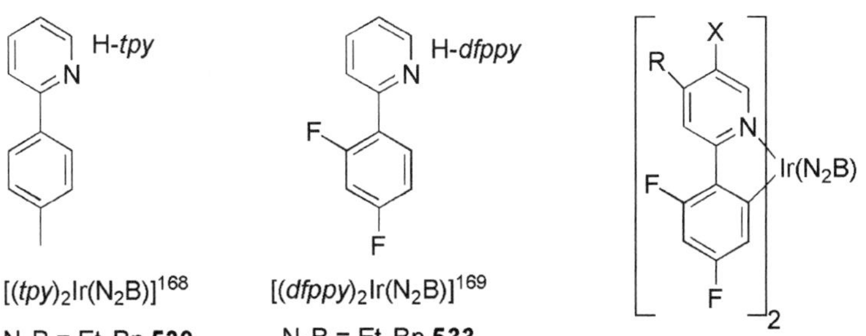

H-*tpy*

$[(tpy)_2Ir(N_2B)]$[168]

$N_2B = Et_2Bp$ **530**,
Ph_2Bp **531**,
κ^2-pzTp **532**

H-*dfppy*

$[(dfppy)_2Ir(N_2B)]$[169]

$N_2B = Et_2Bp$ **533**,
Ph_2Bp **534**,
κ^2-pzTp **535**

$N_2B = \kappa^2$-pzTp[170]
$R = C_8H_{17}$, X = Br **536**
R = H, X = carbazole **537**

CHART 8. Chelating biaryl complexes.

D. *Complexes with σ-Donor/π-Acceptor Ligands*

1. Alkenyl and Alkynyl Complexes

a. Alkenyls

Metal alkenyls have received relatively little attention beyond the ubiquitous vinyl complexes that are obtained either thermally or photochemically from simple alkene complexes of the type $Tp^xM(\eta^2\text{-}C_2H_4)(L)$ ($L = C_2H_4$, PR_3, NCMe, CO, Scheme 49, see also Section III-B.1). The reactivity of these species largely mirrors that of their π-alkene precursors, in which their intermediacy has often been implicated. Most of this chemistry, and studies establishing the relative stability and accessibility of the π-alkene and hydridovinyl forms, are discussed in detail in Section III-B.1; however, some merit further attention.

A disparity between iridium and rhodium reactivity is illustrated by the relative stabilities of **289** and $Tp^*RhH(CH{=}CH_2)(CO)$ (**290**) versus their π-alkene isomers, **289** being thermodynamically preferred, whereas **290** rapidly converts to $Tp^*Rh(CO)(\eta^2\text{-}C_2H_4)$ (**148**) under ambient conditions in darkness.[95] It is thus unsurprising that **294** has never been discretely isolated. Rather, its existence is inferred in the formation of **156**, **293** and **386** from the bis(ethylene) complex **118**, which is believed to arise by ethylene insertion into the Rh–H bond of complexes such as **294**, with subsequent coordination of the Lewis base.[70] It is also noted that both **156** and **293** can be thermally induced to lose ethylene and generate $Tp^*Rh(\eta^2\text{-}C_2H_4)(L)$,[118] while the iridium complexes $Tp^*Ir(\eta^2\text{-}C_2H_4)(L)$ ($L = PMe_3$ (**224**), PMe_2Ph (**226**)) thermolytically rearrange to the hydridovinyls $Tp^*IrH(CH{=}CH_2)(L)$ ($L = PMe_3$ (**540**), PMe_2Ph (**541**)).[92]

Both **224** and its hydridovinyl isomer **540** are protonated by $[H(OEt_2)_2][BAr^f_4]$ to afford the salt $[Tp^*IrH(\eta^2\text{-}C_2H_4)(PMe_3)]BAr^f_4$ (**541 · BAr$^f_4$**).[179] For **540** this was established to occur at the β-vinylic carbon, the initial product (**542**$^+$, Scheme 50), which exists as a 1:1 isomeric mixture, being identified by low temperature (–90 °C) NMR spectroscopic monitoring. Detailed mechanistic studies revealed that conversion of **542**$^+$ to **541**$^+$ proceeds by hydride migration to the ethylidene

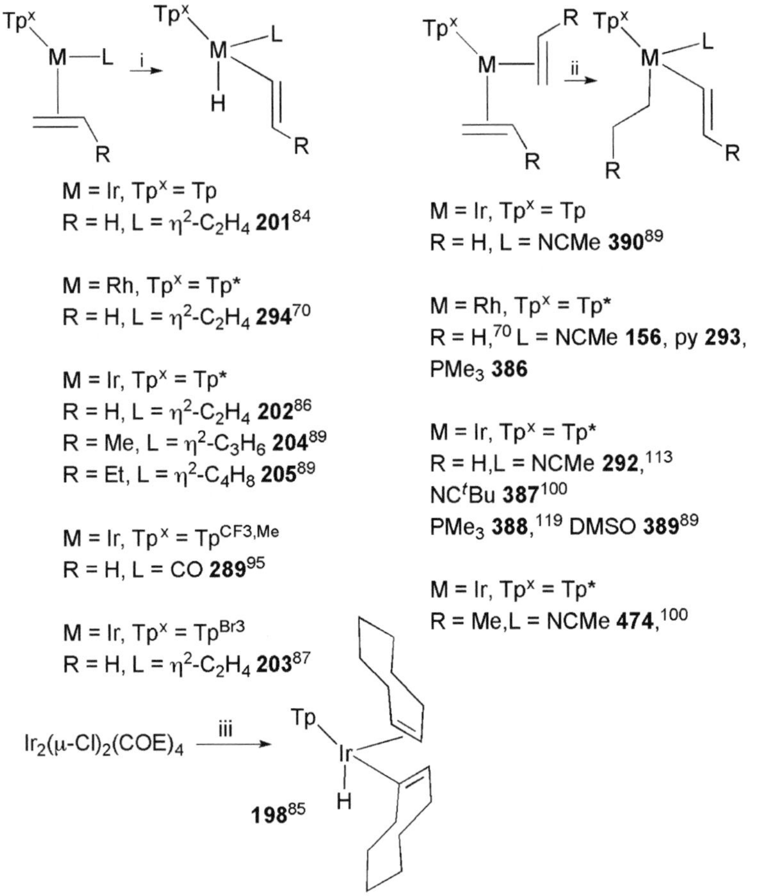

Tp^x M—L →(i) Tp^x M L H R R

Tp^x M R →(ii) Tp^x M L R R R

M = Ir, Tp^x = Tp
R = H, L = η^2-C_2H_4 **201**[84]

M = Rh, Tp^x = Tp*
R = H, L = η^2-C_2H_4 **294**[70]

M = Ir, Tp^x = Tp*
R = H, L = η^2-C_2H_4 **202**[86]
R = Me, L = η^2-C_3H_6 **204**[89]
R = Et, L = η^2-C_4H_8 **205**[89]

M = Ir, Tp^x = $Tp^{CF3,Me}$
R = H, L = CO **289**[95]

M = Ir, Tp^x = Tp^{Br3}
R = H, L = η^2-C_2H_4 **203**[87]

M = Ir, Tp^x = Tp
R = H, L = NCMe **390**[89]

M = Rh, Tp^x = Tp*
R = H,[70] L = NCMe **156**, py **293**,
PMe_3 **386**

M = Ir, Tp^x = Tp*
R = H,L = NCMe **292**,[113]
NC^{t}Bu **387**[100]
PMe_3 **388**,[119] DMSO **389**[89]

M = Ir, Tp^x = Tp*
R = Me,L = NCMe **474**,[100]

$Ir_2(\mu\text{-}Cl)_2(COE)_4$ →(iii) Tp Ir H **198**[85]

SCHEME 49. Alkenyl complexes derived from parent π-alkene complexes. Conditions: (i) Δ or $h\nu$; (ii) Δ, L; (iii) KTp.

Tp* Ir PMe_3 R ⇌(i) [Tp* + Ir PMe_3 C Me R H ⇌ Tp* + Ir PMe_3 C Me R H] →(ii) Tp* + Ir PMe_3 R

R = H **540**
Et **386**

R = H **542**$^+$; Et **544**$^+$

R = H **541**$^+$
Et **543**$^+$

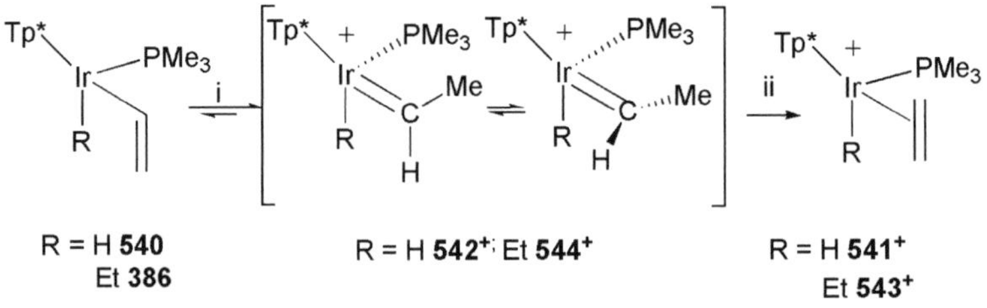

SCHEME 50. Conditions: (i) H^+, $-60\,°C$; (ii) $-47\,°C$.

carbon, with reversible ethylene deinsertion occurring at ambient temperature. Comparable results were obtained with **386**.

A further significant feature of the ethyl–vinyl complexes $Tp^xIr(C_2H_4R)(CH{=}CHR)(NCR')$ (R' = Me (**292**), (**473**), (**390**), R' = ^{t}Bu (**387**)) and, in acetonitrile, **198**, is their propensity for intramolecular [3 + 2] cycloaddition of the alkenyl and nitrile ligands.[113] This affords iridapyrrole complexes **474**, **475**, **545**, **546** and **547**[113] (Chart 9; discussed in Section II-E) for which fully delocalised aromatic structures have been assigned on the basis of NMR spectroscopic data. Besides this study the hydrides **201** and **204** were found to react with a series of aldehydes, resulting in formation of the alkoxide–alkene chelates **244**–**247** (Scheme 13, Section III-B.1).

The rhodium vinyls $Tp^*Rh(CH{=}CH_2)X(CNCH_2{}^tBu)$ (X = Br (**548**), Cl (**549**)) have been prepared, commencing from $[Tp^{Me_2}Rh(CNCH_2{}^tBu)Cl_2]$ (**373**) and $BrMgCH{=}CH_2$ to afford **548**, which is converted to **549** by a 2-step procedure (Scheme 51, see also Section II-C.1).[142] The analogous hydride (**550**) was similarly obtained by treating **549** with Cp_2ZrH_2 (and also photolytically from carbodiimide complex **395**),[73] but is thermally unstable, rearranging slowly under ambient conditions to afford $Tp^*Rh(CNCH_2{}^tBu)(\eta^2\text{-}C_2H_4)$ (**152**). The 3,3-dimethylbutentyl hydride complex **551** is also formed by C–H activation with **395** as a single isomer that is thermally unstable toward the elimination of 3,3-dimethylbut-1-ene.

One more, simple, group 9 vinyl is known, complex **174**, which is obtained by the addition of HCl to the η^2-DMAD complex **172** (Scheme 52), a process believed to result from protonation of the metal centre with subsequent insertion of the alkyne into the Rh–H linkage, given the *trans-β* geometry of the resulting vinyl ligand.[79]

A more complex series of iridium vinyls have been derived from DMAD, directly from $Tp^*Ir(\eta^2\text{-}C_2H_4)_2$ (**194**), $Tp^*Ir(\eta^2\text{-}C_4H_4\text{-}2,3\text{-}Me_2)$ (**234**) or $Tp^*Ir(C_6H_5)_2(N_2)$

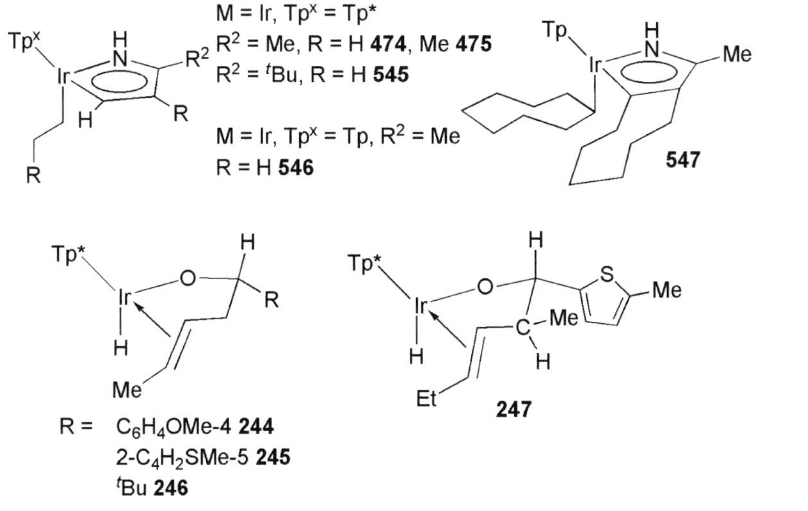

CHART 9. Iridapyrroles and related compounds.

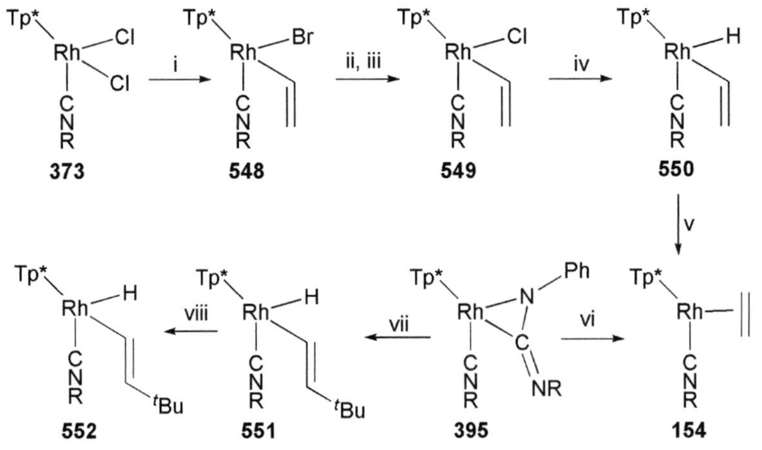

SCHEME 51. R = $CH_2{}^tBu$. Conditions: (i) CH_2=CHMgBr; (ii) AgOTf, THF, rt, (iii) Bu_4NCl, THF, reflux; (iv) Cp_2ZrH_2; (v) 22 °C (vi) *hv*, CH_2=$CH(CMe_3)$; (vii) CCl_4, hexane, −20 °C.

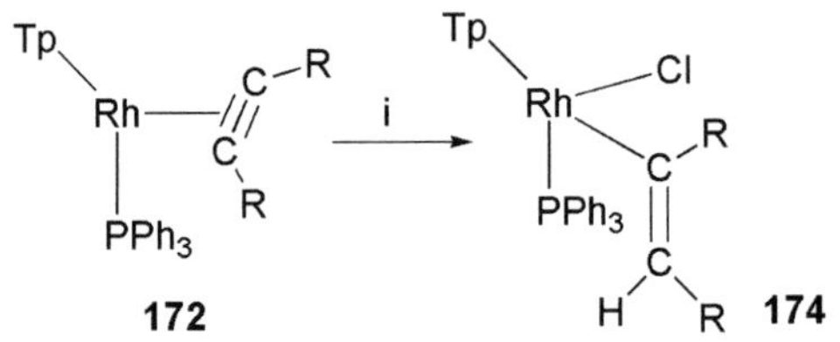

SCHEME 52. R = CO_2Me. Conditions: (i) HCl.

(**211**). Thus, iridacyclopent-2-ene complexes **335** and **336** have been obtained from **194** and one equivalent of DMAD, while **336** will then react further to afford the iridacyclopenta-2,3-diene complex **337** (Scheme 53, see also Scheme 22, Section III-B.1).[132] An analogue of **337** (**339**) is similarly obtained from **234** and two equivalents of DMAD (Scheme 23, Section III-B.1),[134] but reacts further to afford the iridacycloheptatriene complex **338**.[133,134] Alternatively, **339** reacts with but-2-yne to afford an equilibrium mixture of the iridacycloheptatriene **340** and the bicyclic complex **341**, the latter being disfavoured by excess water or other extraneous donors, which will also displace H_2O from both complexes.[134]

Complexes **503** and **504** (analogues of **338**) were similarly prepared from **211**, in an effort to probe the reaction mechanisms. Both complexes were subsequently treated with PMe_3 to afford **555** and **556** (Scheme 54, see also Scheme 46, Section II-C.1),[133] which then extrude the organic fragments upon oxidation with H_2O_2. In contrast, **503**, **504** and indeed **338** are oxidised to the respective chelating ketone complexes **505**, **506** and **557**, the latter two of which are susceptible to further oxidation by ${}^tBuO_2H$ to afford the iridacyclic carboxylato complexes **507** and **558**, both of which have fully delocalised structures (see also Section II-E).[166]

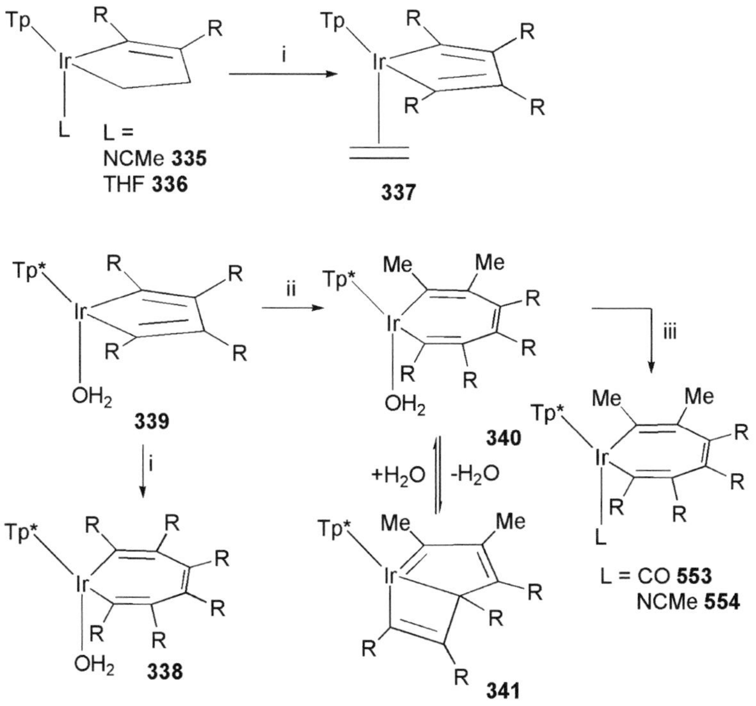

SCHEME 53. R = CO_2Me. Conditions: (i) DMAD, (ii) MeC≡CMe; (iii) L.

b. Alkynyls

No relevant iridium alkynyls exist to date. However, the rhodium complexes Tp*RhH(C≡CPh)(PR_3) (R = C_6H_5 (**559**), C_6H_4F-4 (**560**)) have been prepared by the addition of phenylacetylene to the respective bis(phosphine) complexes,[180] and the ethyl propiolate complex Tp*RhH(C≡CCO_2Et)(PPh_3) (**561**) has very recently been prepared in the same way.[181]

2. Carbene Complexes

Relatively few such materials are known, particularly of rhodium for which only four discrete examples are documented, the first, TpRh{=C(SMe)$_2$}X_2 (X = I (**562**), Cl (**563**)), being obtained only in 1998.[79] These were prepared by sequential double alkylation of the carbon disulfide complex TpRh(η^2-SCS)(PPh_3) (**175**) with MeI, yielding **562**, treatment with excess Bu_4NCl effecting quantitative conversion to **563**. Though the latter was structurally characterised, neither compound has been subjected to further chemical study. The thiocarbamoyl salt [TpRh (η^2-$SCNMe_2$)(PPh_3)]Cl (**564 · Cl**) is also relevant here, since it was determined to possess appreciable multiple-bond character in the Rh–C linkage, implying a significant contribution from a rhodathiirene canonical form.

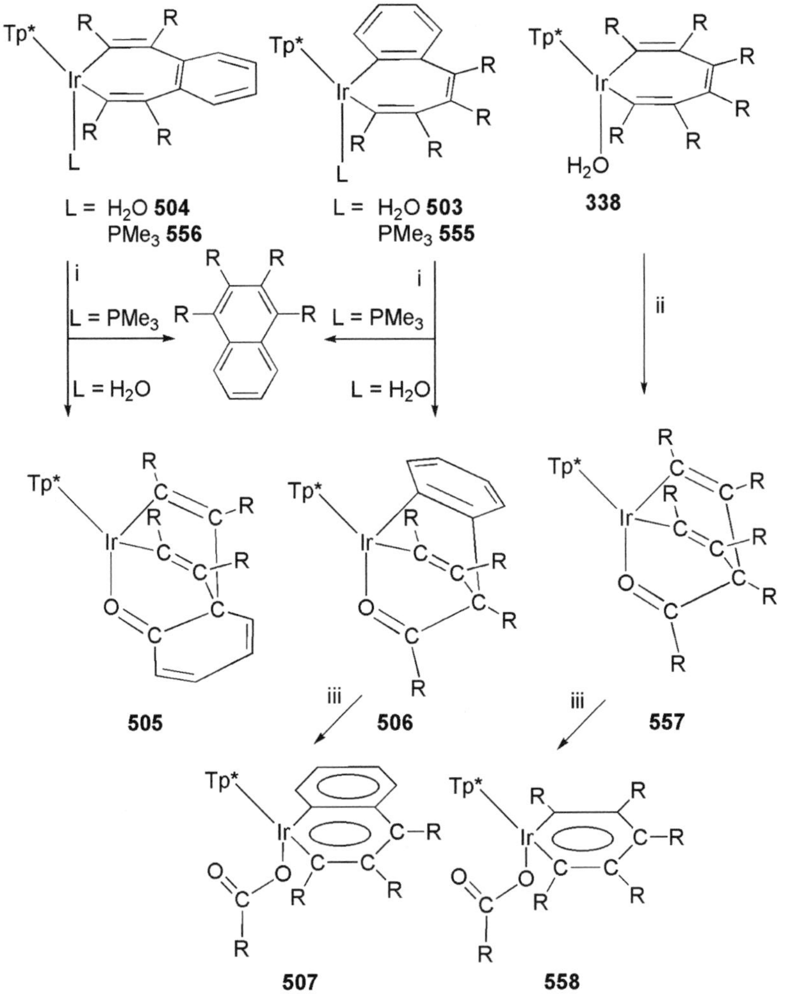

SCHEME 54. Conditions: (i) H_2O_2 (30%) excess, C_6H_{12}, 150 °C, 3d; (ii) $^tBuO_2H$; (iii) excess $^tBuO_2H$, CH_2Cl_2, 20 °C.

Other reports were more fortuitous, as in the unexpected formation of **565** (Scheme 55) upon alkylation of Bp*Rh(CO)(py) (**566**,) with methyl iodide.[182] The mechanism for this was established to be intramolecular (from cross-over experiments with ^{13}C- and ^{2}H-labelled **566**), and proposed to proceed by oxidative addition of MeI, with subsequent migratory insertion of CO. The nature of the hydride migration process from boron remains to be determined. On prolonged heating (45 °C), **565** evolves into **567**; a rare example of reverse α-hydride migration from metal to carbene.

The final example (**568**) was obtained by addition of HN^iPr_2 to $Tp^{Me_2,4Cl}RhCl(CHCl_2)(CO)$ (**569**), the product of chloroform addition to $Tp^{Me_2,4Cl}Rh(CO)_2$

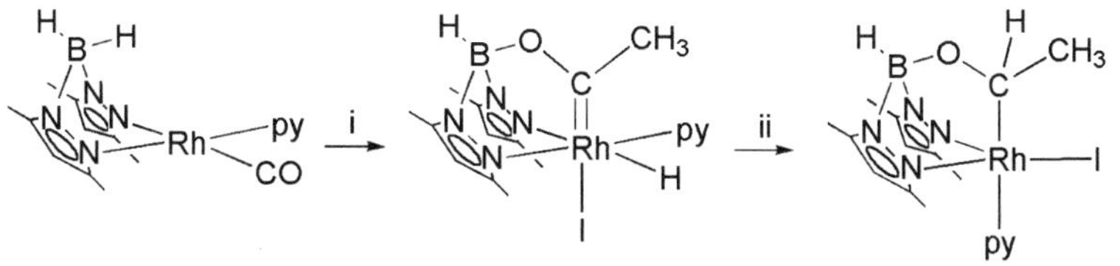

SCHEME 55. Conditions: (i) C_6H_6, CH_3I, r.t.; (ii) 45 °C, 36 h.

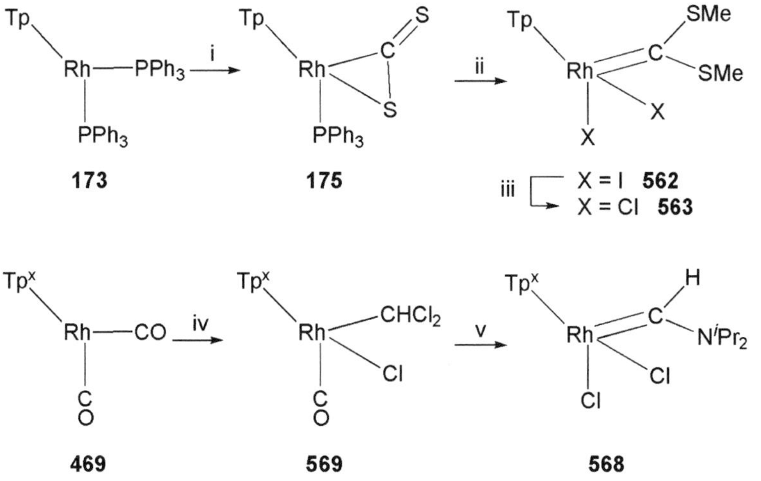

SCHEME 56. $Tp^x = TpMe^{2,4Cl}$. Conditions: (i) CS_2; (ii) MeI; (iii) $CHCl_3$ or $[Bu_4N]Cl$; (iv) $CHCl_3$, 48 h; (v) excess HN^iPr_2.

(**469**, Scheme 56).[183] Though no definitive mechanism was established, it was noted that the analogous reaction between $Tp^*Rh(CO)_2$ and chloroform did not proceed to the corresponding dichloromethyl complex, suggesting that the tridentate coordination mode in the latter might preclude clean oxidative addition.

A larger range of iridium carbenes exists, predominantly resulting from C–H activation of solvent, as in the activation of a range of ethers, and dimethylaniline, by $Tp^{Ph}Ir(\eta^4\text{-}C_4H_5Me\text{-}3)$ (**238**)[98] and $Tp^{Me_2}Ir(C_6H_5)_2(N_2)$ (**211**)[90,117,165] (Scheme 57), some of which were described above.

The operative mechanism is a matter of debate and would seem to be substrate dependant. For instance, it is believed that THF coordinates to the metal prior to successive β-, then α-hydride eliminations,[90] a notion that is also invoked for the activation of MeOPh, EtOPh and Me_2NPh by **238**.[98] It is argued for these cases that carbene formation precedes the cyclometallation step, which is less strained for the carbene-bound ring than for the Tp^{Ph} phenyl group; hence the preferential formation of cyclic carbenes **489**, **490** and **498** over κ^4-*N,N′,N″,C*-Tp^{Ph} analogues of **488**, **491** and **497**. However, the activation of Me_2NPh by **211** was independently

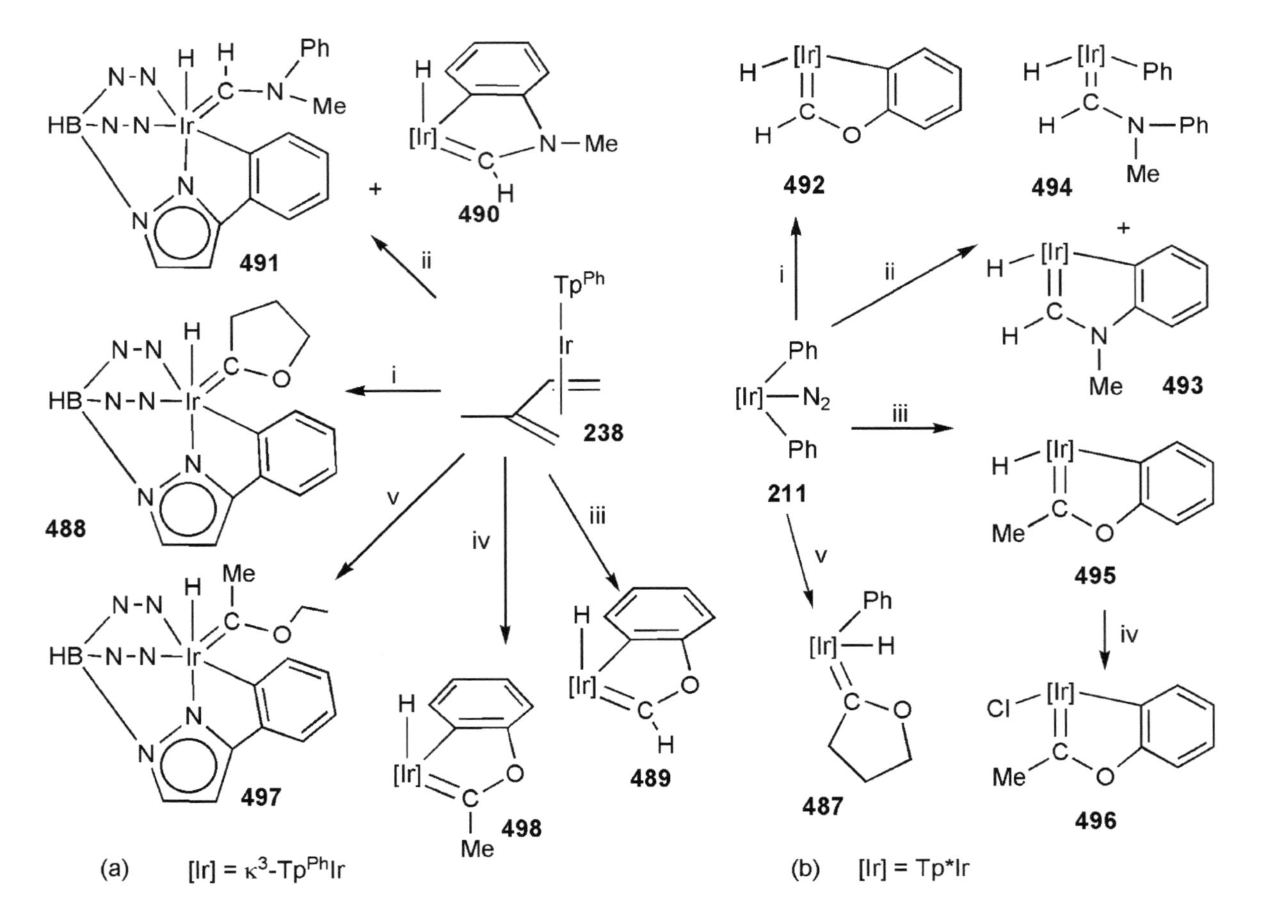

SCHEME 57. Reagents: (a) (i) THF; (ii) Me_2NPh; (iii) MeOPh; (iv) EtOPh; (v) Et_2O. (b) (i) MeOPh; (ii) Me_2NPh; (iii) EtOPh; (iv) CCl_4; (v) THF.

noted to afford **493** *and* **494**, which have been proven to arise from competitive pathways.[165] Since initial coordination of the heteroatom is a prerequisite for formation of **494**, it was concluded that aryl C–H activation must be the initial step *en route* to **493** (i.e. preceding amine coordination). This is credible given the enhanced facility noted for aryl over alkyl C–H activation. Regardless of mechanism, a key observation of both studies is the significantly greater propensity for α-, over β-C–H activation, illustrated by the universal formation of carbenes from Et_2O (**497**)[98] and EtOPh (**495**,[98] **498**),[165] rather than π-alkene complexes.

The activation of THF by $Tp^*Ir(\eta^2\text{-}C_2H_4)_2$ (**194**), and its hydridovinyl isomer $Tp^*IrH(CH{=}CH_2)(\eta^2\text{-}C_2H_4)$ (**202**), has also been reported, the latter complex being deemed the 'active' species.[90,119] In both cases, the activation competes with ethylene coupling, affording a ca. 1:1 mixture of the carbene and Tp*IrH $(\eta^3\text{-}C_3H_4Me)$ (**291**, Section III-B.1). A curious aspect of this system is its inability to activate diethyl ether, though this has not received detailed comment. However, the reactivity of **202** toward a range of cyclic ethers has been explored, and trends established (Scheme 58). It was noted that 2-methyl THF activates more readily than THF itself (presumably due to electronically enhanced coordinative capacity of the ether), a similar, but less pronounced, situation being apparent for diethers. In contrast, pyran derivatives are less prone to activation, presumed to be a result of greater steric hindrance toward coordination.

Mechanistic studies using THF-d_8 revealed a significant kinetic isotope effect (**393**:**393-d_n** ca. 2.2:1), and also implied a complex series of reactions that incorporate deuterium (two to three atoms) into the α and β positions of the ancillary propylene chain; issues that have not been resolved. However, the complexes $Tp^*IrH(R)(\eta^2\text{-}C_2H_4)$ (R = H (**206**), Et (**285**)) were also found to activate THF, affording exclusively the respective carbenes (R = H (**574**), C_2H_5 (**394**)).

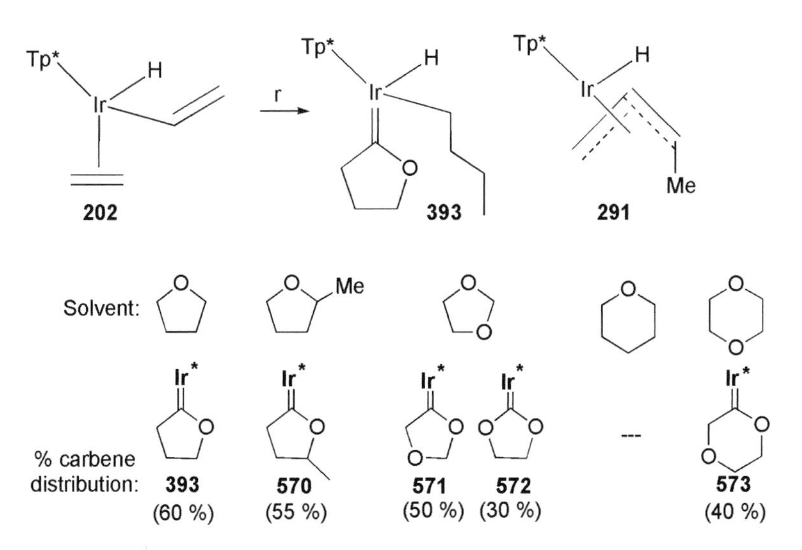

SCHEME 58. Activation of cyclic ethers by **202**; product distribution balance in each case is composed of **291**. Conditions: (i) THF, 60 °C, $t_{1/2}$ ~ 1 h. Ir* = Tp*IrH(^{n}Bu).

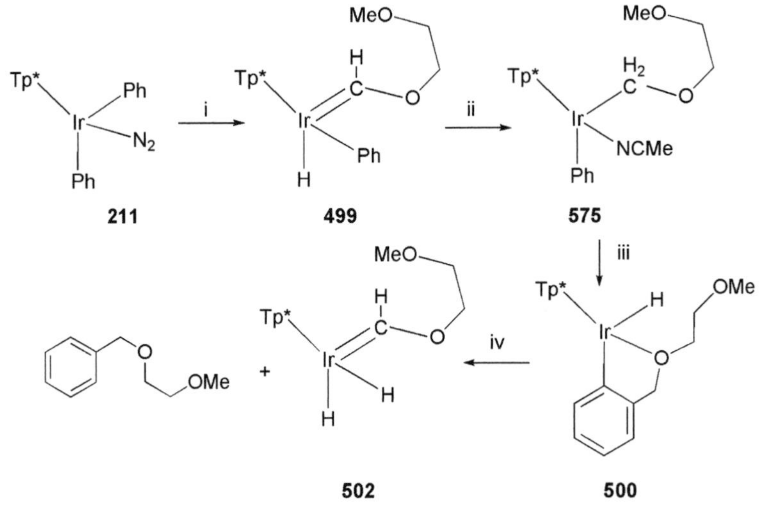

SCHEME 59. Conditions: (i) benzene/dme, 60 °C, 6 h; (ii) NCMe, 80 °C; (iii) NCMe, 80 °C, 24 h; (iv) benzene/dme 80 °C, 24 h.

Moreover, an intermediate in the activation of THF by **211**, formulated as Tp*Ir(C_6H_5)$_2$(THF), was observed spectroscopically, seemingly confirming the supposed pre-coordination of the substrate.

A similar reaction has been reported to arise when **211** is heated in mixtures of benzene and dimethoxyethane, ultimately affording the carbene **502** in admixture with (2-methoxy-ethoxymethyl)benzene, the product of C–C coupling of the solvents (Scheme 59).[118] Though no definitive mechanism is yet available, the initial formation of carbene **499** has been confirmed, and this is presumed to undergo a thermally induced 1,2-hydride shift, the product of which was trapped as **575**. Coupling of the aryl and alkyl groups (*via* a postulated η^2-benzyne complex) would then afford **500**, though how this eliminates the diether and generates **502** is unclear. However, the intermediacy of **500** and **575** has been confirmed, both affording the same products when heated in benzene/dme.

Somewhat more unusual is the recently described formation of the chlorocarbene complex **576**, obtained *via* the fragmentation of CH_2Cl_2 by the dinitrogen complex (κ^5–*N,N′,N″,C,C′*-Tp^{Mes}Ir(N_2) (**577**), which is in turn generated photolytically from **578** (Scheme 60).[184] The identity of **576** was determined spectroscopically, its hydrolytic instability precluding isolation in analytical purity. Nonetheless, treating **576** with one equivalent of 3-mesitylpyrazole cleanly converted it to the pyrazolylcarbene complex **579**, which was isolated and crystallographically characterised.

Several other uncommon activation processes have been reported to afford carbene products, notably that of aromatic aldehydes by Tp*Ir(η^4-C_4H_4-2,3-Me_2) (**234**),[158] which at 140 °C in anisaldehyde affords the carbene **458** *en route* to carbonyl complex **455**. Though **458** cannot be isolated it has been observed spectroscopically, while the analogue derived from activation of

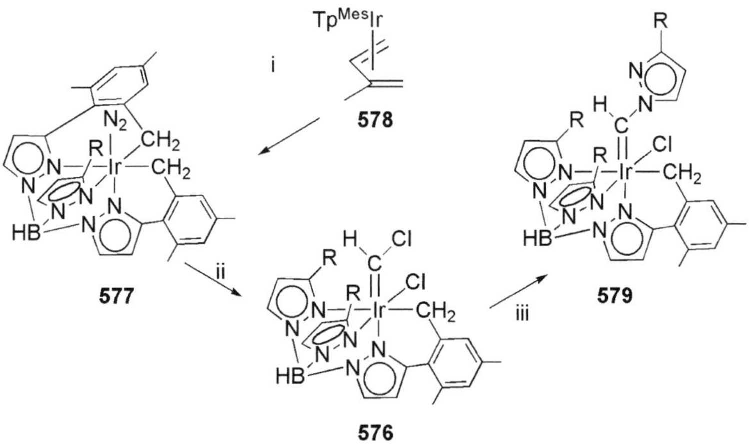

SCHEME 60. $R = C_6H_2Me_3$-2,4,6. Conditions: (i) PhH, $h\nu$; (ii) CH_2Cl_2, 80 °C, 0.5 h; (iii) Hpz^{Mes}.

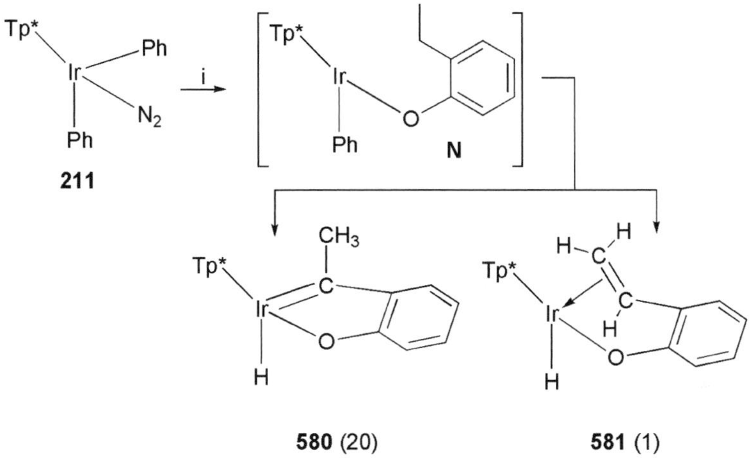

SCHEME 61. Conditions: (i) C_8H_{12} 60 °C.

$Me_2NC_6H_4C(=O)H$ has been isolated and fully characterised. In the presence of excess anisaldehyde **458** reacts further to afford the metallabicyclic carbene **460**, which thermally rearranges to **462** *via* an undetermined mechanism (Schemes 37 and 38).

Rather unique, though related, is the activation of 2-ethylphenol by **211**, which affords predominantly the carbene **580** in admixture with the η^2-alkene complex **581** (ca. 10%, Scheme 61).[185] This is presumed to occur by initial alcoholysis of one Ir–Ph linkage, affording intermediate **N** that then undergoes double activation of the methylene C–H bonds. Significantly, as with ethers (*vide supra*), the first C–H activation is preferentially followed by α-, not β-activation, thus favouring the carbene. Moreover, isolated samples of the π-alkene complex thermally equilibrate to the same 20:1 product mixture, a rare example of a metal–alkene converting to a metal–alkylidene, the reverse being more typical. Indeed, the alkylidenes $[Tp^*Ir(R)\{=C(Me)H\}(PMe_3)]^+$ (R = H (**542**$^+$), Et (**544**$^+$)), obtained by protonation of $Tp^*Ir(R)(CH=CH_2)(PMe_3)$ (R = H (**540**), Et (**386**), Section II-D.1), are observed only at $-90\,°C$, since above this temperature they undergo 1,2-hydride migration to afford the respective π-alkene complexes.[179] It should be noted that **542**$^+$ and **544**$^+$ were the first alkylidenes of iridium(III) ligated by Tp^x ligands.

The protonation of iridapyrrole complex **582** also yields an alkylidene, **583-*syn*** (kinetic product), which in damp solvents rearranged to the thermodynamic product **583-*anti***, *via* an undisclosed mechanism (Scheme 62).[99,100] This process is reversible such that treatment of either isomer with K_2CO_3 regenerates **582**. Interestingly, while the hydride of **583-*syn*** appears to be ideally positioned for migration to the carbene, this is not observed under ambient conditions. However, coordinating solvents (MeOH, NCMe) induce α-migratory insertion of this linkage, affording **584** and **585** respectively, the former being cleanly converted back to **583** *in vacuo*.

It should be noted that while a number iridapyrroles exist (Section II-E) their protonation does not universally afford isolable alkylidenes, particularly those complexes comprising an alkyl ligand in place of the hydride, which typically afford chelating π-alkene complexes.[100]

Finally, the alkylidene **586** (Chart 10) has been obtained by protonation of the chelate complex **256**, which was encountered while studying the activation of 2,5-dimethylthiophene by $Tp^*Ir(C_2H_4)_2$ (**194**, Section III-B.1),[101,102] and results from a complex series of C–H activation and C–C coupling steps. Meanwhile, **341** results from a series of C–C coupling reactions, being the final product of the interaction of but-2-yne with the iridacyclopentadiene complex **339**, obtained from **194** and two equivalents of DMAD (Section III-B.1, see also Scheme 16).[134]

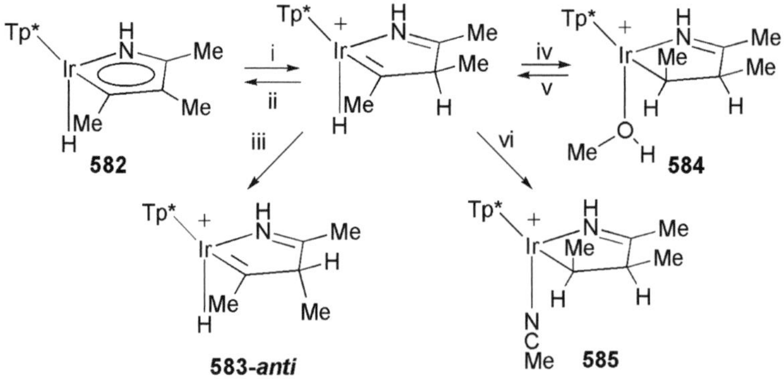

SCHEME 62. Conditions: (i) $[H(OEt_2)][BAr^f_4]$; (ii) K_2CO_3; (iii) $CDCl_3$ (H_2O), 20 °C; (iv) MeOH; (v) –MeOH; (vi) NCMe.

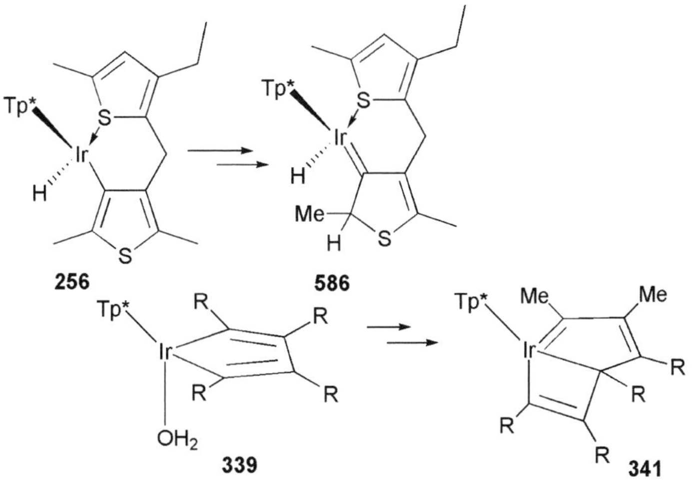

CHART 10. R = CO_2Me. Alkylidenes derived from C–H activation and C–C coupling.

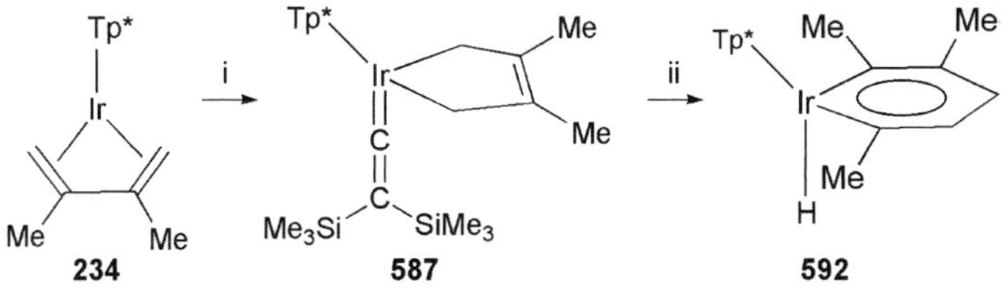

SCHEME 63. Conditions: (i) $Me_3SiC{\equiv}CSiMe_3$, CyH, 80 °C, 12 h; (ii) CF_3COOH, THF, 20 °C, 12 h.

Vinylidenes

A single, very recent, report describes the synthesis of a relevant group 9 vinylidene complex, viz. the bis(trimethylsilyl)vinylidene **587** (Scheme 63, Fig. 7), prepared from Tp*Ir(η^4-2,4-$Me_2C_4H_4$) (**234**) and $Me_3SiC{\equiv}CSiMe_3$.[186] This reaction involves significant electronic reorganisation of the butadiene ligand, comparable to that induced by Lewis acids (Section III-B.1). It is thus believed that **587** forms *via* an intermediate π-alkyne complex that undergoes a 1,2-silyl shift. Evidence in support of this was obtained from a comparable reaction commencing from Tp*Ir(C_6H_5)$_2$(N_2) (**211**) that affords (**588**) (Scheme 64).

The protonation behaviour of **587** and **588** is dependant upon the acid. Thus, treating **588** with excess (six equivalents) $HBF_4 \cdot OEt_2$ affords the cationic alkylidene [Tp*Ir(C_6H_5){=C(Me)Ph}(OEt_2)]$^+$ (**589**$^+$), which can be converted to the neutral complex Tp*Ir(C_6H_5){=C(Me)Ph}Cl (**590**) by reaction with [PPN]Cl. Alternatively, the addition of excess HCl to **588** affords **590** directly, while trifluoroacetic acid (TFA) affords Tp*Ir(C_6H_5)(κ^1-O_2CCF_3){=C(Me)Ph} (**591**) at an appreciably enhanced rate (2 h, r.t.). In contrast, the protonation of **587** is more complex, neither HCl nor HBF_4 affording isolable materials. However,

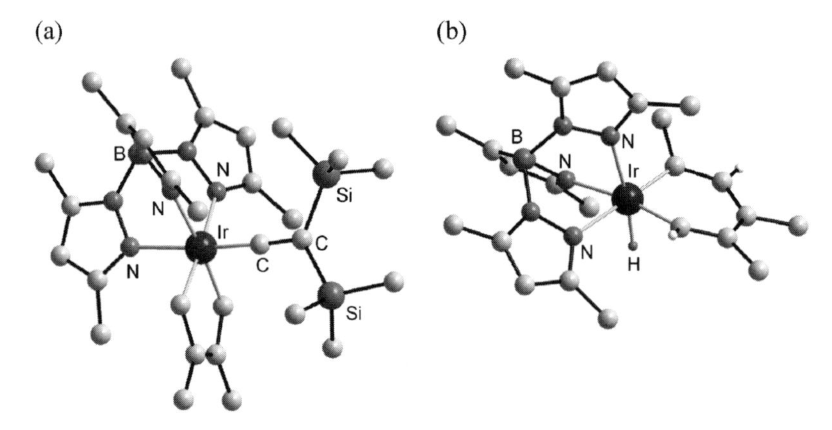

Fig. 7. Molecular structures of (a) the vinylidene complex **587** (hydrogen atoms omitted) and (b) the iridabenzene **587** (Tp* hydrogen atoms omitted).

Scheme 64. Conditions: (i) $Me_3SiC{\equiv}CSiMe_3$, PhH, 90 °C, 5 h; (ii) $HBF_4 \cdot OEt_2$, Et_2O, 10 min; (iii) PPN^+Cl^-; (iv) HCl, PhH, 20 °C, 3 d.

upon treatment with TFA in THF the iridabenzene complex **592** is obtained in ca. 80% yield (Section II-E).

3. Metal Carbonyls

a. M(I) Dicarbonyls

The earliest reported examples were $BpRh(CO)_2$ (**593**) and $Tp^*Rh(CO)_2$ (**430**), which appeared in the patent literature for use as hydroformylation catalysts in 1971,[187] though with no explicit synthetic detail. Several workers subsequently reported the synthesis of **593**,[45,188] and the Bp* analogue (**262**),[188] from $Rh_2(\mu\text{-Cl})_2(CO)_4$ and KBp^x ($Bp^x = Bp$, Bp*) and similarly obtained $BpIr(CO)_2$ (**594**) from '$Na[Ir_2Cl_{4-8}(CO)_4]$'.[188] The analogous reaction of $Rh_2(\mu\text{-Cl})_2(CO)_4$ with KTp*, however, yields only the pyrazole-bridged complex $Rh_2(\mu\text{-pz}^*)_2(CO)_4$,[39] though **430** is obtained using the monomeric precursor 'RhCl(CO)$_2$'.[189] Complications also arose in the attempted synthesis of TpRh(CO)$_2$ and (pzTp)Rh(CO)$_2$ *via* the same route, which instead yielded highly insoluble materials that were formulated as (RTp)$_2$Rh$_2$(CO)$_3$ (R = H (**266**), pz (**477**)).[39,40] These were also obtained from carbonylation of the respective (RTp)Rh(LL) (LL = 2 C_2H_4, 1,5-cod) complexes.[40,44] Though the structures have not been firmly established, they are believed to be polymeric, bridged by exo-polydentate Tp and pzTp ligands, with all three carbonyls adopting μ_2-coordination modes. One report[45] does suggest the existence and transient isolation of the desired dicarbonyls, but also describes their spontaneous decomposition under ambient conditions. It is, nonetheless, surprising that no further attempts have apparently been made to obtain these fundamentally important materials.

In contrast, most of the known dicarbonyl complexes are readily obtained by standard routes (NaTp^x or KTp^x) commencing from $M_2(\mu\text{-Cl})_2(CO)_4$, 'MCl(CO)$_2$', or M(acac)(CO)$_2$ (M = Rh, Ir); or by carbonylation of the respective diene complexes. Current examples are summarised in Table IV (Chart 11) with the respective synthetic route.

The solution and solid-state structures of the dicarbonyls have been discussed at length and the same conclusions reached as for alkene complexes (Section III-B.1); viz. (i) mixtures of 4-(κ^2-Tp^x) and 5-coordinate (κ^3-Tp^x) isomers exist in solution; (ii) dynamic exchange of the pyrazole rings, involving both isomers, is typically facile at ambient temperature; (iii) the electronic and steric nature of the pyrazole

TABLE IV
DICARBONYL COMPLEXES Tp^{x}M(CO)$_2$

Tp^x	Metal	Number	Route	Reference	Tp^x	Metal	Number	Route	Reference
Et$_2$Bp	Rh	**595**	D	138	Tp^{Ph}	Rh	**610**	C	53
Ph$_2$Bp	Rh	**596**	D	65	Tp^{PhCl}	Rh	**611**	D	194
MePhBp^{Me_2}	Rh	**597**	D	64	Tp^{Ph,Me}	Rh	**612**	A	52
pz$_2$BBN	Rh	**598**	D	65	Tp$^{Tol,4'Bu}$	Rh	**613**	C	195
(pz^{Me_2})$_2$BBN	Rh	**599**	D	65	Tp^{Mes}	Rh	**614**	C	63
MeTp^{Me_2}	Rh	**600**	D	64	Tp^{Mes*}	Rh	**615**	C	63
Tp^{Np}	Rh	**601**	D	190	Tp^{An}	Rh	**616**	C	57
Tp^{Me}	Rh	**602**	D	48	(pz^{Me_2})Bp^{Ph,Me}	Rh	**617**^{b}	A	61
Tp^{Me_2,4Cl}	Rh	**469**	–	48	Tp^a	Rh	**618**	C	52
Tp^{iPr_2}	Rh	**603**	C	74	Tp^{a,Me}	Rh	**619**	C	52
Tp^{iPr,4Br}	Rh	**604**	D	48	Tp^{a*}	Rh	**620**	C	55
Tp$^{(CF_3)_2}$	Rh	**605**	D	51	Tp^{a*},3Me	Rh	**621**	C	55
Tp^{CF_3,Me}	Rh	**606**	D	51	Tp^b	Rh	**622**	C	52
Tp^{Br_3}	Rh	**607**	M	191	Tp	Ir	**265**	C	84
Tp^{Menth}	Rh	**608**	D	192	Tp^{Me_2}	Ir	**623**	D	195
Tp^{Mementh}	Rh	**609**	D	193	pzTp	Ir	**624**	C	83

Synthetic routes: A = [M(acac)(CO)$_2$] + M'Tp^x; D = [MCl(CO)$_2$]$_2$ + M'Tp^x; M = [MCl(CO)$_2$] + M'Tp^x; C = Carbonylation of [Tp^{x}M(diene)].

^{a}Synthesis not reported.

^{b}κ^2-*N*(pz^{Me_2})-*N*(pz^{PhMe}).

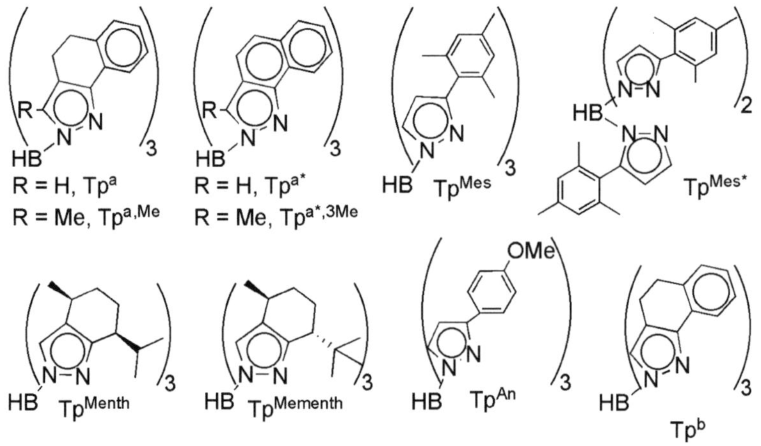

CHART 11. Ligands employed in metal dicarbonyls.

substituents has an, as yet unquantified, influence on the relative isomeric distribution.

More interestingly, during the synthesis of **611** from $Rh_2(\mu\text{-}Cl)_2(CO)_4$, it was found that using a single equivalent of $TlTp^{PhCl}$ did not afford a 50% yield of **611**, but rather the dimeric complex **625**, which was identified crystallographically (Fig. 8).[194] Significantly, the infrared (four CO bands) and ^{1}H-NMR spectroscopic data (3 unique pz environments) for **625** were found to be strikingly similar to those obtained for **610** and **616**, which had previously been interpreted as being indicative of a mixture of two κ^2-Tp^x isomers (**A** and **B**, Scheme 8, Section III-B.1).[53,57] The subsequent isolation of a compound that is spectroscopically comparable to **625** from the 1:1 reaction of $[RhCl(CO)_2]_2$ and $TlTp^{An}$ might thus seem to call this earlier assignment into question, though the authors deferred comment in lieu of crystallographic confirmation.[194] It should, however, be noted that the existence, in solution, of two κ^2-Tp^x isomers for relevant bis(carbonyl) and bis(alkene) complexes has been widely assigned on the basis of IR or ^{1}H-NMR spectroscopic data. As such, if the possible mis-assignment of the Tp^{Ph} and Tp^{An} complexes is verified, the wider implications are considerable. A further striking result from this report was the crystallographic characterisation of three different isomers of (κ^2-*N,N′*-Tp^{PhCl})$Rh(CO)_2$ (Fig. 8).

b. M(I) Mono-carbonyls

The dicarbonyls **262**, **430**, **469**, **593–597** and **612** have each served as precursors to a range of monocarbonyl complexes (Table V) through their interaction with extraneous donors; reactions that are performed under thermal, photochemical or even ambient conditions.

In contrast, Tp*Rh(η^2-C_2H_4)(CO) (**148**) is obtained by carbonylation of the parent bis(ethylene) complex **118**,[70] a notable reaction since for most bis-olefin

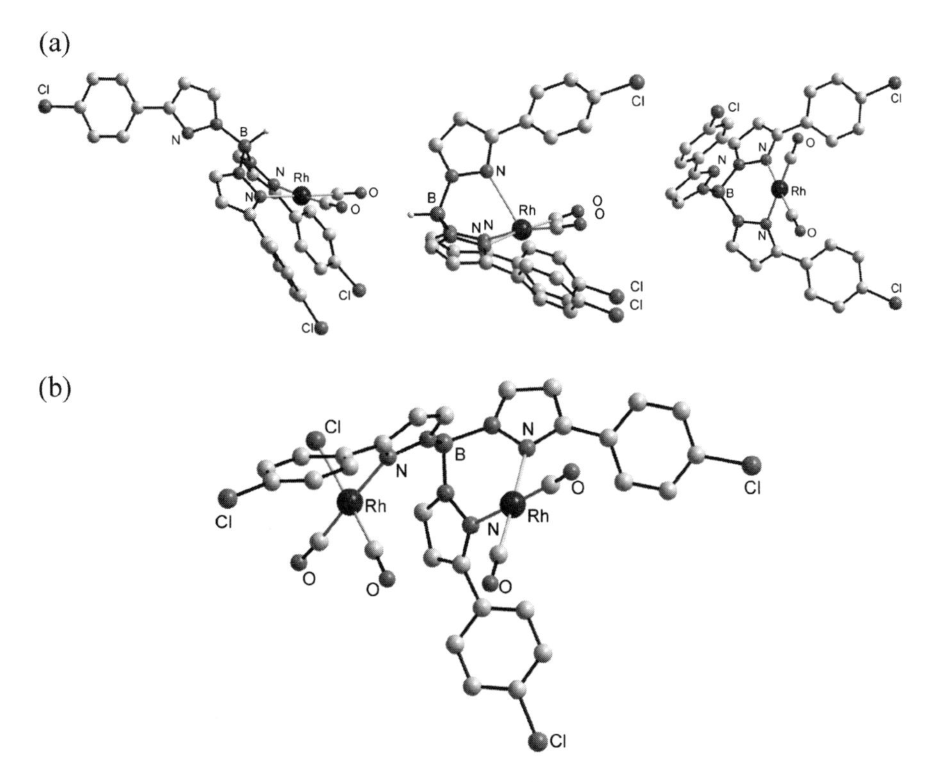

FIG. 8. Molecular structures of (a) three isomers of $(Tp^{PhCl})Rh(CO)_2$ and (b) the binuclear complex $(Tp^{PhCl})Rh_2Cl(CO)_4$ **625** (hydrogen atoms omitted).

TABLE V
COMPLEXES $Tp^xM(CO)(L)$ DERIVED FROM DICARBONYLS

L	Number	Reference	L	Number	Reference
$Tp^{Me_2}Rh(CO)L$			$Tp^{Me_2,4Cl}Rh(CO)L$		
PMe_3	**626**	197,198	PMe_3	**636**	198
PCy_3	**627**	199	PCy_3	**637**	198
PPh_3	**628**	41,198,199	PPh_3	**638**	198
$PMePh_2$	**629**	198	$PMePh_2$	**639**	67,198
PMe_2Ph	**630**	198	PMe_2Ph	**640**	67,198
$PTol_3$	**631**	199	$P(OMe)_3$	**641**	198
$P(m\text{-}Tol)_3$	**632**	199	$P(O^iPr)_3$	**642**	198
$P(OMe)_3$	**633**	198	$P(OPh)_3$	**643**	198
$P(Oph)_3$	**634**	41,198			
$P(NMe_2)_3$	**635**	199			
$BpRh(CO)L$			$Et_2BpRh(CO)(L)$		
PCy_3	**644**	200	CN^tBu	**649**	138
PPh_3	**645**	188,200	$CNCH_2SO_2(Tol\text{-}p)$	**650**	138
$P(p\text{-}An)_3$	**646**	200	$Tp^{Ph,Me}Rh(CO)(PPh_3)$	**651**	202
$P(NC_2H_4)_3$	**647**	200	$BpIr(CO)(PPh_3)$	**652**	188
$Bp^{Me_2}Rh(CO)L$			$TpIr(CO)(\eta^2\text{-}C_2H_4)$	**228**	94
PPh_3	**648**	201			
Py	**566**	182			

complexes carbonylation affords exclusively the respective dicarbonyl. The complex $Tp^{CF_3,Me}Ir(\eta^2\text{-}C_2H_4)(CO)$ (**229**) has also been reported, though without synthetic detail,[95] while **617** is reported to react with one equivalent of PPh_3 to afford the respective monophosphine complex *in situ*, though it has not been isolated.[61]

Coordination of a second equivalent of phosphine has been observed when $Tp^{Me_2,4Cl}Rh(CO)_2$ (**469**) is treated with excess of PR_3, thereby affording $Tp^{Me_2,4Cl}Rh(CO)(PR_3)_2$ ($PR_3 = PMe_3$ (**653**),[104] PMe_2Ph (**654**),[104] $PMePh_2$ (**655**)[67,104]) in which the $Tp^{Me_2,4Cl}$ ligand adopts the rare κ^1-*N* coordination mode.‡ This chemistry is directly analogous to that observed with the bis(ethylene) complex $Tp^*Rh(C_2H_4)_2$ (**118**), which ultimately results in (step-wise) dechelation of the Tp* ligand (Section II-C.1),[104] though this is not observed in the present case. However, the $Tp^{Me_2,4Cl}$ ligand *is* extruded when **469** is treated with excess dppp, affording $[Rh(dppp)_2(CO)][Tp^{Me_2,4Cl}]$ (**656 · Tp^{Me_2,4Cl}**). It is interesting to note that for dicarbonyl complexes only the $Tp^{Me_2,4Cl}$ ligand undergoes such dechelation, Tp^{Me_2} affording just simple mono-phosphine complexes.

c. M(II) Carbonyls

A small number of rhodium(II) carbonyls have been prepared by chemical oxidation of $Tp^*Rh(L)(CO)$ with $[Cp_2Fe][PF_6]$, affording $[Tp^*Rh(L)(CO)]^+$ ($L = PCy_3$ (**627**$^+$), PPh_3 (**628**$^+$),[203] $P(C_6H_4Me\text{-}4)_3$ (**631**$^+$), $P(C_6H_4Me\text{-}3)_3$ (**632**$^+$), $P(OPh)_3$ (**634**$^+$), $P(NMe_2)_3$ (**635**$^+$))[199] as the PF_6 salts. These are rare examples of well characterised, mononuclear rhodium(II) complexes that are believed to be stabilised by coordination of the third pyrazole donor (typically lying pendant in the neutral Rh(I) precursors), affording in each case square-pyramidal coordination geometries. This unprecedented redox-induced $\kappa^2 \rightarrow \kappa^3$ isomerisation, which can also be induced electrochemically,[199,203,204] is undoubtedly a significant contribution to the electronic/steric debate in respect of the favoured geometry in Tp^x complexes.

Chemical oxidation of $Tp^xRh(CO)_2$ (**430**), however, ultimately affords the *N*-protonated Rh(I) complex $[\{\kappa^2\text{-}HB(pz^*)_2(pz^*H)Rh(CO)_2]$ (**657**), which can also be obtained by protonation of **430** with HBF_4. Chemical (Cp_2Co) and electrochemical reduction of **657** have, however, been shown to regenerate **430**, thus demonstrating the intermediacy of **430**$^+$, generated by 1-electron transfer, *en route* to **657**.

d. M(III) Carbonyls

In contrast to rhodium complex **430**, protonation of $Tp^xIr(CO)_2$ ($Tp^x = Tp$ (**265**), Tp* (**623**)) with $HBF_4 \cdot OEt_2$ affords the iridium(III) hydrides $[Tp^xIrH(CO)_2]BF_4$ ($Tp^x = Tp$ (**658 · BF$_4$**), Tp* (**659 · BF$_4$**)),[196,205] and is reversed in the presence of a

‡In the case of **655** a weak 'agostic' B–H–Rh interaction was established on the basis of crystallographic data, making this the first example of a Tp^x complex to exhibit the rare κ^2-*N,H* binding mode.[67]

non-nucleophilic base DBU. In contrast, the strongly nucleophilic bases $^{n}Bu^{-}$ and MeO^{-} attack a carbonyl ligand of **659**$^{+}$ to afford Tp*IrH(C(=O)R}(CO) (R = ^{n}Bu (**660**), OMe (**661**)).[196] Similarly, in methanolic KOH **658**$^{+}$ affords TpIrH{C(=O)OMe}(CO) (**662**),[205] also accessible by heating **265** in methanol, while comparable methods afford TpIrH{C(=O)OR}(CO) (R = Et (**663**), H (**664**)) and the carbamoyl complexes TpIrH{C(=O)NHR}(CO) (R = ^{n}Pr (**665**), C_6H_{11} (**666**), *R*-CH(CH_3)C_6H_{11} (**667-*R***), *S*-CH(CH_3)C_6H_5 (**668-*S***)) derived from the respective primary amines.[206] Each of **662–668** reacts with $HBF_4 \cdot OEt_2$ to afford **658** · $\mathbf{BF_4}$; while **664** will also react with strong base to give the dihydride TpIrH_2(CO) (**669**). The analogous dihydride (pzTp)IrH_2(CO) (**670**) forms directly when **624** is refluxed in wet acetonitrile.

The hydride Tp*IrH_2(CO) (**323**) has been obtained by carbonylation of the thiophene analogue Tp*IrH_2(SC_4H_4) (**521**)[121] and, in admixture with Tp*IrH_4 (**322**), from the photolytic activation of methanol by either Tp*IrH_2(η^2-COE) (**207**) or Tp*Ir(1,5-COD) (**179**);[129,207] carbonylation (3 atm) of the product mixture in each case affording exclusively **323**.[207] In contrast, the rhodium analogue Tp*RhH_2(CO) (**325**) is the sole product from photolysis of Tp*Rh(1,3-COD) (**171**) in methanol.[77,78,130]

Substitution of labile ligands by CO in M(III) precursors has also been employed in several other cases, as in the synthesis of Tp*Ir(2-thienyl)$_2$(CO) (**520**) from the respective thiophene complex[120] and Tp*Ir(O_2CCF_3)$_2$(CO) (**671**) from the water adduct.[41] More typically employed are dinitrogen adducts, from which Tp*Ir (C_6H_5)$_2$(CO) (**486**)[90,117] and [Tp^{Me_2}Ir(Me)(PMe_3)(CO)]BAr^f_4 (**443** · $\mathbf{BAr^f_4}$)[141,155] were prepared directly. The latter complex has also been obtained by thermal activation of acetaldehyde, *via* the isolable adduct [Tp*Ir(Me)(O=CHMe)(PMe_3)]BAr^f_4 (**446** · $\mathbf{BAr^f_4}$), as has the tolyl analogue **448** · $\mathbf{BAr^f_4}$ (from 4-tolualdehyde), while the comparable activation of 4-anisaldehyde has been effected using the butadiene complex **234** (Scheme 37, Section II-C.1).[158]

Thermal C–H activation has also been encountered in the complex $Tp^{CF_3,Me}$Ir (CO)(η^2-C_2H_4) (**229**), which spontaneously rearranges to the hydrido-vinyl **289** under ambient conditions. Under comparable conditions the rhodium analogue Tp*RhH(CH=CH_2)(CO) (**290**), in contrast, preferentially reverts to the π-alkene form (**148**, Section III-B.1).[95] However, **148** cleanly activates benzene under mild thermal conditions (70–100 °C) to afford Tp*RhH(C_6H_5)(CO) (**295**). The photochemical activation of benzene is also effected by **148** (and **430**), affording **295** in admixture with Tp*Rh(Et)(Ph)(CO) (**438**), which results from ethylene insertion into the Rh–H linkage of **295**.[153] Complex **438** will insert CO to afford Tp*Rh {C(=O)Et}(C_6H_5)(CO) (**458**) which liberates phenyl-ethylketone upon treatment with $ZnBr_2$.

Photochemical activation of benzene and a variety of saturated hydrocarbons has also been reported for **430**,[151,152,208,209] though in most instances the resulting hydrides were too labile to isolate, their conversion to the chlorides (stirring with CCl_4) thus being favoured. In this fashion Tp*Rh(R)Cl(CO) (R = ^{n}Pn (**431**), ^{n}Hex (**432**), Cy (**433**), ^{n}Hep (**434**), ^{i}Oct (**435**)) were each isolated in high yield.[151] The methyl analogue (**436**) was obtained by treating Tp*RhH(Cy)(CO), generated *in situ*, with methane prior to chlorination,[152] while several other hydrides

(R = *p*-Xyl, Mes, Tol) were photochemically generated *in situ*.[209] Interestingly, in the presence of Et_3SiH, Si–H activation predominates over C–H, affording exclusively Tp*IrH($SiEt_3$)(CO) (**437**), which is isolable and stable. This greater thermodynamic stability is reiterated by the observation that in *n*-pentane solution, the formation of **437** is competitive with that of **431**, despite the silane being in much lower concentration.[151]

A small number of classical oxidative substitutions have also been employed, the earliest example being addition of $2X_2$ (X = I, Cl) to the dimeric $Tp_2Rh_2(CO)_3$ (**266**), affording $TpRhX_2(CO)$ (X = I (**672**), Cl (**673**)), which remains the only chemistry reported for **266**.[44] The analogous reactions of $(pzTp)_2Rh_2(CO)_3$ (**477**) with Br_2, I_2 and excess C_3F_7I were also reported, affording (pzTp)Rh(X)(Y)(CO) (X = Y = Br (**674**), X = Y = I (**675**), X = C_3F_7, Y = I (**476**)). In spite of this precedent, the later preparation of TpRh(R_f)I(CO) (R_f = $CF_2C_6F_5$ (**273**), C_3F_7 (**275**)) was effected by carbonylation of the respective Rh(III) ethylene complexes **271** and **272**.[106] However, $Tp^xM(R_f)I(CO)$ (Tp^xM = Tp*Rh, R_f = $CF_2C_6F_5$ (**478**), C_3F_7 (**480**); Tp^xM = TpIr, R_f = $CF_2C_6F_5$ (**479**), C_3F_7 (**481**)) were prepared by oxidative substitution of the respective dicarbonyls (Section II-C.1, Scheme 41).

Somewhat less straightforward chemistry was observed upon treatment of Tp*Rh(CO)(L) with MeI. Where L = PMe_3 (**626**) and PMe_2Ph (**630**) the isolated products were identified as the salts [Tp*Rh(Me)(CO)(L)]I (L = PMe_3 (**676 · I**), PMe_2Ph (**677 · I**)) on the basis of spectroscopic data and, for **676 · I**, confirmed crystallographically.[201] These materials undergo slow migratory insertion of CO into the Rh–Me linkage, and coordinate the iodide to afford the acyl complexes Tp*Rh{C(=O)Me}I(L) (L = PMe_3 (**678**), PMe_2Ph (**679**)). Comparable chemistry was reported where L = PPh_3 (**628**) and $PMePh_2$ (**630**), though the products were not isolated. However, with dicarbonyl **430** only the acyl complex Tp*Rh{C(=O) Me}I(CO) (**680**) was obtained, though the intermediacy of an unobserved salt was postulated. The apparently greater facility of migration in the dicarbonyl was attributed to the lower electron density at the metal, a suggestion that correlates well with the observed influence of phosphine basicity upon the kinetics of the reaction.

A handful of more elaborate M(III) monocarbonyls have also been prepared, by a variety of routes. For instance, the H_2O ligand in each of the iridacycloheptatriene complexes **338**, **503**, **504** and **340** (Section III-B.1, Scheme 23, II-C.1, Chart 12) has been displaced by CO to afford **681**–**683** and **553** respectively, the latter also being obtained by carbonylation of the bicyclic carbene **341** (Scheme 23).[133,134]

More exotic is the ketyl-complex **684** and its PPh_3 analogue **687**, which are obtained from the reaction of *o*-chloranil with the carbonyl complexes **430** and **628** respectively (Scheme 65).[210] The isolation of these 'carbonyl insertion products', rather than simple Rh(III) catecholates (as observed in the analogous reactions with Cp*Rh(CO)L),[211] has been attributed to the stabilising influence of the Tp* ligand over that of Cp*. However, decarbonylation to the catecholates can be achieved either by UV irradiation (for **684**) or thermally (for **685**). The catecholates are susceptible to ligand exchange and have also been chemically oxidised with [NO][PF_6] to afford cationic species. These have been extensively studied by

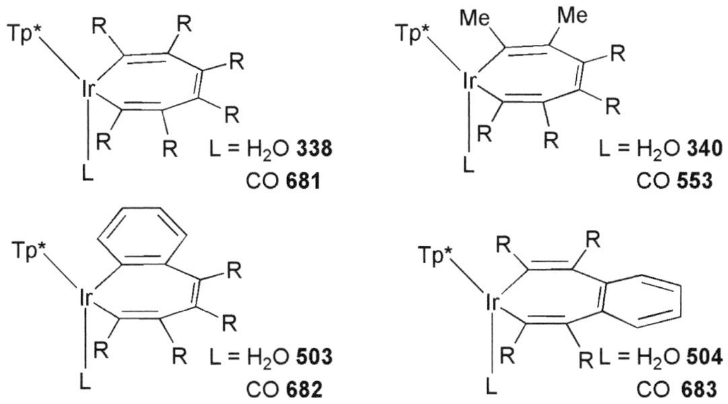

CHART 12. Carbonylation of Iridacycloheptatrienes. R = CO_2Me.

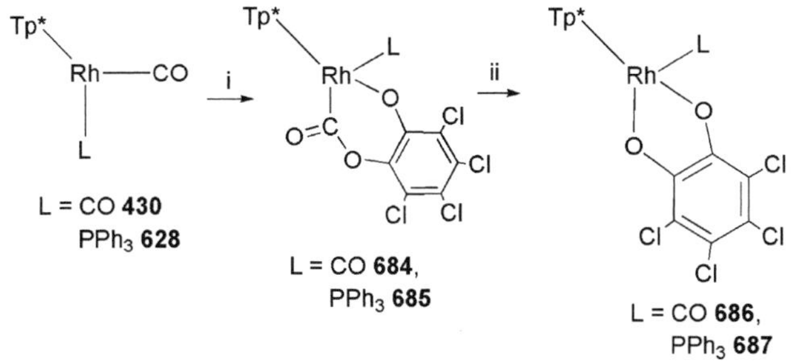

SCHEME 65. Conditions: (i) *o*-$C_6Cl_4O_2$; (ii) *hν*-UV (L = CO), Δ, (L = PPh_3), –CO.

voltammetry and ESR spectroscopy, the results of which are indicative of Rh(III) complexes, in which the unpaired electron is localised to the catechol ligand.

e. Metal Acyls

A limited number of metal-acyl complexes have been reported, most encountered serendipitously while studying the reactivity of the metal-carbonyls. For instance, the complexes $Tp^xIrH\{C(=O)R\}(CO)$ (Tp^x = Tp*, R = ^{n}Bu (**660**), OMe (**661**);[196] Tp^x = Tp, R = OMe (**662**),[205] OEt (**663**), OH (**664**), NH^nPr (**665**), NHC_6H_{11} (**666**), *R*-$CH(CH_3)$-C_6H_{11} (**667-*R***), *S*-$CH(CH_3)C_6H_5$ (**668-*S***))[206] were each obtained from the reaction of the respective cationic complexes $[Tp^xIrH(CO)_2]^+$ (Tp^x = Tp (**658$^+$**) Tp* (**659$^+$**)) with the respective nucleophiles MR (M = Na, Li) or primary amines. Several other examples arose from migratory insertion of CO, as in the formation of Tp*Rh{C(=O)Me}I(L) (L = PMe_3 (**678**), PMe_2Ph (**679**), CO (**680**)) upon oxidative addition of MeI to Tp*Rh(CO)(L),[201] and that of the catecholates **684** and **685** directly from the reaction of the Rh(I) precursors and *o*-$C_6Cl_4O_2$ (Scheme 65).[210]

Similarly, Tp*Rh{C(=O)Et}(C_6H_5)(CO) (**439**) was found to form by migratory insertion when Tp*Rh(Et)(C_6H_5)(CO) (**438**) is treated with CO.[153]

The oxidative addition of aldehydic C–H bonds has also been found to give rise to acyls, as in the formation of Tp*RhH{C(=O)R′}(PR_3) (R′ = $C_6H_4NO_2$-4; R = Ph (**688**), C_6H_4F-4 (**689**)) from O = CHR′ and the respective bis(phosphine) complexes.[180] Interestingly, these acyls exhibit unusually high thermal stability, toluene solutions of **688** exhibiting merely 30% decomposition after 21 h at 70 °C, the break-down product being identified (spectroscopically) as Tp*Rh(CO)(PPh_3) (**628**).

f. Metal Nitrosyls

It is remarkable that not a single example of a rhodium or iridium nitrosyl with a Tp^x ligand has been reported. In a similar manner, analogues based on arenediazenido and thionitrosyl ligands have yet to appear for group 9 metals, despite such compounds being reported for earlier transition metal centres.

4. Metal Isonitriles

The synthesis and study of these materials has been somewhat limited, though many recent advances in C–H activation have relied heavily on the chemistry of such complexes (*vide infra*).

The earliest examples were obtained by treating the dicarbonyl $Et_2BpRh(CO)_2$ (**595**) with one or two equivalents of the isocyanides CNC_6H_4Me-4, $CNCH_2SO_2C_6H_4Me$-4 and CN^tBu.[138] In this way Et_2BpRh(L)(CNR) (L = CO, R = ^tBu (**649**), $CH_2SO_2C_6H_4Me$-4 (**650**); L = CNR, R = ^tBu (**690**), $CH_2SO_2C_6H_4Me$-4 (**691**), C_6H_4Me-4 (**692**)) were prepared, though the CNC_6H_4Me-4 analogue of **649** and **650** remains elusive, bis-substitution being spontaneous. Each of the isonitrile complexes was subsequently found to react with RX (RX = MeI, I_2) (presumably by oxidative addition), and also with $HgCl_2$ to afford 1:1 adducts, the structures of which were not established.

The displacement of more labile ligands has subsequently been applied to the synthesis of a range of complexes of Rh(I) and Rh(III). Thus, sequential treatment of $Rh_2(\mu\text{-}Cl)_2(\eta^2\text{-}C_2H_4)_4$ with MTp* (M = Na, K) and CNR has afforded the complexes Tp*Rh$(CNR)_2$ (R = $CH_2{}^tBu$ (**509**),[167,212] C_6H_4Me-4 (**693**),[150] $C_6H_3Me_2$-2,6 (**694**), Me (**695**)),[212] and similarly BpRh$(CN^tBu)_2$ (**263**).[59] In contrast, Bp*Rh$(CN^tBu)_2$ (**264**) was prepared from the isolated bis(ethylene) complex Bp*Rh$(\eta^2\text{-}C_2H_4)_2$ (**120**); an approach that with Tp*Rh$(\eta^2\text{-}C_2H_4)_2$ (**118**) yields only the mono-substitution products, viz.: Tp*Rh$(\eta^2\text{-}C_2H_4)$(CNR) (R = Cy (**149**),[58] ^tBu (**150**),[58,70] $C_6H_3Me_2$-2,6 (**151**)[72]), while with the Tp ligand only the binuclear complex $Tp_2Rh_2(\mu\text{-}CNCy)_3$ (**696**) has been isolated. It is noted that **696** is a direct analogue of $Tp_2Rh_2(\mu\text{-}CO)_3$ (**266**), the only simple carbonyl complex obtained for the TpRh fragment (Section II-D.2).

The synthesis of the rhodium dichlorides Tp*$RhCl_2$(CNR) (R = ^tBu (**697**),[41] CH_2Ph (**698**),[41] $CH_2{}^tBu$ (**373**)[142]) was effected thermally (toluene reflux) from the precursors Tp*$RhCl_2$(MeOH) (**76**)[41] and Tp*$RhCl_2$(NCMe),[142] while the comparable reaction of Tp*Ir$(C_6H_5)_2(N_2)$ (**211**) with CN^{t}Bu afforded Tp*Ir $(C_6H_5)_2$(CN^{t}Bu) (**486**) the first, and still only, example of an iridium isonitrile complex bearing a Tp^x ligand.[90] More intriguing is the recent report of the rhodium(III) complex Tp*Rh(S_5)(CNC$_6$H$_3$Me$_2$-2,6) (**699**, Fig. 9),[213] which was obtained by ambient-temperature displacement of MeCN from the pentasulphido chelate Tp*Rh(S_5)(NCMe) (**700**), itself prepared from the respective cyclooctene complex (**164**) and S_8.

The majority of work with isonitriles has focused on the use of selected examples as precursors to a wide range of rhodium-alkyl and aryl complexes in which the ancillary isonitrile is typically retained and not directly involved in the reaction of interest. In the simplest case, discussed in Section II-C.1 the dichloride complex **373** was found to react with a variety of Grignard reagents to afford complexes of the form Tp*Rh(R)X(CNCH$_2$ ^{t}Bu) (Table VI, Scheme 31).[142,143] Though scrambling of the metal halide is observed when using bromo- or iodo-magnesium organyls

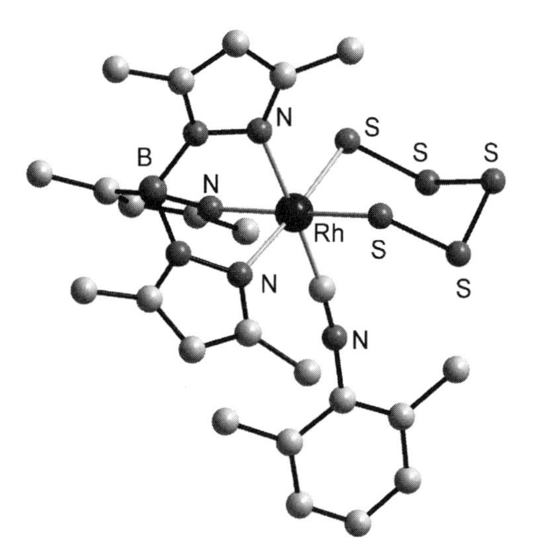

Fig. 9. Molecular structure of the pentasulfide chelate complex Tp*Rh(S_5)(CNC$_6$H$_3$Me$_2$-2,6) (**700**) (hydrogen atoms omitted).

Table VI
Isonitrile Complexes Tp*Rh(R)(X)(CNCH$_2$ ^{t}Bu) Derived from Grignard Reagents

RMgZ	R	X	Number	Reference	RMgZ	R	X	Number	Reference
MeMgCi	Me	Cl	**391**	142	^{n}BuMgCl	^{n}Bu	Cl	**399**	142
(CD$_3$)MgI	CD$_3$	Cl	**392**	142	^{s}BuMgCl	^{s}Bu	Cl	**400**	142
		I	**393**	142	^{n}PnMgCl	^{n}Pn	Cl	**401**	142
EtMgCl	Et	Cl	**394**	143	^{n}HexMgCl	^{n}Hx	Cl	**402**	142
^{n}PrMgCl	^{n}Pr	Cl	**395**	142,143	H$_2$C=CHMgBr	CH=CH$_2$	Cl	**570**	141
^{i}PrMgCl	^{i}Pr	Cl	**396**	142,143		CH=CH$_2$	Br	**569**	141
^{c}PrMgBr	^{c}Pr	Cl	**397**	142					
	^{c}Pr	Br	**398**	142					

quantitative conversion to the chloride is cleanly achieved by sequential treatment with AgOTf and [^{n}Bu$_4$N]Cl (Scheme 31).[142] In several cases (where R = CD$_3$, ^{n}Pr, ^{c}Pr, ^{n}Pn, ^{n}Hex) the respective hydrido complexes (**424**, **378**, **380**, **384** and **701**) have been obtained as isolable materials by treating the chlorides with Cp$_2$ZrH$_2$; a methodology that would seem more widely applicable, though this has not been reported.[143]

The cyclopropyl-hydrido complex **415** exhibits appreciable thermal instability, such that it rearranges quantitatively in ambient temperature benzene solutions to the rhodacyclobutane complex Tp*Rh(σ,σ'-CH$_2$CH$_2$CH$_2$)(CNCH$_2$^{t}Bu) (**154**, Scheme 34, Section II-C.1).[72] Thermal isomerisation of **154** (65 °C, 2.5 h) affords a π-propene complex (**153**) that is able to activate benzene. Alternatively, in the presence of excess CNCH$_2$^{t}Bu, **154** thermally inserts two equivalents of the isonitrile to ultimately afford the rhodacyclohexane complex **427** (Scheme 34).

The bis(isonitrile) complex Tp*Rh(CNCH$_2$^{t}Bu)$_2$ (**509**) has been extensively explored, and found to undergo reversible protonation by HBF$_4$ · OEt$_2$ to afford [Tp*RhH(CNCH$_2$^{t}Bu)$_2$]BF$_4$ (**701 · BF$_4$**).[212] This contrasts the dicarbonyl analogue (**430**), for which *N*-protonation of the Tp* ligand predominates.

A more important reactivity of **509** is its ability to activate benzene under ambient temperature photolysis to afford Tp*RhH(C$_6$H$_5$)(CNCH$_2$^{t}Bu) (**425**).[167] In seeking to enhance the facility of this reaction, the complex was treated with PhN$_3$, resulting in conversion of one isonitrile ligand to an η^2-carbodiimide, thus affording the moderately stable but photolabile complex **395**.[150] This was the first reported example of a stable η^2-carbodiimide complex of rhodium, analogues of which were similarly prepared from the 2,6-xylyl (**694**) and methyl (**695**) isonitrile complexes (Scheme 66). In order to enable crystallographic characterisation a further example was prepared from the *o*-tolylisonitrile complex **693** and N$_3$C$_6$H$_3$Me$_2$-2,4 (Fig. 10).

Even before crystallographic confirmation, spectroscopic data had pointed to the 'metallaaziridine' being the dominant canonical form, since these data were more consistent with this Rh(III) species than a Rh(I) π-complex. However, at elevated temperature ^{1}H-NMR spectroscopic data are consistent with 'propeller-type'

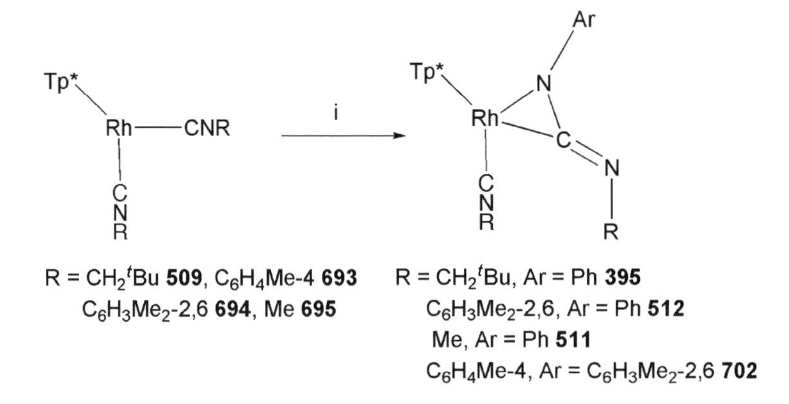

SCHEME 66. Conditions: (i) ArN$_3$, hexane (Ar = Ph, 2,6-Xyl).

rotation of the carbodiimide ligand (ΔE_{act} 20 kcal mol^{-1}; cf. 15 kcal mol^{-1} for the archetypal CpRh(η^2-C_2H_4)$_2$, implying a significant contribution from the π-complex description under these conditions.

As anticipated, the carbodiimide ligands are significantly more labile than the isonitriles, such that photolytic activation of benzene by **395** proceeds several orders of magnitude faster than with **509**, though with **511** and **512** the rate-enhancement is somewhat less. The lability of **395** is further illustrated by its capacity to undergo intramolecular aryl C–H activation in the solid state both thermally (1 month) and more rapidly (14 h) under photolysis. In each case the product is Tp*RhH (*o*-$C_6H_4N{=}C{=}NCH_2{}^tBu$)($CNCH_2{}^tBu$) (**515**); a remarkable outcome given the considerable separation (>4 Å) of Rh and its most proximal C–H within **395**. Though several mechanistic possibilities were suggested, no definitive conclusion was reached.

A further peculiarity of **395** is its reaction with an excess of neopentyl isocyanide, which instead of regenerating **509** (as observed for Tp*RhH(C_6H_5)($CNCH_2{}^tBu$) (**425**)[144]) is believed to result in insertion of $CNCH_2{}^tBu$ into the Rh–N bond, to afford **703**, the precise structure of which has not been determined, though ^{1}H-NMR spectroscopic data would appear consistent with *either* structure **a** *or* **b** as illustrated in Chart 13.

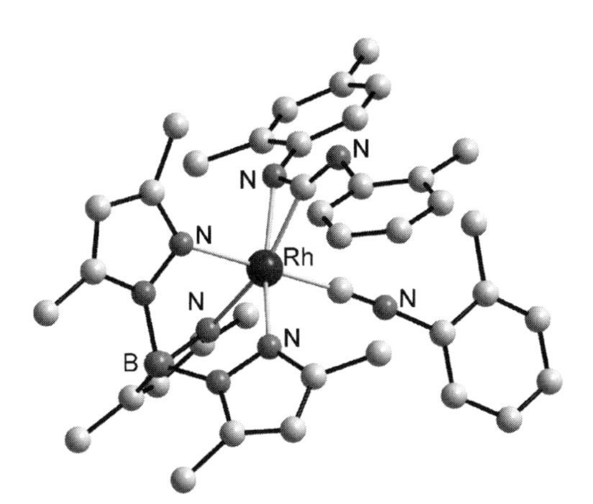

FIG. 10. Molecular structure of the carbodiimide complex **702** (hydrogen atoms omitted).

CHART 13. R = $CH_2{}^tBu$. Possible structures for **703**.

The capacity of **395** to effect hydrocarbon C–H activation has been extensively explored in great detail, with respect to the mechanism, kinetics and selectivity.[143,146,147,149] The photolysis of **395** in neat hydrocarbon solutions generates the complexes Tp*RhH(R)($CNCH_2{}^{t}Bu$) (**396–410**), which were characterised spectroscopically, but isolated as the respective chlorides (generated *via* reaction with CCl_4) since the hydrido complexes themselves are unstable toward purification.[145] Analytically pure samples of the hydrides were again obtained by treating the chlorides with Cp_2ZrH_2.[143]

For the aromatic substrates (toluene and mesitylene) competition is observed between benzylic and aromatic C–H activation. Thus, toluene affords the *meta*- (**402**) and *para* (**403**) tolyl-complexes, alongside the benzyl system **401**. Upon heating, the benzyl complex is lost, and a 2:1 thermodynamic mixture of **402**:**403** obtained. Similarly, mesitylene affords the 3,5-dimethylbenzyl (**405**) and mesityl (**404**) complexes (3:1), though comprehensive characterisation of the latter was precluded by its failure to convert cleanly to the corresponding chloride, the dichloride **373** instead being obtained.

In contrast, chlorocarbons undergo exclusive activation of a terminal methyl C–H, with no evidence for C–Cl addition to the metal.[147] Moreover, with 2-chloropentane only the 4-chloropentyl product (**410**) was obtained, the absence of the 2-chloro isomer being attributed to rapid β-chloride elimination to afford Tp*RhHCl($CNCH_2{}^{t}Bu$) (**422**) (28% by-product) and pent-1-ene.

E. *Delocalised Metallacycles*

Several of these immensely interesting compounds have been prepared for iridium, though the earliest examples date back to only a decade.

1. Iridapyrroles

The original report[113] outlined the formation of an iridapyrrole (**474**, Schemes 40 and 67) by thermally induced intramolecular [3 + 2] cycloaddition of the acetonitrile and vinyl ligands in Tp*Ir(Et)($CH{=}CH_2$)(NCMe) (**292**). An aromatic, rather than localised, structure was established for **474** on the basis of multinuclear 1 and 2-D NMR spectra, which revealed, *inter alia*, the iridium-bound 'methine' unit to have resonances (Δ_H 10.71, δ_C 191.3) intermediate between those typical of metal–carbene and metal–vinyl bonds. This was ultimately supported by a crystallographic study (Fig. 11).[100] Preliminary kinetic and mechanistic studies firmly established that **474** forms intramolecularly (pseudo-first-order behaviour and negligible incorporation of CD_3CN solvent) and is catalysed by small amounts of water.

Comparable [3 + 2] additions have been effected with a range of nitriles (NCR, R = Me, ^{t}Bu, Ph, CH_2-2-C_4H_3S), proceeding either from their pre-isolated adducts, Tp*Ir(Et)($CH{=}CH_2$)(NCR) (R = Me (**292**), ^{t}Bu (**387**)), or in a one-pot reaction from Tp*Ir(η^2-C_2H_4)$_2$ (**118**) and NCR (R = Ph, CH_2-2-C_4H_3S).[100,113] Analogous chemistry has also been explored with the Tp ligand, though in several cases this has proven more complex. For instance, while TpIr(Et)($CH{=}CH_2$)(NCMe) (**390**)

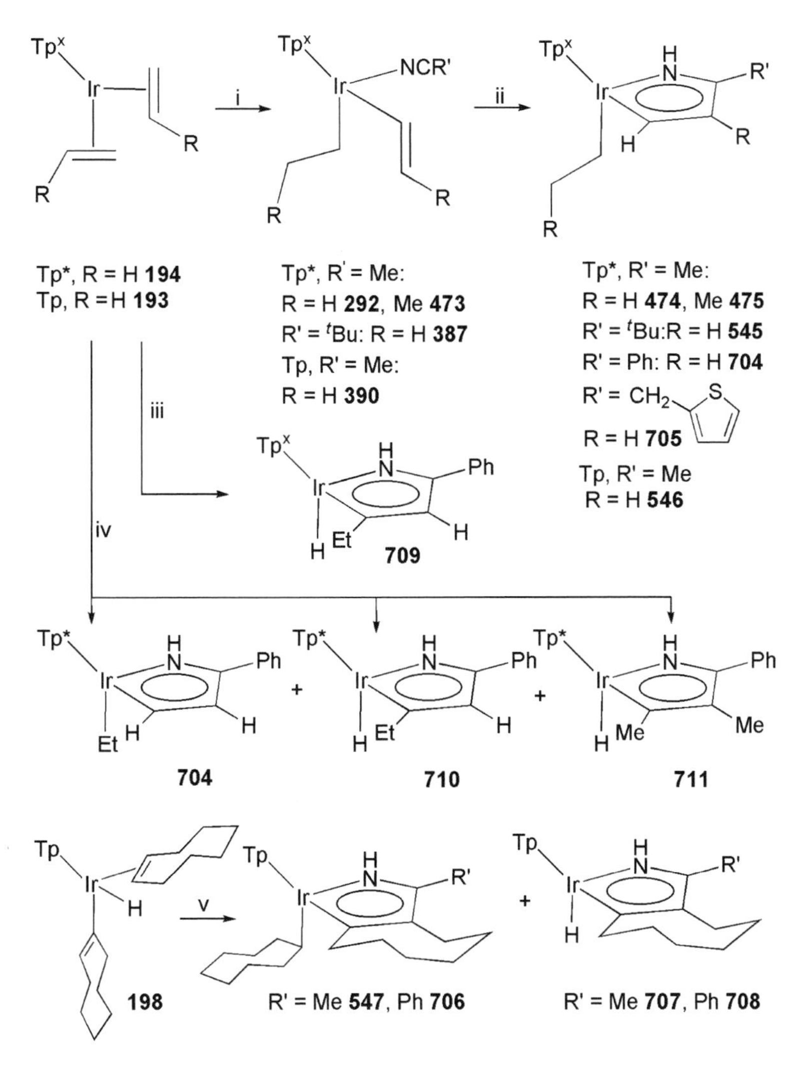

Scheme 67. Conditions: (i) NCR, 60 °C; (ii) 80 °C–100 °C; (iii) C_6H_{12}, NCPh, 100 °C, 18 h.; (iv) C_6H_{12}, NCPh, 100 °C; (v) 80–100 °C, NCR.

undergoes the anticipated coupling to afford **546**,[113] the reaction of TpIrH(σ-C_8H_{13})(η^2-C_8H_{14}) (**198**) with both NCMe and NCPh affords mixtures of the expected products (**547** and **706** respectively) and the metal hydrides **707** and **708**, from which the η^2-COE ligand has been lost.[100] This has been attributed to high steric hindrance within **198** resulting in competition between insertion of cyclooctene into Ir–H the bond (necessary to afford the 'TpIr(σ-C_8H_{15})(σ-C_8H_{14})(NCR)' intermediate) and its elimination, which has been observed on heating **198** in cyclohexane to afford TpIrH(η^3-C_8H_{13}) (**200**).[89] Isolated samples of **200**, when heated (130 °C) with the respective nitrile afford **707**/**708**, implying reversibility of the vinyl→allyl transformation.[100]

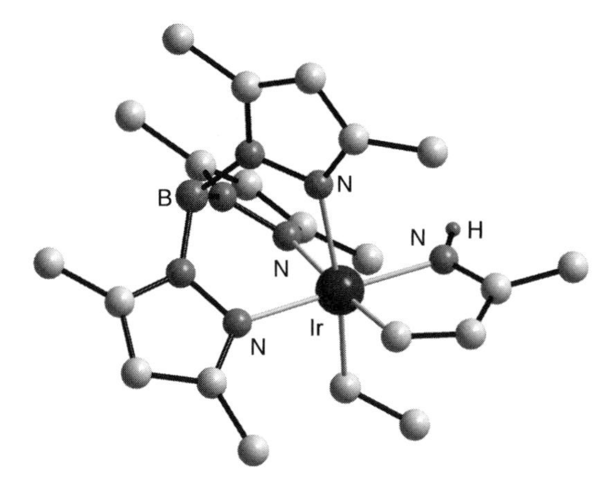

FIG. 11. Molecular structure of iridapyrrole complex **493**. Hydrogen atoms omitted.

Similar anomalies were encountered on heating $Tp^xIr(\eta^2\text{-}C_2H_4)_2$ ($Tp^x = Tp$ (**193**), Tp* (**194**)) with NCPh in cyclohexane (cf. neat reactions, *vide supra*) which for the Tp system resulted in incorporation of the ethyl chain into the iridapyrrole ring, with formation of a metal hydride. With Tp* however, three products were obtained. These outcomes have been attributed to inefficient trapping by NCPh of the 16-electron ethyl–vinyl intermediates, resulting instead in coupling of these ligands to afford the (undetected) unsaturated species 'Tp*IrH {C(Et)=CH_2}'. This is then either trapped directly by NCPh (Tp) or first rearranges *via* the allyl Tp*IrH(η^3-C_3H_4Me) (**291**) (Tp*), the eventual adducts then undergoing intramolecular cycloaddition. This suggestion was supported by the observation that heating **291** in NCR cleanly affords **711** and the methyl analogue **582**.

Despite being preferably described as aromatic, iridapyrroles exhibit appreciable alkene character at the β-carbon. Thus **474** and **475** are both readily protonated at this site to afford the olefin complexes **242** and **243**, which form as kinetic mixtures of *cis* and *trans* isomers, but revert completely to the *trans* isomers over time (3 h for **243**, 12 h at 80 °C for **244**, Scheme 68).[100] In the case of **244**, NMR monitoring revealed the intermediacy of two other species that convert to the aforementioned mixture of ***244-cis/trans*** over several hours. The identity of these species was established indirectly, by protonation of **582**, which instead of a hydrido-olefin complex quantitatively affords alkylidene complex ***583-syn***; reversibly in the presence of K_2CO_3.[99,100] Over time, solutions of ***583-syn*** convert irreversibly to the thermodynamically more stable *trans* isomer, which similarly regenerates **582** in the presence of base. However, in coordinating solvents (MeOH, NCMe) the metal hydride undergoes facile α-migration to the alkylidene, stereospecifically affording adducts **584** and **585**. Thus, the transiently observed intermediates in formation and isomerisation of **243** and **244** were proposed to be analogues of **584** and **585**,[99] resulting from α-migration of the metal–alkyl group to the alkylidene carbon.

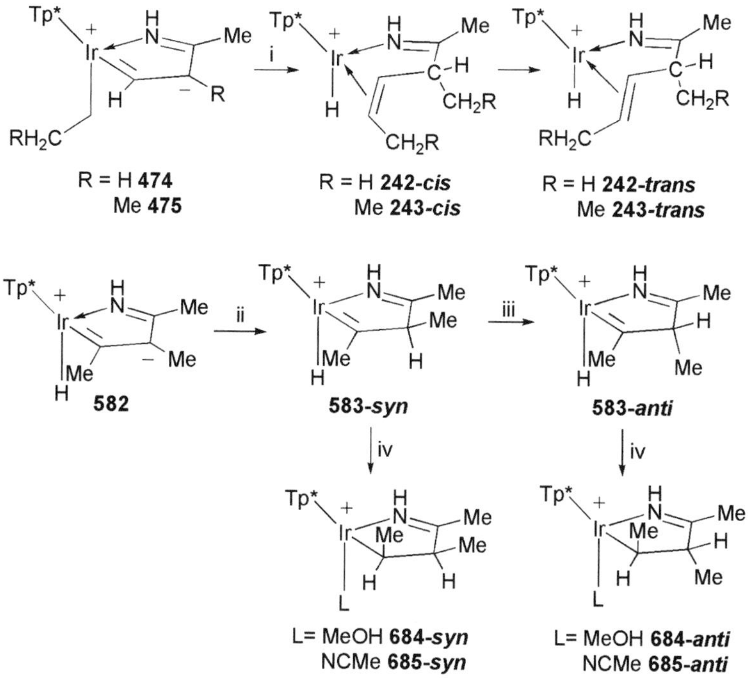

SCHEME 68. Conditions: (i) $[H(OEt_2)][BAr^f_4]$ 20 °C; (ii) $[H(OEt_2)][BAr^f_4]$ 20 °C; (iii) time; (iv) L.

2. Iridabenzenes

A small number of iridabenzenes have also been recently reported. The first (**558**) was obtained by oxidation of the chelating ketone complex **557** with $^t BuO_2H$ under ambient conditions.[166] Similarly, oxidation of **506** afforded the iridanaphthalene **507**, which was the first example of its kind (Scheme 69). For both materials NMR spectroscopic and X-ray diffraction data were deemed consistent with delocalised aromatic structures, though the iridium atom is in each case displaced from the mean ring plane, by 0.74 and 0.76 Å respectively, which has been attributed to steric factors. Though an interesting chemistry is claimed for both **557** and **507**, only for the latter has this been exemplified, its conversion to the benzannelated iridacyclohexadiene **712** being observed upon elution from silica gel (Scheme 69). A mechanism has not been established, but is believed to involve hydrolysis of the ester function and 'subsequent nucleophilic attack at the γ-carbon of the electron-poor iridanaphthalene'. This conversion is reversed upon treatment with $ClC(O)CO_2Me$.

The precursor **558** is derived from the iridacyclopentadiene **339** by insertion of DMAD to afford iridacyclohexatriene (**338**), followed by oxidation.[132–134] More recently it has been reported that **357** also reversibly inserts propene at room temperature, though at 60 °C this becomes irreversible and affords the iridabenzene **713**.[214] Similarly, **714** gives rise to the iridabenzene **715**, which contains a methyl,

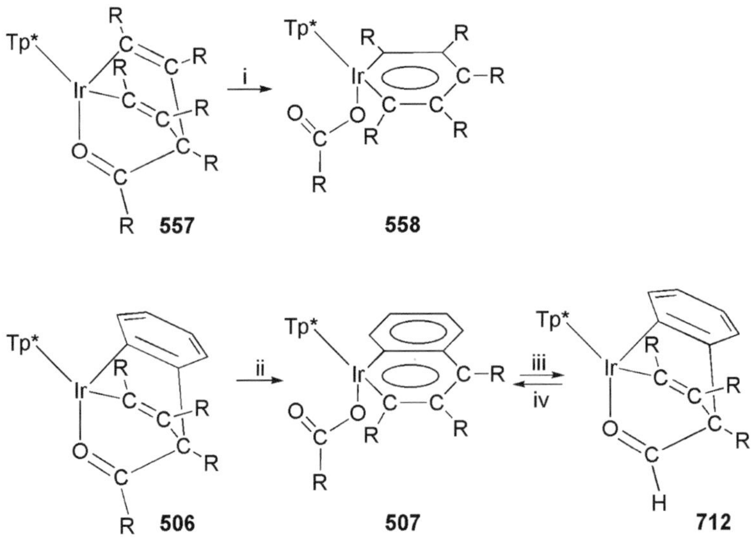

SCHEME 69. R = CO_2Me. Conditions: (i) $^tBuO_2H$, 20 °C, DCM; (ii) $^tBuO_2H$, 20 °C, DCM; (iii) Et_2O, silica gel; (iv) ClC(=O)CO_2Me, py, 20 °C.

rather than hydride, ancillary ligand. The mechanism of these reactions is not well understood, but is believed to involve: (i) π-propene to alkylidene rearrangement, (ii) migratory insertion into the Ir–C linkage, and (iii) a β-elimination step (Scheme 70). This would necessitate, in the case of **714**→**715**, a dimethylcarbene unit (=CMe_2) being favoured over a linear propylidene (=C(H)Et), for which there remains no clear explanation. Preliminary reactivity studies of the two iridabenzenes (**713** and **715**) have been briefly described, demonstrating that interaction with acetonitrile destroys the aromatic integrity of both compounds. Thus, **715** extrudes propene, reverting to **714**, while **713** undergoes hydride migration to the two α-carbon centres.

Finally, a very recent report describes the synthesis of the iridabenzene **592**, by protonation of the vinylidene complex **587** with CF_3CO_2H.[186] The mechanism for this conversion remains to be determined, but is postulated to involve acid-cleavage of the $SiMe_3$ groups with subsequent protonation of the β-carbon affording a cationic ethylidyne complex. Migration of one alkyl terminus of the σ^2-but-2-ene ligand onto the carbyne carbon followed by two proton rearrangements (β-elimination and abstraction) would then yield **592** (Scheme 71).

F. *Metallacarboranes*

Despite the recent proliferation of transition metal carbollide (metallacarboranes) chemistry merely three such compounds comprising a poly(pyrazolyl)borate ligand are known within group 9, viz. *closo*-3-(κ^3-Tp)-3,1,2-$RhC_2B_9H_{11}$ (**716**),[215] *closo*-2-(κ^3-Tp)-2,1,2-$RhC_2B_9H_{11}$ (**717**)[215] and Tp*Rh(η^5-7-$NH_2{}^tBu$-7-$CB_{10}H_{10}$) (**718**),[216]

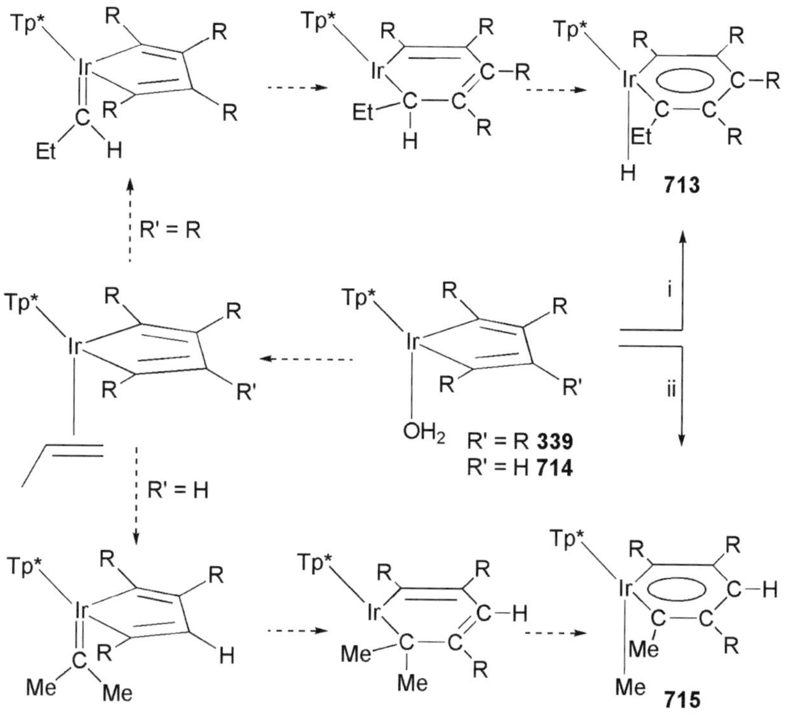

SCHEME 70. Conditions: (i) $MeCH{=}CH_2$, DCM, 60 °C; (ii) $MeCH{=}CH_2$, DCM, 20–60 °C.

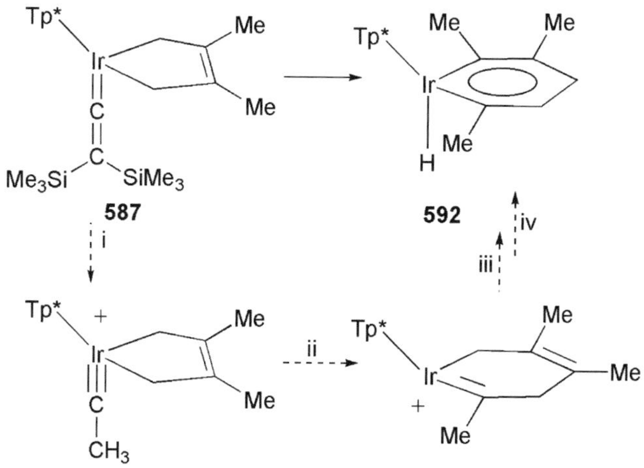

SCHEME 71. (i) H^+ (xs); (ii) migration; (iii) α-H migration; (iv) deprotonation.

each obtained from phosphinohalide precursors and KTp^x (Scheme 72, Fig. 12). The tris(pyrazolyl)methane analogue of **718** was similarly prepared in the presence of $TlPF_6$, though as the salt **719 · PF_2O_2**, presumed to result from hydrolysis of PF_6^- during crystallisation from acetone.[216] Each of these compounds was fully

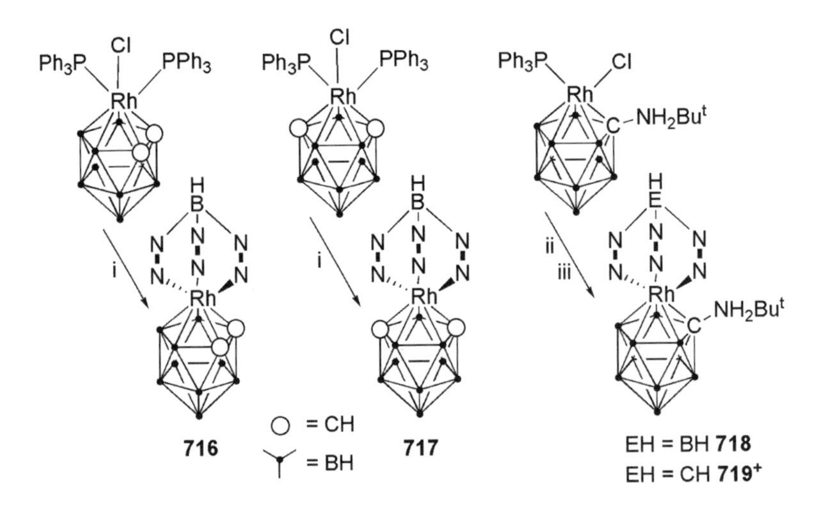

SCHEME 72. Conditions and reagents: (i) KTp′ (ii) KTp*; (iii) $TlPF_6$, Tpm.

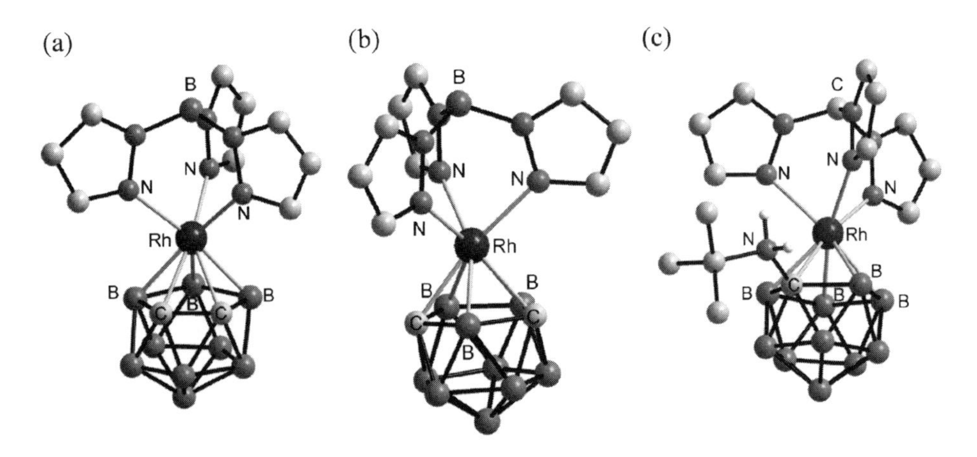

FIG. 12. Molecular structures of (a) **716**; (b) **717**; (c) **719**$^+$. Hydrogen atoms omitted.

characterised by NMR spectroscopy, mass spectrometry and X-ray crystallography. It should be noted that attempts to prepare the cobalt analogue of **716** directly from $CoCl_3$ with KTp^x and Na[carbollide] was unsuccessful, affording instead $[CoTp_2]^+$[*commo*-3,1,2-Co(3,1,2-$C_2B_9H_{11})_2]^-$.[215]

G. *Metal Hydrides*

As one would expect, metal hydrides have been frequently encountered during organometallic studies, often as the result of inter- or intramolecular C–H activation, or α-/β-elimination processes. These are well documented elsewhere (Sections III-B.1, II-C and III-D.1) and will not be discussed here unless significant chemistry has been observed for the hydride ligand, or indeed during its formation.

The earliest reported examples were a series of rhodium(III) monohydrides, prepared by hydrogenation (toluene reflux, 1 atm) of Tp*RhX_2(MeOH) (X = Cl (**76**), O_2CCF_3),[41] yielding, in the presence of triethylamine, [Tp*$RhHX_2$][Et_3NH] (X = Cl, **720 · Et_3NH**, O_2CCF_3 (**721 · Et_3NH**)). These salts were then treated with a series of group 5 donors to afford the neutral complexes Tp*RhHCl(L) (L = PPh_3 (**722**), $PMePh_2$ (**723**), PPh_2Me (**724**), PEt_3 (**725**), $PMe(C_6H_4Me\text{-}2)_2$ (**726**), $AsMe_2Ph$ (**727**)), which were fully characterised but have not received further study.

The related dihydrides $Tp^xMH_2(PR_3)$ (M = Rh, Tp^x = Tp, PR_3 = PPh_3 (**728**),[68] PMe_2Ph (**729**),[58] PEt_3 (**730**);[58] Tp^x = Tp*, PR_3 = PMe_3 (**731**),[58] PMe_2Ph (**732**),[58] PEt_3 (**733**),[58] M = Ir, Tp^x = Tp, PR_3 = PPh_3 (**734**),[68] PMe_3 (**735**);[69,215] Tp^x = Tp*, PR_3 = PMe_3 (**736**),[92] PMe_2Ph (**737**),[92] 1/2 dmpe (**738**)[92]) were also obtained by low-pressure (1–2 atm) hydrogenation, in this case of the respective $Tp^xM(\eta^2\text{-}C_2H_4)$ (PR_3) complexes. The dideuterides **728-d_2** and **734-d_2** were similarly obtained,[68] while $Bp^xRhH_2(PPh_3)$ (Bp^x = Bp (**739**), Bp* (**740**)) were prepared by hydrogenating the respective bis(phosphine) complexes (Section III-H).[59] Interestingly, the iridium complexes **734–736** can alternatively be prepared from the salts [Cp*IrH_3 (PR_3)]OTf, which in acetonitrile form [$(MeCN)_3IrH_3(PR_3)$]OTf under mild conditions; these then readily react with NaTp or KTp*.[69]

The reactivities of **734–736** have received limited study, their treatment with HBF_4 (one equivalent) being reported to afford the cationic dihydrogen complexes $[Tp^xIrH(H_2)(L)]^+$ (Tp^x = Tp, PR_3 = PPh_3 (**741$^+$**), PMe_3 (**742$^+$**), Tp^x = Tp* PMe_3 (**743$^+$**)) as BF_4 salts, each of which has been fully characterised.[69,217] The gradual (several hours) decomposition of **743 · BF_4** has been noted at ambient temperature, though its solution lifetime is extended as the BAr^f_4 salt. Similarly, while protonation of **728** with HBF_4 results in immediate decomposition, [TpRhH(H_2) (PPh_3)BAr^f_4] (**744 · BAr^f_4**) is solution-stable below 250 K. Each of **741$^+$–743$^+$** exhibits dynamic exchange of the H_2 and hydride ligands, exemplified by detailed NMR investigations of the mono- and di-deuterated and tritiated analogues, which were prepared by exposing the parent solutions to an atmosphere of D_2 or T_2 for several hours.[69]

The reactivity of **731** toward excess PMe_3 has also been explored, and demonstrated to effect step-wise dechelation of the Tp* ligand, directly analogous to that observed for Tp*$Rh(\eta^2\text{-}C_2H_4)_2$ (**118**) and $Tp^{Me2,4Cl}Rh(CO)_2$ (**469**) (Section III-B.1 and II-D.3). Thus, heating **731** (κ^3-Tp*) with PMe_3 affords initially (50 °C, 2 h) (κ^2-Tp*)$RhH_2(PMe_3)_2$ (**745**) and ultimately (8 h) [$RhH_2(PMe_3)_4$][Tp*] (**746 · Tp***).[103,104] A comparable reaction proceeds with the Tp analogue of **731**, though its synthesis has not been reported.

The same dechelation chemistry has been observed for the Ir(V) polyhydride Tp*IrH_4 (**322**), prepared by hydrogenation of Tp*$Ir(\eta^2\text{-}C_2H_4)_2$ (**194**), which upon prolonged heating with PMe_3 affords exclusively [*cis*-$IrH_2(PMe_3)_4$][Tp*] (**747 · Tp***), again *via* the intermediacy of a simple substitution product, Tp*IrH_2 (PMe_3) (**736**).[207,218] This reaction, and that with CO (90 °C, 3 atm, 24 h) to afford Tp*IrH_2(CO) (**323**) proceed only under forcing conditions, which has been cited as evidence in support of **322** adopting a classical Ir(V) tetrahydride structure. This is in contrast to the analogous rhodium complex

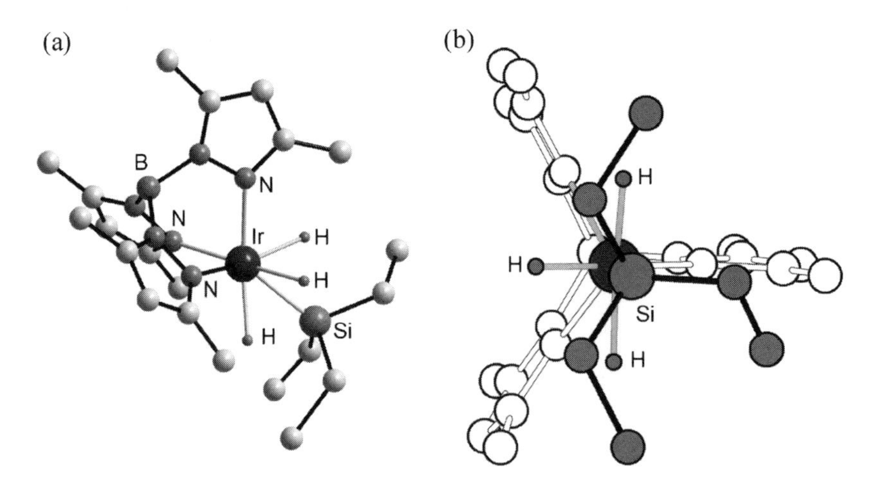

FIG. 13. (a) Molecular structure of [$Tp^{Me_2}IrH_3(SiEt_3)$] (**749**); (b) projection along the Si–Ir–B axis.

$Tp^*RhH_2(H_2)$ (**748**)§, prepared from [PPh_4][Tp^*RhCl_3] and $NaBH_4$,[219] which was the first example of a 'non-classical' (η^2-H_2) hydride stabilised only by nitrogen donors. The 'non-classical' structure of this material was assigned on the basis of extensive multinuclear and multidimensional NMR investigations,[219,220] and was supported by inelastic neutron-scattering experiments[221] and DFT investigations of the fictitious '$TpRhH_4$' which exhibits no propensity for a classical tetrahydrido structure.[222,223]

The assignment of a classical structure to **322** was made predominantly on the basis of NMR spectroscopic studies, the measured T_1(min) of ca. 400 ms (500 MHz) being deemed consistent with other classical hydrides (e.g. 390 ms for **731**, 360 ms for **736** at 400 MHz), as opposed to the non-classical hydrides **748** (42 ms, 400 MHz)[219,220] and **742**$^+$ (22 ms, 500 MHz).[69] However, **322** is unusual among classical hydrides in exhibiting resolvable J_{DH} couplings and an isotope perturbation equilibrium, deuteration in C_6D_6, D_2O or CD_3OD affording all of the possible isomers $Tp^*IrH_{4-n}D_n$.[207] This latter observation was attributed to the presence of two different types of Ir–H bond (a corollary of **322** adopting a hydride-capped, distorted octahedral structure) as calculated from NMR parameters. Moreover, for $Tp^*IrH_3(SiEt_3)$ (**749**), synthesised from $HSiEt_3$ and $Tp^*IrH_2(SC_4H_4)$ (**521**), an X-ray crystallographic study (Fig. 13) confirmed a distorted octahedron with capping $SiEt_3$ group.[207]

It should, however, be noted that theoretical investigations[224] of $TpIrH_4$ (**750**), which has been prepared in low (< 10%) yield but not fully characterised,[207] suggest the C_{3v} capped-octahedron is a high-energy saddle point (Fig. 14). Two isoenergetic

§The H–H interaction in **748** is weak (*i.e.* close to the limit of oxidative addition of H_2 to the metal). Hence, a slight increase in electron-donation from the Tp ligand (*e.g.* [$Tp^{Me_2}Rh(H_2)H_2$]) destabilises the η^2-H_2 form, while a reduction (*e.g.* [$Tp^{CF_3,Me}Rh(H_2)H_2$]) significantly increases stability. These results are documented in: Bucher, U. E. Dissertation No. 10166, ETH Zurich, 1993, but have yet to appear in the literature.

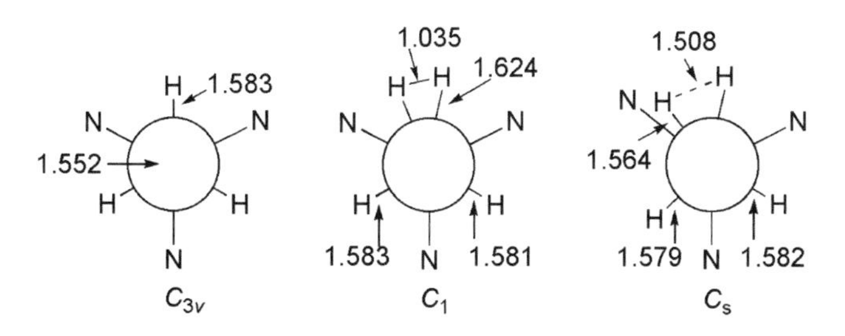

FIG. 14. Schematic representation of the possible molecular geometries for $TpIrH_4$ (**750**) projected along the Ir–B axis with the pyrazole structure omitted for clarity.

minima are calculated that correspond to a C_s 'edge-bridged' octahedron (Ir(V)) and the C_1 η^2-dihydrogen dihydride,[223] though in lieu of structural data for **322** or **750**, or indeed theoretical studies of the former, a definitive conclusion in this respect cannot be reached.

The Ir(V) tetrahydride **322**, Ir(II) carbonyl-dihydride **323** and their fully deuterated congeners have also been encountered as the result of photolytic degradation of methanol by Tp*IrH_2(η^2-COE) (**207**) and Tp*Ir(η^4-1,5-COD) (**179**).[128] The latter has also served as precursor to **207** under high-pressure hydrogenation (500 atm),[81] while at lower pressure (200 atm) the related dihydride Tp*IrH_2(η^2-1,5-COD) (**209**) is instead obtained.

The photochemistry of **207** has received much attention, as a result of Tp*IrH(C_6H_5){P(OMe)$_3$} (**321**)[91] being formed upon photolysis of benzene solutions in the presence of P(OMe)$_3$. Having explored the photochemical cycle of this conversion it was concluded that the initial photoproduct is likely 'Tp*IrH_2', since in the absence of benzene the phosphite complex Tp*IrH_2{P(OMe)$_3$} (**751**) is obtained, while photolysis in *tert*-butylacrylate affords Tp*IrH_2(η^2-CH_2=$CHCO_2$ ^{t}Bu) (**218**), neither of which activate benzene. Thus, 'Tp*IrH_2' presumably activates benzene to afford the Ir(V) trihydride Tp*IrH_3(C_6H_5), which then undergoes photolytic elimination of H_2, to yield a 16-electron complex that is trapped by P(OMe)$_3$.

Comparable photochemistry has been reported for Tp*Rh(1,3-COD) (**171**) with dissociation of the alkene again proposed as the initial step.[77,78,130] Thus, in the presence of P(OMe)$_3$ the complex Tp*RhH(C_6H_5){P(OMe)$_3$} (**324**) is obtained, while *tert*-butylacrylate is believed to afford Tp*RhH(C_6H_5)(η^2-CH_2=$CHCO_2$ ^{t}Bu), which rapidly undergoes alkene insertion into the Rh–H linkage, so that only Tp*Rh(C_6H_5)($CH_2CH_2CO_2$ ^{t}Bu) (**326**) is isolated.[78] The activation of diethylether has also been reported (with P(OMe)$_3$) to afford Tp*RhH(OEt) {P(OMe)$_3$} (**327**);[77] a reaction that is believed to occur by trapping of the initial 'Tp*Rh' photoproduct by P(OMe)$_3$, followed by C–O oxidative addition and elimination of ethylene. Complex **327** is itself reactive and thermally extrudes EtOH in benzene to ultimately afford **324**, while prolonged selective photolysis generates small amounts of Tp*RhH_2{P(OMe)$_3$} (**752**). This is presumed to occur by reductive elimination and dehydrogenation of EtOH, in much the same way that

MeOH is dehydrogenated by **207**[129] and **171**[77] to afford the hydrides **322**, **323** and $Tp^*RhH_2(CO)$ (**325**) respectively.

Other Notable Hydrides

While the products of C–H activation are numerous, far fewer examples exist of E–H bond activation for the heavier group 14 elements (E = Si, Ge), the first being the photolytic activation of Et_3SiH by $Tp^*Rh(CO)_2$ (**430**) to afford $Tp^*RhH(SiEt_3)(CO)$ (**437**).[151] The complex $Tp^{^iPr_2}Rh(\eta^2\text{-COE})(NCMe)$ (**163**) has also been subjected to oxidative substitution by H_2SiEt_2 and H_3SiPh under ambient conditions, affording $Tp^{^iPr_2}RhH(SiH_{3-n}R_n)(NCMe)$ ($n = 2$, R = Et (**753**); $n = 1$, R = Ph (**754**)).[74] It was, however, noted that primary silanes failed to add to **163**, a fact attributed to their higher steric hindrance. In contrast, the primary stannane $HSnPh_3$ adds readily to $Tp^*Rh(PR_3)_2$ (R = Ph (**755**), C_6H_4F-4 (**756**)) under comparable conditions to afford $Tp^*RhH(SnPh_3)(PR_3)$ (R = Ph (**757**), C_6H_4F-4 (**758**)).[180]

Precursors **755** and **756** have also been found to react with the terminal alkynes $HC{\equiv}CPh$[180] and $HC{\equiv}CCO_2Et$[181] to similar effect (Section III-D.1), and with the aldehyde $O{=}CHC_6H_4NO_2$-4 to afford the acyl hydride complexes $Tp^*RhH\{C({=}O)C_6H_4NO_2\text{-}4\}(PR_3)$ (R = Ph (**688**), C_6H_4F-4 (**689**)), which are remarkably stable, undergoing merely 30% decomposition over 21 h at 70 °C.[180]

One other phosphine–hydride complex has been prepared, specifically **759 · OTf**, which was obtained by treating methanol solutions of $[Rh(NCMe)_3(triphos)](OTf)_3$ with NaTp (Scheme 73).[225] The identity of **759 · OTf** was established both spectroscopically and by a single crystal X-ray diffraction study (Fig. 15), but the material has received no more elaboration.

Disparity in the chemistry of rhodium and iridium complexes has been illustrated by the facile generation of $[Tp^*IrH(CO)_2]BF_4$ (**659 · BF$_4$**)[196] and its Tp analogue **658 · BF$_4$**[205] by protonation of the respective dicarbonyls (**623** and **265**) with HBF_4, while $Tp^*Rh(CO)_2$ (**430**) instead undergoes N-protonation of the uncoordinated pyrazole group.[196] Protonation of the iridium systems is reversible in the presence of DBU, and they also interact with hard nucleophiles (e.g. BuLi and NaOMe) to afford acyl complexes as described in Section II-D.3.

The allylic hydride complex $Tp^*RhH(\eta^3\text{-}C_3H_5)$ (**349**) is noteworthy, not for its existence, but because it is formed from $Tp^*RhBr(\eta^3\text{-allyl})$ (**344**) and the

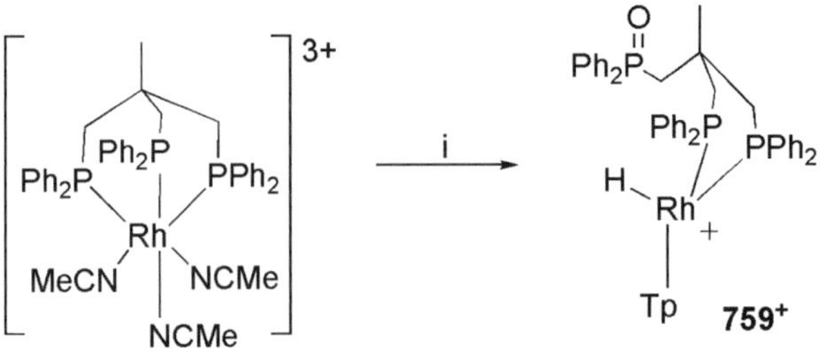

SCHEME 73. Conditions: (i) NaTp, MeOH, Δ, 3 h.

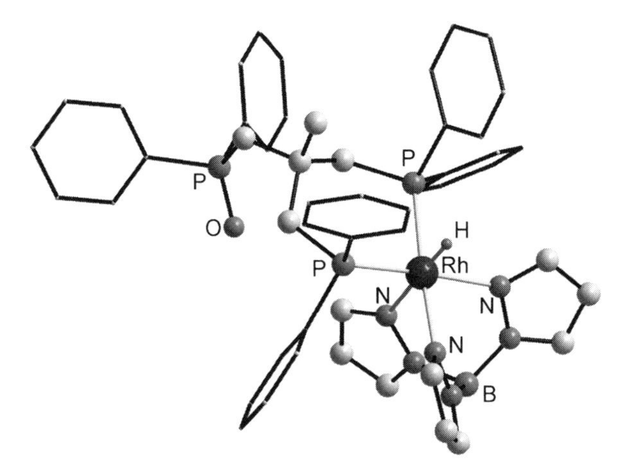

FIG. 15. Molecular structure of **759**$^+$. Phenyl groups simplified, hydrogen atoms omitted.

nucleophilic hydride donor Li[HBEt$_3$], which attacks not at the metal, but rather the C2 of the allyl group, affording a rhodacyclobutane intermediate, which undergoes hydride migration to the metal (Section II-C.2).[75] Thus, the reaction of **344** with Li[DBEt$_3$] affords Tp*RhH(η^3-H$_2$C=CDCH$_2$) (**350**). The metal hydride can then be exchanged for bromide by treatment with allyl bromide.

Also noteworthy are some alkylidenes that exemplify rare reactivity for metal hydrides. The first is the cyclic carbene complex **565**, the formation of which is itself unusual, proceeding as it does from the interaction of Bp*Rh(CO)(py) (**566**) and methyl iodide.[182] This is proposed to involve the oxidative addition of MeI and subsequent migratory insertion of CO, though at what stage the B–H activation occurs remains to be determined. More significant, however, is that on heating to 45 °C, **565** irreversibly evolves into the alkyl complex **567** *via* a rare 'reverse α-hydride migration' onto the alkylidene carbon (Scheme 55, Section II-D.2).

Finally, the iridium carbene complex **580**, obtained from the double C–H activation of 2-ethylphenol by Tp*Ir(C$_6$H$_5$)$_2$(N$_2$) (**211**), illustrates the rare incidence of an equilibrium between alkylidene hydride and alkene hydride complexes. The alkylidene forms in admixture with ca. 5% of the alkene hydride isomer **581**, illustrating a preference for the α- over β- hydrogen in the second activation step. However, in isolation **581** is observed to re-establish the same equilibrium mixture (i.e. 20:1 **580**:**581**, Scheme 61); a rare example of a metal–alkene converting to a metal–alkylidene, the reverse reaction being more typical.

H. *Phosphine Complexes*

As with hydrides, phosphines feature prominently as ancillary ligands, and also as archetypal donors employed to induce rearrangements and migrations, or to stabilise coordinately unsaturated species. In these spectator roles the nature of the phosphine and its chemistry are in general of secondary importance, other than allowing electronic and steric tuning of the metal centre and will largely be ignored here since they are well documented elsewhere. For instance, numerous complexes of the type $Tp^xM(\eta^2\text{-}C_2H_4)(PR_3)$ (for Rh with Tp, Tp*, Bp*; **142–147**, **158** Section III-B.1; for Ir with Tp, Tp*; **220–227**,), $Tp^xM(CO)(PR_3)$ (for Rh with Tp*, $Tp^{Me_2,Cl}$, $Tp^{Ph,Me}$, Bp, Bp*; **626**–**648**, **651** and for Ir with Bp **652**, Section II-D.2) have been prepared by thermal or photochemical ligand substitution from the parent bis(ethylene) or bis(carbonyl) complexes respectively. Low-pressure hydrogenation of the mono-phosphino ethylene complexes has been employed to generate the respective dihydrides $Tp^xMH_2(PR_3)$ (for Rh with Tp, Tp*, Bp, Bp*; **728–733**, **739**, **740**, for Ir with Tp, Tp* (**734–738**), Section II-G) while the chlorohydrido complexes $Tp^xRhHCl(PR_3)$ (**722–726** and with $AsMe_2Ph$ **727**) were obtained by treating $[Tp^*RhHCl_2][Et_3NH]$ (**720** · $\mathbf{Et_3NH}$) with the respective donor.

A wider range of M(III) complexes has been prepared *via*: (i) ligand substitution (e.g. fluoroalkyl complexes **274**, **276**, $\mathbf{278}^+$ (Section II-C.1)) 2-thienyls **517–520**, **526**, $\mathbf{528}^+$ (Section III-D.1) $Tp^*Ir(C_6H_5)_2(PMe_3)$ (**392**) (Section II-D.2) and complex metallacycles **464**, **555**, **556** (Section II-E), (ii) trapping coordinately unsaturated species (e.g. the photoproducts $Tp^*MH(R)\{P(OMe)_3\}$ **321**, **324**, **327** (Sections III-B.1 and II-C.2)) or the ethyl–vinyls $Tp^*M(Et)(CH{=}CH_2)(PMe_3)$ **386**, **388** (Section III-B.1 and III-D.1), (iii) intramolecular rearrangement of M(I) phosphine complexes (e.g. $Tp^*IrH(CH{=}CH_2)(PR_3)$ **540**, **541** (Section III-D.1), (iv) rearrangement induced by extraneous phosphine (e.g. the σ^1:σ^1 butadiene complexes **302**, **309**, **314** (Section III-B.1) and $Tp^*Rh(\sigma\text{-allyl})Br(PMe_2Ph)$ (**346**), obtained from $Tp^*Rh(\eta^3\text{-}C_3H_5)Br$ (**344**, Section III-B.2); (v) oxidative addition of MeI to a range of $Tp^*Rh(CO)(PR_3)$ complexes ($PR_3 = PMe_3$ **626**, PPh_3 **628**, $PMePh_2$ **629**, PMe_2Ph **630**) to afford initially the salts $[Tp^*Rh(Me)(CO)(PR_3)]I$ that rearrange to $Tp^*Rh\{C({=}O)Me\}I(L)$ (Section II-D.3).

A range of complexes $[Tp^xRh(PR_3)Cl_2]$ ($Tp^x = Tp$, $PR_3 = PPh_3$ (**760**), PPh_2Me (**761**), $PPhMe_2$ (**762**), PEt_3 (**763**); $Tp^x = Tp^*$, $PR_3 = PPh_3$ (**764**), PPh_2Me (**765**), $PPhMe_2$ (**766**), $PMe(C_6H_4Me\text{-}2)_2$ (**767**), PMe_3 (**768**)[142]) have also been prepared from the respective $Tp^xRhCl_2(NCMe)$ complex and donors, as have the arsine complexes $TpRhCl_2(AsR_3)$ ($AsR_3 = AsPh_3$ (**769**), $AsPhMe_2$ (**770**)) and $Tp^*IrCl_2(AsPhMe_2)$ (**771**).[41] The complex $Tp^*IrBr_2(PMe_3)$ (**772**) has also been prepared, but by bromination of the respective dihydride **732**.[155] The complex $(pzTp)RhI_2(PPh_3)$ (**773**) has been prepared by ligand substitution from the respective carbonyl complex **675**[44] and the Tp analogue has also, apparently, been prepared and treated with MeI to afford $TpRh(Me)I(PPh_3)$ (**371**).[139,140] Particularly notable among these is **760**, the reactions of which with AgX ($X = BF_4$, NO_3, SbF_6) have been reported to afford the remarkable heterotrinuclear complex salts $[\{TpRh(PPh_3)(\mu\text{-}Cl)_2\}_2Ag]X$ (**774** · **X**),[226] the structure of which was determined crystallographically for $X = BF_4$, revealing an unusual distorted square-planar geometry for Ag(I) (Fig. 16).

More significant are a range of bis(phosphine) complexes obtained typically by ligand substitution of η^2-alkene precursors, or on occasion from $RhCl(PR_3)_3$ or $Rh_2(\mu\text{-}Cl)_2(LL)_2$ ($LL = 2\ PR_3$, $R_2P\text{-}PR_2$) with the respective pyrazolylborate anion. Thus, the complexes $Bp^xRh(L_2)$ ($Bp^x = Bp$, $L_2 = 2\ PPh_3$ (**261**); $Bp^x = Bp^*$, $L_2 = 2\ PPh_3$ (**259**), $2\ PMe_3$ (**260**), depe (**257**), dppe (**258**)),[59] $Tp^xRh(L_2)$ ($Tp^x = Tp$, $L_2 = 2$

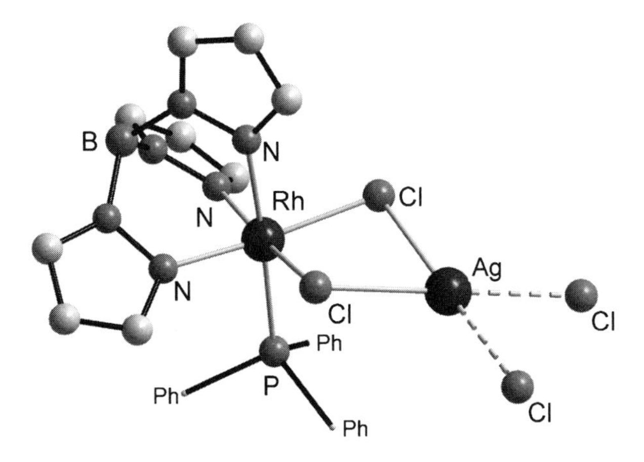

FIG. 16. Truncated molecular structure of $[\{TpRh(PPh_3)(\mu\text{-}Cl)_2\}_2Ag]^+$ (**774**$^+$).

PPh_3 (**173**),[79] $Tp^x = Tp^*$, $L_2 = 2\ PPh_3$ (**755**),[180] $2\ P(C_6H_4F\text{-}4)_3$ (**756**),[180] $2\ PMe_3$ (**268**),[104] dppe (**775**);[199] $Tp^x = pzTp$, $L_2 = 2\ PPh_3$ (**776**);[199] $Tp^x = Tp^{iPr_2}$, $L_2 =$ dppm (**777**), dppe (**471**), dppp (**778**);[74] dppene (**779**)[227]) and TpIr(dppip) (**780**)[105] have each been prepared, isolated and comprehensively characterised.¶ Complexes **173**, **774** and **755** have also been chemically oxidised with $[Cp_2Fe]PF_6$ (also BF_4 for **755**) to the respective Rh(II) cations,[199,204] the electrochemistry of which has been studied to some length, as was the case for a series of $Tp^xRh(CO)(PR_3)$ complexes (Section II-D.3).

In several instances more intriguing outcomes have been observed, as in the attempted synthesis of dppm and dppe analogues of **779** through the treatment of $TpIr(\eta^2\text{-}C_2H_4)_2$ (**193**) with the respective diphosphine.[105] In the case of dppm this affords the novel iridium hydride complex **334** (Fig. 17), in which ligand displacement has been accompanied by hydrogen transfer from the methylene-bridge to iridium, the connectivity being confirmed crystallographically. Protonation of **334** with HBF_4 thus affords $[TpIrH(dppm)]BF_4$ (**781** · **BF$_4$**), while the (much slower) reaction with MeI has also been observed, but the product is not reported. The dppip analogue of **781**$^+$ can be prepared by direct protonation of **780** with HBF_4.

Treatment of **193** with dppe, in contrast, effects extrusion of the Tp ligand, affording the salt $[Ir(dppe)_2][Tp]$ (**270** · **Tp**). Though the mechanism of this reaction has not been studied, the formally related extrusions of Tp* from $Tp^*Rh(\eta^2\text{-}C_2H_4)_2$ (**118**) and $Tp^*RhH_2(PMe_3)$ (**731**) upon their treatment with excess PMe_3 (ultimately affording $[Rh(PMe_3)_4][Tp^*]$ (**269** · **Tp***) and $[Rh(H)_2(PMe_3)_4][Tp^*]$ (**746** · **Tp***) respectively) have, in considerable detail.[103,104] This has revealed a step-wise dechelation of the Tp* ligand ($\kappa^3 \rightarrow \kappa^2 \rightarrow \kappa^1 \rightarrow \kappa^0$), demonstrated in the former case by the isolation of the intermediate complexes $(\kappa^2\text{-}Tp^*)Rh(PMe_3)_2$ (**268**) and $(\kappa^1\text{-}N\text{-}Tp^*)Rh(PMe_3)_3$ (**267**). A comparable reaction sequence has been proposed

¶dppene = cis-bis(diphenylphosphino)ethene; dppip = 2,2-bis(diphenylphosphino)propane.

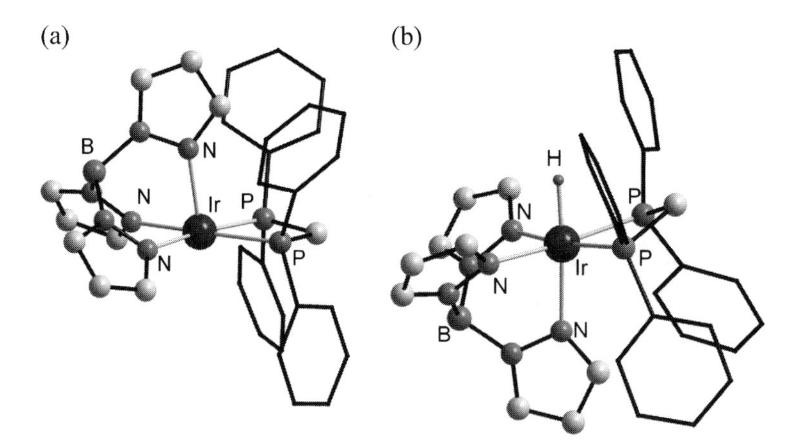

FIG. 17. Molecular structure of (a) **334** (Ir–H not shown) and (b) the protonated form **781**$^{+}$. Phenyl groups simplified.

for Tp*Ir(η^2-C_2H_4)$_2$ (**194**), which upon prolonged heating with PMe_3 affords exclusively [*cis*-$IrH_2(PMe_3)_4$][Tp*] (**747** · **Tp***), *via* the intermediacy of Tp*IrH_2(PMe_3) (**736**).[207,218] Also, $Tp^{Me_2,4Cl}Rh(CO)_2$ (**470**) reacts with PR_3 to afford (κ^1-$Tp^{Me_2,4Cl}$)Rh(CO)(PR_3)$_2$ ($PR_3 = PMe_3$ (**653**),[104] PMe_2Ph (**654**),[104] $PMePh_2$ (**655**)[67,104]). In this instance, however, complete extrusion is effected only with dppp, which yields initially [Rh(CO)(dppp)$_2$][$Tp^{Me_2,4Cl}$] (**656** · **$Tp^{Me_2,4Cl}$**) and ultimately [Rh(dppp)$_2$][$Tp^{Me_2,4Cl}$] (**782** · **$Tp^{Me_2,4Cl}$**).

The chemistry of several of the bis(phosphine) complexes has been developed, most notably that of TpRh(PPh_3)$_2$ (**173**), which despite its simplicity was prepared only in 1998, commencing from Wilkinson's catalyst and KTp.[79] The reported reactions of **173** are summarised in Scheme 74, and include generation of the first rhodium–carbene complexes (Section II-D.2) bearing a 'Tp^{x}' ligand, viz. TpRh{=C(SMe)$_2$}X_2 (X = I (**562**), Cl (**563**)), by double alkylation of TpRh(η^2-SCS)(PPh_3) (**175**) with MeI, monoalkylation with [Et_3O][BF_4] instead affording the dithioethoxycarbonyl salt **176** · **BF_4**. It is also noted that **173** effects the catalytic demercuration of Hg(C≡CR)$_2$ (R = C_6H_4Me-4) to afford the 1,3-diyne RC≡C–C≡CR, though the complex cannot be recovered after the reaction, a mixture of unidentified species instead being obtained. It is, however, surprising that no further work concerning **173** has been described, given the importance other authors have attributed to this seminal demonstration of a versatile entry into the organometallic chemistry of the 'TpRh' fragment.

Subsequently the syntheses and limited reactivity studies were reported for the complexes Tp*Rh(PPh_3)$_2$ (**755**) and Tp*Rh{P(C_6H_4F-4)$_3$}$_2$ (**756**), both of which have been demonstrated to undergo oxidative substitution by PhC≡CH, Ph_3SnH and O=$CHC_6H_4NO_2$-4 to afford Tp*RhH(X)(PR_3) (R = C_6H_5, *C_6H_4F-4* X = C≡CPh (**559**), **560**, $SnPh_3$ (**757**), **758**), C(=O)$C_6H_4NO_2$ (**688**), **689**).[180] Significantly, the acyl complex **688** exhibits remarkable thermal stability, over 70% of the material remaining unchanged after 21 h at 70 °C.

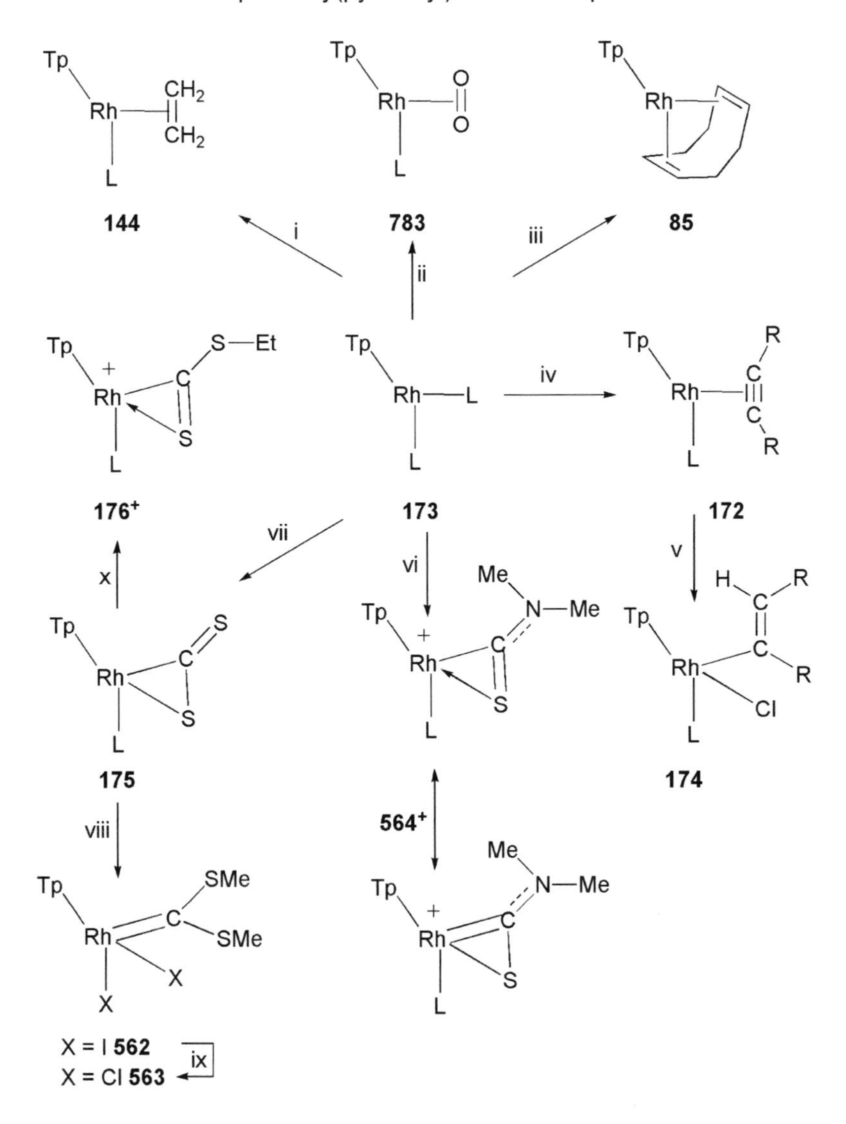

SCHEME 74. Conditions: (L = PPh_3, R = CO_2Me) (i) C_2H_4; (ii) O_2; (iii) 1,5-COD, toluene (110 °C); (iv) RC≡CR; (v) HCl; (vi) $Me_2NC(=S)Cl$; (vii) CS_2; (viii) MeI; (ix) $CHCl_3$; (x) $[Et_3O][BF_4]$.

The reaction of **755** with Ph_3SiH at 50 °C in toluene has also been reported, affording the dihydride Tp*$RhH_2(PPh_3)$ (**784**). However, the same reagents combined in dichloromethane results in fragmentation of the Tp* ligand, affording $RhH_2Cl(PPh_3)_2$(Hpz*), or over longer periods $RhHCl_2(PPh_3)_2$(Hpz*).[228] This reaction also proceeds, though more slowly, in the absence of the silane, and can be accelerated by addition of radical scavengers, leading the authors to suggest an ionic mechanism that is accelerated by proton donors. The same work demonstrated the reaction of **755** with $HgCl_2$, which affords a mercury bridged dirhodium complex that is devoid of pyrazole fragments (**785**) and also with C_6F_5SH, which affords the sulphido-bridged Rh(I)–Rh(III) complex **786** (Scheme 75). This demonstration of the capacity of **755** to activate S–H bonds, albeit in this instance uncontrolled, inspired a recent study that has successfully employed **755** as the catalyst in alkyne hydrothiolation.[229]

Ligand fragmentations have also been observed to arise from certain oxygenation reactions of $Tp^{^iPr_2}Rh(dppe)$ (**471**), in contrast to the cases of **173** and Tp*Rh(η^2-C_2H_4)(PR_3) ($PR_3 = PMe_3$ (**145**), PEt_3 (**147**)) that oxygenate readily, affording TpRh(η^2-O_2)(PPh_3) (**782**)[79] and Tp*Rh(η^2-O_2)(PR_3) ($PR_3 = PMe_3$ (**787**), PEt_3 (**788**)) respectively.[58] However, for **471** (Scheme 76) no simple dioxygen adduct is obtained, but rather low yields (< 20%) of $Tp^{^iPr_2}Rh(\eta^2\text{-}O_2)(Hpz^{^iPr_2})$ (**789**), wherein the pyrazole ligand derives from fragmentation of $Tp^{^iPr_2}$;[160,161] indeed, the presence

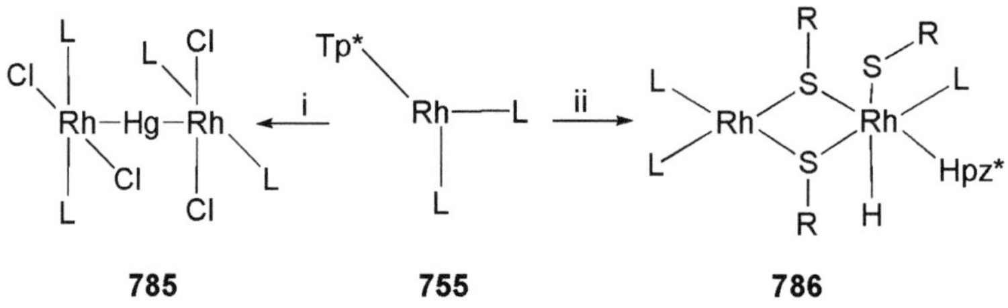

SCHEME 75. $L = PPh_3$, $R = C_6F_5$. Conditions: (i) C_6F_5SH; (ii) $HgCl_2$.

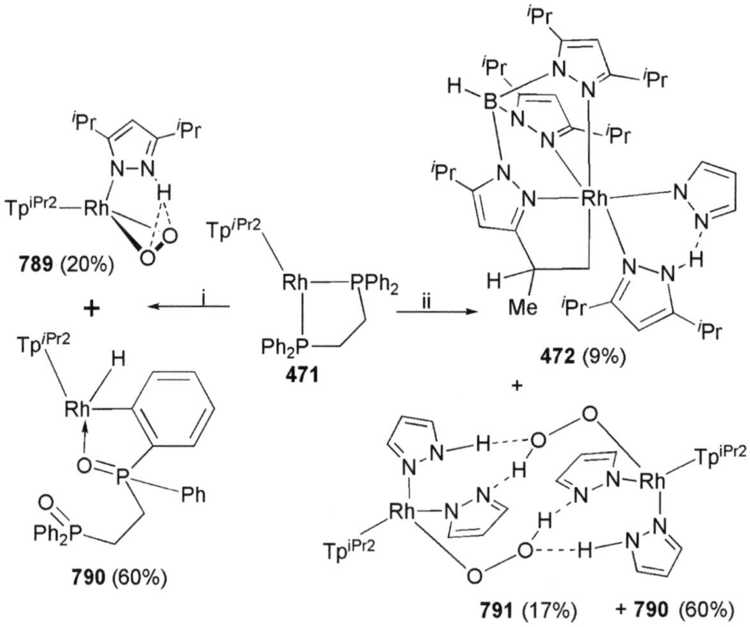

SCHEME 76. Conditions: (i) O_2, $Hpz^{^iPr_2}$; (ii) O_2, Hpz.

of free 3,5-diisopropylpyrazole enhances yields of **789** to >20%. The major product in each case is the ortho-metallated dppeO$_2$ complex **790**. Oxygenation of **471** in the presence of pyrazole continues to afford small amounts (<10%) of **789**, while **790** is again the major product, but now in admixture with the hydroperoxide complex **791** and the curious complex **472**, in which one 3-isopropyl substituent has undergone C–H activation at the metal. The formation of these products has been investigated and discussed at length,[161] though no definitive mechanism has been established.

Intramolecular C–H activation has also been observed upon treatment of TpIr(η^2-C$_2$H$_4$)(PPh$_3$) (**220**) with excess PPh$_3$, giving rise to an equilibrium between **220** and (κ^3-*N,N′,C*^{5} –Tp)IrH(PPh$_3$)$_2$ (**333**, Scheme 21).[93,131]

A range of other intra- and intermolecular C–H activation processes have been described in preceding sections that involve ancillary phosphine ligands, and these will not be repeated here given the negligible significance of the ligands in this context. However, particularly noteworthy among these are the range of complexes derived from Tp*Ir(Me)Br(PMe$_3$) (**372**), which upon treatment with AgOTf affords Tp*Ir(Me)(OTf)(PMe$_3$) (**440**). This in turn can be converted to [Tp*Ir(Me)(N$_2$)(PMe$_3$)]BAr$^f_4$ (**441 · BAr$^f_4$**),[141,157] which enters into a variety of intriguing C–H activation processes (Section III-D.1). Similarly notable are the vinyl complexes Tp*Ir(R)(CH=CH$_2$)(PMe$_3$) (R=H (**561**), Et (**540**)), which at −60 °C undergo protonation by HBF$_4$ at the β-carbon to afford the ethylidene complexes [Tp*Ir(R)(=CHMe)(PMe$_3$)]$^+$ (R=H (**542**$^+$), Et (**544**$^+$)), which constitute the *first* alkylidenes of iridium(III) bearing Tp^x ligands (Section III-D.1).[179]

Complexes of Polydentate Phosphines

The formation of 'Tp^{x}Rh' complexes with the polydentate phosphines MeC(CH$_2$PPh$_2$)$_3$ (triphos) and P(C$_7$H$_7$)$_3$ (tris(cyclohepta-2,4,6-trienyl)phosphine) have been briefly and independently explored. In the former case, [Rh(NCMe)$_3$(triphos)](OTf)$_3$ was found to react with NaTp in methanol under reflux (Scheme 73) to afford the octahedral complex TpRhH(κ^2-*P,P′*-triphosO)OTf (**759 · OTf**), in which one, pendant, arm of the triphos ligand is oxidised.[225] Given that [TpRu(κ^3-triphos)]OTf can be prepared from the isoelectronic Ru(II) precursor [Ru(NCMe)$_3$(triphos)]$^{2+}$, the authors reasoned that steric factors alone cannot induce the 'arm-off' reaction with rhodium, implicating oxidation of the pendant arm as a key step in the formation of **759**$^+$, the hydride being subsequently abstracted from the solvent. However, whether this is impeded under non-oxidising conditions was not reported.

Somewhat more significant is the formation of (κ^2-*N,H*-Tp*)Rh{κ^3-*P*-η^2:η^2-(C$_7$H$_7$)$_2$P(C$_7$H$_7$)} (**140a**) either from Tp*Rh(1,5-COD) and P(C$_7$H$_7$)$_3$ or from [RhCl{κ^4-*P*-η^2:η^2:η^2-P(C$_7$H$_7$)$_3$}] and KTp*.[66] This represented the first unequivocal proof of a κ^2-*N,H* binding mode for a Tp^x ligand. The structure was confirmed crystallographically in the solid state whilst its retention in solution was established on the basis of comprehensive multinuclear NMR spectroscopic studies (i.e. ^{1}H, ^{11}B, ^{11}B{^{1}H}, ^{13}C, ^{31}P, ^{103}Rh). That **140a** is the kinetically controlled product has been demonstrated by the observation, at trace levels in the solid and in solution, of two coordination isomers, assigned as (κ^2-N,N'-Tp*)Rh{κ^2-P-η^2-$C_7H_7P(C_7H_7)_2$} (**140b**) and the coordinately unsaturated, 14-electron (κ^2-N,N'-Tp*)Rh{$P(C_7H_7)_2$} (**140c**). These have been implicated in the slow (months) conversion of **140a** into the dioxygen adduct (κ^2-N,N'-Tp*)Rh(η^2-O_2){$P(C_7H_7)_2$} (**141**).

Hybrid Poly(pyrazolyl)(phosphino)borates

An exciting recent development has been the synthesis and preliminary investigation of the first members of a new generation of poly(pyrazolyl)borate ligand that bear a hydrocarbyl chain in place of the hydride, and are also bifunctionalised with a phosphine donor. The first such ligand, $[H_2C{=}CHCH_2B(CH_2PPh_2)(pz)_2]^-$ (herein abbreviated (σ-allyl)(P)Bp) was prepared (Scheme 77) as

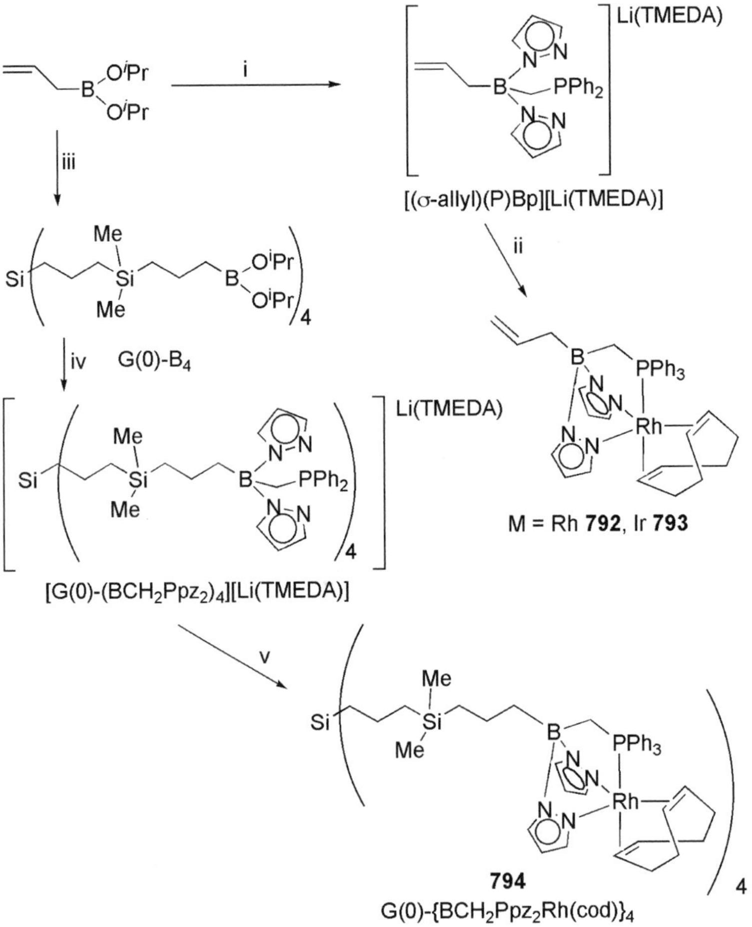

[(σ-allyl)(P)Bp][Li(TMEDA)]

G(0)-B_4

[G(0)-$(BCH_2Ppz_2)_4$][Li(TMEDA)]

M = Rh **792**, Ir **793**

794
G(0)-{$BCH_2Ppz_2Rh(cod)$}$_4$

SCHEME 77. Reagents: (i) 2 Hpz, [Li(TMEDA)(CH_2PPh_2)]; (ii) $M_2(\mu\text{-Cl})_2(cod)_2$; (iii) Si[$(CH_2)_3SiMe_2H$]$_4$, 'Pt', Toluene; (iv) 2 Hpz, [Li(TMEDA)(CH_2PPh_2)]; (v) 2 $Rh_2(\mu\text{-Cl})_2(cod)_2$.

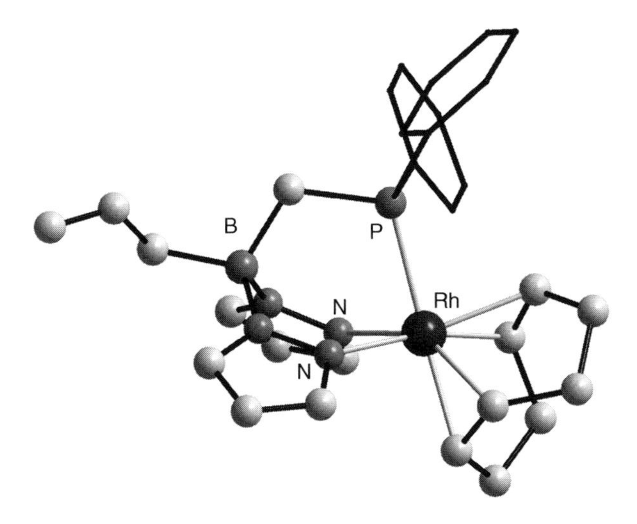

FIG. 18. Molecular structure of [{(σ-allyl)(P)Bp}Rh(1,5-cod)] (**792**). Phenyl groups simplified. Hydrogen atoms omitted.

the Li(TMEDA) salt, which reacts with $M_2(\mu\text{-Cl})_2(1,5\text{-COD})_2$ (M = Rh, Ir) to afford the novel complexes **792** (Fig. 18) and **793**, which adopt slightly distorted trigonal-bipyramidal geometries with one pyrazole donor lying axial.[230]

In an extension of this basic chemistry, the precursor (σ-allyl)B(O^{i}Pr)$_2$ was combined (*via* Pt-catalysed hydrosilylation) with the carbosilane $Si[(CH_2)_3SiMe_2H]_4$ to afford the dendrimer $Si[(CH_2)_3SiMe_2(CH_2)_3B(O^iPr)_2]_4$ (**G(0)-B$_4$**), from which the polyanionic, dendrimeric ligand $Si[(CH_2)_3SiMe_2(CH_2)_3B(CH_2PPh_2)(pz)_2]_4$ (**G(0)-(BCH$_2$Ppz$_2$)$_4$**) was generated as before. Treatment of this ligand with two equivalents of $Rh_2(\mu\text{-Cl})_2(1,5\text{-COD})_2$ afforded the metallodendrimer $Si[(CH_2)_3SiMe_2(CH_2)_3B(CH_2PPh_2)(pz)_2Rh(cod)]_4$ (**G(0)-(BCH$_2$Ppz$_2$Rh)$_4$; 794**), the identity of which is supported by analytical, NMR spectroscopic and MALDI-TOF mass spectrometric data.

IV

PERSPECTIVES IN C–H ACTIVATION AND CATALYSIS

A. *C–H Bond Activation Processes*

A recurrent theme in the chemistry of rhodium and iridium poly(pyrazolyl)borate complexes is their capacity to activate hydrocarbon C–H bonds, both intra and intermolecularly, under photochemical and/or thermal conditions. This has increasingly become the subject of detailed investigations aimed at elucidating the salient mechanistic features, not least the extent to which the 'Tp^x' ligand participates, through its capacity for variable denticity. This body of work (to 2001) was recently reviewed by Slugovc and Carmona,[231] and many details are described elsewhere herein. This section thus seeks to provide a concise overview and reference resource, emphasising more recent advances and key mechanistic points, rather than comprehensively summarising all relevant reactions.

1. Overview

The activation of hydrocarbon C–H bonds by complexes of the type $Tp^xM(LL)$ (M = Rh, Ir; LL = η^4-diene, 2 η^2-C_2H_4, 2 CO, 2 PR_3, or combinations thereof) or $Tp^xMH(CH{=}CH_2)(L)$ (M = Ir, Rh; L = η^2-C_2H_4, CO, PR_3) has become a ubiquitous route to alkyl (Section II-C.1), aryl (Section II-C.1)[72,143,145–147,149,151–153] and alkenyl (Section III-D.1)[70,84–87,89,95,100,113,119] compounds, as described in the relevant sections. Some examples have proven more serendipitous, e.g. the numerous examples of intramolecular cyclometallation of the pyrazole rings[93,191] or their substituents.[88,97,98,160,161,165,184] In other cases outcomes are more complex, as in the activation of variously substituted thiophenes by $Tp^*Ir(\eta^2\text{-}C_2H_4)_2$ (**194**) and $Tp^*Ir(\eta^4\text{-}C_4H_4Me_2\text{-}2,3)$ (**234**) which not only afford the simple 2-thienyl products, but also, *via* a series of successive C–H activations and C–C bond formations, a number of complex bicyclic chelates.[120,121] In contrast, the rhodium complexes $Tp^*Rh(\eta^2\text{-}C_2H_4)(PR_3)$ (R = Me (**145**), Et (**147**)) thermally activate thiophene with appreciable control to afford $Tp^*RhH(2\text{-thienyl})(PR_3)$ (>95%), though under photochemical conditions the competitive insertion of rhodium into the C–S bond predominates.[71]

The double α-C–H activation of cyclic and acyclic ethers and amines, affording Fischer-type, i.e. hetero-atom substituted carbene complexes, has also been well documented (Section III-D.1) for the complexes $Tp^{Ph}Ir(\eta^4\text{-}C_4H_5Me\text{-}3)$ (**238**),[98] $Tp^*Ir(C_6H_5)_2(N_2)$ (**211**),[90,117,118,165] $Tp^*Ir(\eta^2\text{-}C_2H_4)_2$ (**194**) and $Tp^*IrH(CH{=}CH_2)(\eta^2\text{-}C_2H_4)$ (**202**).[90,119] Definitive mechanisms for these conversions could not be established, deuteration studies revealing a significant kinetic isotope effect and implying a complex series of reactions.[90,119] However, some evidence was obtained in support of ether coordination being the initial step, while a universal preference for α- over β-C–H activation was noted, given the formation of carbenes rather than π-alkene complexes, and tested with a range of substrates. Several related, but less explored, conversions have been reported, including the fragmentation of CH_2Cl_2 by $Tp^{Mes}Ir(\eta^4\text{-}C_4H_5Me\text{-}3)$ (**578**) to afford a rare chlorocarbene complex,[184] and the activation of aryl-aldehydes by $Tp^*Ir(\eta^4\text{-}C_4H_4Me_2\text{-}2,3)$ (**234**).[158]

It should be noted that there is also a report of intramolecular C–H activation by a relevant cobalt complex,[232] the identity of which was inferred to be the cobalt imido compound $Tp^{^tBu,Me}Co{=}NSiMe_3$. However, this imido species has not been discretely observed.

2. Mechanistic and Kinetic Aspects of Intermolecular C–H Activation

Though observed with many complexes, C–H activation has been most widely studied for $Tp^*Rh(\eta^4\text{-}1,3\text{-COD})$ (**171**), $Tp^*IrH_2(\eta^2\text{-COE})$ (**207**), $Tp^*Rh(CO)_2$ (**430**) and $Tp^*Rh(\eta^2\text{-}Ph\text{-}N{=}C{=}NR)(CNR)$ (**395**, R = $CH_2{}^tBu$), which photochemically activate a range of hydrocarbons. On this basis (detailed below) the generic mechanism has been determined to involve: (i) initial photo-dissociation of a labile ligand, to afford a coordinately unsaturated complex that becomes solvated by the hydrocarbon substrate; (ii) oxidative addition of a C–H linkage (a thermal process with a low activation barrier); (iii) coordination of a donor (either extraneous or a 'pendant' pyrazole group) to afford coordinative saturation and stabilise the product.

a. Activation by Tp*Rh(η^4-1,3-COD) (171) and Tp*IrH$_2$(η^2-COE) (207)

Both complexes photolytically activate benzene, in the presence of P(OMe)$_3$, to afford Tp*MH(C$_6$H$_5$){P(OMe)$_3$} (M = Rh (**324**), Ir (**321**)). For **207**, this was found to involve initial dissociation of cyclooctene, since in the absence of benzene the photoproduct is trapped by P(OMe)$_3$ as Tp*IrH$_2${P(OMe)$_3$} (**751**).[91] Moreover, since **751** itself fails to activate benzene, it was reasoned that the active species is the 'Tp*IrH$_2$' fragment. This presumably oxidatively adds benzene to afford Tp*IrH$_3$(C$_6$H$_5$), which has not been observed, but is expected to be unstable toward photolytic elimination of H$_2$. The resulting 16-electron complex would then be stabilised by binding of P(OMe)$_3$. In a similar fashion, the comparable photolysis of **171** was determined to afford initially 'Tp*Rh' *en route* to Tp*RhH(C$_6$H$_5$){P(OMe)$_3$} (**324**).[77,78,130] Both **171** and **207** were also found to activate a small number of other substrates, including diethyl ether, and degrade methanol to afford mixtures of dihydrido-carbonyl,[77,78,130] and polyhydrido-[129] complexes.

b. Activation by Tp*Rh(CO)$_2$ (430)

The photochemical activation of hydrocarbons by **430** was the first such report for any group 9 Tp^x complex,[151,152] and has subsequently been the subject of numerous independent mechanistic studies. The earliest of these involved matrix isolation (Ar, CH$_4$, N$_2$, ^{13}CO) infrared and electronic spectroscopic studies of **430**, and its analogue Bp*Rh(CO)$_2$ (**262**).[233] On this basis it was established that the initial photoreaction results in loss of CO and mono-dechelation of the, initially κ^3, Tp* ligand to afford (κ^2-Tp*)Rh(CO), which in N$_2$ and ^{13}CO matrices was rapidly trapped. It was also noted that at 12 K no evidence of C–H activation was observed in either the methane matrix, or in Nujol mulls. However, at 77 K activation of Nujol was observed, thus establishing a thermal requisite for the C–H activation step, i.e. while photolysis generates the 'active species', subsequent C–H oxidative addition is a thermal process.

More detailed solution studies revealed that C–H activation by **430** is extremely efficient, with no interference from thermal processes or secondary photoreactions,[208,234] (cf. Cp and Cp* analogues).[235] This allowed for the determination of absolute quantum efficiencies for C–H activation (ϕ_{CH}),[151,208,234–236] which exhibit an appreciable dependence on the excitation-wavelength. Thus, in *n*-pentane, a maximum efficiency of 0.34 is attained upon irradiation at 313 nm (or $\phi_{CH} = 0.31$ at $\lambda = 366$ nm for *n*-pentane, hexane, heptane and isooctane), dropping to $\phi_{CH} = 0.01$ at $\lambda = 458$ nm. This was interpreted in terms of two distinct, highly reactive excited states, the lower-energy of which (populated by visible excitation, $\lambda = 458$ nm) is inert toward C–H activation, while this is facile for the upper state.[235] The lower-energy state was thus correlated with the $\kappa^3 \rightarrow \kappa^2$ interconversion of the Tp* ligand, since although a facile thermal pathway exists for this process, no C–H activation is observed in the dark. In turn, excitation at 366 nm was deemed responsible for carbonyl dissociation to afford the 'active complex'.[151] It should be noted that thermal excitation to this upper level, from the lower, was ruled out on the basis of the temperature dependence observed for ϕ_{CH} at both wavelengths, from which activation barriers of 5.4 ($\pm$2.5) kJ mol^{-1} (λ=458 nm) and 8.3 ($\pm$2.5) kJ mol^{-1} (λ=366 nm) were calculated. These are too low for thermal excitation ($\sim$60 kJ mol^{-1}) and were thus attributed to the R–H bond dissociation step.[151]

Deeper insight into this mechanism was afforded by femtosecond time-resolved infrared studies[237–239] that enabled observation of intermediates and the calculation of relative energy barriers. Thus, upon UV irradiation **430** loses CO (<100 fs) to afford a 16-electron monocarbonyl complex[237] that is rapidly solvated (barrier-less process) to afford Tp*Rh(CO)(RH), which vibrationally cools in 20 ps.[238] Thermal mono-dechelation ($\kappa^3 \rightarrow \kappa^2$) of the Tp* ligand ($\Delta G$=4.2 kcal mol^{-1}) proceeds within 200 ps to afford a second intermediate, which was formulated as (κ^2-Tp*) Rh(CO)(Xe) during kinetic studies in liquid Xenon.[240] Both of these intermediates were also predicted on the basis of DFT studies.[241] The C–H activation step proceeds with a time-constant of 230 ns (ΔG $\sim$ 8.3 kcal mol^{-1})[238] and is followed rapidly (<< 200 ns) by rechelation ($\kappa^2 \rightarrow \kappa^3$) of the Tp* ligand.

The kinetics of this C–H activation process have been studied in a series of hydrocarbons,[242] establishing that: (i) linear hydrocarbons are activated faster than their cyclic congeners, a fact attributed to preferential activation of primary, over secondary C–H functions, and (ii) aryl C–H activation is significantly slower, due to the intermediate being a strongly bound π-arene complex.

c. Activation by Tp*Rh(η^2-Ph–N=C=NR)(CNR) (395, R=CH$_2$ ^{t}Bu)

Photolysis of the carbodiimide complex **395** in hydrocarbon solvents similarly effects C–H activation, *via* initial dissociation of the carbodiimide ligand to afford 'Tp*Rh(CNCH$_2$ ^{t}Bu)' as the active species.[144] The reaction proceeds in a variety of neat aliphatic hydrocarbons,[145] with apparently exclusive selectivity for the primary C–H of terminal methyl groups. This includes the case of chloroalkanes,[147] for which no oxidative addition of the C–Cl bond is observed. Competition is, however, noted for aromatic substrates bearing methyl substituents,[145] with both benzylic and aromatic activation proceeding under kinetic control, though the aromatic activation products are thermodynamically favoured. Thus, toluene affords a kinetic mixture of aryl and benzyl complexes, which thermally redistributes to a 2:1 mixture of *meta* and *para* isomers of the aryl complex, while mesitylene yields a 3:1 mixture of the mesityl and benzyl complexes.

The involvement of alkane σ-complexes of the type Tp*Rh(CNCH$_2$ ^{t}Bu)(RH) was again established, both in C–H activation and subsequent reductive elimination processes, the kinetics and energetics of which were explored in detail.[143,146,149] The reductive elimination of methane from Tp*RhD(CH$_3$)(CNCH$_2$ ^{t}Bu) (**423**) and

Tp*RhH(CD_3)($CNCH_2{}^t Bu$) (**424**) provided indirect evidence for a σ-complex, since in both cases scrambling of the metal hydride and methyl ligands was observed prior to elimination of CH_2D and CHD_2 respectively.[149] Moreover, in benzene/hexafluorobenzene solvents the rate of elimination was found to depend upon the concentration of benzene, indicative of an associative component to the reductive elimination. These processes were deemed consistent with the reversible formation of a σ-complex. Similar observations were also noted during the investigation of benzene elimination from Tp*RhH(C_6H_5)($CNCH_2{}^t Bu$) in the presence of $CNCH_2{}^t Bu$, which was also found to be reversible and associative.[144]

On the basis of these studies, and others focussing on alkane reductive elimination and skeletal rearrangement within Tp*RhH(R)($CNCH_2{}^t Bu$) complexes,[143] and the kinetic selectivity of activation in mixed solvents,[143,145,146] a series of mechanistic conclusions were reached. Specifically, primary alkane σ-complexes are involved in both activation and reductive elimination of C–H bonds, and are significantly more prone to activation than dissociation. Migration to a secondary alkane complex is also favoured over dissociation, though once formed the secondary complex is more likely to dissociate than is the primary. However, more likely still is further migration, either back to the primary alkane complex, or preferably to an alternative secondary alkane complex. Nonetheless, secondary activation remains the slowest possible process, rearrangement or dissociation being over 100 times more likely, thus this is rarely encountered. Indeed, near exclusive selectivity for primary over secondary C–H activation is observed.

3. Intramolecular C–H Activation of Alkenes

The involvement of alkene complexes in C–H activation reactions, both thermal and photochemical, is widespread and well documented. However, given the diversity of reagents and outcomes encountered, mechanistic studies have necessarily focussed on individual reaction schemes, with emphasis upon optimising and accounting for the observed outcomes. One exception, that has been explored in some detail, is the intramolecular C–H activation of olefin complexes (typically of ethylene) giving rise to hydridovinyl species. This reaction is, in fact, of fundamental importance, since it has been directly implicated in many intermolecular activations, the hydridovinyl being often considered the active species.

The seminal report in this area documented the thermally-induced isomerisation of $Tp^{CF_3,Me}Ir(CO)(\eta^2\text{-}C_2H_4)$ (**229**) to $Tp^{CF_3,Me}IrH(CH{=}CH_2)(CO)$ (**289**),[95] which proceeds cleanly at 60 °C. This thermodynamic preference is the reverse of that observed for the Cp and Cp* analogues, which favour the η^2-alkene complex, as does the photochemically generated Tp*RhH($CH{=}CH_2$)(CO) (**290**), which rapidly reverts to Tp*Rh(η^2-C_2H_4)(CO) (**148**) under ambient conditions.[95] This disparity between the rhodium and iridium π-alkene complexes has been widely observed. However, while the rhodium hydridovinyls are rarely isolated, the kinetic accessibility of Tp*RhH($CH{=}CH_2$)(η^2-C_2H_4) (**294**) has been established.[70] It was found that on treatment with NCMe or pyridine, Tp*Rh(η^2-C_2H_4)$_2$ (**118**) cleanly converts to Tp*Rh(CH_2CH_3)($CH{=}CH_2$)(L) (L = NCMe (**156**), py (**293**)), presumably *via* migratory insertion of ethylene into a Rh–H linkage, thus implicating the intermediacy of **294**.

Early investigations of the stability of these π-alkene and hydridovinyl complexes, and in particular the influence of the 'Tp^x' ligand, concluded that: (i) methyl substituents (e.g. in Tp*, $Tp^{CF_3,Me}$) are essential to the thermal stability of both π-alkene and hydridovinyl isomers,[86] and (ii) hapticity change (κ^2-$Tp^x \rightarrow \kappa^3$-$Tp^x$) stabilises the M(III) product of C–H activation (d^6-octahedral), thus serving as a driving-force.[114] The hapticity change was subsequently dismissed as an inadequate driving force,[95] and it should also be noted that many M(I) $Tp^xM(\pi$-alkene)(L) complexes preferentially engage in κ^3-Tp^x coordination in the ground state.

A greater insight into both influences was afforded by *ab initio* studies of intramolecular C–H activation in a series of model compounds, including the 'real' complexes $Tp^xM(\eta^2$-$C_2H_4)_2$ (M = Rh, Ir; $Tp^x = Tp^{Me}$, Tp^{CF_3}).[115] It was thus concluded that 3-substituents on the pyrazole rings exert a destabilising steric influence upon the π-complex (i.e. greater steric bulk in $Tp^xM(\eta^2$–$C_2H_4)_2$ (κ^3- or κ^2-Tp^x) weakens the M–C_{alkene} bond). Moreover, for rhodium κ^2-chelation is favoured over κ^3, and (κ^2-Tp^x)Rh(η^2-$C_2H_4)_2$ are 20–28 kcal mol^{-1} more stable than the respective hydridovinyl isomers. In contrast, for iridium all κ^2 isomers are unstable with respect to κ^3, while the relative stabilities of (κ^3-Tp^x)Ir(η^2-$C_2H_4)_2$ and (κ^3-Tp^x)IrH(CH=CH_2)(η^2-C_2H_4) are governed by the sterics of the Tp ligand; i.e. greater bulk favours the hydridovinyl. Thus, the conversion $Tp^{CF_3,Me}$Ir(CO)(η^2-C_2H_4) (**229**) → $Tp^{CF_3,Me}$IrH(CH=CH_2)(CO) (**289**) is exothermic. Moreover, all such isomerisations are less endothermic for iridium than rhodium, this being attributed to the preference of late 2nd-row elements to adopt a d^{n+1} ground state with high-energy d^ns^1 excited state, cf. d^ns^1 ground state and low-lying d^ns^1 excited state for the 3rd row; i.e. oxidative addition is easier for 3rd row elements.

B. *Catalysis*

There are relatively few reports of group 9 poly(pyrazolyl)borates having been employed in catalytic applications. The first mention in this respect is a patent of 1971 pertaining to hydroformylation catalysts,[187] which lists BpRh(CO)$_2$ (**593**) and Tp*Rh(CO)$_2$ (**430**), though details are limited. While no further reports have elaborated upon this, the use of BpRh(CO)(L) (L = PCy$_3$ (**644**), PPh$_3$ (**645**), P(p-An)$_3$ (**646**) P(NC$_4$H$_4$)$_3$ (**647**)) to catalyse hydroformylation of hex-1-ene has been recently described.[200] All four materials exhibited high catalytic activity, affording 80% of the aldehyde and 20% hex-2-ene, though optimal results require a high catalyst concentration.

The regioselective catalytic hydrogenation of quinoline has been effected with several complexes generated *in situ* from NaTp and $M_2(\mu$-Cl$)_2$(1,5-COD)$_2$, (M = Rh, Ir) and $Ir_2(\mu$-Cl$)_2$(COE)$_4$ and also with isolated samples of TpIr(η^2-$C_2H_4)_2$ (**193**).[243] These studies revealed the rhodium/1,5-COD system to be most active, a result that was replicated using an isolated sample of TpRh(1,5-COD) (**85**), though the catalysts are decomposed during the reaction. In contrast, the iridium/COE system catalyses the reaction under mild conditions without decomposition, though with reduced activity, while with TpIr(η^2-C_2H_4)$_2$ (**193**) lower conversions are achieved. A slightly enhanced rate was observed upon use of the Tp* ligand in place of Tp.

In a similar study, the regioselective hydrogenation of *trans*-cinnamaldehyde was effected using $Ir_2(\mu$-$Cl)_2(COE)_4$ stabilised by Tp, Tp^{Me}, Tp* and Bp, and using isolated samples of Tp*Ir(1,5-COD) (**179**).[244] This revealed that the presence of methyl substituents on the pyrazole rings of the tris(pyrazolyl)borates enhances selectivity for the C=O bond. Thus, while with Tp and Bp* dihydrocinnamaldehye is obtained in 60% yield with only traces (<5%) of cinnamol, this product distribution is reversed by Tp*. Moreover, with Tp^{Me} cinnamol alone isolated in excess of 90% yield.

The complexes $Tp^{Ph,Me}$Rh(CO)(L) (L=CO (**612**), PPh_3 (**651**)) have been successfully applied in the hydrosilylation of 1-octene with triethoxysilane, with **612** exhibiting a particularly high level of activity and selectivity.[245] However, a lower selectivity was noted for the hydrosilylation of allyl alcohol ethoxylate. It should be noted that attempts to employ TpIr(η^2-C_2H_4)$_2$ (**193**) in the hydrosilylation of phenylacetylene were unsuccessful.[84,246] However, Tp*Rh(η^2-C_2H_4)$_2$ (**118**) has apparently been used to hydrosilylate ethylene,[231,247] and TpRh(1,5-COD) (**86**) catalyses the hydrosilylation of *trans* 1,4-bis(trimethylsilyl)-3-buten-1-yne though with little selectivity.[248] More recently, the hydrothiolation of alkynes by aromatic and aliphatic thiols has been catalysed using Tp*Rh(PPh_3)$_2$ (**755**), which exhibits outstanding activity (yields 63–94%) and selectivity.[229] Moreover, when **755** is used to catalyse the reaction between phenylacetylene and thiophenol, the distribution of regioisomers is comparable to that obtained through palladium catalysis (the inverse of most rhodium systems), but is obtained under ambient conditions (cf. 80 °C for Pd).

Finally, the polymerisation of phenylacetylene has been explored using the rhodium complexes Tp^xRh(1,5-COD) (Tp^x=Tp (**85**), Tp* (**90**), $Tp^{^iPr_2}$ (**102**), Tp^{Ph_2} (**104**)), Tp*Rh(NBD) (**99**), Tp*Rh(η^2-C_2H_4)$_2$ (**118**), Tp*Rh(η^2-C_2H_4)(NCMe) (**155**), and Bp^xRh(1,5-COD) (Bp^x=Bp (**87**), Bp* (**91**), $Bp^{(CF_3)_2}$ (**121**)).[54] This revealed that the 1,5-COD complexes are significantly more active catalysts than their norbornadiene congeners, and that in dichloromethane solution the substituted Tp^x or Bp^x complexes afford high yields of polyphenylacetylene with high stereoregularity (exothermically for **90**, **102** and **104**) while the Tp and Bp complexes are essentially inactive. However, independent studies have revealed that ionic liquids, tetraalkylammonium salts and alcohols serve as co-catalysts for phenylacetylene polymerisation by TpRh(1,5-COD) (**85**) which affords over 90% polyphenylacetylene in these media.[249]

It is also noted that the polymerisation of methylmethacrylate has been observed with the cobalt complex $Tp^{^tBu,Me}$CoMe (**28**) (70 °C, benzene), though the efficiency of this process has not been investigated. Similarly, acrylonitrile has been polymerised in the presence of $Tp^{^tBu,Me}$CoEt (**29**), giving 23% polyacrylonitrile with a molecular mass of 51,000.[250]

Finally, it is noted that the iridium complex Tp*Ir(η^4-$C_4H_4Me_2$-2,3) (**234**) has been found to act as an effective H/D exchange catalyst,[165] by virtue of its capacity to activate C–H and C–D bonds. This complex has been applied to the deuteration of norbornadiene derivatives, in which context the mechanism and regioselectivity of the reaction have been explored in some detail.[251] The complex $Tp^{*}IrBr_2(PMe_3)$ **772** has also been reported in the patent literature as a catalyst for deuterium and tritium exchange.[252]

V
CONCLUDING REMARKS

The organometallic chemistry of group 9 poly(pyrazolyl)borate complexes has in many respects been comprehensively investigated. This is particularly so for rhodium and iridium, for which a wealth of complexes comprising π-alkene, σ-alkyl/aryl and carbonyl ligands have been prepared and thoroughly characterised. Nonetheless, numerous gaps remain in current knowledge, including several notable omissions, and compounds that remain elusive.

Thus, while numerous carbonyl complexes have been prepared, $TpRh(CO)_2$ remains unknown due to its apparent propensity to convert to $Tp_2Rh_2(CO)_3$, despite the stability of less sterically encumbered analogues. More surprisingly, the only example of a nitrosyl complex is $Tp^{^tBu,Me}Co(NO)$, whilst more conventional iridium or rhodium examples remain unknown.

More frustrating has been the quest for a definitive understanding of the influence exerted by steric and electronic factors on the hapticity of the Tp^x ligand, many solution-phase structures remaining unclear. Indeed, despite the wealth of novel Tp^x ligands that have been explored there would seem to be few systematics in respect of whether a κ^3 or κ^2 structure is adopted in the solid state, the solution-phase being further complicated by a proliferation of dynamic processes. Nonetheless, the elucidation of the particular coordination mode within specific complexes has become a relatively well-defined science, though recent observations of Tp^x bridged dirhodium polycarbonyls whose spectroscopic data closely resemble those previously assigned to coordination isomers of $Tp^xRh(CO)_2$ may have potentially far-reaching implications.

While many of these issues remain unresolved, the greater importance for the future of this field seems to lie with the further development of rhodium and iridium poly(pyrazolyl)borate complexes in the role of catalysts and/or reagents for C–H activation and C–C coupling. In these roles several detailed mechanistic investigations have already been undertaken, thus the impetus may well now shift to the application of these data in industrially viable processes; a goal that is already being realised with the ever-expanding patent literature in this area.

ACKNOWLEDGEMENTS

The author thanks Prof. A. F. Hill for the opportunity to contribute to this volume, and Dr N. A. Barnes, University of Manchester, and Dr B. R. Clare, Monash University, for their comments and suggestions during manuscript preparation. The Australian Research Council (DP0342701) is gratefully acknowledged for funding.

References

(1) Cano, M.; Heras, J. V.; Trofimenko, S.; Monge, A.; Gutierrez, E.; Jones, C. J.; McCleverty, J. A. *J. Chem. Soc. Dalton Trans.* **1990**, 3577.
(2) Kitajima, N.; Hikichi, S.; Tanaka, M.; Moro-oka, Y. *J. Am. Chem. Soc.* **1993**, *115*, 5496.
(3) Shirasawa, N.; Nguyet, T. T.; Hikichi, S.; Mora-oka, Y.; Akita, M. *Organometallics* **2001**, *20*, 3582.
(4) Calabrese, J. C.; Domaille, P. J.; Thompson, J. S.; Trofimenko, S. *Inorg. Chem.* **1990**, *29*, 4429.
(5) Bondi, A. *J. Phys. Chem.* **1964**, *68*, 441.
(6) Rheingold, A. L.; Liable-Sands, L. M.; Golen, J. A.; Yap, G. P. A.; Trofimenko, S. *Dalton Trans.* **2004**, 598.
(7) Belderrain, T. R.; Paneque, M.; Carmona, E.; Gutierrez-Puebla, E.; Monge, M. A.; Ruiz-Valero, C. *Inorg. Chem.* **2002**, *41*, 425.
(8) Ghosh, P.; Bonanno, J. B.; Parkin, G. *J. Chem. Soc. Dalton Trans.* **1998**, 2779.
(9) Ruman, T.; Ciunik, Z.; Mazurek, J.; Wolowiec, S. *Eur. J. Inorg. Chem.* **2002**, 754.
(10) Ruman, T.; Ciunik, Z.; Goclan, A.; Lukasiewicz, M.; Wolowiec, S. *Polyhedron* **2001**, *20*, 2965.
(11) Trofimenko, S.; Calabrese, J. C.; Thompson, J. S. *Angew. Chem. Int. Ed. Engl.* **1989**, *28*, 205.
(12) Trofimenko, S.; Calabrese, J. C.; Thompson, J. S. *Inorg. Chem.* **1992**, *31*, 974.
(13) Trofimenko, S.; Hulsbergen, F. B.; Reedijk, J. *Inorg. Chim. Acta.* **1991**, *183*, 203.
(14) Egan, J. W. Jr.; Haggerty, B. S.; Rheingold, A. L.; Sendlinger, S. C.; Theopold, K. H. *J. Am. Chem. Soc.* **1990**, *112*, 2445.
(15) Jewson, J. D.; Liable-Sands, L. M.; Yap, G. P. A.; Rheingold, A. L.; Theopold, K. H. *Organometallics* **1999**, *18*, 300.
(16) Akita, M.; Shirasawa, N.; Hikichi, S.; Moro-oka, Y. *Chem. Commun.* **1998**, 973.
(17) Trofimenko, S. *Chem. Rev.* **1972**, *72*, 497.
(18) O'Sullivan, D. J.; Lalor, F. J. *J. Organomet. Chem.* **1973**, *57*, C58.
(19) Brunker, T. J.; Barlow, S.; O'Hare, D. *Chem. Commun.* **2001**, 2052.
(20) Brunker, T. J.; Cowley, A. R.; O'Hare, D. *Organometallics* **2002**, *21*, 3123.
(21) Brunker, T. J.; Green, J. C.; O'Hare, D. *Inorg. Chem.* **2003**, *42*, 4366.
(22) Shirasawa, N.; Akita, M.; Hikichi, S.; Moro-oka, Y. *Chem. Commun.* **1999**, 417.
(23) Trieu, T. N.; Vu, D. D.; Trinh, N. C.; Akita, M.; Morooka, Y. *Tap Chi Hoa Hoc* **1999**, *37*, 41; *Chem. Abstr.* 132:347715.
(24) King, R. B.; Bond, A. *J. Am. Chem. Soc.* **1974**, *96*, 1334.
(25) Yoshimitsu, S.-I.; Hikichi, S.; Akita, M. *Organometallics* **2002**, *21*, 3762.
(26) Hikichi, S.; Yoshizawa, M.; Sasakura, Y.; Komatsuzaki, H.; Moro-oka, Y.; Akita, M. *Chem. Eur. J.* **2001**, *7*, 5011.
(27) Reinaud, O. M.; Theopold, K. H. *J. Am. Chem. Soc.* **1994**, *116*, 6979.
(28) Detrich, J. L.; Konecny, R.; Vetter, W. M.; Doren, D.; Rheingold, A. L.; Theopold, K. H. *J. Am. Chem. Soc.* **1996**, *118*, 1703.
(29) Detrich, J. L.; Reinaud, O. M.; Rheingold, A. L.; Theopold, K. H. *J. Am. Chem. Soc.* **1995**, *117*, 11745.
(30) Carreon-Macedo, J.-L.; Harvey, J. N. *J. Am. Chem. Soc.* **2004**, *126*, 5789.
(31) Doren, D. J.; Konecny, R.; Theopold, K. H. *Catal. Today* **1999**, *50*, 669.
(32) Thyagarajan, S.; Incarvito, C. D.; Rheingold, A. L.; Theopold, K. H. *Inorg. Chim. Acta* **2003**, *345*, 333.
(33) Thyagarajan, S.; Incarvito, C. D.; Rheingold, A. L.; Theopold, K. H. *Chem. Commun.* **2001**, 2198.
(34) Uehara, K.; Hikichi, S.; Akita, M. *Organometallics* **2001**, *20*, 5002.
(35) Uehara, K.; Hikichi, S.; Inagaki, A.; Akita, M. *Chem. Eur. J.* **2005**, *11*, 2788.
(36) Uehara, K.; Hikichi, S.; Akita, M. *J. Chem. Soc. Dalton Trans.* **2002**, 3529.
(37) Gorun, S. M.; Hu, Z.; Stibrany, R. T.; Carpenter, G. *Inorg. Chim. Acta* **2000**, *297*, 383.
(38) Shay, D. T.; Yap, G. P. A.; Zakharov, L. N.; Rheingold, A. L.; Theopold, K. H. *Angew. Chem. Int. Ed.* **2005**, *44*, 1508.
(39) Borkett, N. F.; Bruce, M. I. *J. Organomet. Chem.* **1974**, *65*, C51.
(40) O'Sullivan, D. J.; Lalor, F. J. *J. Organomet. Chem.* **1974**, *65*, C47.
(41) May, S.; Reinsalu, P.; Powell, J. *Inorg. Chem.* **1980**, *19*, 1582.

(42) Albinati, A.; Bucher, U. E.; Gramlich, V.; Renn, O.; Ruegger, H.; Venanzi, L. M. *Inorg. Chim. Acta* **1999**, *284*, 191.
(43) Trofimenko, S. *J. Am. Chem. Soc.* **1969**, *91*, 588.
(44) Cocivera, M.; Desmond, T. J.; Ferguson, G.; Kaitner, B.; Lalor, F. J.; O'Sullivan, D. J. *Organometallics* **1982**, *1*, 1125.
(45) King, R. B.; Bond, A. *J. Organomet. Chem.* **1974**, *73*, 115.
(46) Cocivera, M.; Ferguson, G.; Kaitner, B.; Lalor, F. J.; O'Sullivan, D. J.; Parvez, M.; Ruhl, B. *Organometallics* **1982**, *1*, 1132.
(47) Cocivera, M.; Ferguson, G.; Lalor, F. J.; Szcsecinski, P. *Organometallics* **1982**, *1*, 1139.
(48) Bucher, U. E.; Currao, A.; Nesper, R.; Ruegger, H.; Venanzi, L. M.; Younger, E. *Inorg. Chem.* **1995**, *34*, 66.
(49) Northcutt, T. O.; Lachicotte, R. J.; Jones, W. D. *Organometallics* **1998**, *17*, 5148.
(50) Akita, M.; Ohta, K.; Takahashi, Y.; Hikichi, S.; Moro-oka, Y. *Organometallics* **1997**, *16*, 4121.
(51) Ministro, E. D.; Renn, O.; Ruegger, H.; Venanzi, L. M.; Burckhardt, U.; Gramlich, V. *Inorg. Chim. Acta* **1995**, *240*, 631.
(52) Rheingold, A. L.; Ostrander, R. L.; Haggerty, B. S.; Trofimenko, S. *Inorg. Chem.* **1994**, *33*, 3666.
(53) Sanz, D.; Santa Maria, M. D.; Claramunt, R. M.; Cano, M.; Heras, J. V.; Campo, J. A.; Ruiz, F. A.; Pinilla, E.; Monge, A. *J. Organomet. Chem.* **1996**, *526*, 341.
(54) Katayama, H.; Yamamura, K.; Miyaki, Y.; Ozawa, F. *Organometallics* **1997**, *16*, 4497.
(55) Rheingold, A. L.; Haggerty, B. S.; Yap, G. P. A.; Trofimenko, S. *Inorg. Chem.* **1997**, *36*, 5097.
(56) Rheingold, A. L.; Liable-Sands, L. M.; Trofimenko, S. *Inorg. Chem.* **2000**, *39*, 1333.
(57) Santa Maria, M. D.; Claramunt, R. M.; Campo, J. A.; Cano, M.; Criado, R.; Heras, J. V.; Ovejero, P.; Pinilla, E.; Torres, M. R. *J. Organomet. Chem.* **2000**, *605*, 117.
(58) Nicasio, M. C.; Paneque, M.; Perez, P. J.; Pizzano, A.; Poveda, M. L.; Rey, L.; Sirol, S.; Taboada, S.; Trujillo, M.; Monge, A.; Ruiz, C.; Carmona, E. *Inorg. Chem.* **2000**, *39*, 180.
(59) Baena, M. J.; Reyes, M. L.; Rey, L.; Carmona, E.; Nicasio, M. C.; Perez, P. J.; Gutierrez, E.; Monge, A. *Inorg. Chim. Acta* **1998**, *273*, 244.
(60) Rheingold, A. L.; Incarvito, C. D.; Trofimenko, S. *Inorg. Chem.* **2000**, *39*, 5569.
(61) Ruman, T.; Ciunik, Z.; Trzeciak, A. M.; Wolowiec, S.; Ziolkowski, J. J. *Organometallics* **2003**, *22*, 1072.
(62) Ruman, T.; Ciunik, Z.; Wolowiec, S. *Polyhedron* **2004**, *23*, 219.
(63) Rheingold, A. L.; White, C. B.; Trofimenko, S. *Inorg. Chem.* **1993**, *32*, 3471.
(64) Bucher, U. E.; Fassler, T. F.; Hunzinker, M.; Nesper, R.; Ruegger, H.; Venanzi, L. M. *Gazz. Chim. Ital.* **1995**, *125*, 181; *Chem. Abstr.* 123:112383.
(65) Bortolin, M.; Bucher, U. E.; Ruegger, H.; Venanzi, L. M.; Albinati, A.; Lianza, F.; Trofimenko, S. *Organometallics* **1992**, *11*, 2514.
(66) Herberhold, M.; Eibl, S.; Milius, W.; Wrackmeyer, B. Z. *Anorg. Allg. Chem.* **2000**, *626*, 552.
(67) Malbose, F.; Kalck, P.; Daran, J.-C.; Etienne, M. *J. Chem. Soc. Dalton Trans.* **1999**, 271.
(68) Oldham, W. J. Jr.; Heinekey, D. M. *Organometallics* **1997**, *16*, 467.
(69) Oldham, W. J. Jr.; Hinkle, A. S.; Heinekey, D. M. *J. Am. Chem. Soc.* **1997**, *119*, 11028.
(70) Perez, P. J.; Proveda, M. L.; Carmona, E. *Angew. Chem. Int. Ed. Engl.* **1995**, *34*, 231.
(71) Paneque, M.; Taboada, S.; Carmona, E. *Organometallics* **1996**, *15*, 2678.
(72) Wick, D. D.; Northcutt, T. O.; Lachicotte, R. J.; Jones, W. D. *Organometallics* **1998**, *17*, 4484.
(73) Wick, D. D.; Jones, W. D. *Organometallics* **1999**, *18*, 495.
(74) Ohta, K.; Hashimoto, M.; Takahashi, Y.; Hikichi, S.; Akita, M.; Moro-oka, Y. *Organometallics* **1999**, *18*, 3234.
(75) Ikeda, S.; Maruyama, Y.; Ozawa, F. *Organometallics* **1998**, *17*, 3770.
(76) Seino, H.; Yoshikawa, T.; Hidai, M.; Mizobe, Y. *J. Chem. Soc. Dalton Trans.* **2004**, 3593.
(77) Ferrari, A.; Merlin, M.; Sostero, S. *Helv. Chim. Acta* **1999**, *82*, 1454.
(78) Boaretto, R.; Ferrari, A.; Merlin, M.; Sostero, S.; Traverso, O. *J. Photochem. Photobiol. A* **2000**, *135*, 179.
(79) Hill, A. F.; White, A. J. P.; Williams, D. J.; Wilton-Ely, J. D. E. T. *Organometallics* **1998**, *17*, 3152.
(80) Albinati, A.; Bovens, M.; Ruegger, H.; Venanzi, L. M. *Inorg. Chem.* **1997**, *36*, 5991.
(81) Bovens, M.; Gerfin, T.; Gramlich, V.; Petter, W.; Venanzi, L. M.; Haward, M. T.; Jackson, S. A.; Eisenstein, O. *New J. Chem.* **1992**, *16*, 337.

(82) Sowa, J. R. Jr.; Angelici, R. J. *J. Am. Chem. Soc.* **1991**, *113*, 2537.
(83) Fernandez, M. J.; Rodrigues, M. J.; Oro, L. A. *Polyhedron* **1991**, *10*, 1595.
(84) Tanke, R. S.; Crabtree, R. H. *Inorg. Chem.* **1989**, *28*, 3444.
(85) Fernandez, M. J.; Rodriguez, M. J.; Oro, L. A.; Lahoz, F. J. *J. Chem. Soc. Dalton Trans.* **1989**, 2073.
(86) Perez, P. J.; Poveda, M. L.; Carmona, E. *J. Chem. Soc. Chem. Commun.* **1992**, *8*.
(87) Rodriguez, P.; Diaz-Requejo, M. M.; Belderrain, T. R.; Trofimenko, S.; Nicasio, M. C.; Perez, P. J. *Organometallics* **2004**, *23*, 2162.
(88) Slugovc, C.; Mereiter, K.; Trofimenko, S.; Carmona, E. *Chem. Commun.* **2000**, 121.
(89) Alvarado, Y.; Boutry, O.; Gutierrez, E.; Monge, A.; Nicasio, M. C.; Poveda, M. L.; Perez, P. J.; Ruiz, C.; Bianchini, C.; Carmona, E. *Chem. Eur. J.* **1997**, *I*, 860.
(90) Gutierrez-Puebla, E.; Monge, A.; Nicasio, M. C.; Perez, P. J.; Poveda, M. L.; Carmona, E. *Chem. Eur. J.* **1998**, *4*, 2225.
(91) Ferrari, A.; Polo, E.; Ruegger, H.; Sostero, S.; Venanzi, L. M. *Inorg. Chem.* **1996**, *35*, 1602.
(92) Gutierrez-Puebla, E.; Monge, A.; Nicasio, M. C.; Perez, P. J.; Poveda, M. L.; Rey, L.; Ruiz, C.; Carmona, E. *Inorg. Chem.* **1998**, *37*, 4538.
(93) Heinekey, D. M.; Oldham, W. J. Jr.; Wiley, J. S. *J. Am. Chem. Soc.* **1996**, *118*, 12842.
(94) Ciriano, M. A.; Fernandez, M. J.; Modrego, J.; Rodriguez, M. J.; Oro, L. A. *J. Organomet. Chem.* **1993**, *443*, 249.
(95) Ghosh, C. K.; Hoyano, J. K.; Krentz, R.; Graham, W. A. G. *J. Am. Chem. Soc.* **1989**, *111*, 5480.
(96) Boutry, O.; Poveda, M. L.; Carmona, E. *J. Organomet. Chem.* **1997**, *528*, 143.
(97) Slugovc, C.; Mereiter, K.; Trofimenko, S.; Carmona, E. *Helv. Chim. Acta* **2001**, *84*, 2868.
(98) Slugovc, C.; Mereiter, K.; Trofimenko, S.; Carmona, E. *Angew. Chem. Int. Ed.* **2000**, *39*, 2158.
(99) Alias, F. M.; Poveda, M. L.; Sellin, M.; Carmona, E. *Organometallics* **1998**, *17*, 4124.
(100) Alias, F. M.; Daff, P. J.; Paneque, M.; Poveda, M. L.; Carmona, E.; Perez, P. J.; Salazar, V.; Alvarado, Y.; Atencio, R.; Sanchez-Delgado, R. *Chem. Eur. J.* **2002**, *8*, 5132.
(101) Paneque, M.; Poveda, M. L.; Salazar, V.; Carmona, E.; Ruiz-Valero, C. *Inorg. Chim. Acta* **2003**, *345*, 367.
(102) Paneque, M.; Poveda, M. L.; Carmona, E.; Salazar, V. *Dalton Trans.* **2005**, 1422.
(103) Paneque, M.; Sirol, S.; Trujillo, M.; Gutierrez-Puebla, E.; Monge, M. A.; Carmona, E. *Angew. Chem. Int. Ed.* **2000**, *39*, 218.
(104) Paneque, M.; Sirol, S.; Trujillo, M.; Carmona, E.; Gutierrez-Puebla, E.; Monge, M. A.; Ruiz, C.; Malbosc, F.; Serra-Le Berre, C.; Kalck, P.; Etienne, M.; Daran, J. C. *Chem. Eur. J.* **2001**, *7*, 3868.
(105) Wiley, J. S.; Heinekey, D. M. *Inorg. Chem.* **2002**, *41*, 4961.
(106) Bowden, A. A.; Hughes, R. P.; Lindner, D. C.; Incarvito, C. D.; Liable-Sands, L. M.; Rheingold, A. L. *J. Chem. Soc. Dalton Trans.* **2002**, 3245.
(107) Jones, W. D.; Feher, F. J. *J. Am. Chem. Soc.* **1984**, *106*, 1650.
(108) Wenzel, T. T.; Bergman, R. G. *J. Am. Chem. Soc.* **1986**, *108*, 4856.
(109) Baker, M. V.; Field, L. D. *J. Am. Chem. Soc.* **1986**, *108*, 7433.
(110) Baker, M. V.; Field, L. D. *J. Am. Chem. Soc.* **1986**, *108*, 7436.
(111) Stoutland, P. O.; Bergman, R. G. *J. Am. Chem. Soc.* **1985**, *107*, 4581.
(112) Stoutland, P. O.; Bergman, R. G. *J. Am. Chem. Soc.* **1988**, *110*, 5732.
(113) Alvarado, Y.; Daff, P. J.; Perez, P. J.; Poveda, M. L.; Sanchez-Delgado, R.; Carmona, E. *Organometallics* **1996**, *15*, 2192.
(114) Ghosh, C. K.; Rodgers, D. P. S.; Graham, W. A. G. *J. Chem. Soc. Chem. Commun.* **1988**, 1511.
(115) Jimenez-Catano, R.; Niu, S.; Hall, M. B. *Organometallics* **1997**, *16*, 1962.
(116) Paneque, M.; Perez, P. J.; Pizzano, A.; Poveda, M. L.; Taboada, S.; Trujillo, M.; Carmona, E. *Organometallics* **1999**, *18*, 4304.
(117) Gutierrez, E.; Monge, A.; Nicasio, M. C.; Poveda, M. L.; Carmona, E. *J. Am. Chem. Soc.* **1994**, *116*, 791.
(118) Paneque, M.; Poveda, M. L.; Santos, L. L.; Salazar, V.; Carmona, E. *Chem. Commun.* **2004**, 1838.
(119) Boutry, O.; Gutierrez, E.; Monge, A.; Nicasio, M. C.; Perez, P. J.; Carmona, E. *J. Am. Chem. Soc.* **1992**, *114*, 7288.
(120) Paneque, M.; Poveda, M. L.; Rey, L.; Taboada, S.; Carmona, E.; Ruiz, C. *J. Organomet. Chem.* **1995**, *504*, 147.

(121) Paneque, M.; Poveda, M. L.; Salazar, V.; Taboada, S.; Carmona, E. *Organometallics* **1999**, *18*, 139.
(122) Paneque, M.; Poveda, M. L.; Salazar, V. *Organometallics* **2000**, *19*, 3120.
(123) Ogilvy, A. E.; Draganjac, M.; Fauchfuss, T. B.; Wilson, S. R. *Organometallics* **1988**, *7*, 1171.
(124) Jones, W. D.; Chin, R. M. *J. Organomet. Chem.* **1994**, *472*, 311.
(125) Dong, L.; Duckett, S. B.; Ohman, K. F.; Jones, W. D. *J. Am. Chem. Soc.* **1992**, *114*, 151.
(126) Chirsholm, M. H.; Haubrich, S. T.; Martin, J. D.; Streib, W. D. *J. Chem. Soc. Chem. Commun.* **1994**, 683.
(127) Myers, A. W.; Jones, W. D.; McClements, S. M. *J. Am. Chem. Soc.* **1995**, *117*, 11740.
(128) Bianchini, C.; Jimenez, M. V.; Melt, A.; Moneti, S.; Vizza, F. *J. Organomet. Chem.* **1995**, *504*, 27.
(129) Ferrari, A.; Merlin, M.; Sostero, S.; Traverso, O.; Ruegger, H.; Venanzi, L. M. *Helv. Chim. Acta* **1998**, *81*, 2127.
(130) Boaretto, R.; Paolucci, G.; Sostero, S.; Traverso, O. *J. Mol. Catal. A* **2003**, *204–205*, 253.
(131) Wiley, J. S.; Oldham, W. J. Jr.; Heinekey, D. M. *Organometallics* **2000**, *19*, 1670.
(132) O'Conner, J. M.; Closson, A.; Gantzel, P. *J. Am. Chem. Soc.* **2002**, *124*, 2434.
(133) Alvarez, E.; Gomez, M.; Paneque, M.; Posadas, C. M.; Poveda, M. L.; Rendon, N.; Santos, L. L.; Rojas-Lima, S.; Salazar, V.; Mereiter, K.; Ruiz, C. *J. Am. Chem. Soc.* **2003**, *125*, 1478.
(134) Paneque, M.; Poveda, M. L.; Rendon, N.; Mereiter, K. *J. Am. Chem. Soc.* **2004**, *126*, 1610.
(135) Restivo, R. J.; Ferguson, G.; O'Sullivan, D. J.; Lalor, F. J. *Inorg. Chem.* **1975**, *14*, 3046.
(136) Carmona, E.; Cingolani, A.; Marchetti, F.; Pettinari, C.; Pettinari, R.; Skelton, B. W.; White, A. H. *Organometallics* **2003**, *22*, 2820.
(137) Herberich, G. E.; Buschges, U. *Chem. Ber.* **1989**, *122*, 615.
(138) Clark, H. G.; Goel, S. *J. Organomet. Chem.* **1979**, *165*, 383.
(139) Chambron, J.-C.; Elchhorn, D. M.; Franczyk, T. S.; Stearns, D. M. *Acta Crystallogr. Sect. C* **1991**, *47*, 1732.
(140) McCallum; Bergman, R. G. *unpublished results.*
(141) Tellers, D. M.; Bergman, R. G. *J. Am. Chem. Soc.* **2000**, *122*, 954.
(142) Wick, D. D.; Jones, W. D. *Inorg. Chem.* **1997**, *36*, 2723.
(143) Northcutt, T. O.; Wick, D. D.; Vetter, A. J.; Jones, W. D. *J. Am. Chem. Soc.* **2001**, *123*, 7257.
(144) Jones, W. D.; Hassell, E. T. *J. Am. Chem. Soc.* **1992**, *114*, 6087.
(145) Jones, W. D.; Hassell, E. T. *J. Am. Chem. Soc.* **1993**, *115*, 554.
(146) Vetter, A. J.; Flaschenriem, C.; Jones, W. D. *J. Am. Chem. Soc.* **2005**, *127*, 12315.
(147) Vetter, A. J.; Jones, W. D. *Polyhedron* **2004**, *23*, 413.
(148) Berry, M.; Elmin, K.; Green, M. L. H. *J. Chem. Soc. Dalton Trans.* **1979**, 1950.
(149) Wick, D. D.; Reynolds, K. A.; Jones, W. D. *J. Am. Chem. Soc.* **1999**, *121*, 3974.
(150) Hessell, E. T.; Jones, W. D. *Organometallics* **1992**, *11*, 1496.
(151) Purwoko, A. A.; Lees, A. J. *Inorg. Chem.* **1996**, *35*, 675.
(152) Ghosh, C. K.; Graham, W. A. G. *J. Am. Chem. Soc.* **1987**, *109*, 4726.
(153) Ghosh, C. K.; Graham, W. A. G. *J. Am. Chem. Soc.* **1989**, *111*, 375.
(154) Burger, P.; Bergman, R. G. *J. Am. Chem. Soc.* **1993**, *115*, 10462.
(155) Tellers, D. M.; Bergman, R. G. *Organometallics* **2001**, *20*, 4819.
(156) Tellers, D. M.; Bergman, R. G. *Can. J. Chem.* **2001**, *79*, 525.
(157) Tellers, D. M.; Bergman, R. G. *J. Am. Chem. Soc.* **2001**, *123*, 11508.
(158) Gutierrez-Puebla, E.; Monge, A.; Paneque, M.; Poveda, M. L.; Salazar, V.; Carmona, E. *J. Am. Chem. Soc.* **1999**, *121*, 248.
(159) Teuma, E.; Malbozc, F.; Pons, V.; Serra-Le Berre, C.; Jaud, J.; Etienne, M.; Kalck, P. *J. Chem. Soc. Dalton Trans.* **2001**, 2225.
(160) Takahashi, Y.; Hashimoto, M.; Hikichi, S.; Akita, M.; Moro-oka, Y. *Angew. Chem. Int. Ed.* **1999**, *38*, 3074.
(161) Takahashi, Y.; Hashimoto, M.; Hikichi, S.; Moro-oka, Y.; Akita, M. *Inorg. Chim. Acta* **2004**, *357*, 1711.
(162) Hughes, R. P.; Lindner, D. C.; Rheingold, A. L.; Yap, G. P. A. *Organometallics* **1996**, *15*, 5678.
(163) Hughes, R. P.; Smith, J. M.; Liable-Sands, L. M.; Concolino, T. E.; Lam, K.-C.; Incarvito, C.; Rheingold, A. L. *J. Chem. Soc. Dalton Trans.* **2000**, 873.
(164) Barrientos, C.; Ghosh, C. K.; Grahame, W. A. G.; Thomas, M. J. *J. Organomet. Chem.* **1990**, *394*, C31.

(165) Santos, L. L.; Mereiter, K.; Paneque, M.; Slugovc, C.; Carmona, E. *New. J. Chem.* **2003**, *27*, 107.
(166) Paneque, M.; Posada, C. M.; Poveda, M. L.; Rendon, N.; Salazar, V.; Onate, E.; Mereiter, K. *J. Am. Chem. Soc.* **2003**, *125*, 9898.
(167) Jones, W. D.; Duttweiler, R. P. Jr.; Feher, F. J.; Hessell, E. T. *New J. Chem.* **1989**, *13*, 725.
(168) Li, J.; Djurovich, P. I.; Alleyne, B. D.; Tsyba, I.; Ho, N. N.; Bau, R.; Thompson, M. E. *Polyhedron* **2004**, *23*, 419.
(169) Li, J.; Djurovich, P. I.; Alleyne, B. D.; Yousufuddin, M.; Ho, N. N.; Thomas, J. C.; Peters, J. C.; Bau, R.; Thompson, M. E. *Inorg. Chem.* **2005**, *44*, 1713.
(170) Mak, C. S. K.; Hayer, A.; Pascu, S. I.; Watkins, S. E.; Holmes, A. B.; Kohler, A.; Friend, R. H. *Chem. Commun.* **2005**, 4708.
(171) D'Andrade, B. W. D.; Datta, S.; Forrest, S. R.; Djurovich, P.; Polikarpov, E.; Thompson, M. E. *Org. Electron.* **2005**, *6*, 11.
(172) Holmes, R. J.; D'Andrade, B. W.; Forrest, S. R.; Ren, X.; Li, J.; Thompson, M. E. *Appl. Phys. Lett.* **2003**, *83*, 3818.
(173) Ren, X.; Li, J.; Holmes, R. J.; Djurovich, P. I.; Forrest, S. R.; Thompson, M. E. *Chem. Mater.* **2004**, *16*, 4743.
(174) Sergey, L.; Thompson, M. E.; Adamovich, V.; Djurovich, P. I.; Adachi, C.; Baldo, M. A.; Forrest, S. R.; Kwong, R. C. *WO 2002015645*, **2002**, *Chem. Abstr.* 136:191506.
(175) Thompson, M. E.; Djurovich, P. I.; Li, J. *WO 2004017043*, **2004**, *Chem. Abstr.* 140:225477.
(176) Ren, X.; Holmes, R.; Forrest, S.; Thompson, M. E. *US 2004209116*, **2002**, *Chem. Abstr.* 141:372557.
(177) Tung, Y.-J.; Ngo, T. *US 2005006642*, **2005**, *Chem. Abstr.* 142:103508.
(178) Oshiyama, T.; Katoh, E.; Kita, H.; Oi S.; Inoue, Y. *WO 2005097943*, **2005**, *Chem. Abstr.* 143:413208.
(179) Alias, F. M.; Poveda, M. L.; Sellin, M.; Carmona, E. *J. Am. Chem. Soc.* **1998**, *120*, 5816.
(180) Circu, V.; Fernandes, M. A.; Carlton, L. *Inorg. Chem.* **2002**, *41*, 3859.
(181) Cao, C.; Wang, T.; Patrick, B. O.; Love, J. A. *Organometallics* **2006**, *25*, 1321.
(182) Bergman, R. G.; Yeston, J. S. *Organometallics* **2000**, *19*, 2947.
(183) Teuma, E.; Malbosc, F.; Etienne, M.; Daran, J.-C.; Kalck, P. *J. Organomet. Chem.* **2004**, *689*, 1763.
(184) Lopez, J. A.; Mereier, K.; Paneque, M.; Poveda, M. L.; Serrano, O.; Trofimenko, S.; Carmona, E. *Chem. Commun.* **2006**, 3921.
(185) Paneque, M.; Poveda, M. L.; Santos, L. L.; Carmona, E.; Lledos, A.; Ujaque, G.; Mereiter, K. *Angew. Chem. Int. Ed.* **2004**, *43*, 3708.
(186) Ilg, K.; Paneque, M.; Poveda, M. L.; Rendon, N.; Santos, L. L.; Carmona, E.; Mereiter, K. *Organometallic* **2006**, *25*, 2230.
(187) Lawrenson, M. J. *DE 2058814*, **1971**, *Chem. Abstr.* 75:109845.
(188) Bonati, F.; Minghetti, G.; Banditelli, G. *J. Organomet. Chem.* **1975**, *87*, 365.
(189) Trofimenko, S. *Inorg. Chem.* **1971**, *10*, 1372.
(190) Calabrese, J. C.; Trofimenko, S. *Inorg. Chem.* **1992**, *31*, 4810.
(191) Rheingold, A. L.; Liable-Sands, L. M.; Incarvito, C. L.; Trofimenko, S. *J. Chem. Soc. Dalton Trans.* **2002**, 2297.
(192) LeCloux, D. D.; Keyes, M. C.; Osawa, M.; Reynolds, V.; Tolman, W. B. *Inorg. Chem.* **1994**, *33*, 6361.
(193) Keyes, M. C.; Young, V. G. Jr.; Tolman, W. B. *Organometallics* **1996**, *15*, 4133.
(194) Criado, R.; Cano, M.; Campo, J. A.; Heras, J. V.; Pinilla, E.; Torres, M. R. *Polyhedron* **2004**, *23*, 301.
(195) Rheingold, A. L.; Liable-Sands, L. M.; Golan, J. A.; Trofimenko, S. *Eur. J. Inorg. Chem.* **2003**, 2767.
(196) Ball, R. G.; Ghosh, C. K.; Hoyano, J. K.; McMaster, A. D.; Graham, W. A. G. *J. Chem. Soc. Chem. Commun.* **1989**, 341.
(197) Chauby, V.; Serra-Le Berre, C.; Kalck, Ph.; Daran, J.-C.; Commenges, G. *Inorg. Chem.* **1996**, *35*, 6354.
(198) Malbosec, F.; Chauby, V.; Serra-Le Berre, C.; Etienne, M.; Daran, J.-C.; Kalck, P. *Eur. J. Inorg. Chem.* **2001**, 1689.

(199) Connelly, N. G.; Emslie, D. J. H.; Geiger, W. E.; Hayward, O. D.; Linehan, E. B.; Orpen, A. G.; Quayle, M. J.; Rieger, P. H. *J. Chem. Soc. Dalton Trans.* **2001**, 670.
(200) Trzeciak, A. M.; Borak, B.; Ciunik, Z.; Ziolkowski, J. J.; Guedes da Silva, M. F. C.; Pombeiro, A. J. L. *Eur. J. Inorg. Chem.* **2004**, 1411.
(201) Chauby, V.; Daran, J.-C.; Serra-Le Berre, C.; Malbosc, F.; Kalck, P.; Gonzalez, O. D.; Haslam, C. E.; Haynes, A. *Inorg. Chem.* **2002**, *41*, 3280.
(202) Moszner, M.; Wolowiec, S.; Trosch, A.; Vahrenkamp, H. *J. Organomet. Chem.* **2000**, *595*, 178.
(203) Connelly, N. G.; Emslie, D. J. H.; Metz, B.; Orpen, A. G.; Quayle, M. J. *Chem. Commun.* **1996**, 2289.
(204) Geiger, W. E.; Ohrenberg, N. C.; Yeomana, B.; Connelly, N. G.; Emslie, D. J. H. *J. Am. Chem. Soc.* **2003**, *125*, 8680.
(205) Fernandez, M. J.; Rodriguez, M. J.; Oro, L. A. *J. Organomet. Chem.* **1992**, *438*, 337.
(206) Fernandez, M. J.; Modrego, J.; Rodriguez, M. J.; Santamaria, M. C.; Oro, L. A. *J. Organomet. Chem.* **1992**, *441*, 155.
(207) Gutierrez-Puebla, E.; Monge, A.; Paneque, M.; Poveda, M. L.; Taboada, S.; Trujillo, M.; Carmona, E. *J. Am. Chem. Soc.* **1999**, *121*, 346.
(208) Purwoko, A. A.; Less, A. J. *Inorg. Chem.* **1995**, *34*, 424.
(209) Purwoko, A. A.; Tibensky, S. D.; Lees, A. J. *Inorg. Chem.* **1996**, *35*, 7049.
(210) Connelly, N. G.; Emslie, D. J. H.; Hayward, W. D.; Orpen, A. G.; Quayle, M. J. *J. Chem. Soc. Dalton Trans.* **2001**, 875.
(211) Connelly, N. G.; Freeman, M. J.; Manners, I.; Orpen, A. G. *J. Chem. Soc. Dalton Trans.* **1984**, 2703.
(212) Jones, W. D.; Hessell, E. T. *Inorg. Chem.* **1991**, *30*, 778.
(213) Nagao, S.; Suito, N.; Kojima, A.; Seino, H.; Mizobe, Y. *Bull. Chem. Soc. Jpn.* **2005**, *78*, 1641.
(214) Alvarez, E.; Paneque, M.; Poveda, M. L.; Rendon, N. *Angew. Chem. Int. Ed.* **2006**, *45*, 474.
(215) Schubert, D. M.; Knobler, C. B.; Trofimenko, S.; Hawthorne, M. F. *Inorg. Chem.* **1990**, *29*, 2364.
(216) Jeffery, J. C.; Jelliss, P. A.; Lebedev, V. N.; Stone, F. G. A. *Organometallics* **1996**, *15*, 4737.
(217) Heinekey, D. M.; Oldham, W. J. Jr.; *J. Am. Chem. Soc.* **1994**, *116*, 3137.
(218) Panequ, M.; Poveda, M. L.; Taboada, S. *J. Am. Chem. Soc.* **1994**, *116*, 4519.
(219) Bucher, U. E.; Lengweiler, T.; Nanz, D.; von Philipsborn, W.; Venanzi, L. M. *Angew. Chem. Int. Ed. Engl.* **1990**, *29*, 548.
(220) Nanz, D.; von Philipsborn, W.; Bucher, U. E.; Venanzi, L. M. *Magn. Reson. Chem.* **1991**, *29*, S38.
(221) Eckert, J.; Albinati, A.; Bucher, U. E.; Venanzi, L. M. *Inorg. Chem.* **1996**, *35*, 1292.
(222) Gelabert, R.; Moreno, M.; Lluch, J. M.; Lledos, A. *Organometallics* **1997**, *16*, 3805.
(223) Eckert, J.; Webster, C. E.; Hall, M. B.; Albinati, A.; Venanzi, L. M. *Inorg. Chim. Acta* **2002**, *330*, 240.
(224) Webster, C. E.; Singleton, D. A.; Szymanski, M. J.; Hall, M. B.; Zhao, C.; Jia, G.; Lin, Z. *J. Am. Chem. Soc.* **2001**, *123*, 9822.
(225) Brandt, K.; Sheldrick, W. S. *Inorg. Chim. Acta* **1998**, *267*, 39.
(226) Carmona, D.; Viguri, F.; Lanoz, F. J.; Oro, L. A. *Inorg. Chem.* **2002**, *41*, 2385.
(227) Akita, M.; Hashimoto, M.; Hikichi, S.; Moro-oka, Y. *Organometallics* **2000**, *19*, 3744.
(228) Circu, V.; Fernandes, M. A.; Carlton, L. *Polyhedron* **2003**, *22*, 3293.
(229) Cao, C.; Fraser, L. R.; Love, J. A. *J. Am. Chem. Soc.* **2005**, *127*, 17614.
(230) Casado, M. A.; Hack, V.; Camerano, J. A.; Ciriano, M. A.; Tejel, C.; Oro, L. A. *Inorg. Chem.* **2005**, *44*, 9122.
(231) Slugovc, C.; Padilla-Martinez, I.; Sirol, S.; Carmona, E. *Coord. Chem. Rev.* **2001**, *213*, 129.
(232) Thyagarajan, S.; Shay, D. T.; Incarvito, C. D.; Rheingold, A. L.; Theopold, K. H. *J. Am. Chem. Soc.* **2003**, *125*, 4440.
(233) Bloyce, P. E.; Mascetti, J.; Rest, A. J. *J. Organomet. Chem.* **1993**, *444*, 223.
(234) Lees, A. J.; Purwoko, A. A. *Coord. Chem. Rev.* **1994**, *132*, 155.
(235) Purwoko, A. A.; Drolet, D. P.; Lees, A. J. *J. Organomet. Chem.* **1995**, *504*, 107.
(236) Lees, A. J. *J. Organomet. Chem.* **1998**, *554*, 1.
(237) Lian, T.; Bromberg, S. E.; Yang, H.; Prouls, G.; Bergman, R. G.; Harris, C. B. *J. Am. Chem. Soc.* **1996**, *118*, 3769.

(238) Bromberg, S. E.; Yang, H.; Asplund, M. C.; Lian, T.; McNarnara, B. K.; Kotz, K. T.; Yeston, J. S.; Wilkens, M.; Frei, H.; Bergman, R. G.; Harris, C. B. *Science* **1997**, *278*, 260.
(239) Asplund, M. C.; Yang, H.; Kotz, K. T.; Bromberg, S. E.; Wilkens, M. J.; Harris, C. B. *Laser Chem.* **1999**, *19*, 253.
(240) Yeston, J. S.; McNarnara, B. K.; Bergman, R. G.; Moore, C. B. *Organometallics* **2000**, *19*, 3442.
(241) Zaric, S.; Hall, M. B. *J. Phys. Chem. A* **1998**, *102*, 1963.
(242) Asplund, M. C.; Snee, P. T.; Yeston, J. S.; Wilkens, M. J.; Payne, C. K.; Yang, H.; Kotz, K. T.; Frei, H.; Bergman, R. G.; Harris, C. B. *J. Am. Chem. Soc.* **2002**, *124*, 10605.
(243) Alvarado, Y.; Busolo, M.; Lopez-Linares, F. *J. Mol. Catal. A* **1999**, *142*, 163.
(244) Lopez-Linares, F.; Agrifoglio, G.; Labrador, A.; Karam, A. *J. Mol. Catal. A* **2004**, *207*, 115.
(245) Ganicz, T.; Mizerska, U.; Moszner, M.; O'Brien, M.; Perry, R.; Stanczyk, W. A. *Appl. Catal. A-Gen.* **2004**, *259*, 49.
(246) Tanke, R. S.; Crabtree, R. H. *J. Am. Chem. Soc.* **1990**, *112*, 7984.
(247) Trujillo, M. Ph.D. thesis, University of Sevilla, **1999**.
(248) Maruyama, Y.; Yoshiuchi, K.; Ozawa, F.; Wakatsuki, Y. *Chem. Lett.* **1997**, 623.
(249) Trzeciak, A. M.; Ziolkowski, J. J. *Appl. Organomet. Chem.* **2004**, *18*, 124.
(250) Tsuchihara, K.; Hagiwara, H.; Takeuchi, K.; Asai, M.; Shiono, T.; Ban, H. *JP 2004099730*, **2004**, *Chem. Abstr.* 140:271404.
(251) Slugovc, C.; Perner, B. *Inorg. Chim. Acta* **2004**, *357*, 3104.
(252) Bergman, R. C.; Klei, S. R. *US 2002173666*, **2002**, *Chem. Abstr.* 137:370219.

Index

B

C

D

E

F

G

H

I

N

O

P

S

T

Cumulative List of Contributors for Volumes 1–36

Cumulative Index for Volumes 37–56